METAMORPHOSIS
A Problem in Developmental Biology

2ND EDITION

METAMORPHOSIS
A Problem in Developmental Biology
2ND EDITION

Edited by
LAWRENCE I. GILBERT
The University of North Carolina
Chapel Hill, North Carolina

and

EARL FRIEDEN
Florida State University
Tallahassee, Florida

PLENUM PRESS · NEW YORK AND LONDON

Library of Congress Cataloging in Publication Data

Main entry under title:

Metamorphosis, a problem in developmental biology.

Includes bibliographies and index.
1. Metamorphosis. I. Gilbert, Lawrence Irwin, 1929-   . II. Frieden, Earl.
[DNLM: 1. Metamorphosis, Biological. QL 981 M587]
QL981.M47   1981   591.3′34   81-17691
ISBN 0-306-40692-2   AACR2

© 1981 Plenum Press, New York
A Division of Plenum Publishing Corporation
233 Spring Street, New York, N.Y. 10013

Printed in the United States of America

Contributors

Burr G. Atkinson, Department of Zoology, Cell Science Laboratories, University of Western Ontario, London, Ontario, Canada

Walter E. Bollenbacher, Department of Anatomy and Department of Zoology, University of North Carolina, Chapel Hill, North Carolina 27514

Robert H. Broyles, Department of Biochemistry and Molecular Biology, University of Oklahoma at Oklahoma City, Health Sciences Center, Oklahoma City, Oklahoma 73190

Virginia Carver, Department of Chemistry, The Florida State University, Tallahassee, Florida 32306

Ernest S. Chang, Department of Animal Science, University of California, Davis, California 95616, and Bodega Marine Laboratory, P.O. Box 247, Bodega Bay, California 94923

Deborah Ann Check, Physiology Group, T. H. Morgan School of Biological Sciences, University of Kentucky, Lexington, Kentucky 40506

Harold Fox, Department of Zoology, University College, London WC1E 6BT, United Kingdom

Earl Frieden, Department of Chemistry, Florida State University, Tallahassee, Florida 32306

James W. Fristrom, Department of Genetics, University of California, Berkeley, California 94720

Lawrence I. Gilbert, Department of Zoology, University of North Carolina, Chapel Hill, North Carolina 27514

Walter Goodman, Department of Entomology, University of Wisconsin, Madison, Wisconsin 53706

Noelle A. Granger, Department of Anatomy and Department of Zoology, University of North Carolina, Chapel Hill, North Carolina 27514

K. C. Highnam, Department of Zoology, The University, Sheffield S10 2TN, United Kingdom

John J. Just, Physiology Group, T. H. Morgan School of Biological Sciences, University of Kentucky, Lexington, Kentucky 40506

Jerry J. Kollros, Department of Zoology, University of Iowa, Iowa City, Iowa 52242

Jeanne Kraus-Just, Physiology Group, T. H. Morgan School of Biological Sciences, University of Kentucky, Lexington, Kentucky 40506

M. Locke, The Cell Science Laboratories, Zoology Department, University of Western Ontario N6A 5B7, Canada

Charles S. Nicoll, Department of Physiology–Anatomy, University of California, Berkeley, California 94720

John D. O'Connor, Department of Biology, University of California, Los Angeles, California 90024

Sandra J. Smith-Gill, Department of Zoology, University of Maryland, College Park, Maryland 20742

S. Sridhara, Department of Biochemistry, University of Mississippi Medical Center, Jackson, Mississippi 39216

George Wald, Biological Laboratories of Harvard University, Cambridge, Massachusetts 02138

Bruce A. White, Department of Physiology–Anatomy, University of California, Berkeley, California 94720. Present address: Laboratory of Cellular Gene Expression and Regulation, Memorial Sloan-Kettering Cancer Center, New York, New York 10021

A Personal Foreword

"The old order changeth, yielding place to new." When Tennyson wrote this, he was unfamiliar with the pace of modern science else he would have said the new is displaced by the newer. When Gilbert and I gathered the papers for the first edition of this overview of metamorphosis, we aimed to provide a broad basis upon which the experimental analysis of the developmental changes called metamorphosis could proceed. We were both aware then that with the new techniques of biochemistry and with the revolutionary breakthrough to the nature of the gene, countless new possibilities were being opened for the exploration of the molecular basis of development. The resources offered by metamorphic changes offered unique opportunities to trace the path from gene to phenotype. Our expectations were high. I visited Larry Gilbert and Earl Frieden in their laboratories and saw with envy how far advanced they were then in the use of molecular methods of analysis. I had started on a different approach to develop an *in vitro* test for thyroid action on amphibian tissue. But circumstances limited my own progress to the initial delimitation of the technical possibilities of the *in vitro* system. Only from the sidelines could I watch the steady if slow progress of biology in penetrating the maze of molecular events by which animal tissues respond to hormonal and other developmental factors. In retirement, I succumbed to the inevitable and turned to an old love—social behavior and the new speculations of sociobiology. Hence, I now look forward as a novice in a strange land to reading this new text. However, unlike Sir Bedivere, who was reluctant to throw the sword Excalibur into the lake, we in science think that a thirteen-year-old review of so rich a topic as metamorphosis has waited, perhaps too long, to be displaced by the new. I congratulate Larry and Earl on the splendid talents they have persuaded to contribute to this new edition and, unlike Sir Bedivere, I look forward to a new Camelot.

WILLIAM ETKIN

Preface

Amphibian and insect metamorphosis has excited the interest and imagination of man since the time of Aristotle. Metamorphosis exemplifies the most profound change in form found during the postembryonic development of all animals. As early as the late 1800s the process was recognized as reflecting significant steps in biological evolution both vertebrate and invertebrate. More recently a wide spectrum of life scientists ranging from developmental biologists to biochemists have addressed the fundamental questions related to metamorphosis. The commanding role of both invertebrate and vertebrate hormones was quickly recognized and exploited by endocrinologists and other biologists. Currently, molecular biologists—who were formerly preoccupied with phage and *Escherichia coli*—have begun to focus their attention on the problems of cellular transitions and to utilize animal forms that undergo metamorphosis as a model system for the study of development and differentiation.

The presence of several authoritative reviews on specific facets of this subject does not mitigate the need for a thorough and systematic work that brings together the many disparate strands of knowledge and research on metamorphosis. This work presents in a single volume the numerous aspects of metamorphosis, as did the first volume of *Metamorphosis: A Problem in Developmental Biology*, published in 1968. Except for the title and subject matter, there is a minimum of overlap between the two editions. The subject has been arranged to permit several points of view. A new group of contributors has been encouraged to use their own unique approaches and styles to keep the volume diverse and stimulating.

The book has been designed primarily for research scientists and biology graduate students. It should be useful in any college course encompassing development, endocrinology, physiology, or biochemistry. We recommend it for scholars interested in growth and development, from undergraduate juniors to senior researchers.

Earl Frieden
Lawrence I. Gilbert

Tallahassee, Florida
Chapel Hill, North Carolina

Contents

Chapter 3

Cell Structure during Insect Metamorphosis

M. Locke

Chapter 4

Hormonal Control of Insect Metamorphosis

Noelle A. Granger and Walter E. Bollenbacher

Chapter 5

Chemistry, Metabolism, and Transport of Hormones Controlling Insect Metamorphosis

Lawrence I. Gilbert and Walter Goodman

Chapter 6

Macromolecular Changes during Insect Metamorphosis

S. Sridhara

PART II: VERTEBRATES

Chapter 9
Survey of Chordate Metamorphosis
John J. Just, Jeanne Kraus-Just, and Deborah Ann Check

Chapter 10
Cytological and Morphological Changes during Amphibian Metamorphosis
Harold Fox

Chapter 11

Hormonal Control of Amphibian Metamorphosis

Bruce A. White and Charles S. Nicoll

Chapter 12
Biological Basis of Tissue Regression and Synthesis
Burr G. Atkinson

Chapter 13
Transitions in the Nervous System during Amphibian Metamorphosis
Jerry J. Kollros

Chapter 14

Changes in the Blood during Amphibian Metamorphosis

Robert H. Broyles

Chapter 15

**Biochemical Characterization of Organ Differentiation
and Maturation**

Sandra J. Smith-Gill and Virginia Carver

Chapter 16

The Dual Role of Thyroid Hormones in Vertebrate Development and Calorigenesis

Earl Frieden

CHAPTER 1

Metamorphosis: An Overview

GEORGE WALD

1. INTRODUCTION

With metamorphosis it is easy to know where to start, but hard to know where to stop. The big metamorphoses make a great impression. Ovid must have known of them in naming his great poem that is the source of most of our classic mythology. The term itself is a misnomer, for much more is going on than the changes of form that it literally means. We realize now that along with the form the biochemistry is profoundly altered; and both kinds of change prepare for large changes in ecology and behavior. There are also always aspects of reminiscence, evocations of the metamorphosing animal's evolutionary past. Altogether a profound change of life, a *metabiosis*. And once one has explored a complete metamorphosis, one begins to recognize its bits and pieces all around.

For of course life is all changes, big and little. Not only is every living organism, even the simplest, unique, different from every other, even of its own kind; it also differs from its former self from moment to moment. In a sense, as every Hindu or Buddhist is taught, one dies each moment to be reborn to live the next moment, yet not altogether the same since one was changed by having lived the moment before. And as backdrop for this individual inconstancy, Life itself, the great tide of evolution, residues of which keep floating up from the depths. There is a lot of remembering.

GEORGE WALD · Biological Laboratories of Harvard University, Cambridge, Massachusetts 02138.

And no wonder. For we humans—like every other creature now alive—carry within ourselves the entire reach of our evolution. Every organism now alive, animal or plant, represents a continuous line of life that goes back to the first living cells to appear upon this planet, some three billion years ago; for if that line had ever broken, how could we be here? That virtual immortality is in the germ cells. All that time germinal cells, eventually eggs and sperm, have gone on living as single-celled organisms, Protists, reproducing their kind by simple fission, and now and then engaging in cell fusion, syngamy, as do numbers of other Protists, having got ready for that by halving their chromosome numbers.

But then those fused cells, now fertilized eggs, dividing repeatedly produce not only more germ cells to propagate the endless line, but bodies, to carry them about, feed them, protect them, on occasion warm them; to represent them in the arena of natural selection; and finally to mingle them in another fertilization. Then, having completed their primary function, the bodies can be discarded, at once or after some delay. That is what death is about; it came upon the scene in evolution with the first segregation of germ cells from bodies (cf. Wald, 1973). When the germ cells change, that is evolution; when bodies change, that is metamorphosis.

Our own human eggs and sperm have been through all of that. Some 300 million years ago they were making fishes, then amphibians, then reptiles, then other mammals, and now men. If we only exercise the good sense and restraint to let them go on unhindered, heaven knows what they may make in ages to come. But will we?

Big metamorphoses abound. They occur in every large animal grouping, as Table 1 shows. Once one grasps what is happening in a drastic metamorphosis, one recognizes the components as they appear singly or in smaller constellations much more widely. For example, the hemoglobins change in structure during the very striking metamorphoses of lampreys and frogs; but then one finds such changes in the hemoglobins of embryos or larvae and adults in every other vertebrate that has been examined.

What one calls metamorphosis is all a matter of how greatly and abruptly animals partition their lives. There are always the two dominant themes: feeding, for growth and maintenance; and sex, for reproduction. The feeding and growth necessarily come in early, the sexual phase later. The animal may change minimally, or may assume very different forms for each of these phases, metamorphosing from one to the other.

It is curious how obsessive each of these phases can be, how exclusively they may occupy the animal. The business of a fly larva or a caterpillar is to eat and grow; the business of the metamorphosed adult is to couple and lay the eggs. It may go on eating, or not (e.g., the mayfly). Salmon, alewives, and eels setting out on their spawning migrations apparently have stopped feeding. Their digestive systems have degenerated and would probably be unable to deal with food even if ingested.

TABLE 1. METAMORPHOSIS[a]

Phylum	Class	Larval form
Porifera		Amphiblastula
Coelenterata		Planula
Platyhelminthes	Turbellaria Polycladida	Müller's larva
	Trematoda	Miracidium, cercaria, redia, sporocyst
	Cestoda	Onchosphere
Nemertinea		Pilidium
Bryozoa	Ectoprocta	Cyphonautes larva
Annelida		Trochophore
Mollusca		Trochophore, veliger
Arthropoda	Crustacea	Nauplius in Entomostraca
		Zoea in Malacostraca
	Insecta	Nymphs of insects with incomplete metamorphosis
		Caterpillars, grubs, etc. of insects with complete metamorphosis
Echinoderma		Pluteus, Bipinnaria, Auricularia, etc.
Chordate subphyla		
Hemichordata		Tornaria
Urochordata		Tadpole
Cephalochordata		Amphioxus larva
Vertebrata	Agnatha (Cyclostomata)	Ammocoete
	Teleost fishes	Leptocephalus (eel), parr (salmon), etc.
	Amphibia	Tadpoles (frog), various aquatic larvae of salamanders

[a] From Balinsky (1975).

The jaws of male king salmon are often deformed so that they no longer meet. The two species of North Atlantic "freshwater" eel (*Anguilla*), American and European, make journeys of up to thousands of miles to overlapping areas in the Sargasso Sea without feeding, there to spawn and die (Schmidt, 1924). For all such fishes the spawning migration and reproduction are the last acts of life, and the metamorphosis that prepares for them prepares these animals also to die. Of course the same is true of the metamorphosed mayfly's single day in the sun.

Hence, there is a strange wisdom planted in the young, left to fend for themselves. A free-swimming sea squirt larva (*Tunicate*) has a very simple structure, serving almost exclusively for motility, and it may exist as a larva for only minutes or hours; yet it performs the complex job of finding a place of attachment for the sessile adult. How much active choice is involved we do not know; the larva does possess sense organs— eye spots, a statolith—that the adult lacks. Nor do we know how often it misses, since we are left only with the successes.

How do the larval eels, bred in the Sargasso Sea, find their way back alone to the American and European coasts? How do the two species sort themselves out? It takes the American species about 15 months to get back, the European eels about 3 years. What impels them, still as larvae, to leave the Gulf Stream to journey to the coasts? We still not only have no answers, but not even a plausible guess. It has been alleged that after metamorphosing only the female elvers ascend the rivers into fresh water, the males staying out at the river mouths (Bertin, 1956); but I wonder. For the eels remain almost wholly undifferentiated sexually during the next 5–15 years, totally occupied with feeding and growing, having put off all sexual manifestations until after a second metamorphosis (of which more below), when they will set out again for the Sargasso.

And how do the prematurely born kangaroo joeys, still essentially embryonic, find their way unassisted by the mother into the pouch, and there fuse to a nipple?

Those great drives, feeding and sex, are of course universal. They involve large metamorphoses only when the animals pursue them, not together, but consecutively, dividing their lives into stages, each occupied primarily or exclusively with one or the other function. Then metamorphosis enters like a rite of passage, closing one phase and preparing for the next.

So a free-swimming echinoderm or urochordate larva, specialized almost wholly for motility and hence dispersal, metamorphoses into a sedentary or sessile adult, specialized for feeding and reproduction. The winged insects reverse this order: the sedentary larvae, specialized for feeding and growth, metamorphose into highly motile, winged forms, specialized for coupling and reproduction.

It would be a mistake to assess the importance of these stages in the life cycle by the lengths of time they occupy. Ascidians may remain as free-swimming larvae from a few minutes (*Molgula*) to a few hours, up to a maximum of several days (*Styela*), before metamorphosing to the long-lived adult. At the other extreme, mayflies (Ephemeroptera) develop gradually as aquatic nymphs for many months to over a year, to burst forth as winged insects with admirable sensory and motor equipment, prepared to couple, lay their eggs in water, and perish, all within a day.

These transformations recall fundamental evolutionary progressions. The wormlike caterpillar or grub in metamorphosing to a winged insect kaleidoscopes the evolutionary sequence that is thought to derive the Arthropods from polychaete worms, and so binds together the Annelid–Arthropod superphylum. Similarly, the striking resemblances between echinoderm larvae and larvae of tunicates and acorn worms (Hemichordata) offer the most substantial evidence that the chordate stock shares ancestry with the echinoderms to form the Echinoderm–Chordate superphylum. And, of course, the tadpole metamorphosing to a frog or toad seems to compress within a few weeks the changes that marked the emergence of freshwater fishes to the land.

Even in such an obvious instance as this last, there are occasional surprises. Offered two newts, one with lungs, the other without, surely one would guess the former terrestrial, the latter aquatic. Frequently a wrong guess: our New England spotted newt *Diemyctylus*, when sexually mature, has lungs and is highly aquatic, whereas its lungless neighbor *Desmognathus* is one of the most persistently land-dwelling salamanders we know. Throw both into a tank that offers both environments, and the newt with lungs seeks the water, the lungless one the land (Wilder, 1912–1913). Indeed most lungless salamanders are more terrestrial in habit than aquatic.

That does not represent an aberration in the animals, but a misunderstanding by biologists. Wet-skinned amphibians are primarily skin-breathers. Their lungs are mainly not for respiration, but hydrostatic organs to help the animal to float, particularly its foresection, in the water. When that now favorite object of metamorphic studies, the bullfrog, bursts forth with legs and lungs, one thinks—and writes—Ah, yes! The aquatic tadpole has now become a terrestrial frog. Tell that to the bullfrogs! I have been watching them in a nearby pond all summer. They rarely go ashore unless forced to. They spend most of their lives floating quietly in the water with the upper part of the head just breaking the surface. The clawed toad *Xenopus* metamorphoses completely, then never leaves the water. And, as for legs—don't forget the webbed feet!—which is more agile in the water, a tadpole or a frog? I think a frog. Tadpoles are vegetarian, frogs carnivorous. The legs may be more useful for catching prey in the water than for walking about on the land.

Again, *Amphioxus*, the lancelet, after metamorphosis is a beautifully streamlined, spindle-shaped creature, looking admirably adapted for swimming. Yet the highly asymmetrical larva gets around much more than the adult, which buries itself in sand or gravel up to its mouth, and remains semisessile like other filter feeders.

The most striking features of classic instances of metamorphosis—their abruptness and extent of transformation—hide what is often an underlying continuity. The larva frequently possesses rudiments of adult structures, the adult vestiges of larval structures. One is dealing in fact with a dual organism, the larva and adult developing not so much in succession as side by side. In each phase of the life cycle, the one set of structures is realized, the other latent. One is tempted to invent a term for this: *schizobiosis*, meaning split life.

With this underlying contiguity goes an extraordinary independence. Each form, larval and adult, pursues its own program of development relatively independent of the other. Each is controlled by its own constellation of genes, that advance and recede in their levels of activity. Each has established independently its own ecological relationships and adaptations; and looked at in longer perspective, each has pursued its own evolution. We may take it for granted that whenever the modes of life of young and adult begin to differ markedly, selective factors will operate differently upon them, they will diverge genetically more and more, the development of each will become more necessarily independent, and the metamorphosis from one stage to the other more complete.

An ascidian larva, for example, presents sharply demarcated from each other its own transitory organs, some vestiges of which will remain in the adult, and rudiments of the adult organs that will not develop further until metamorphosis (cf. Barrington, 1968, pp. 225–229). So also one finds in the larvae of many types of winged insects the clusters of cells called imaginal discs, which have been set aside to form the adult organs, and indeed represent in visible form in the larva much of the adult organism (cf. Wigglesworth, 1954, pp. 7–8).

Either the larva or the adult can develop genetic changes apart from the other. Thus, larvae of the silkworm or of the gypsy moth *Lymantria* exist in many different racial varieties, all of which metamorphose to the same adult. Conversely, very different varieties of some species of butterfly (e.g., *Papilio polytes*) may have indistinguishable caterpillars (cf. Wigglesworth, 1954, p. 6).

In some instances such relationships can be traced back into the egg. Thus, in certain flies (*Musca, Drosophila*) the egg as laid is a mosaic. Local injuries produced by ultraviolet irradiation immediately after laying can cause local defects in the larvae without seeming to affect the adults; but by 7 hours after laying, injuries to the egg can cause highly localized defects in the adults, often without visible effect on the larvae (Geigy, 1931). Similarly, in the clothes moth *Tineola* irradiating different points in the egg can cause purely larval or purely adult defects (Lüscher, 1944; cf. Wigglesworth, 1954, p. 4).

Could the argument above for the persistence of both larval and adult structures throughout life be made also for the molecules? That is an interesting question that needs pursuing much further. The very observation that brought me into this field, the metamorphosis of visual pigments in the bullfrog from porphyropsin in the tadpole to rhodopsin in the froglet, had a surprising aftermath many years later in the discovery that fully mature bullfrogs still retain some porphyropsin, though withdrawn to the dorsal borders of their retinas (Reuter *et al.*, 1971; see below). Similarly, the human fetus, though 65–95% of its hemoglobin is of the fetal varieties, also possesses minor amounts of the adult hemoglobins. The fetal hemoglobins diminish rapidly for a few months after birth, yet persist permanently in the adult at levels under 1% (Lehmann and Kynoch, 1976).

This persistence of larval or fetal characters throughout life makes much more readily understandable the phenomena of *second metamorphosis*, the return of larval· characters, as in certain fishes and amphibians, at sexual maturity in the late adult (Wald, 1958; Grant, 1961). It is perhaps generally the case that what changes in metamorphosis is only the balance of activity in the sets of genes most concerned with establishing the successive stages in the life cycle. Those most active in the early embryos or larvae are perhaps never entirely turned off, but remain active at depressed levels throughout the adult stages.

And as the genes, the hormones. In both urodele and anuran amphibians, both the thyroid hormones and prolactin from the anterior hypophysis interact as antagonists throughout life, in both larval and adult stages, and in both first and second metamorphoses. The thyroid is already functioning in the larva, but is overbalanced by prolactin, which acts as a larval growth hormone, meanwhile staving off metamorphosis (Etkin and Gona, 1967; Bern et al., 1967). Then the thyroid hormones rise in concentration, stimulating the metamorphosis of tadpoles to frogs, and larval newts to red efts, including the land drive in the latter animals that sends them ashore. Finally prolactin again becomes predominant, stimulating in newts changes in skin texture and the water drive that brings them back into fresh water to spawn (Grant and Grant, 1958). Yet, even after that last transformation, an injection of thyroxine can drive these animals back to land, returning the skin also toward the eft condition (Grant and Cooper, 1964).

This two-hormone control is strikingly similar to the interplay throughout insect development, again as antagonists, of the juvenile hormone (neotenin), which maintains larval or nymphal development, and ecdysone, the molting hormone, which stimulates metamorphosis and adult development. Again as in amphibians, the larval hormone may return in the adult in specific association with reproduction, for the juvenile hormone has been reported to act as a gonadotropin in some adult insects.

It is the extraordinary good fortune of biologists that a group of organisms that displays complete metamorphosis, the flies (Diptera), also possesses giant chromosomes, great bundles of DNA strands lined up in phase, in which banding makes visible the positions of individual genes. Here and there one sees at times a swelling—a so-called puff—at a gene locus, which appears to represent an area of active transcription, synthesis of messenger RNA. At any one time such puffs tend to appear at the same points in the chromosomes of all cells of a specific type, though at other points in other types of cells, even within the same organ. The "puff spectrum" also differs in each tissue at various stages in the life cycle. Thus, during early molting cycles, when both neotenin and ecdysone are present, the puff patterns differ regularly from those during metamorphosis, when neotenin is in decline. That the latter puffing is induced by ecdysone can be shown in vitro as well as in vivo, correlated closely in both situations with the dosage (Clever, 1962, 1965).

And now I should like to go over to my own special absorption with metamorphosis.

One of the most exciting experiences of my scientific life was the discovery of the molecular metamorphosis of visual systems in the bullfrog (Wald, 1945). McCutcheon some years before had found hemoglobin metamorphosing in the same species (1936). It made a beginning with what was then a new and heady realization: that when the anatomy changes in metamorphosis, so does the biochemistry. That might have been anticipated; still, it had to be found. Soon there was much more, coming from many quarters. I was led eventually to try to think through the significance of metamorphosis in the lives of vertebrates, involving as it always does simultaneous changes in anatomy, biochemistry, and ecology; and always in profound association with phylogeny (Wald, 1958).

What follows is essentially a revision of that 1958 essay, of which, I am glad to note, much remains cogent.

2. PRIMACY OF THE SPAWNING ENVIRONMENT

Two kinds of visual systems are found in the rods of vertebrate retinas. One is based upon the red visual pigment, rhodopsin, formed by the combination of the protein opsin with retinal$_1$ (retinaldehyde), the aldehyde of vitamin A$_1$. The other is based upon the purple pigment, porphyropsin, formed from the same type of opsin combined with retinal$_2$, the aldehyde of vitamin A$_2$. Retinal$_2$ and vitamin A$_2$ differ from retinal$_1$ and vitamin A$_1$ only in possessing an added double bond in the ring in the 3,4 position (Fig. 1) (Wald, 1956, 1958; Morton and Pitt, 1957).

The porphyropsin system was first discovered in freshwater fishes. Marine fishes and land vertebrates characteristically possess the rhodopsin system (Figs. 2 and 3) (Wald, 1939a,b).

What of the fishes that are neither freshwater nor marine, but migrate between both environments? It would be well before discussing them to clarify somewhat their biological position.

Most fishes are restricted throughout their lives to narrow ranges of salinity. Such forms are called "stenohaline" and are of two kinds, freshwater and marine. A much smaller group of fishes can live as adults in a wide range of salinities. They are called "euryhaline," and, again, are of two kinds, anadromous and catadromous, meaning "upstream" and "downstream." These terms refer to the direction of the spawning migration. Salmon, for example, are typically anadromous forms, coming upstream to spawn, whereas the "freshwater" eels are catadromous, going downstream to the sea on their spawning migration.

It is probably true, however, that no euryhaline fish has to leave its spawning environment to complete a normal life cycle. Many instances are known in which anadromous fishes remain permanently in fresh water. The same is true of such an anadromous cyclostome as the sea lamprey, which colonized the Great Lakes and virtually destroyed the freshwater fisheries there.

So far as we know, the spawning environment is always fixed. The eggs, the sperms, or the embryos, perhaps sometimes all three, are stenohaline. Euryhalinity develops later in life and permits, though does not compel, these animals to migrate to the other environment. Migration

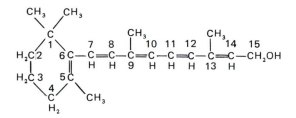

Vitamin A₁ (Retinol) C₁₉H₂₇CH₂OH

Retinal₁ C₁₉H₂₇CHO

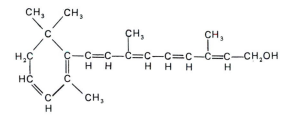

Vitamin A₂ (3,4-dehydro-retinol) C₁₉H₂₅CH₂OH

Retinal₂ C₁₉H₂₅CHO

FIGURE 1. Structures of vitamin A₁, retinal₁, vitamin A₂, and retinal₂.

is only a potentiality, which some of these forms exploit regularly and others rarely. The salmons are essentially freshwater fishes with the privilege of going to sea as adults; the freshwater eels are marine fishes with the capacity of coming as adults into fresh water.

The significant biological statement concerning such fishes is not that they migrate but that, being fixed in spawning environment, they are euryhaline as adults. I should like on this basis to redefine the terms applied to them. An anadromous fish is a euryhaline form that spawns in fresh water; a catadromous fish, one that spawns in the sea (Wald, 1939a).

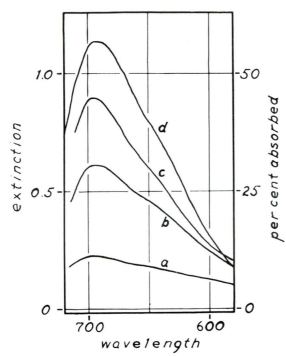

FIGURE 2. The vitamin A_2 of the retinas of freshwater fishes. Antimony chloride tests with extracts of wholly bleached retinas display only the absorption band maxima at 690–696 nm characteristic of vitamin A_2. This result has been obtained invariably in about 12 widely distributed species of freshwater teleosts. In almost all cases the visual pigment has also been extracted, and this pigment has been found to be porphyropsin, with λ_{max} about 522 nm. Extinction = log I_0/I, in which I_0 is the incident and I the transmitted intensity at each wavelength. Republished by permission of the *Journal of General Physiology*; Wald, 1939a.

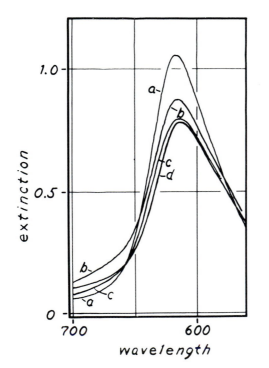

FIGURE 3. The vitamin A_1 of the retinas of marine fishes. Spectra of the antimony chloride tests with extracts of bleached retinas reveal the λ_{max} at 615–620 nm characteristic of vitamin A_1. This result has been obtained with a great variety of bony, and a few elasmobranch, fishes and is characteristic also of land vertebrates. However, a number of wrasse fishes (Labridae) and closely related Coridae, though wholly marine, are exceptional in having a predominance of vitamin A_2 in their retinas. Republished by permission of the *Journal of General Physiology*; Wald, 1939a.

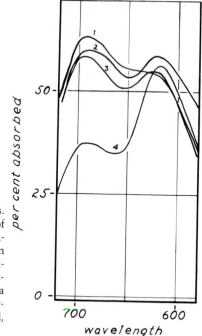

FIGURE 4. Retinal vitamins A in euryhaline fishes. Spectra of antimony chloride tests with extracts of bleached retinas from (1) chinook salmon, (2) rainbow trout, (3) brook trout, and (4) the American "freshwater" eel. All these tissues contain both vitamins A_1 and A_2, the anadromous salmonids a predominance of vitamin A_2, the catadromous eel a higher proportion of vitamin A_1. Republished by permission of the *Journal of General Physiology*; Wald, 1939a.

On examining the visual systems of several genera of salmonids, I found that all of them possess mixtures of the rhodopsin and porphyropsin systems, yet primarily the latter, characteristic of the spawning environment. Conversely, the American freshwater eel possesses a mixture of both visual pigments, in which rhodopsin—again the spawning type— predominates (Fig. 4) (Wald, 1939a).[*] Certain other anadromous fishes— alewife, white perch—possess porphyropsin almost alone (Fig. 5). All the euryhaline fishes examined follow a simple rule: all of them possess, either predominantly or exclusively, the type of visual system characteristically associated with the spawning environment (Wald, 1939a, 1941).

To a first approximation these patterns are genetic and independent of the immediate environment. The salmonids that were found to possess mixtures of both visual systems had spent their entire lives in fresh water.

---

[*] Later experiments in our laboratory and at the Stazione Zoologica in Naples by P. K. Brown and P. S. Brown demonstrated considerable variation in the proportions of vitamins A_1 and A_2 in the retinas of individual eels of both the American and the European species. These proportions vary between 65:35 and 25:75, with a mean value of approximate equality. The visual pigments, rhodopsin and porphyropsin, are present in approximately the same *molar* ratios; but since rhodopsin possesses a higher specific extinction than porphyropsin, extinctionwise rhodopsin tends to predominate.

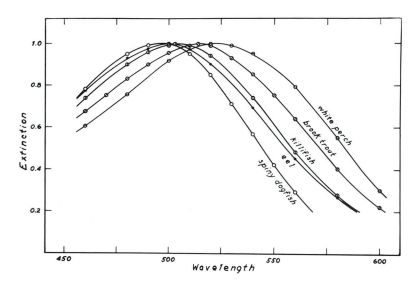

FIGURE 5. Spectra of the visual pigments of various fishes, illustrating the transition from an exclusively rhodopsin to an exclusively porphyropsin system, correlated with salinity relations. The permanently marine dogfish possesses rhodopsin alone (λ_{max} about 497 nm); the catadromous eel and brackish-water killifish, predominantly rhodopsin; the anadromous brook trout, predominantly porphyropsin; and the anadromous white perch, wholly porphyropsin (λ_{max} about 522 nm). Republished by permission of the *Journal of General Physiology*; Wald, 1941.

Alewives just in from the sea on their spawning migration possess porphyropsin almost exclusively. Most striking of all, the cunner and tautog, members of the wholly marine family of Labridae, the wrasse fishes, and some closely related Coridae, possess porphyropsin (Wald, 1939a, 1941; Kampa, 1952).

Since the distribution of visual systems among fishes is genetic, one may ask whether it fits into some evolutionary pattern. Many paleontologists are convinced that the vertebrate stock originated in fresh water. It is from such freshwater ancestors that our freshwater fishes were ultimately derived. The observation that these animals characteristically have the porphyropsin system suggests that this may have been the ancestral vertebrate type—a view that receives some support from the observation that the sea lamprey, *Petromyzon marinus*, a member of the most primitive living group of vertebrates, possesses porphyropsin as an adult (see below) (Wald, 1942). Subsequently vertebrates undertook two great evolutionary migrations, one into the sea, the other to land. Both led them to the use of rhodopsin in rod vision, for this is the pigment we find characteristically in marine fishes and land vertebrates. The euryhaline fishes are intermediate between freshwater and marine forms both in life history and in the composition of their visual systems. In this regard one can arrange the fishes in such an ordered sequence as shown in Fig. 6.

I have said above that freshwater fishes "characteristically" have vitamin A_2 and porphyropsin in their retinas, marine fishes "characteristically" vitamin A_1 and rhodopsin. When one makes such a generalization, it immediately becomes a target. There is little elation in confirming it, much more however in refuting, perhaps destroying it. In the process the attacker paradoxically begins by puffing up the importance of the generalization, to imply that he is bringing down big game. He may accord it the ultimate insult of calling it "classical," meaning it's wrong, but ancient enough to be forgiven.

So I had no sooner announced this partition of visual systems between freshwater and marine fishes than a hunt began to find exceptions. Happily, I had found the first exceptions myself: two wrasses (Labridae), which, though permanently marine, possess mainly vitamin A_2 and porphyropsin. Kampa (1952) found the same condition in the closely related Coridae. These exceptions were no embarrassment. On the contrary, they made the highly positive contribution of helping to show that the visual chemistry is determined genetically, independently of the environment. I think that is always the proper use of exceptions—to lead to more complete, more accurate generalizations. To use them merely to cast doubt or destroy is an act of vandalism. The whole business of science is to find meaningful generalizations; and every exception becomes an invitation to find the wider generalization.

After a lapse of some 20 years, how has this one fared? A whole little literature developed around alleged exceptions. I would feel better about

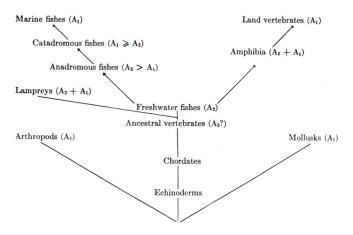

FIGURE 6. Distribution of vitamins A_1 and A_2 in vertebrate retinas. The observations, all made on contemporary animals, are here correlated with the present ecology. They may also, however, represent evolutionary sequences, and in that case they convey the suggestion that primitive vertebrate vision was based on vitamin A_2.

it if, as here, the estimation of what visual system or what mixture of visual systems is in use rested mainly on or at least included the antimony chloride tests with the vitamins A. The point is simple enough. The antimony chloride bands of vitamins A_1 and A_2 are sharp and about 75 nm apart; of retinals 1 and 2 about 40 nm apart; whereas the direct spectra of vitamins A_1 and A_2, or of the retinals or retinal oximes 1 and 2, or of rhodopsin and porphyropsin, are only about 20 nm apart, and are nearly twice as broad as the antimony chloride bands. I think therefore that the only really sensitive and accurate way to determine which systems are present and in what proportions is through the antimony chloride spectra with the vitamins A, with the antimony chloride bands of the retinals a second choice. Those are all too often lacking.

There is no doubt that a few marine fishes display the porphyropsin system in their retinas, and a few freshwater fishes the rhodopsin system. Nevertheless, I think that workers in this field might by now agree that if one picks a permanently marine fish out of the sea, no other questions asked, the chance of finding only rhodopsin and vitamin A_1 in its retinas is about 90%; and if a permanently freshwater fish, the chance of finding only porphyropsin and vitamin A_2 is about the same.

3. INTERNAL OR EXTERNAL SALINITY—WHICH DETERMINES THE VISUAL PATTERN?

The phylogenetic diagram in Fig. 6 suggests that the ancestral vertebrates used vitamin A_2 in vision, and passed this habit on to the freshwater fishes. Then the evolutionary migration to the sea brought the marine fishes, over euryhaline intermediate forms, over to the use of vitamin A_1. In reality this has probably not been a one-way transition. Except for the African lungfish (A_2), all the freshwater fishes examined were modern teleosts. The lungfish lies close to the line of fleshy-finned fishes that led through amphibians to the land vertebrates; but the teleosts seem to have come relatively recently out of another line, the ray-finned fishes. These again are believed to have arisen primitively in fresh water, and some (e.g., the carp: A_2) to have remained continuously in that environment. Others entered the sea to yield our marine fishes (mainly A_1); and some of them, including most of the freshwater fishes examined in this regard, are believed later to have left the sea to reenter fresh water. So there has been a lot of going back and forth; and if the euryhaline fishes are intermediate in this regard, there is no way to be sure in which direction they are evolving. An anadromous fish such as the salmon could be getting ready to become permanently marine, or could as well have just arrived in fresh water from the sea.

Our observations imply that whenever such evolutionary changes of habitat have occurred, the visual biochemistry changed with it. Whatever the complex of genetic changes that prompted the change of environment somehow carried this seemingly irrelevant biochemical transformation with it.

A similar problem involves freshwater turtles. They seem to represent an evolutionary return to freshwater existence after reptiles had already adapted to a wholly terrestrial life. All reptile eggs are laid on land, except for a few forms in which they are retained in the body until hatching. The few marine turtles and land tortoises that have been examined use vitamin A_1 in vision; but a freshwater turtle had A_2 exclusively. That fact seems to violate directly our thesis that the spawning environment principally determines which vitamin A is used in vision. A freshwater turtle is "catadromous" as we have used that term, and I would have expected it to use primarily vitamin A_1.

There is a way out of this dilemma for both the fishes and the turtles. It is to consider the blood salinity—the plasma osmotic pressure—rather than the salinity relations in the external environment to be perhaps the more important variable.

One must begin by realizing that vertebrates regulate their blood osmotic pressure relatively independent of the environment. The freezing point depression—a measure of blood osmotic pressure—clusters around 0.6°C in mammals and birds (Fig. 7). In amphibians it lies well below, from about 0.4 to 0.55°C, exactly the same range as in freshwater fishes. The lungfish *Protopterus* and the lampreys also fall in this range. On the other hand, marine fishes, as also marine and land turtles, fall distinctly higher, the latter at 0.6–0.7°C, the marine fishes from about 0.67 to over 1°C (cf. Wald, 1952).

One would not be far off the mark in stating that those vertebrates with plasma freezing point depressions above 0.6°C are likely to use vitamin A_1 in vision, those in which it is well below 0.6°C are likely to use vitamin A_2. That simple rule takes very good care of all the fishes, the lungfish included, and the lamprey. It makes us wonder a little why any amphibians, as low as is their blood osmotic pressure, develop rhodopsin. But it really helps with the freshwater turtles; surprisingly, they have blood osmotic pressures as low as any freshwater fishes.

The euryhaline fishes for which I would find such measurements also come out very interesting on this basis. Both anadromous and catadromous fishes—salmons and eels—have plasma freezing point depressions clustering around 0.6°C when in fresh water, and close to 0.8°C, like marine fishes, when at sea.

There is even evidence that the wrasses (Labridae), almost unique among permanently marine fishes in having a great predominance of the porphyropsin system, tend to be low also among marine fishes in the freezing point depression of their blood plasma.

To sum up: Perhaps the fundamental adjustment as animals evolve from freshwater environments to the sea and land—i.e., from hydrating to dehydrating environments—is in the change of blood osmotic pressure;

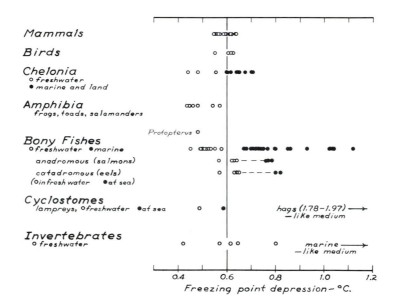

FIGURE 7. Osmotic pressures of the internal fluids of animals, assembled from various sources. The osmotic pressure is expressed in terms of the depression of the freezing point of water, to which it is proportional. Seawater possesses freezing point depressions ranging from about 1.8 to 2.4°C; similar freezing point depressions are found in the blood of hags and marine invertebrates. All the values shown are for adult animals. Note especially that the freshwater fishes divide completely from the marine fishes; that the euryhaline fishes are intermediate when in fresh water; and that the amphibians and the freshwater turtles fall in the same range as the freshwater fishes. From Wald, 1952.

and the pattern of visual systems may tend to follow this change in the internal rather than the external environment.

There was a heyday in measuring blood osmotic pressures early in this century, that has lapsed long since. I think it would be a fruitful line to take up again, not only in association with habitat relations of all sorts, but particularly with metamorphosis.

4. EURYHALINE FISHES AND AMPHIBIANS

One can hardly develop such an argument as outlined above without raising questions regarding the amphibians. These animals come between freshwater fishes and land vertebrates, just as the euryhaline fishes do between freshwater and marine fishes. Most amphibians, like most euryhaline fishes, spawn in fresh water. Indeed, the life cycle of the common frog runs strikingly parallel with that of such an anadromous fish as the salmon. Both originate and go through a larval period in fresh water.

Both, after undergoing deep-seated anatomical and physiological changes that can in both be described as metamorphosis, migrate for the growth phase, the salmon to sea, the frog to land. Both return to fresh water at sexual maturity to spawn. What land is to the frog, the sea is to the salmon. *The euryhaline fishes are the amphibians among the fishes.*

In this sense one might speak of almost all amphibians as "anadromous," meaning that they spawn in fresh water and are free, as adults, to go back and forth between fresh water and the land. [One would like to commit further etymological atrocities. The essence of the euryhaline condition is the capacity to migrate, not so much between low and high salt concentrations as between hydrating (freshwater) and dehydrating (sea and land) conditions. In this sense the amphibians are "euryhaline." Obviously one needs new terms, firmly grounded in ecological essentials rather than in trivialities.] A few amphibians (red-backed and tree salamanders) have developed special devices for living permanently ashore. As for a "catadromous" amphibian—that is, one that spawns on land and goes through its growth phase in the water—the common, lungless American salamander *Desmognathus fusca* comes close (Wilder, 1912–1913). This lives in and on moist earth, where it mates and lays its eggs, and where the larvae spend their first few days before migrating into shallow running water. There they remain nearly a year, before metamorphosing and establishing themselves permanently ashore. A number of aquatic reptiles also (alligators, freshwater snakes and turtles) could be thought of as "catadromous."

If these are substantial parallels, and if the spawning environment mainly decides the pattern of visual pigments, then one should expect such an "anadromous" amphibian as the common frog to possess mainly porphyropsin, like a salmon. Yet rhodopsin was originally discovered in the rods of frogs, and for a long period all that we knew of this pigment was learned with frogs.

Finding that the African lungfish *Protopterus* has vitamin A_2 in its retinas pointed up the problem, for this animal lies close to the line of fleshy-finned fishes from which the amphibians seem to be derived (Wald, 1945).

In this dilemma I turned to a tailed amphibian with the thought that it might display more primitive properties than the tailless types. Adults of the common New England spotted newt, *Diemyctylus* (formerly *Triturus*) *viridescens*, were found to possess porphyropsin exclusively (Wald, 1952). This brought a first amphibian into the same fold with certain anadromous fishes but left the frog in a more aberrant position than ever.

On examining bullfrogs in metamorphosis, however, I found that tadpoles just entering the metamorphic climax possess porphyropsin almost entirely, whereas newly emerged frogs have changed almost entirely to rhodopsin (Wald, 1945). The anatomical metamorphosis, which in this species takes about 3 weeks, is accompanied by this biochemical metamorphosis of visual systems. The bullfrog enters metamorphosis with porphyropsin, like a freshwater fish, and emerges with rhodopsin, like a land vertebrate (Fig. 8).

These observations have recently been confirmed (Wilt, 1959; Crescitelli, 1958); and Wilt has shown that the metamorphosis of visual pigments in the bullfrog can be stimulated to occur prematurely by treatment with thyroxine. A similar metamorphosis has also been observed in the Pacific tree frog, *Hyla regilla* (Crescitelli, 1958). On the other hand, several instances have been recorded in which the visual pigment does *not* appear to change at anatomical metamorphosis: the frogs *Rana esculenta* and *R. temporaria* (Collins *et al.*, 1953) and the toad *Bufo boreas halophilus* (Crescitelli, 1958).

Certain amphibians, therefore, like euryhaline fishes, may display both the rhodopsin and the porphyropsin systems. It seemed for a time that one difference between both groups might be that in euryhaline fishes the patterns of visual system are fixed, whereas in amphibians they change abruptly with metamorphosis. I shall have more to say of this later.

The discovery of the metamorphosis of visual systems in the bullfrog, made originally in 1942, had an astonishing aftermath almost 30 years later. Tom Reuter, in our laboratory, found that mature bullfrogs still possess the porphyropsin system, but confined to the dorsal quarter, sometimes only the dorsal border, of the retina. There, however, it can

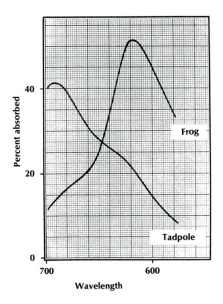

FIGURE 8. Biochemical metamorphosis of visual systems in the bullfrog, *R. catesbeiana*. The tadpole just entering metamorphic climax has in its retina vitamin A_2 (that is, porphyropsin) with only a trace of vitamin A_1 and rhodopsin, whereas the newly emerged froglet has just the reverse pattern. Republished by permission of the *Harvey Lectures*; Wald, 1945.

be in considerably higher density than rhodopsin in the remainder of the retina (Reuter *et al.*, 1971).

This observation changes our whole notion of what is going on. As Fig. 8 shows, the vitamin A_2 from tadpole retinas is accompanied by a small amount of A_1, and conversely, the A_1 from froglet retinas has with it a trace of A_2. I thought at the time that was only because I had bracketed the metamorphosis so closely. My tadpole was just entering the metamorphic climax, my froglet just emerging. I thought that younger tadpole retinas would have had vitamin A_2 alone, older frogs A_1 alone.

But now we know that the adult bullfrog retains the porphyropsin system throughout life. That means it retains the capacity to make vitamin A_2. No known plant carotenoid contains the 3,4-double bond that would make it a direct precursor of vitamin A_2. It must be formed from vitamin A_1 (or retinal$_1$), presumably through the operation of a 3,4-vitamin A_1 (or retinal$_1$) dehydrogenase. The blood and liver of the tadpole contain vitamin A_1, but we could find in them no measurable A_2. Apparently the conversion of A_1 to A_2 occurs in the eye itself (Ohtsu *et al.*, 1964), and our experiments showed that it occurs in the pigment epithelium, the single layer of cells in contact with the retinal photoreceptors. Apparently the 3,4-vitamin A_1 dehydrogenase, and hence the gene that determines its presence, remains active throughout the life of the bullfrog, though withdrawing in the adult to the most dorsal zone of the retinal pigment epithelium. Metamorphosis in this instance therefore involves not the turning off but only the anatomical segregation of a larval character.

5. BIOCHEMISTRY OF METAMORPHOSIS

At the time the metamorphosis of visual systems was discovered in the bullfrog, another such change in the same species had already been described. McCutcheon (1936) had found that the properties of hemoglobin in this animal change markedly at metamorphosis. The oxygen equilibrium curve of hemoglobin, measured at one temperature and pH, goes through a remarkable transition between tadpoles and adults. The hemoglobin of tadpoles has a high affinity for oxygen and it seemed from McCutcheon's measurements that the shape of its oxygen equilibrium curve might be hyperbolic, whereas the hemoglobin of young adults has a relatively low affinity for oxygen, and its equilibrium curve is distinctly S-shaped.

Riggs (1951) reexamined this situation in our laboratory. He confirmed McCutcheon's finding of a striking loss of oxygen affinity at metamorphosis. He found, however, that the *shape* of the oxygen equilibrium curve does not alter at metamorphosis; it is equally sigmoid throughout development. He found another important change: tadpole hemoglobin exhibits almost no loss of oxygen affinity on acidification (that is, no

Bohr effect), whereas frog hemoglobin has a very large Bohr effect (Fig. 9).

Though these statements hold for the whole hemoglobin content, further details have recently emerged (Broyles and Frieden, 1973; Watt and Riggs, 1975). Bullfrog tadpole hemoglobin now appears to have four major components. I and II predominate in very young tadpoles, which seem to synthesize them in the kidney. In older tadpoles III and IV largely replace I and II, and are synthesized in the liver. After metamorphosis, the adult hemoglobin is synthesized in bone marrow. Throughout this progression the hemoglobin keeps declining in oxygen affinity, a type of change that has tended to accompany the general evolution of hemoglobin in the animal kingdom (cf. Wald, 1952).

It is clear therefore that hemoglobin, like the pigment of rod vision, metamorphoses in the bullfrog at the time of anatomical metamorphosis. Both substances are conjugated proteins. In the rod pigment, it is the prosthetic group, retinal, that changes; the protein opsin, so far as known, remains unaltered. In hemoglobin it is the protein, globin, that changes; the prosthetic group, heme, is the same always.*

Frieden *et al.* (1957) have described a third change in the proteins of this species at metamorphosis (Fig. 10). In the bullfrog tadpole the predominant proteins of the blood plasma are globulins. At metamorphosis, the protein concentration of the plasma doubles, and albumins become predominant. These changes can be induced prematurely, just as can anatomical metamorphosis, by administering triiodothyronine.

Hahn (1962) has found similar changes in paedogenic tiger salamanders (*Ambystoma tigrinum*) metamorphosed with triiodothyronine administration.

Still another type of biochemical change has been shown to accompany metamorphosis in frogs and salamanders. Fishes excrete most of their nitrogen as ammonia, whereas land vertebrates excrete their nitrogen primarily as urea or uric acid. Munro (1939) showed some years ago that whereas the tadpoles of the frog *R. temporaria* excrete the great bulk of their nitrogen as ammonia, at metamorphic climax this animal goes over to excreting its nitrogen primarily as urea. At this time, also, arginase, the last in the chain of enzymes that forms urea, makes its first appearance in the liver (Fig. 11). Munro (1953) has demonstrated similar

---

* Such changes in hemoglobin seem to represent a fundamental property of vertebrates, for they penetrate to the most primitive forms. Adinolfi *et al.* (1959) have reported that larvae of the lamprey *P. planeri* possess two hemoglobins, and on metamorphosis change to two others. Sometimes they found all four hemoglobins together in metamorphosing animals. Manwell (1957) has reported finding different hemoglobins in postlarval and adult California sculpins (*Scorpaenichthys marmoratus*, a marine teleost). He states also that preliminary experiments on a live-bearing surf perch (*Embiotica lateralis*) reveal a fetal form of hemoglobin. Recently, Giles and Vanstone (1976) have studied the metamorphosis of hemoglobins during the parr–smolt transition in coho salmon.

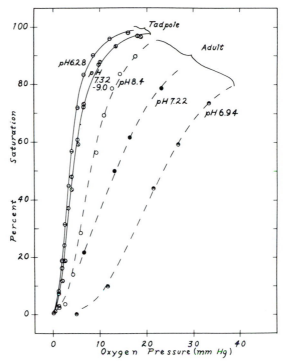

FIGURE 9. Biochemical metamorphosis of hemoglobin in the
bullfrog, *R. catesbeiana*. Tadpole
hemoglobin has a high affinity
for oxygen and virtually no loss
of affinity on acidification (the
Bohr effect); adult hemoglobin
has a much lower affinity for oxygen and a large Bohr effect. Republished by permission of the
Journal of General Physiology;
Riggs, 1951.

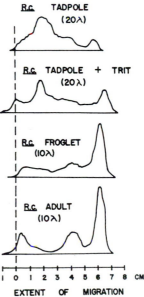

FIGURE 10. Biochemical metamorphosis of plasma proteins in the blood of the bullfrog, *R. catesbeiana*. Tadpole
plasma contains a predominance of globulins that migrate
slowly on electrophoresis at pH 8.6; in the froglet a change
to a predominance of rapidly migrating albumin takes
place. This change is also induced prematurely by treatment of tadpoles with triiodothyronine (T_3). The volume
of serum used is indicated in parentheses. Republished
by permission of *Science*; Frieden *et al.*, 1957.

changes accompanying metamorphosis in the toad *Bufo bufo*, the salamanders *Triturus vulgaris* and *T. cristatus*, and the axolotl *Siredon
mexicanum*.

These observations made a first beginning with the biochemistry of
metamorphosis. They showed that just as animals in metamorphosis
undergo radical alterations in anatomy, so their biochemistry is funda-
mentally revised at the same time. Indeed both kinds of change, anatomical and biochemical, herald an ecological transition, for they are followed
by radical changes of habitat. They mark also an evolutionary transition,
for these changes offer the most striking instances we know of recapitulation. The amphibian in metamorphosis seems to repeat in rapid summary the changes that accompanied the emergence of vertebrates from
fresh water onto land. The transformations of visual systems and of the
patterns of nitrogen excretion seem to provide clear instances of *biochemical recapitulation*. The changes in hemoglobin also seem to involve
aspects of recapitulation (Wald, 1952; Wald and Allen, 1957). Whether
the changes in serum proteins have this character, it is too early to say.
In any case, in metamorphosis the anatomy, the biochemistry, and,

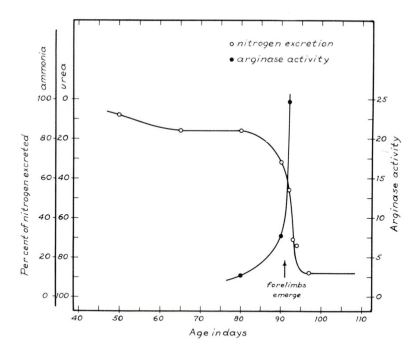

FIGURE 11. Biochemical metamorphosis of nitrogen excretion in the European frog, *R.
temporaria*. At metamorphic climax, this animal changes from excreting about 90% of its
nitrogen as ammonia to excreting about 90% as urea. Simultaneously, the activity of arginase, the enzyme that hydrolyzes arginine to form urea, rises sharply in the liver. Drawn
from the data of Munro, 1939; republished by permission of Academic Press; Wald, 1952.

shortly afterward, the ecology all are transformed, and frequently in some degree of accord with the animal's evolutionary history.

It is interesting to realize how closely these patterns hold together. An aberration in one of them seems to call forth appropriate aberrations in the others. The mud puppy, *Necturus maculosus*, for example, remains to some degree a permanent larva, never losing its external gills and never emerging from the water. Some years ago I found that adult mud puppies have porphyropsin alone, like a freshwater fish (Wald, 1946; Brown *et al.*, 1963).

The clawed toad, *Xenopus laevis*, a member of the peculiar family Aglossa, which possesses neither tongue nor teeth, is a purely aquatic form, which, though it metamorphoses, ordinarily never emerges from the water. Adults of this species possess in their retinas vitamin A_2 and porphyropsin almost exclusively (Dartnall, 1954; Wald, 1955; Dartnall, 1956). Underhay and Baldwin (1955) showed that this species also exhibits peculiar changes in the pattern of its nitrogen excretion. As a tadpole it excretes nitrogen primarily as ammonia. At metamorphosis, like other amphibians, it begins to change over toward urea excretion, so that at the height of metamorphosis it excretes a little more nitrogen as urea than as ammonia. Toward the end of metamorphosis, however, it swings back again, so that the adult excretes about three times as much ammonia as urea nitrogen. It is as though this animal, having got ready to leave the water, changed its mind; and both the getting ready and the change of mind are reflected in the nitrogen excretion. Indeed, *Xenopus* can change its mind again, for if kept moist and out of water it accumulates huge amounts of urea, perhaps as a device for conserving water such as is practiced by the elasmobranch fishes. Its return to water is attended by a massive excretion of urea accompanied by very little ammonia, after which it goes back to excreting its nitrogen primarily as ammonia (Cragg, 1953).

We see, therefore, that even the aberrations of amphibian metamorphosis, anatomical and ecological, are paralleled closely by the biochemistry. It is probably true that in all cases in which the anatomy or the ecology changes, the biochemistry also changes. Indeed the biochemistry may have a primary status; the visible alterations in anatomy and ecology may only reflect prior biochemical changes.

6. SECOND METAMORPHOSIS

The first requirement of a life cycle is that it be *circular*. Any organism that leaves its natal environment to explore, or grow up in, another must return at maturity to reproduce its kind. The spawning environment is fixed, whatever excursions animals may make as adults, and it is a truism that all animals must return to their natal environments to spawn.

For this reason, any animal that undergoes profound changes preparatory to migrating from its natal environment is likely to undergo a second series of changes *in the reverse direction* before returning. *Every metamorphosis invites a second metamorphosis.*

Let us begin with the common spotted newt mentioned above. This animal begins its life as an olive-green, gilled larva, living wholly in the water. After several months it metamorphoses to a lung-breathing, land-dwelling eft. The color changes to a brilliant orange-red, the skin becomes rough and dry, the lateral line organs recede. The newt now lives 2–3 years wholly on land, growing meanwhile almost to full size. Then it undergoes a second metamorphosis: the color returns approximately to that of the larva, and the newt regains the wet, shiny, mucus-covered skin, the keeled tail, the functional lateral line organs, though not, of course, gills. In this mature state it reenters the water to spawn and live out the remainder of its life (Noble, 1926, 1929; Dawson, 1936).

Many anatomical and behavioral aspects of the second metamorphosis can be induced prematurely in red efts by injection or implantation of anterior pituitary preparations (Dawson, 1936; Reinke and Chadwick, 1940); and a significant part of this complex of changes—the drive to reenter the water and the molt to a smooth, wet skin—is stimulated in hypophysectomized red efts by injections of prolactin, the lactogenic hormone of the anterior pituitary (Grant and Grant, 1958).

I have already said that the mature animal possesses porphyropsin, like a freshwater fish. These animals, however, had already undergone the second metamorphosis. Red efts on examination were found to possess mixtures of rhodopsin and porphyropsin, predominantly rhodopsin (Fig. 12). The second metamorphosis in this species is accompanied therefore by the biochemical metamorphosis of its visual system from a predominantly land type to that characteristic of freshwater types (Wald, 1952).

In extension of the present argument, Nash and Fankhauser (1959) examined the nitrogen excretion of this animal. Like other amphibians already mentioned, the larval newt excretes about 90% of its total nitrogen as ammonia. At the first metamorphosis, it goes over to excreting urea, and the red eft excretes almost 90% of its nitrogen in this form. Then, at the second metamorphosis, it turns back again, so that in the adults about one-fourth of the total nitrogen is excreted again as ammonia.

This, in turn, brings us back to the sea lamprey. This animal has a life cycle much like that of a salmon (Fig. 13). After passing 4–5 years as a blind ammocoete larva, living buried in the sand or mud of its natal stream, it undergoes, while still in that position, a profound metamorphosis, preparatory to migrating downstream to the ocean or a lake for its growth phase. This lasts $1\frac{1}{2}$–$3\frac{1}{2}$ years. Then the sea lamprey undergoes a second metamorphosis, to the sexually mature adult. The sexes differentiate visibly for the first time: the gonads mature, secondary sex char-

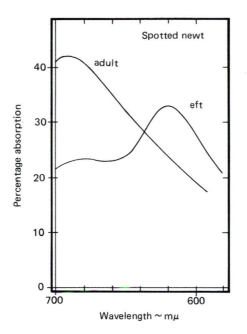

FIGURE 12. Second metamorphosis of visual systems in the New England spotted newt, *Diemyctylus viridescens*. Retinas of the land-living red eft contain a preponderance of vitamin A_1, with a minor admixture of A_2; retinas of water-phase, sexually mature adults contain vitamin A_2 predominantly or exclusively. (See Wald, 1952.)

acteristics appear, the males develop a ropelike ridge along the back, and either sex may assume golden mating tints. Then these animals migrate upstream to spawn (Gage, 1927).

Some years ago I found that the sea lamprey, taken on its spawning migration, has almost exclusively vitamin A_2 in its retina, and I concluded from this that it probably possesses porphyropsin (Wald, 1942). Since lampreys are members of the ancient class Agnatha, which includes the most primitive living vertebrates, I took this observation to support the view that porphyropsin is the ancestral type of visual pigment in vertebrates.

Later, however, Crescitelli (1956) reported that he had extracted rhodopsin from the retinas of this species and pointed out that this goes better with the opposed view, that rhodopsin is the primitive vertebrate pigment.

The specimens of sea lamprey examined by Crescitelli had just metamorphosed from the larval condition and had begun to migrate downstream, whereas the ones I had examined were at the other end of their life cycle, migrating upstream to spawn. On obtaining downstream migrants like Crescitelli's, I confirmed his observations exactly (Fig. 14). The retinas of such animals contain vitamin A_1 and rhodopsin alone. The upstream migrants, however, possess vitamin A_2 and porphyropsin

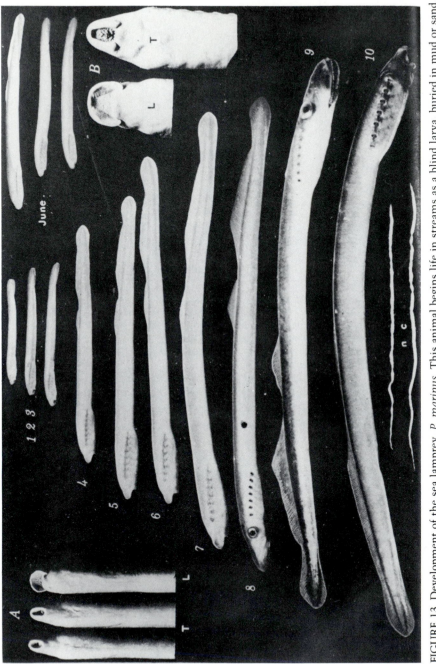

FIGURE 13. Development of the sea lamprey, *P. marinus*. This animal begins life in streams as a blind larva, buried in mud or sand (stages 1–7). Then it undergoes a first metamorphosis while still in this position (stages 8–10), preparatory to migrating downstream. Several years later it undergoes a second metamorphosis, to the sexually mature adult, and migrates upstream to spawn and die. (A, B) Transformation of the mouth at first metamorphosis from the larval, hooded form (L) to the contracted circular form (T). (nc) Notochords of decayed adults found in streams after spawning. Republished by permission of the New York State Conservation Department; Gage, 1927.

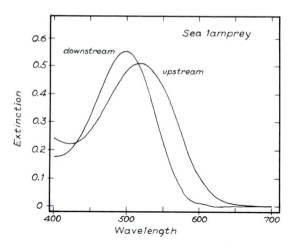

FIGURE 14. Second metamorphosis of visual systems in the
sea lamprey, *P. marinus*. The
downstream migrants, having
recently undergone the first metamorphosis from the larval stage,
possess rhodopsin. The sexually
mature adults, migrating upstream from the sea to spawn,
have porphyropsin alone. From
Wald, 1957.

virtually alone (Wald, 1957). We find therefore in this most primitive
group of vertebrates another biochemical example of second metamorphosis, like that previously observed in the newt.

Beatty (1966) has made an extensive study of changes in the proportions of rhodopsin and porphyropsin during the spawning migrations of
Pacific salmon. These are therefore further examples of second metamorphosis. In adult, migrating coho, king, pink, and chum salmon, the
retinas went from containing predominantly rhodopsin to predominantly
porphyropsin. In sockeye salmon, though the proportion of porphyropsin
increased, rhodopsin always predominated. On the other hand, no evidence was found in these fishes of visual pigment changes during the
first metamorphosis from the freshwater parr to the smolt about to migrate seaward.

Such second metamorphoses expose fundamental characteristics of
the metamorphic process.

1. Both the first and the second metamorphoses *anticipate* changes
in environment. Ordinarily they occur in the old environment and are
completed there. They are *preparations* for the new environment, not
responses to it.

2. Striking hormonal relationships are associated with these events.
It has been known for many years that the first metamorphosis is stimulated in many instances by thyroid hormones. A number of instances
are known in which phenomena associated with the second metamorphosis are stimulated by hormones of the anterior pituitary, including
specifically the lactogenic hormone, prolactin (Grant and Grant, 1958).
It is now recognized that in both tailed and tailless amphibians, both
prolactin and the thyroid hormones act antagonistically to each other
throughout life. Prolactin stabilizes the larval condition until overborne
by a more active thyroid, stimulating the first metamorphosis. Then, at
sexual maturity, prolactin may regain the upper hand as in the newt,

bringing on second metamorphosis and the reappearance of many larval traits (cf. Bern and Nicoll, 1968).

3. Just as the first metamorphosis prepares the animal to leave its natal environment, so the second metamorphosis prepares it to return, completing the life cycle. It is of the essence of a second metamorphosis to reverse in part the changes that accompanied the first metamorphosis. The two metamorphoses tend to be opposed in direction, anatomically and biochemically.

4. Just as the changes in the first metamorphosis tend to have the character of recapitulations—that is, to coincide somewhat with the animal's evolutionary history—so the changes that occur in a second metamorphosis are likely to be antirecapitulatory, to reverse in direction the sequence of changes that accompanied the animal's evolution.

The last consideration involves a potential source of confusion. As I have already said, a life cycle is circular. If one section of it runs parallel with the course of evolution, another section is likely to run counter to that course. Just as every metamorphosis invites a second metamorphosis, so every associated recapitulation invites a subsequent antirecapitulation. This is only proper, provided it occurs at the point in the animal's history when it is being prepared for the return to the natal environment.

7. DEEP-SEA FISHES; EELS

Heretofore I have discussed only changes in the visual pigments that involve their prosthetic groups. I should like now to discuss another type of change, involving the other component of a visual pigment, the protein opsin.

Denton (1957), and Denton and Warren (1956) have reported that the visual pigments of deep-sea fishes, instead of having absorption maxima (λ_{max}) near 500 nm, as do the rhodopsins of surface forms, have λ_{max} near 480 nm. In consequence, they are orange in color rather than red, and Denton and Warren proposed that they be called chrysopsins, or visual gold. For reasons that appear below, I prefer to call them deep-sea rhodopsins.

This observation has since been confirmed by Munz (1957) and by Wald et al. (1957). It makes good ecological sense; for the surface light that penetrates most deeply into clear sea water is blue, and made up of wavelengths near 480 nm. This is also the waveband in which, below the photic zone, bioluminescence by marine bacteria is concentrated (cf. Wald, 1967). Hence, the rhodopsins of deep-sea fishes are more effective with their maximal absorption in this region of the spectrum.

As might be expected, the transition from surface to deep-sea rhodopsin is not sudden. A preliminary exploration shows that the absorption spectra of the rhodopsins shift more or less systematically with depth from the surface to about 200 fathoms (Fig. 15). We find that throughout such a series the prosthetic group—the retinal—remains the same. It is the opsin that alters. We have here a relationship comparable with that familiar in the hemoglobins, all of which possess the same heme joined with a variety of globins, different in every species.

Disregarding the relatively few rhodopsins and porphyropsins that lie in exceptional positions, one sees, therefore, a major transition from λ_{max} 480 to λ_{max} 500 nm in the rhodopsins of marine fishes, correlated with depth, and depending on a systematic change of opsins; this connects with a further major transition from rhodopsin to porphyropsin (from λ_{max} 500 to λ_{max} 522 nm) correlated with the transfer to fresh water, and depending on the change of chromophore from retinal$_1$ to retinal$_2$.

With this we can return to the freshwater eel (*Anguilla*). Carlisle and Denton (1959) confirmed our observation that this animal, when taken in fresh water, ordinarily possesses the mixture of rhodopsin and porphyropsin described earlier; but they found that toward the beginning of its spawning migration it goes over to deep-sea rhodopsin (Fig. 16). Whereas the absorption peak of its usual mixture of visual pigments, when the eel is in fresh water, lies at about 505 nm, that of the animal about to migrate lies close to 485 nm. Indeed, the rhodopsin of such a

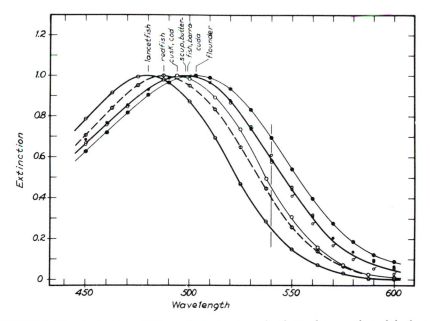

FIGURE 15. The rhodopsins of fishes taken at various depths in the sea. That of the lancet fish, found ordinarily below 200 fathoms, has λ_{max} about 480 nm; those of surface forms (scup, butterfish, barracuda, flounder) have λ_{max} 498–503 nm. The cusk and cod (from summer depths of 40–50 fathoms) have λ_{max} 494–496 nm, and the redfish (from a depth of about 100 fathoms) has λ_{max} 488 nm. Republished by permission of *Nature (London)*; Wald *et al.*, 1957.

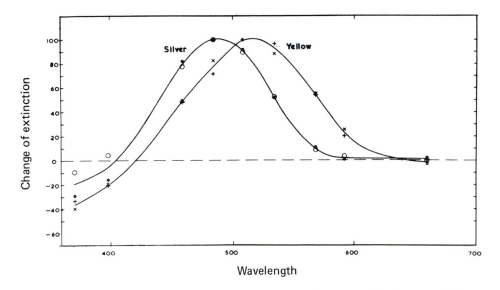

FIGURE 16. Second metamorphosis of visual pigments in the European freshwater eel. The yellow eel, prior to second metamorphosis, has a mixture of rhodopsin and porphyropsin (net λ_{max} 500–505 nm) corresponding to the mixture of vitamins A shown in Fig. 4. The sexually mature silver eel, about to begin its spawning migration to the Sargasso Sea, has changed over to deep-sea rhodopsin (λ_{max} about 487 nm). Republished by permission of the *Journal of the Marine Biological Association of the United Kingdom*; Carlisle and Denton, 1959.

freshwater eel preparatory to migration is virtually identical in spectrum with that of the permanently deep-sea conger eel (Denton and Walker, 1958).

This is another instance of a second metamorphosis (Fig. 17). The eel, having been spawned in the depths of the Sargasso Sea, journeys as a larva (leptocephalus) to the shores of America and Europe (Schmidt, 1922). There it metamorphoses to the adult form and usually, though probably not always, migrates into fresh water for its growth phase. Eventually it metamorphoses again: its color changes, the eyes approximately double in diameter, the digestive system deteriorates. As though getting ready for its return to the Sargasso Sea, it changes also to deep-sea rhodopsin.

My co-workers, Paul and Patricia Brown (1958) have examined such animals at the Stazione Zoologica in Naples. The European eel about to migrate seaward seems first to lose its retinal vitamin A_2 and retinal$_2$ and then to begin to combine vitamin A_1 and retinal$_1$ with a new, deep-sea opsin. The animal has already progressed far with this, as with the anatomical changes of the second metamorphosis, while still in fresh water.

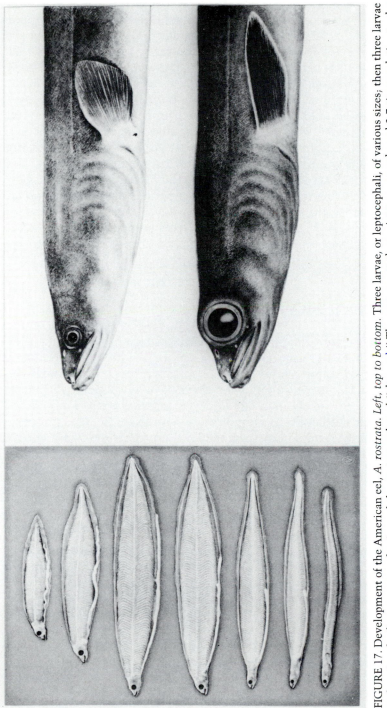

FIGURE 17. Development of the American eel, *A. rostrata. Left, top to bottom.* Three larvae, or leptocephali, of various sizes; then three larvae undergoing the first metamorphosis; and the metamorphosed "glass eel." The uppermost three pictures are enlarged 2.7 times relative to the others. *Right.* The second metamorphosis: (*top*) American eel in the "green" stage; (*bottom*) a European eel in the "silver" stage, ready for its spawning migration. Note particularly in the latter the approximate doubling in diameter of the eye. This large-eyed stage has not yet been observed in the American species. Republished by permission of the Department of Fisheries, Province of Quebec; Vladykov, 1955.

Such observations can tell us something concerning the larval condition. To my knowledge, no one has yet examined the visual pigment of the leptocephalus larva, but the foregoing discussion suggests strongly that the pigment is deep-sea rhodopsin. Similarly, I noted in 1958 that though no one had as yet examined the retinal pigment of the larval New England newt, our observation that the adult at maturity metamorphoses to porphyropsin implied that this is also the larval pigment. A few years later we confirmed that supposition (D. O. Carpenter, P. K. Brown, and G. Wald, unpublished observations). Again, since the blind ammocoete larva of the sea lamprey metamorphoses to an eyed adult possessing rhodopsin, this is the first visual pigment to appear in this species. Yet the fact that in the second metamorphosis the pigment changes to porphyropsin implies that the latter represents the true, albeit missing, larval type. That is, since the second metamorphosis involves some measure of return to the larval condition, it can tell us something of the larval state, even of larval properties that have been lost in the course of evolution.

8. LAND VERTEBRATES

Land vertebrates still pursue their embryogeny in water, but they have brought the water ashore. In a sense they are erstwhile amphibians that have carried water ashore in which their embryos go through the larval stages and first metamorphosis. They have developed two special devices for this: the boxed-in or cleidoic egg, and viviparity. Amphibians still experiment with both. Certain of them—for example, the American red-backed, slimy, and worm salamanders—lay eggs on land within which the larvae complete their entire development. Others—such as the European black salamander, *Salamandra atra*—retain the eggs in the body until the young are fully formed. The European spotted salamander, *S. maculosa*, ordinarily lays its eggs in streams, but if it cannot reach water, permits them to develop internally.

One might hope, therefore, to find residues of metamorphosis in the embryogeny of land vertebrates, and in this one is not disappointed. Anatomical residues abound; they were the original source of the idea of recapitulation and were principally responsible for its early overexuberance. The embryo of a land vertebrate undergoes an anatomical metamorphosis approaching that of an amphibian. Unlike a larval amphibian, it never has functional gills; but for a time it does of course have gill slits, as well as other evidences of earlier aquatic life.

One finds biochemical metamorphosis also in the embryos of land vertebrates, and it includes some of the same changes with which metamorphosis in amphibians has already made us familiar.

An example: In general, vertebrates hold the osmotic pressures of their body fluids within narrow limits. In the various groups of vertebrates the blood osmotic pressure takes characteristic values, correlated to a degree with the ecology, and perhaps also with the phylogeny (see Wald, 1952). Thus, the freezing point depression (Δfp) of the plasma in freshwater fishes and amphibians lies at 0.45–0.55°C, whereas adult birds and mammals exhibit values of 0.55–0.65°C. Measurements in the developing chick (Howard, 1957) show that the fluids of the early embryo have Δfp values of about 0.47°C (Fig. 18). The fluid osmotic pressure rises throughout development, with a final spurt at hatching that brings it to the adult value (Δfp = 0.58°C). That is, the embryo begins with a fluid osmotic pressure characteristic of freshwater fishes and amphibians and ends with that characteristic of mature birds.

This may be one example of a much more general phenomenon. In the frog *R. temporaria*, the ovarian eggs have an osmotic pressure like that of adult blood (Δfp = 0.41°C). Within a few hours after fertilization this has fallen to about 0.33°C, and in the gastrula stage reaches the extraordinarily low minimum of 0.275°C. Then it rises again, so that toward the end of the first week of development, it again approaches the adult level (Krogh *et al.*, 1938; Backman and Runnström, 1909; Bialaszewicz, 1912). I hardly know whether these changes in frogs and chicks are properly to be described as "metamorphoses." They may come too early in development and may be too continuous for that. I include them tentatively in this discussion in the hope that further examination will clarify their status.

A second example: Hall (1934b) has shown that during the embryonic development of the chick, its hemoglobin changes radically, continuously

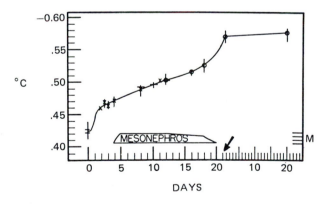

FIGURE 18. Biochemical metamorphosis of fluid osmotic pressure in the developing chick. *Ordinates*: freezing point depression, a measure of osmotic pressure. *Abscissae*: days of incubation, to hatching on the 21st day (*arrow*), and days thereafter; (=) unincubated egg white, (X) subgerminal fluids, (−) amniotic fluids, (O) blood. The embryo begins with osmotic pressures characteristic of freshwater fishes and amphibians and ends with the much higher osmotic pressure characteristic of mature birds and mammals. The duration of functional activity of the mesonephros (M) is also indicated. Republished by permission of the *Journal of Cellular and Comparative Physiology*; Howard, 1957.

losing affinity for oxygen, so that in an adult chicken more than twice the oxygen pressure is needed for half-saturation as is needed in a 10-day-old chick (Fig. 19). These changes persist for some time after hatching. They are similar in direction to the change in hemoglobin that accompanies metamorphosis in the bullfrog.

Comparable changes in hemoglobin accompany the embryonic development of all mammals so far examined. It is now well recognized that in mammals generally, man included, fetal hemoglobin is a molecular species different from maternal or adult hemoglobin (Fig. 20) (Hall, 1934a). Always—with the possible exception of man—the change in oxygen affinity is in the same direction, a loss of affinity as development progresses. The fetal and adult hemoglobins of mammals differ also in many other ways: in electrophoretic mobility, sedimentation rate, resistance to alkali, immunological specificity, solubility, crystal shape, and amino acid composition (for references see Wald, 1952). All these changes involve the globin moiety of hemoglobin; the heme is always the same.

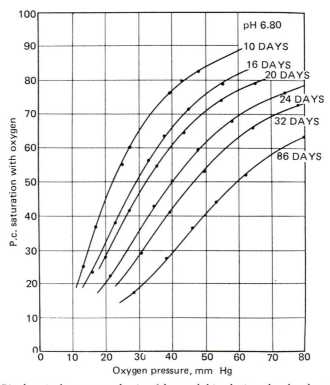

FIGURE 19. Biochemical metamorphosis of hemoglobin during the development of the chick. Measurements on dilute solutions of hemoglobin, buffered at pH 6.80, and equilibrated with oxygen at 37°C. The affinity for oxygen decreases regularly from the 10th day of incubation, and this change continues for some time after hatching. Republished by permission of the *Journal of Physiology*; Hall, 1934b.

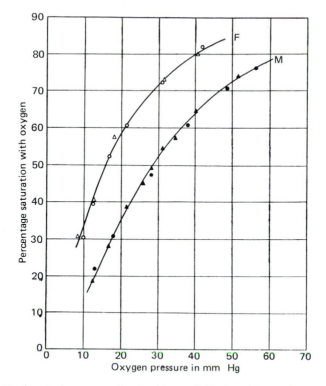

FIGURE 20. Biochemical metamorphosis of hemoglobin in a placental mammal. Oxygen equilibrium curves of hemoglobin from a goat fetus (F) and from the mother (M). Both hemoglobins were obtained at 15–18 weeks' gestation, and measured in solution at 37°C and pH 6.8. The fetal hemoglobin has almost twice as high an affinity for oxygen as the maternal hemoglobin. Republished by permission of the *Journal of Physiology*; Hall, 1934a.

The phenomenon of metamorphosis, biochemical as well as anatomical, extends therefore beyond the amphibians and fishes to include the land vertebrates, both egg-laying and placental.

Do land vertebrates exhibit vestiges of a second metamorphosis? I suppose that puberty is to be so regarded. To be sure, this does not prepare a land vertebrate to migrate, for the natal environment is now segregated, and puberty prepares the animal only to mate. Here only one representative cell—the spermatozoon—completes the return to the natal environment, and this, of course, undergoes a profound metamorphosis before being launched upon a migration as formidable, relative to its size, as that of any salmon.

It seems significant also that prolactin, a dominant hormone in larval development and at second metamorphosis in amphibians, continues to play roles associated with sexual maturity in land vertebrates, stimulating the crop gland that provides nourishment for newly hatched pigeons, and, of course, stimulating lactation in mammals.

9. CONCLUSION

Metamorphosis is a basic and general phenomenon, common to the whole vertebrate stock. It includes anatomical, physiological, and—perhaps prior to these—biochemical components, all designed to prepare the animal to leave its natal environment. Necessarily, in order to reproduce, the animal must eventually return, so completing its life cycle; and its return may be prepared for by a second metamorphosis, in some aspects the reverse of the first.

Our history as vertebrates is not dust to dust but water to water. From this point of view Nicodemus's question to Jesus, "How can a man be born when he is old? Can he enter the second time into his mother's womb and be born?" (John 3:4), can be given a broad and positive biological answer. Every animal can and must return to the "womb"—not, indeed, to be born again, but to bear the next generation. For a catadromous fish, the "womb" is the sea; for anadromous fishes and amphibians, a pond or stream; for land vertebrates, a uterus or egg.

10. EPILOGUE

In the opening pages of this chapter, I spoke of the fact that we human beings, along with all other organisms now alive, are the products of some 3 billion years of continuous life; and that that continuity is conveyed in our germ cells, eggs and sperm. I should like now to carry this argument a further step, involving the genes that, carried in the chromosomes of the eggs and sperm, largely determine what we are. They do that mainly by specifying the structure of our proteins.

We now realize that over virtually the entire range of evolution, the genes have displayed a degree of conservatism of structure, and hence have retained a closeness of relationship to one another that no one would have thought possible a generation ago.

Take the familiar respiratory enzyme cytochrome c, which occurs in every organism that respires oxygen. It is a heme enzyme, its prosthetic group heme joined to a protein that is a single chain of about 104 amino acids in vertebrates, to which invertebrate and plant cytochromes c add numbers of further amino acids at the $-NH_2$ end. Between man and the rhesus monkey, one amino acid in the chain of 104 has changed; between man and horse, 12; between man and chicken, 14; between man and tuna fish, 22; and between man and the yeast cell, 44.

Since each amino acid is coded in the genes by a triplet of nucleotides, the gene that specifies cytochrome c has at least 312 nucleotides. Interspecies differences almost always involve a change in only one nucleotide in a triplet, the minimal change needed to code for a different amino acid. What has been said, therefore, is that between man and the rhesus monkey one nucleotide in 312 has changed, and between man and yeast

44 nucleotides in 312. (That is somewhat of a simplification. When that many changes have occurred, some may have occurred more than once; but that hardly affects the argument.)

There was a time ages ago—perhaps 2 billion years ago, perhaps longer—when man and yeast shared a common ancestry. Some of those mutual ancestors went one way, to become yeasts; some took another road and eventually became men. Two diverging pathways lead from that remote time; and their double journey has resulted in the change of only 44 nucleotides in 312.

A further observation helps to show what that double journey means. Between the cytochrome c of yeast and that of the bread mold *Neurospora*, there are 39 changes of amino acid—a different group of changes, to be sure, yet almost the same number as divide yeast and man. Between *Neurospora* and man there are again 43 changes.

That tells us that yeast and *Neurospora* have been evolving apart from each other almost as long as either of them from man. One readily makes the mistake of thinking that such simpler organisms as yeasts and bread molds are in our remote ancestry; but no, they are our contemporaries, products of just as long an evolution: not our ancestors, but our cousins, several orders removed.

Matters are not very different when the genes reside in the same animal. The α and β chains of our adult blood hemoglobin, for example, and the γ chain of our fetal hemoglobin probably all evolved from a parent type resembling myoglobin, our muscle hemoglobin, by gene duplication followed by translocation in the chromosome (Ingram, 1961). Our α chain has 141 amino acids, the β and γ chains each 146. Between the α and β chains there are 78 amino acid changes, between α and γ chains 83. (We cannot make any simple comparison with the lengths of time occupied by the cytochrome c changes, for different genes mutate at very different rates.) Apparently the evolutionary progression has been α to γ to β. All three started as one, but α and γ have evolved apart from each other a little longer (25–50 million years?) than α and β. For even such highly mutable genes as these seem to take on the order of 10 million years on the average to establish a single change of amino acid in a protein sequence as the species norm, whether in the same or different organisms.

The glory of metamorphosis, particularly when "complete," is that one is dealing not just with two genes or two protein chains, but with two animals occupying the same skin, the one—larval, fetal—maintaining some degree of ancestral relation to the adult. For though they were once one, before setting out upon their schizoid course, the larval member has held more closely to the ancestral condition. Both develop and evolve in an astonishing degree of independence of each other, the constellations of genes that principally govern each of them drawing further and further apart, as all separated genes do with time; yet never so greatly as to lose large structural and functional contacts with one another and with the unity that was in their remote ancestry.

REFERENCES

Adinolfi, M., Chieffi, G., and Siniscalco, M., 1959 *Nature (London)* **184**:1325.

Backman, E. L., and Runnström, J., 1909, *Biochem. Z.* **22**:290.

Balinski, B. I., 1975, *An Introduction to Embryology*, p. 535, Saunders, Philadelphia.

Barrington, E. J. W., 1968, in: *Metamorphosis: A Problem in Developmental Biology* (W. Etkin and L. I. Gilbert, eds.), pp. 223-270, Appleton–Century–Crofts, New York.

Beatty, D. D. 1966, *Can. J. Zool.* **44**:429.

Bern, H. A., and Nicoll, C. S., 1968, *Recent Progr. Horm. Res.* **24**:681.

Bern, H. A., Nicoll, C. S., and Strohman, R. C., 1967, *Proc. Soc. Exp. Biol. Med.* **126**:518.

Bertin, L., 1956, *Eels: A Biological Study*, Cleaver–Hume, London.

Bialaszewicz, K., 1912, *Arch. Entwicklungsmech. Org.* **34**:489.

Brown, P. K., Gibbons, I. R., and Wald, G., 1963, *J. Cell. Biol.* **19**:79.

Brown, P. K., and Brown, P. S., 1958, unpublished observations cited in Wald, G., in: *Comparative Biochemistry*, Volume 1 (M. Florkin and H. S. Mason, eds.), Academic Press, New York (1960).

Broyles, R. H., and Frieden, E., 1973, *Nature New Biol.* **241**:207.

Carlisle, D. B., and Denton, E. J., 1959, *J. Mar. Biol. Assoc. U.K.* **38**:97.

Clever, U., 1962, *Gen. Comp. Endocrinol.* **2**:604.

Clever, U., 1965, in: *Mechanisms of Hormone Action* (P. Karlson, ed.), pp. 142–148, Academic Press, New York.

Collins, F. D., Love, R. M., and Morton, R. A., 1953, *Biochem. J.* **53**:632.

Cragg, M. M., 1953, Ph.D. thesis, cited in Underhay and Baldwin, 1955.

Crescitelli, F., 1956, *J. Gen. Physiol.* **39**:423.

Crescitelli, F., 1958, *Ann. N.Y. Acad. Sci.* **74**:230.

Dartnall, H. J. A., 1954, *J. Physiol. (London)* **125**:25.

Dartnall, H. J. A., 1956, *J. Physiol.* **134**:327.

Dawson, A. B., 1936, *J. Exp. Zool.* **74**:221.

Denton, E. J., and Warren, F. J., 1956, *Nature (London)* **178**:1059.

Denton, E. J., 1957, *J. Mar. Biol. Assoc. U.K.* **36**:651.

Denton, E. J., and Walker, M. A., 1958, *Proc. R. Soc. London Ser. B* **148**:257.

Etkin, W., and Gona, A., 1967, *J. Exp. Zool.* **165**:249.

Frieden, E., Herner, A. E., Fish, L., and Lewis, E. J. C., 1957, *Science* **126**:559.

Gage, S. H., 1927, in: *Biological Survey of the Oswego River System*, p. 158, Suppl. 17th Annu. Rep. N.Y. State Conserv. Dept.

Geigy, R., 1931, *Wilhelm Roux Arch. Entwicklungsmech. Org.* **125**:406.

Giles, M. A., and Vanstone, W. E., 1976, *J. Fish. Res. Board Can.* **33**:1144.

Grant, W. C., 1961, *Am. Zool.* **1**:163.

Grant, W. C., and Cooper, G., 1964, *Am. Zool.* **4**:413.

Grant, W. C., and Grant, J. A., 1958, *Biol. Bull. (Woods Hold)* **114**:1.

Hahn, W. E., 1962, *Comp. Biochem. Physiol.* **7**:55.

Hall, F. G., 1934a, *J. Physiol* **82**:33.

Hall, F. G., 1934b, *J. Physiol.* **83**:222.

Howard, E., 1957, *J. Cell. Comp. Physiol.* **50**:451.

Ingram, V. M., 1961, *Nature (London)* **189**:704.

Kampa, E. M., 1952, *Yearb. Am. Philos. Soc.* p. 161.

Krogh, A. K., Schmidt-Nielsen, K., and Zeuthen, E., 1938, *Z. Vgl. Physiol.* **26**:230.

Lehmann, H., and Kynoch, P. A. M., 1976, *Human Hemoglobin Variants*, p. 3, North-Holland, Amsterdam.

Lüscher, M., 1944, *Rev. Suisse Zool.* **51**:531.

McCutcheon, F. H., 1936, *J. Cell. Comp. Physiol.* **8**:63.

Manwell, C., 1957, *Science* **126**:1175.

Morton, R. A., and Pitt, G. A. J., 1957, *Fortschr. Chem. Org. Naturst.* **14**:244.

Munro, A. F., 1939, *Biochem. J.* **33**:1957.

Munro, A. F., 1953, *Biochem. J.* **54**:29.

Munz, F. W., 1957, *Science* **125**:1142.

Nash, G., and Fankhauser, G., 1959, *Science* **130**:714.

Noble, G. K., 1926, *Am. Mus. Novit.* No. 228.

Noble, G. K., 1929, *Am. Mus. Novit.* No. 348.

Ohtsu, K., Naito, K., and Wilt, F. H., 1964, *Dev. Biol.* **10**:216.

Reinke, E. E., and Chadwick, C. S., 1940, *J. Exp. Zool.* **83**:224.

Reuter, T. E., White, R. H., and Wald, G., 1971, *J. Gen. Physiol.* **58**:351.

Riggs, A. F., 1951, *J. Gen. Physiol.* **35**:23.

Schmidt, J., 1922, *Philos. Trans. R. Soc. London Ser. B* **211**:179.

Schmidt, J., 1924, *Annu. Rep. Smithsonian Inst.* p. 279.

Underhay, E. E., and Baldwin, E., 1955, *Biochem. J.* **61**:544.

Vladykov, V. D., 1955, in: *Fishes of Quebec*, Album No. 6, Department of Fisheries, Quebec, Canada.

Wald, G., 1939a, *J. Gen. Physiol.* **22**:391.

Wald, G. 1939b, *J. Gen. Physiol.* **22**:775.

Wald, G., 1941, *J. Gen. Physiol.* **25**:235.

Wald, G., 1942, *J. Gen. Physiol.* **25**:331.

Wald, G., 1945, *Harvey Lect.* **41**:117.

Wald, G., 1946, *Biol. Bull.* **91**:239.

Wald, G., 1952, in: *Modern Trends in Physiology and Biochemistry* (E. S. G. Barron, ed.), p. 337, Academic Press, New York.

Wald, G., 1955, *Nature (London)* **175**:390; see also Dartnall, 1956.

Wald, G., 1956, in: *Enzymes: Units of Biological Structure and Function* (O. H. Gaebler, ed.), p. 355, Academic Press, New York.

Wald, G., 1957, *J. Gen. Physiol.* **40**:901.

Wald, G., 1958, *Science* **128**:1481; see also *Circulation* **21**:916 (1960).

Wald, G., 1967, *Wistar Inst. Symp. Monogr.* No. 5, p. 59.

Wald, G., 1973, in: *The End of Life* (J. D. Roslansky, ed.), p. 3, North-Holland, Amsterdam.

Wald, G., and Allen, D. W., 1957, *J. Gen. Physiol.* **40**:593.

Wald, G., Brown, P. K., and Brown, P. S., 1957, *Nature (London)* **180**:969.

Watt K. W. K., and Riggs, A., 1975, *J. Biol. Chem.* **250**:5934.

Wigglesworth, V. B., 1954, *The Physiology of Insect Metamorphosis*, Cambridge University Press, London.

Wilder, I. W., 1912–1913, *Biol. Bull. (Woods Hole)* **24**:251.

Wilt, F. H., 1959, *Dev. Biol.* **1**:199.

PART I

INSECTS

CHAPTER 2

A Survey of Invertebrate Metamorphosis

K. C. HIGHNAM

1. INTRODUCTION

Because of the great variety and numbers of insects, together with their undoubted economic importance, it is sometimes forgotten that they are rather atypical invertebrates. The majority of the invertebrate phyla are marine, some with representatives in fresh water and on land; many are sessile as adults, or are burrowers in the sea bottom, or lead otherwise more or less sedentary lives. The reproductive strategies of such invertebrates commonly include free-living larval stages, which may be of some developmental importance as feeding stages to build up sufficient reserves for the production of the adult body, or this role may be secondary to the more important functions of dispersal and site selection. Insects, on the other hand, have been preeminently terrestrial creatures from the Devonian onwards, and the majority of modern forms have highly mobile flying adults. Consequently, insect larvae have a more profound developmental role than their counterparts in other phyla, since the functions of habitat selection and dispersal have been transferred to the adult generation.

In marine invertebrates, the planktonic larvae may be lecithotropic, utilizing contained yolk for their energy requirements, or planktotrophic, feeding upon other planktonic organisms. The equivalents of lecithotropic larvae are not found in insects, again a reflection of their devel-

K. C. HIGHNAM · Department of Zoology, The University, Sheffield S10 2TN, United Kingdom.

opmental rather than dispersal or site selection role. The adults of many invertebrate phyla may brood their young externally, so that the larval stages are condensed or eliminated; yet others, and this includes many insect species, show various degrees of ovoviviparity and viviparity. Closely related species may show quite different developmental patterns, and the reasons for this are usually obscure. In terms of fecundity, species with a long planktotrophic larval stage produce by far the largest numbers of eggs; viviparous species produce dramatically fewer numbers (Thorson, 1950). But it is difficult to estimate the energy input by the parent for the production of large numbers of more or less alecithal eggs compared with that for the production of a smaller number of heavily yolked eggs. It is even more difficult to relate such energy input to reproductive success, i.e., the net production of surviving reproductive adults of the next generation. But the enormous wastage of planktotrophic larvae must be balanced by the advantages of dispersal, such as the ability to colonize an impermanent habitat (Thorson, 1950), or greater genetic exchange, or the potential to extend the geographical range of a species (Crisp, 1974). With the exception of certain colonial species, the fecundity of female insects is rarely very great, and never approaches that of many sedentary or sessile marine invertebrates. This again emphasizes the importance of the adult as the dispersal and habitat selection stage: the presumed advantages of an enormous larval wastage in planktotrophic species simply do not obtain in the insects. This is not to say that the reproductive potential of an insect is not as great as that of many marine invertebrates; in many insect species, the combination of a relatively high reproductive rate with a short life cycle can produce very large numbers of offspring in a breeding season, and the overall wastage due to predation and disease must be equally large. But the individual female insect, although very efficient in converting stored reserves or ingested food into eggs, is not nearly as fecund as some of her counterparts in other phyla.

In spite of the role reversal between larval and adult insect, compared with the larvae and adults of other invertebrate phyla, metamorphosis remains as the process that transforms the earlier into the later stage, and many similarities exist between metamorphosis in insects and that in other invertebrates. This is a view not always held by authors of texts on embryology and developmental biology. Where metamorphosis is treated more than perfunctorily, the morphogenetic changes and controlling mechanisms in the insect transformation are described in detail. This is perhaps a necessary consequence of the wealth of information that exists about insect metamorphosis, to which this volume testifies. But it is surely unnecessary to suggest that the morphogenetic changes associated with metamorphosis in the various invertebrate phyla differ so much in their nature and causation that it is impossible to describe them in common terms (e.g., Balinsky, 1970, p. 609). Moreover, metamorphosis in invertebrates is often described as traumatic, a radical change in form associated with a correspondingly extensive change in the organism's way of life (Berrill, 1961; Ede, 1978; Grant, 1978; Barrington, 1979). When comparisons are subsequently made with insect metamorphosis, it is frequently the holometabolous insects that are used as examples, since the apparently dramatic nature of the transformation equates more nearly with the process in some other invertebrates. This implies a rather narrow view of the nature of the insect larva and the whole process of insect metamorphosis. The similarities between hemi- and holometabolous insect metamorphoses have often been stressed (Wigglesworth, 1954, 1959, 1970). It is believed that these views allow constructive comparisons to be made between metamorphosis in insects and the process in other invertebrates.

Metamorphosis may be a more or less dramatic change, depending upon the period of time over which it takes place, upon the change in habitat, or of physiology, which may take place between larva and adult, and also upon the degree to which the larval tissues are transformed into those of the adult. The last is of particular importance, since natural selection can produce change and adaptation in the larva independently of adult evolution. The resulting specialization of parts of the larval body may make them unsuitable for the adult transformation and they may be discarded at metamorphosis; or dedifferentiation may occur and the cells reused for building the adult body; or the larval tissues may be relatively unspecialized, and metamorphosis may then appear as no more than a more or less uncomplicated spurt of further growth and differentiation to produce the adult form. All such kinds of metamorphoses are found in the insects, and also in other invertebrate phyla. The different kinds of change must have evolved many times in the invertebrates; that such a variety exists also in the insects is once more a reflection of the diversity and adaptability of the phylum Arthropoda.

This survey of invertebrate metamorphosis will attempt to compare the morphogenetic changes underlying the various transformations with apparently similar changes to be found in different insect groups. Our greater knowledge of the underlying genetic programming involved in insect metamorphosis, together with its capacity for change and adaptation, may then shed more light upon the essential principles involved in the metamorphosis of other invertebrates.

2. INVERTEBRATE METAMORPHOSES

The structure and metamorphosis of "typical" or "characteristic" larvae are described in the account that follows: it should be remembered that the larvae in any taxon will have attributes that are species specific, and their metamorphoses will also differ in detail. But it is hoped that sufficient detail has been given to illustrate the general features involved.

2.1. Porifera

The amphiblastula larva of the Calcarea consists of two differentiated groups of cells: tall columnar flagellate cells that are directed forwards during swimming, and large nonflagellate granular cells. Shortly before attachment, the long axis of the larva shortens, the granular cells forming a saucer-shaped structure into which the flagellate cells invaginate. The larva attaches by the "blastopore" end and elongates to form the olynthus stage. The granular cells of the larva give rise to the epidermal pinacocytes and to the porocytes and scleroblasts. The flagellate cells become choanocytes, amoebocytes, and archaeocytes. Other sponges have a solid parenchymula larva that has an almost complete covering of flagellate cells around a mass of granular cells. At metamorphosis, inversion of the layers occurs. Larval settlement and metamorphosis in both types of larvae takes only a few minutes.

Sponge larvae therefore seem to undergo a kind of gastrulation before or during settlement, although the process is quite different from that in the Metazoa. But perhaps the larvae are best considered a free-living embryonic stage. In any event, there is a complete transformation of larva into adult. The larvae of some sponges also exhibit gregarious settlement, and fusion between individuals may occur after settlement. This may reduce individual mortality because it confers some trophic advantage; in a hermaphrodite, a genetic advantage will also accrue from the mixture of types (see Crisp, 1974, for references).

2.2. Cnidaria

The planula larva of hydrozoans has a ciliated ectoderm surrounding an initially solid mass of endoderm cells. Nematocysts, muscle processes, sensory, nervous, and secretory elements are all present. After a variable period of time the coelenteron develops within a layer of endoderm cells, and the larva settles on a suitable substrate (Section 4.1). In some species, the larva may "crawl" over the substrate before attachment. Settlement occurs at what was the anterior end of the larva and attachment is brought about by glandular secretions and the action of nematocysts. The cilia are lost, and a mouth and tentacles develop at what was the posterior end of the larva. In colonial forms, the colony is produced by budding from the first polyp. In many hydroid hydrozoans, dispersal is aided by the free-swimming medusoid generation, which bears the sex organs; the medusae do not become detached from the hydroid colony, and may develop no further than sporosacs, which may be considered almost the equivalent of sex organs on the hydroid. In such species, the planulae are as important for dispersal as for site selection. In *Tubularia*, the planula develops a mouth and tentacles while still within the sporosac: this actinula larva has a short free existence before attaching at its proximal end. The freshwater *Hydra* has complete direct development. In the trachyline hydrozoans, the free-swimming planula develops into a form very like the actinula, which develops into the adult medusa.

In the Scyphozoa, there is again a planula larva. This has a brief free existence and settles as a small polyp, the scyphistoma, which produces ephyrae larvae by strobilation. The ephyrae develop gradually into the adult medusae. In some species, there is direct transformation of the planula into an ephyra.

The Anthozoa also possess a planula larva, but there is never a medusoid generation as in the Hydrozoa. The mesenteries and tentacles develop in a particular and regular order, so that metamorphosis is more gradual but more extensive than in the Hydrozoa.

2.3. Ctenophora

The free-swimming larvae of ctenophores possess the basic structure of the adult, although their general form may be different. Metamorphosis here involves a gradual period of differential growth until the adult body is developed.

2.4. Platyhelminthes

2.4.1. Turbellaria

The majority of Turbellaria have direct development, but in the suborder Cotylea of the Polycladida, and also in a few Acotylea, the embryo hatches as the small ciliated Müller's larva. Its distinctive feature is the development of eight posteriorly directed lappets edged with long cilia. There is an apical sensory tuft and sometimes a caudal one also; frontal gland, eyes, brain, and mouth leading into the intestine are also present. Mesenchyme fills the space between the intestine and the surface ectoderm. The larva has a free-swimming life of a few days. During metamorphosis, the ciliated lappets together with their muscles are absorbed, and the sensory tufts disappear. Extra eyes develop, and the larva flattens out to become the young polyclad. The adult pharynx arises as a ring-shaped evagination of the innermost part of the larval stomodaeum with which it communicates by a narrow slit. During metamorphosis, the outer part of the larval stomodaeum is everted, and its walls become part of the external surface of the adult body; the ring-shaped evagination becomes the adult pharyngeal sheath surrounding the pharynx. The larval muscles, other than those within the lappets, persist into the adult. The metamorphosis of Müller's larva thus presents an excellent example of disposing of those structures adapted for larval life, but retaining or developing others necessary for adult existence. In the genus *Stylochus*, some species have direct development while others produce a modified Müller's larva with only four lappets (Götte's larva). A modified Müller's larva develops within the egg capsule of *Planocera reticulata*.

2.4.2. Trematoda

The parasitic flatworms have highly modified life histories compared with those of the free-living forms. The larvae of monogenetic trematodes possess ciliated bands, eyes, and a hooked opisthaptor. After attachment to the host, the cilia and eyes are lost, and the larval opisthaptor either develops into the adult type, or is shed and the adult structure develops *de novo*.

With more than one host in the life cycle, development is more complicated. The digenetic trematodes have a series of larval stages, each involving the multiplication of individuals and usually each with a distinct host. Typically, the embryo develops into a ciliated larva, the miracidium, in which the epidermis consists of a definite number of cells arranged in four or five tiers, the number per tier being characteristic of different families. An apical papilla, without cilia but armed with a protrusible stylet and gland openings, is used for host penetration, and an eyespot and nerve ganglion are also present. The posterior part of the body contains propagatory cells or germ balls.

Miracidia have a free-living existence of only a few days. After finding a suitable host—a snail or less often a bivalve mollusc—the larva uses its apical papilla to attach to and burrow through the skin. Inside the host, the larval features are lost and transformation into a sporocyst occurs. The propagatory cells or germ balls multiply and form either daughter sporocysts or rediae, which are self-exclusive (see below). The rediae escape from the sporocyst, and move about in and feed upon the host tissue: their body plan resembles that of the adult, but they too contain germ balls that develop either into a further generation of rediae or into cercariae. If the sporocyst produces daughter sporocysts, then the redial generation is omitted and cercariae develop from the sporocyst germ balls. The cercariae are swimming larvae, with a simple or forked tail.

The cercaria again has a short free-living existence during which it has to be eaten by the second intermediate host, a member of a number of invertebrate phyla, sometimes a fish or amphibian. The cercaria may bore through the skin of the host. Within the host, the cercaria loses its tail and then encysts as a metacercaria, surrounded by the cyst wall produced either by the secretions of special glands, or composed partly or entirely of host connective tissue. The purely larval tissues degenerate and disappear during encystment, and the adult features are developed. When the second intermediate host is eaten by the final, definitive host, the cyst is digested and the young fluke migrates to its species-specific location within the host.

The seven phases of egg–miracidium–sporocyst–redia–cercaria–metacercaria–adult, typically occurring in species where a bird is the final host, can be varied considerably in different fluke species. The miracidium may never be free-living, or redial generations may be omitted;

the sporocyst may be lacking, and rediae develop within the miracidia. Consequently, the first or second intermediate hosts may be omitted from some life cycles. It has been suggested that the digenetic trematodes were originally parasites of molluscs, particularly snails, before the evolution of the vertebrates, and adapted and evolved with their new hosts. But as Hyman (1951) points out, it is difficult to accept that all modern flukes have followed the same evolutionary pattern, leaving none as sexually reproducing forms in the original mollusc hosts.

Although the free-living stages of the digenetic trematodes are called larvae, and they fulfill the functions of dispersal and site (host) selection, the sequential changes that they undergo should not be considered as hypermetamorphosis (Section 3.4), a term that should be restricted to the changes an individual undergoes during its life cycle. Strictly, metamorphosis in these flukes should be restricted to the cercaria–metacercaria–adult transformation, initiated by the loss of the cercarial tail and continuing through the encysted metacercaria, which can be considered functionally analogous to the holometabolous insect pupa. The previous stages are adaptive mechanisms for the sequential multiplication of individuals through polyembryony, to counteract the enormous wastage that occurs in such complicated life cycles.

2.4.3. Cestoda

In the pseudophyllid cestodes, the larva is the coracidium, and shows little cellular differentiation except for three pairs of hooks at the posterior pole, some muscle fibers, and a pair of excretory flame cells, surrounded by an ectodermal membrane that may be ciliated in some species. Ingested by the first intermediate host, a crustacean, the body elongates and the posterior end with its hooks constricts off as a rounded or elongated tail, used as an attachment organ. This is the procercoid stage, reminiscent of the trematode cercaria. When eaten by the second intermediate host, usually a fish, the larva penetrates the intestinal wall and develops to the plerocercoid in various parts of the body. The plerocercoid grows and elongates, develops the scolex and other adult features, and becomes a juvenile tapeworm, except that it does not strobilate. Instead it encysts in the host tissues until eaten by the final host, a larger carnivorous fish, or less often a bird or mammal, where it attaches to the intestinal wall and grows to the adult worm.

In taenioid cestodes, there is only one intermediate host. The taenioid onchosphere differs from the pseudophyllid coracidium in being surrounded by one to three acellular membranes. On ingestion by the intermediate host, the onchosphere develops either into a cysticercoid or a cysticercus. The former consists of an anterior vesicle containing the scolex and a posterior, tail, region on which the larval hooks persist for some time; the tails may take a variety of forms in different species. The

cysticercus is a fluid-filled vesicle into which is invaginated the intro-
verted scolex. Cysticercoids encyst in a variety of invertebrate hosts; the
cysticerci, or bladder worms, are found in reptilian and mammalian,
rarely avian, intermediate hosts. When eaten by the final host, the scolex
emerges from both cysticercoid and cysticercus and attaches to the intestinal wall.

Although asexual multiplication is found in some tapeworm species,
the important difference between cestode and trematode life histories is
that in the former one egg produces one tapeworm, while in the latter
one egg may give rise to thousands or even millions of individuals. Cestode multiplication depends upon the very large numbers of eggs produced
in the proglottides continuously strobilated by the adult. In pseudophyllid
cestodes, metamorphosis can be considered the transformation procercoid–plerocercoid–adult; in taenioids, from onchosphere–cysticercoid/
cysticercus–adult, the shift in major differentiation being consequent
upon the number of intermediate hosts in the life history.

2.5. Nemertinea

In many nemertines, development is direct and some species are
viviparous. But many heteronemertines have an indirect development
with a free-swimming larva, the pilidium. This consists of an upper bell-
shaped region, with two pendent lobes on either side of the wide mouth.
The body is ciliated, with a band of longer cilia fringing the periphery of
the umbrella and the margins of the lateral lobes. An apical thickening
of ectoderm bears one or two particularly stout groups of fused cilia.
Internally, between the ectodermal body wall and the gut, is a gelatinous
mass in which variously shaped mesenchyme cells occur, some of which
develop into muscle bands that traverse the larva. The gut is blind ending,
with no anus.

Compared with what occurs in the apparently rather similar larvae
of some other invertebrate phyla, metamorphosis of the pilidium is extraordinarily complex. Discs of ectoderm invaginate from the body wall
of the larva: anterior cephalic, lateral cerebral, and posteroventral discs
are all paired; a dorsal disc and, in some forms, a proboscis disc are
unpaired. The invaginated paired discs initially form small sacs or vesicles, the inner walls of which develop a columnar epithelium while the
outer walls become flattened as they separate from the larval ectoderm,
to form the amnion of each paired disc. The single dorsal disc does not
form an amnion; neither does the proboscis disc when this is present.
The discs grow and spread around the larval gut, including some larval
mesenchyme as they do so. They eventually fuse to form a complete
covering around the larval gut and mesenchyme. As the discs fuse, the
amnion separates from each and fusion between the amnia also occurs.
Consequently, the ectodermal covering derived from the inner layers of
the original invaginations is itself covered by a complete, very thin am-

niotic membrane derived from the flattened outer cells of the original paired discs.

The cephalic discs form the epidermis of the anterior end of the adult worm; from them ectodermal masses move inwards to form the cerebral ganglia, which also give rise to the nerve cords. The cerebral discs form the canals of the cerebral organs, while the posteroventral and dorsal discs develop into the epidermis of the rest of the body. The proboscis disc, where present, forms the proboscis; otherwise this structure develops from the median region of the fused cephalic discs. The larval gut is retained, but replacement of the yolk-filled larval cells occurs by the development of particular cells in its wall. The anus develops from a proctodaeal invagination that connects with the intestine. The originally larval mesenchyme elements develop into the muscles and other mesodermal structures. The remains of the pilidium larva, including its ectoderm and apical plate, are cast off, the amnion is shed, and the young worm emerges.

Presumably as an adaptation to a shallow-water habitat, Desor's larva of *Lineus* develops within the egg membranes and lacks the apical tuft, oral lobes, and ciliated band of the pilidium. Its metamorphosis, via the agency of invaginated ectodermal discs, is, however, very similar to that of the pilidium.

2.6. Acanthocephala

The larva of the spiny-headed worms is the acanthor, contained within a two-valved shell. It possesses a rostellum with three pairs of hooks and is covered with spines. A pair of retractor muscles and circular fibers are associated with the rostellum, and a subepidermal muscle fiber network develops, surrounding a central syncytial core of highly condensed nuclei.

After ingestion by the intermediate host, usually a larval insect, or an isopod or amphipod, the shell splits and the acanthor penetrates the gut wall by means of its rostellum. The rostellar hooks, their associated muscles, and the body spines degenerate. The larva assumes a spherical shape and is now called the acanthella: it lives in the host hemocoele. The primordia of the adult structures—proboscis, ganglion, musculature, ligament sacs, gonads, and urinogenital system—differentiate in a particular order from the inner nuclear mass. A membrane develops on the surface of the acanthella. After differentiation of the inner nuclear mass, the juvenile remains quiescent until ingested by the final host, usually a fish, bird, or mammal, when the membrane ruptures and the adult worm develops in the digestive tract. The juveniles may also be ingested by fish, frogs, reptiles, etc., which are not the definitive hosts; they remain quiescent in these "transport hosts," which may be a more acceptable source of food to the final host that the intermediate host.

2.7. Aschelminthes

2.7.1. Nematoda

Nematodes, whether free-living or parasitic in plants or animals, all have the same basic life cycle: four larval stages separated by molts and an adult stage. Many species, however, may undergo the first one or two molts within the egg capsule, perhaps as an adaptation to parasitism. The larval stages are little different in form from the adult, except for size and mouth parts, and the lack of gonads and copulatory structures. It has been suggested that the term "larva" is a misnomer, and "juvenile" should be used instead. However, particularly in the parasitic forms, changes in habit and physiology must occur between the free-living and the parasitic stages, and in all species, the differences between immature and mature forms can be compared with those found in the life cycles of some insects (Section 3.3.2) so the term larva is considered to be appropriate. At each molt, the cuticle covering the body and lining the pharynx and rectum is shed; the basal part of the stylet is dissolved in some species and its remains are attached to the old cuticle. In other species, particularly animal parasites, the third-stage larva is ensheathed in a protective covering formed from the cuticle of the second stage that is not cast off. Exsheathment occurs by rupture of a particular ring of weakness around the anterior end of the sheath, and the tip of the sheath breaks off in the form of a cap.

2.7.2. Nematomorpha

The gordioid larva of the Nematomorpha has a spiny eversible proboscis and its body is divided into presoma and trunk. During its parasitic existence in the hemocoele of various insects, it grows into the juvenile worm without a dramatic metamorphosis, although the proboscis hooks and stylets, together with their associated epidermis and muscles, are lost when the larva leaves its host and molts to the adult.

2.8. Rotifera

Development is usually direct in the rotifers, although in sessile forms the females hatch as free-swimming juveniles, often called larvae. When they attach, the eyes are lost together with the cilia of the foot; the foot elongates to form the stalk.

2.9. Kinorhyncha

The eggs hatch into simple larvae, which molt several times with the addition of body joints until the adult form is attained. During each molt, the entire cuticle is shed, including the mouth styles and procto-daeal lining, the new cuticle being preformed beneath the old.

2.10. Priapulida

The minute larva has its trunk encased in a cuticularized armor, the lorica, made up of several plates. After a year or two of bottom living, the lorica is molted and the juvenile emerges.

2.11. Entoprocta

The larva of *Pedicellina* has a superficial resemblance to a trochophore, with an apical organ composed of a ciliary tuft springing from an ectodermal sac with underlying ganglion. A similar structure is present in front of the mouth. There is an equatorial girdle of long cilia. A vestibular invagination is present between mouth and anus. The gut is U-shaped. The larva of *Loxosoma* has an oval, rather flattened shape, with an anteriorly placed apical organ on the arched dorsal surface.

Metamorphosis begins when the larva attaches itself by its ciliary rim to a suitable substrate. The cilia are lost, and the rim becomes an attachment disc, with a more or less well-developed pedal gland. By closure of the rim, the vestibule with the mouth and anus loses connection with the outside. Mesenchyme accumulates between the vestibule and the attachment disc, and this region elongates to form the stalk. The apical and preoral organs are lost. The vestibule with mouth and anus now rotates through 180°; tentacles develop from the vestibular margin and the covering epidermis breaks down. The remaining larval structures pass into the adult. The originally ventral surface of the larva thus comes to face upwards in the adult, opposite to the stalk.

2.12. Polyzoa

The cyphonautes larva can be considered characteristic of gymnolaematous ectoprocts, although it may be considerably changed in those many species that brood their eggs and in which the young consequently have only a short free-living existence. The cyphonautes is triangular in shape, very compressed laterally, with a pair of shell plates covering the body, exposing ciliated ectoderm along the anterior and posterior margins. The aboral pole bears an apical nervous organ with sensory bristles, and the oral surface carries a ciliated ring. The U-shaped alimentary tract is complete, although this may be lacking in egg-brooding species. A glandular invagination, the pyriform organ, is situated anteriorly on the oral surface, associated with a ciliated cleft and a sensory vibratile plume of cilia. An adhesive sac is situated close to the anus. The shell plates are connected by an adductor muscle, and other muscles are associated with the alimentary canal. The larva swims with the aboral pole forward.

Metamorphosis is similar in all gymnolaemate larvae, whether of the cyphonautes type or modified. When a suitable substrate is found, by means of the sensory vibratile plume, the adhesive sac is suddenly

everted by muscular contractions and adheres as a flat disc with its secretion. All projecting larval parts are retracted into the interior. In the
cyphonautes, the adductor muscle between the shell plates is ruptured,
and the plates open out and cover the other tissues; later the plates are
discarded. The larval tissues are histolysed, with the exception of part
of the epidermis, which extends over the disintegrating tissues and secretes a cuticle. At about the center of the free surface of the mass, an
ectodermal vesicle is invaginated and becomes covered with mesenchyme. The vesicle constricts to form inner and outer portions: the latter
eventually forms the pharynx with associated tentacles, the former elongates, curves distally, and forms the remainder of the digestive tract. The
ganglion forms as an ectodermal invagination and muscles develop in the
mesenchyme. The primary zooid so formed—the ancestrula—takes from
1 to 6 days to develop after the initial larval attachment, and buds off
more zooids, which also bud to form the colony.

2.13. Brachiopoda

In the Inarticulata, the individual becomes free-swimming when
three pairs of tentacles have developed and the main parts of the adult
body are already defined. The larval shell, the protegulum, covers the
mantle lobes. The tentacles increase in number, the mantle lobes enlarge,
and the shell valves also enlarge by deposition of material along the sides
and anterior edge of the protegulum. The pedicle arises as an evagination
at the posterior end of the ventral mantle lobe, and as it grows, coils up
in the posterior part of the mantle cavity. At a particular stage of tentacle
number, the pedicle is uncoiled and the individual attaches for its sessile
life.

In the Articulata, the free-swimming larva has the same general plan
as that of the inarticulates, with an anterior lobe that develops into the
greater part of the body; a median lobe that becomes the mantle; and a
posterior lobe, the pedicle. The anterior lobe is ciliated, and often has an
apical tuft with underlying ganglion: these are soon lost. After attachment, metamorphosis consists largely of the reversal of the mantle, which
was previously directed backwards around the pedicle; the mantle turns
forwards around the body and begins to secrete the shell.

2.14. Annelida

Leeches and freshwater and terrestrial oligochaetes have direct development, but in the archiannelids and polychaetes, the trochophore
larva is the characteristic free-swimming stage. The trochophore has an
upper bell-shaped region, with an apical plate consisting of an epidermal
thickening with sensory cilia and a ganglion beneath. The lower part of
the body tapers conically. Around the equator of the larva is a ciliated
band, the prototroch, with sometimes another ciliated band, the meta-

troch, beneath it. In some species, another band of cilia surrounding the anus occurs; this is the paratroch. The mouth opens beneath the prototroch, on the future ventral surface, and the anus is at the lower pole. A pair of protonephridia, muscles, and mesenchyme are present internally, together with a pair of mesodermal bands, developed from particular cells originally forming part of the hinder end of the intestine.

Metamorphosis is initiated by the multiplication of the cells of the mesodermal bands, together with the cells of the trunk blastema, which furnishes new ectoderm. The cells of the gut also multiply, and the anus is carried backwards at the end of the posttrochal projection so formed. The mesodermal bands become several cells thick, and coelomic cavities appear within them. The mesoderm grows dorsally and ventrally around the gut until the layers meet and fuse and the coelom becomes continuous. Sequential divisions appear and the segmentation of the adult worm is initiated. The growth zone is just anterior to the anus, so the most anterior primitive segments are the oldest, the posterior ones younger. The upper half of the original larva becomes the prostomium of the adult worm, and the original lower half becomes incorporated into the segmented region. The larval cilia and sensory apparatus disappear.

Variations in the trochophore larva are not only confined to the number and position of the ciliated bands. In the trochophore of *Polygordius lacteus*, the thin ventral surface develops a ring-shaped invagination, forming a circular fold surrounding the trunk blastema. The cavity extends forwards and upwards, and when further growth begins in the trunk blastema, the new posttrochal body is folded up within the cavity. At metamorphosis, the body is straightened and thrust out as the segmented trunk of the worm. In some species, the trochophore may lack a mouth and anus—a protrochophore stage; in others, the posterior half of the larva is segmented, with parapodia and chaetae, the nectochaetous larva—which not only swims but can creep over the substrate. In the mitraria larva, the bell becomes considerably enlarged relative to the posterior region, and the ciliated band with extensive outfoldings comes to lie more anteriorly. Large provisional larval chaetae develop from protuberances on the posterior region. The trunk of the worm develops as an invagination between the mouth and the anus. At metamorphosis, which is extremely rapid, the trunk is everted and the larval gut pulled into it. The metamorphosed worm may eat the remaining disintegrated larval tissues (Wilson, 1932).

2.15. Sipunculida

Sipunculid larvae are of the trochophore type, with an apical plate and prototroch, but with paired ocelli in addition. Metamorphosis involves the development of the trunk from the region posterior to the prototroch. The apical plate degenerates, but the larval ocelli are carried inwards by the developing brain and become the adult ocelli. The trochoblasts disintegrate and are absorbed. The larva does not feed, the anus not developing until the adult trunk is established.

2.16. Echiurida

In *Echiurus*, the larva is a fairly typical trochophore type, with apical plate and prototroch, but with several ciliated bands at its posterior end. Metamorphosis is gradual, with loss of cilia, development of the trunk from the posterior region, and transformation of the oral and preoral parts into the prostomium. In *Bonellia*, the trochophore stage is succeeded by one in which the ciliation largely disappears, remaining only on the ventral surface. The anterior region grows considerably in length. The individual is flattened, and rather resembles a planarian. Metamorphosis into a female adult progresses gradually. However, if the ciliated larva contacts the proboscis of a female, it adheres and develops into a male; no metamorphosis occurs, and the larva itself becomes sexually mature.

2.17. Arthropoda

The special characteristic of metamorphosis in the various arthropod groups in which it occurs is the consecutive series of molts that the larval stages undergo. Superficially, the metamorphic process may appear to be very different in different arthropods: in the crustaceans and pycnogonids, for example, the young larvae hatch with far fewer body segments than the adult, and metamorphosis proceeds through the larval molts in which body segments are progressively added; in the insects, on the other hand, the addition of segments rarely occurs through the series of larval molts. Such differences can be explained in terms of the particular functions of larvae in the different arthropod groups: the principles underlying the metamorphic process are similar in all arthropods, and indeed can be compared with those of other invertebrates (Section 3.3.2). Insect metamorphosis is omitted from this section, since it is treated overall in Section 3, and is examined in considerable detail in other chapters in this volume.

2.17.1. Xiphosura

The young king crab emerges from the egg as the so-called trilobite stage. It has a wide, flat cephalothoracic shield, with two longitudinal furrows demarcating a keeled central region and two lateral regions. The median and lateral eyes are situated in the furrows. Nine abdominal segments can be distinguished, the last being the rudiment of the caudal spine. The first pair of abdominal limbs develops into the operculum, and branchial lamellae develop on the succeeding limbs. Metamorphosis is gradual, with greater fusion of the abdominal segments, and lengthening of the caudal spine during successive molts until the adult form is attained.

2.17.2. Pycnogonida

Most sea spider larvae hatch with only three pairs of limbs, although some can be in a more advanced stage of development. The body is usually square or rounded, very rarely elongated. The three-jointed anterior limbs are chelate, the second and third pairs are hooked. Paired eyes, closely set together, lie behind the conical proboscis.

Metamorphosis involves the addition of posterior segments; the limbs either pass into the adult, or some or all may degenerate, adult limbs developing from the same sites. Limbs develop with the additional segments until the full complement is attained. The abdomen, which is very short, arises as a posterior saclike swelling upon which the anus opens.

2.17.3. Crustacea

The nauplius larva is characteristic of many crustaceans, usually of an oval shape, more rounded anteriorly, although it can be laterally or dorsoventrally compressed, elongated, or broad in form. Although body segmentation is not obvious, the presence of three pairs of appendages shows it to be three-segmented. The appendages are anterior uniramous antennules, middle biramous antennae, and posterior biramous mandibles, which are all involved in locomotion. In the branchiopods, for example, there is no sudden metamorphosis: segments and limbs are added during successive molts, through a series of metanauplius stages, until the adult form is reached. But in other groups, for example the copepods, there are six nauplius stages, the final one still bearing the original three pairs of appendages but with the rudiments of five further pairs. This metamorphoses to the first copepodid state, in which the body is no longer flexed ventrally but is straight; a constriction marks off anterior and posterior regions and the furcal appeandages and the rudiment of the fourth thoracic segment develops. This stage has essentially the form of the adult, and may be called a juvenile; in subsequent copepodid stages, the full segmentation and form of the adult is attained. In the cirripedes, a number of nauplius stages eventually produce the metanauplius, with rudiments of the abdominal limbs; this then molts into the free-swimming cypris larva, so-called because of the bivalved shell that envelops the whole body. This was at one time called a pupa, although unlike the insect analog, it is freely motile. When a suitable substrate has been found (Section 4.1), the cypris attaches by its modified antennules and by a sticky fluid secreted by the cement glands at their bases. Metamorphosis involves loss of the shell valves, and development of the anterior part of the head, particularly in the goose barnacles, together with the production of the shell plates of the adult.

In the penaeids (Decapoda), the nauplius progresses through a series of molts to the protozoea stage, with fully segmented thorax and a small carapace, but still lacking compound eyes. The protozoea molts successively to the zoeal stages in which the remainder of the segments and appendages are developed. In the true crabs, nauplius and protozoeal stages are omitted and the eggs hatch as zoea larvae that progress to a megalopa, with all its segments and limbs present. The megalopa uses its abdominal swimmerets to swim, but finally moves to the bottom to molt to the juvenile adult with folded abdomen. Lobster larvae hatch at a later stage of development, with all the thoracic appendages present; abdominal swimmerets and the uropods appear at succeeding molts. The spiny lobster and its relatives have the phyllosoma larva: with three pairs of long bristled legs and a flattened thorax, it is well adapted to planktonic life. The true shrimps and prawns hatch as zoea larvae, with three pairs of functional maxillipedes; the remaining legs appear at successive molts.

There is little doubt that the various developmental stages found in the "higher" crustaceans are related to their methods of locomotion, from largely antennal in the nauplius, through thoracic swimming appendages and finally thoracic plus abdominal (Gurney, 1942). Correlated with adult size and the yolkiness of the eggs, preliminary larval stages can be omitted. In the cirripedes, the intervention of the cypris stage allows prior development of adult features, so that the process of attachment and metamorphosis can take place with great rapidity.

2.18. Mollusca

The molluscan trochophore larva bears a close resemblance to that of the annelids, with the exception that the stomodaeum often has a posterior outgrowth, the rudiment of the radula sac, the foot primordium occurs as a postoral protrusion, and there is a shell gland on the dorsal surface opposite to the mouth. The posterior end of the gut usually ends blindly, but two protruding anal cells indicate where the anus will break through; a telotroch is sometimes present in this region.

In the polyplacophorans, metamorphosis is gradual, the posttrochal region elongating to form the trunk of the adult chiton. The shell gland produces the shell plates in its transverse grooves, and the mantle groove appears along the border of the shell area. A proctodaeum develops posteriorly and connects with the hindgut. As with the annelid trochophore, the pretrochal region undergoes the most marked changes: the prototrochal ciliary cells lose their nuclei and disintegrate; the apical tuft disappears; the pretrochal region becomes conical in shape and ventrally flattened; the proboscis is delimited when the mantle groove extends into this region, and the mouth moves to the proboscis. In the wormlike aplacophorans, the adult trunk region grows out of the posterior part of the trochophore and the anterior part forms a "test" covering the developing anterior region of the mollusc. During metamorphosis, the larval test is either shed, or covered with ectodermal folds from the developing adult.

The gastropods, lamellibranchs, and scaphopods also have trochophore larvae, but another stage, the veliger, is interposed between trochophore and adult. In the veliger larva, the prototroch has developed into a pair of powerfully ciliated semicircular folds, often subdivided into two or three lappets. The trochophore apical tuft may disappear; in its place, or beside it if it remains, a special apical plate develops. Lateral to the apical plate are the cephalic plates from which develop the cerebral ganglia, tentacles, and eyes. The foot rudiment grows out and begins to develop the shape characteristic of each group; an operculum may be present on its posterior face. The larval musculature is well developed, and includes velar retractors that can pull the velum partly or completely into the larval shell, the latter produced by the shell gland on the originally dorsal side of the trochophore. The mantle is produced by cells around the periphery of the shell gland. In the gastropod trochophore, the shell has a more or less symmetrical position; in the veliger, it begins to surround the body dorsally and posteriorly, displaced to one side, and is spirally coiled by unequal growth along its margins. In the lamellibranchs, the shell becomes bivalved, and in the scaphopods lateral mantle folds grow around the body and fuse along their ventral margins; the cylindrical mantle tube secretes the shell on its outer surface.

Metamorphosis of the veliger involves the breakdown and loss of its particular larval structures, since the basic organization of the adult is already present. In the scaphopods, the velum cells are autolysed, lose their cilia, and are finally expelled. In lamellibranchs, loss of the velum is more dramatic as it is cast off in fragments; the apical plate connects secondarily with the epidermis above the mouth and grows out on either side, contributing to the formation of the labial palps. In gastropods, torsion occurs in the veliger, whereby the visceropallial complex rotates counterclockwise through 180° so that the animal can be retracted into the protective mantle cavity and shell—sensitive head and velum first, followed by the tougher foot. In many gastropods, torsion occurs in two stages: an initial twisting of 90° caused by contraction of the asymmetrical larval retractor muscles, and the final torsion caused by differential growth. In others, differential growth alone brings about torsion. Torsion can therefore take from a few hours to several days, according to the relative importance of the mechanisms involved. The velum of the metamorphosing gastropod veliger is expelled in a manner similar to that of the scaphopod. The larval shell is cast off in some gastropod veligers, but more usually the adult shell develops around its outer margin. Where the adult shell is different in form from the larval, the transition from one to the other is very obvious.

In the primitive protobranch lamellibranchs, the adult body develops within the trochophore, which acts as a test. Metamorphosis is sudden and rapid, with the expulsion of prototroch, apical plate, and part of the stomodaeum.

Dreissenia is probably a recent invader of brackish and freshwater habitats and retains its veliger larva; all other freshwater lamellibranchs retain the developing embryos in a marsupium in the exhalent part of the mantle cavity. In the Unionidae, the glochidium larva may be considered a very much modified veliger. It has a bivalved shell armed with hooks and hinged spines, connected by a larval adductor muscle. The larvae attach to fish or urodeles, probably species specifically, and are enclosed in a tumor formed by the host tissues. The larval gut is rudimentary, without mouth or anus, and the larva feeds on the host tissue through enzyme secretion and assimilation through its mantle. The glochidium undergoes a gradual metamorphosis involving complete histolysis of the larval tissues and the development of adult structures. The young adult then breaks out of the host tumor and falls to the bottom. Other freshwater lamellibranchs, and also freshwater pulmonates and prosobranchs, have direct development, as do all the cephalopods.

2.19. Echinodermata

The various larval forms of echinoderms can all be referred to the dipleurula stage, in which the mouth is situated in a saddle-shaped depression bordered by a ciliated band; the alimentary canal is U-shaped, with the anus opening posterior to the ciliated band; and with paired coelomic enterocoelic sacs.

In the pluteus larvae of echinoids and ophiuroids, the ciliated band is extended by development of long, paired arms, those of the echinopluteus differing in shape and position from those of the ophiopluteus. In some genera, additional well-developed cilia, the epaulettes, develop between certain of the arms. The ciliated arms of the pluteus contain calcareous rods, and are capable of slow approximation and divarication due to muscular strands connecting the rods near the aboral pole.

In the bipinnaria larva of asteroids, the ciliated band is thrown into folds by differential growth and becomes divided into pre- and postoral bands. Short, stubby arms are developed that extend the ciliated areas.

The auricularia larva of many holothuroids resembles the early bipinnaria, but the anterior and posterior ciliated folds break up into sections that are rearranged to form from three to five transverse ciliated bands—the doliolaria stage. In crinoids, the doliolaria larva (= vitellaria) has four or five ciliated bands.

Metamorphosis of the echinopluteus involves the development of the oral surface of the adult from the left side of the larva, the adult aboral surface from the larval right side. The orientation of the future adult thus involves a 90° displacement of axis relative to that of the larva. Metamorphosis begins with an ectodermal vestibular invagination that grows inwards and becomes flattened against the left hydrocoele. The invagination is cut off externally, and further development of the echinus rudiment takes place in the vestibular cavity. The right hydrocoele of the larva disappears, and the left hydrocoele grows around the gut to form the water vascular ring. Five blunt lobes project into the vestibular cavity as the rudiments of the five radial water vascular canals, establishing the adult pentamerous symmetry. The adult stomodaeum breaks through the ectoderm of the floor of the vestibular cavity and connects inwardly with the larval stomach. Pedicellariae and particular calcareous plates develop on the oral surface, together with their associated spines. Rudiments of the tube feet appear. The larva now sinks to the bottom and the tube feet adhere to the substratum. The larval arms are rapidly absorbed and their supporting rods protrude as naked spines, which are soon broken off. The larval stomodaeum becomes disconnected from the stomach and forms a shallow pit, in which the stumps of the preoral and anterolateral arms persist for a short time before being absorbed. The final transformation may take as little as 1 hour to achieve, although, of course, preparation for the event may have taken weeks or even months of planktonic life.

Metamorphosis of the ophiopluteus follows the same general pattern as that of the echinopluteus, although there are many differences in detail. It is again characterized by the preponderant growth of the organs of the left side, the left hydrocoele and left posterior coelom. The latter extends forwards and its outer wall develops conical protuberances that raise corresponding humps in the overlying ectoderm, from which develop the adult arms. The anterior part of the larva, anus, arms, and supporting rods are absorbed or discarded. The posterolateral arms are the last to be discarded, and they may continue to swim for some time, and may even show signs of regeneration. Whether a further metamorphosis can occur is debatable.

Metamorphosis of the asteroid bipinnaria often involves a period of temporary fixation. Three nonciliated clubbed arms develop, with adhesive cells, together with an adhesive disc between their bases. The larva is now called a brachiolaria. After attachment, metamorphosis is essentially similar to that of the pluteus larvae. The remains of the larva are absorbed to a greater or lesser extent, even in the same species, and is probably related to the finding of a suitable substrate for metamorphosis (Wilson, 1978). The exceptionally large bipinnaria of *Luidia sarsi* can continue a separate existence for a considerable time after the young sea star has become detached; the larval bodies have a mouth and anus, but as the gut between them is severed at metamorphosis, they cannot feed (Wilson, 1978).

Metamorphosis of the holothuroid auricularia larva includes shrinkage of the preoral region, which causes the mouth, lying at the bottom of a deep atrium and originally on the left side of the larva, to become positioned at the anterior pole of the larva; this stage is sometimes called

the barrel-shaped larva or pupa. Further development of the adult structures is gradual in the free-swimming larva. The atrium opens out and
the adult tentacles develop and protrude freely to the exterior.A kneelike
bend develops in the intestine and one loop grows forward until it reaches
almost to the hydrocoele, establishing the ascending and descending intestinal limbs. The ciliated bands disappear, and the individual sinks to
the bottom as a young adult. Metamorphosis in the holothurian is thus
less drastic than in the other echinoderm groups, with little or no dis-
carding of larval tissues.

The oval vitellaria larva of the crinoid, *Antedon*, has a free-swimming
life of a few hours to 4 or 5 days. The larva has four to five ciliated bands,
an apical plate with a tuft of cilia, and a preoral fixation disc lies on the
ventral surface. After fixation by the disc, the originally anterior portion
of the larva becomes the stalk of the young crinoid. At first it lies with
its length parallel to the substrate, but it soon becomes erect. The stomodaeum becomes closed to the exterior, resembling the vestibular cavity of the metamorphosing echinopluteus. The stomodaeum rotates backwards along the ventral surface until it comes to occupy what was the
original posterior pole of the larva. The developing internal organs, including the hydrocoele and its associated tentacles protruding into the
stomodaeum, also rotate. The cilia are shed, and the ciliated band cells
secrete a cuticle and then retreat from the surface. The oral vestibule
opens to the exterior, the tentacles are protruded, and the young crinoid
is now able to feed. After about 6 weeks, the arms of the crinoid develop
as vertical upgrowths of the calyx: this stage is the pentacrinoid larva.
After further months of sessile life, the adult breaks away from the stalk
to begin its free-living existence.

2.20. Phoronidea

The actinotrocha larva is divided into three body regions: a preoral
lobe with a sensory organ in front of a neural plate, sometimes with
associated eyespots; a postoral region carrying a crown of tentacles projecting backwards and covering the posterior part of the body like an
apron; and a posterior or anal section, with a circumanal ciliated ring.
An ectodermal invagination develops on the ventral surface of the posterior region of the body, and this grows into a coiled tube within the
larval body cavity. Small truncated processes develop at the base of the
crown of tentacles.

The larva sinks to the bottom, and metamorphosis is completed
within about 15 minutes. This consists largely of the evagination of the
tube described above, and since the alimentary canal with its mesentery
is attached to the inner end of the tube, this also is drawn out so that the
alimentary canal becomes U-shaped, with mouth and anus close to-
gether.The larval preoral lobe is thrown off, together with the larval
tentacles and circumoral ciliated ring. The small truncated processes at

the base of the original crown of tentacles increase in size and take the form of the horseshoe-shaped adult lophophore.

Thus the adult is derived very largely from the ventral invagination of the larval body, the dorsal part of the latter forming only the short tract between adult mouth and anus. Parts of the larval body are cast off, but others, such as the gut, are transformed into the adult body.

2.21. Chaetognatha

Although resembling the adult form, the young arrowworm has a free-swimming life and can be considered a larval stage. The young *Sagitta* lacks external ciliation, has a rounded head containing the pharynx, and a slender trunk with closed gut. There is no dramatic metamorphosis, the gut cavity appears and the larval body develops into that of the adult.

2.22. Hemichordata

Development of the enteropneusts may be direct or indirect with the presence of a tornaria larva. This resembles the echinoderm dipleurula, but with an additional posterior ciliated band, the telotroch. The tornaria is planktotrophic, and the circumoral ciliated band becomes increasingly sinuous as the larva grows. Metamorphosis is less dramatic than in the echinoderms: the anterior part of the body grows in length, becomes conical, and gradually takes on the form of the adult proboscis. The collar and trunk regions develop posterior to the proboscis. The neural plate develops dorsally at the position of the grooves which mark the hinder end of the collar region, and sinks in to become the neural tube. The notochord develops as a median dorsal, forwardly directed pouch from the anterior end of the esophagus. The larvae of *Saccoglossus horsti* are lecithotropic, and have a much shorter free-swimming life than tornariae; they can, however, exercise the function of site selection, deferring metamorphosis for a short time if a suitable habitat is not found.

The hemichordates have been included briefly in this survey of invertebrate metamorphosis, as they can be considered to be an independent phylum (Hyman, 1959). The urochordates and cephalochordates are then considered subphyla of the phylum Chordata, and their metamorphoses are considered later (see Just *et al.*, this volume).

3. COMPARISON OF INSECT METAMORPHOSIS WITH THAT OF OTHER INVERTEBRATES

Insects are usually considered to exhibit three major patterns of postembryonic development: ametabolous, in which the development of the adult form is attained gradually through successive molts; hemimetabolous, in which the change from last-instar larva to adult at the final molt is more dramatic; and holometabolous, in which a pupal stage is interpolated between the last larval instar and the adult, where larval tissues are histolysed to a greater or lesser degree, and adult structures are developed, again to a greater or lesser degree, from particular cells set aside for this purpose at an earlier stage. The insects can thus duplicate within the limits of a single class many of the developmental strategies employed by a variety of invertebrate phyla. The similarities are sometimes obscured by the necessity of insects to discard their cuticles as they increase in size, so that their changes in form are revealed only at each ecdysis. But increase in cell number by mitosis is essentially the same in insects as in other animals that do not have an exoskeleton, although it is subordinated to the particular exigencies of the molt cycle.

3.1. Ametabolous Development

In insects, this term is usually restricted to the apterygotes, where the juvenile forms closely resemble the adults, lacking only developed gonads and genitalia; the Protura develop the full number of abdominal segments only after several molts. But ametabolous development differs only in degree from the hemimetabolous type. Growth is largely harmonic, so that the proportions of the adult body are similar to those of earlier stages. It is the absence of wings and their associated thoracic musculature and skeletal modification that makes apterygote metamorphosis seem slight or wanting: in some hemimetabolous insects that have secondarily lost their wings, e.g., female Embioptera, metamorphosis can be equally slight. Progressive adult differentiation can occur: in the Thysanura, for example, scales and coxal styles appear in the third instar, and gonapophyses are present in the fourth.

3.2. Hemimetabolous Development

In the exopterygote insects, metamorphosis is often described as simple or direct, with a gradual development of external genitalia and wing buds during the larval stages and with usually a more marked transformation at the final molt when the winged adult emerges. But metamorphosis in these insects can vary a great deal more than such a description implies. In many exopterygote orders, the larval stages may resemble the adults in general form, but metamorphosis through the successive molts can involve the expansion of some parts of the body, reduction of others; certain cell lines may be discontinued, others, for example those that will form the wings and genitalia, initiated; the cuticular products of the epidermal cells can change both in form and in structure as metamorphosis proceeds; and pigmentation patterns can differ markedly between larval and adult stages. In the Plecoptera, Ephemeroptera, and Odonata, the larvae have respiratory and locomotory adaptations for their aquatic existence that are lost in the aerial adults. In the Thysanoptera, and in the Aleyrodidae and male Coccoidea in the Hemiptera, there are quiescent stages preceding that of the adult that are reminiscent of the holometabolous pupal stage, and in which a similar greater or lesser replacement of larval tissues can occur. The degree of divergence between larval and adult stages in the hemimetabola, with concomitant differences in form and habit, can make metamorphosis in these insects a more or less striking process.

3.3. Holometabolous Development

In the endopterygote insects, a pupal stage is interposed between the last larval and the adult stages. Transformation of the larval to the adult tissues occurs through the pupa, and these insects are said to undergo a complete or indirect metamorphosis. Emphasis is often laid upon the endopterygote orders such as the Lepidoptera, Hymenoptera, and Diptera, in which metamorphosis is the most spectacular: the larval tissues are broken down to a greater or lesser extent, and much of the adult body is developed from imaginal buds or discs, nests of cells that are not functional in the larva, serving only to produce adult structures at metamorphosis.

The degree to which the larval tissues are histolysed and imaginal discs produce the adult body can vary in different endopterygote orders. In the Coleoptera, for example, much of the larval epidermis passes without destruction into the adult, and in general where the appendages of the larval head and thorax are not greatly different from those of the adult, the latter develop, often as folded structures, within the larval appendages. Even where there may be spectacular changes in the form of the body at metamorphosis, these are brought about by the descendants of the cells that formed the larval body. Imaginal discs may also appear at different times during development: in the higher Diptera they are set aside toward the end of embryonic life; in other orders, the imaginal discs that develop into particular structures may appear as ectodermal invaginations at different times during larval life. If traced back far enough, a stage is reached when the imaginal discs are indistinguishable from the general epidermis of the larva or the embryo.

3.3.1. Insect Metamorphosis

Insect metamorphosis thus covers a wide spectrum of transformation, from the various degrees found in the Hemimetabola to the equal variety in the Holometabola. But at the cellular level, all insect metamorphoses are essentially similar: larval cells produce the body form and all associated structures of the larval stages, but they also have within them the potential for producing the adult body, and the pupal body where this exists. Metamorphosis is the realization of the potential for adult development within the cells (Wigglesworth, 1954, 1959). Insect development can be considered as the manifestation of sequential polymorphism, and as with other, parallel, polymorphisms the ultimate causation must reside within the genome (Wigglesworth, 1967, 1970).

3.3.2. Metamorphosis in Noninsect Phyla

Nematode development most closely resembles that of the ametabolous apterygote insects, with generally harmonic growth, and with the consequent difficulty of deciding whether the early stages should be called juveniles rather than larvae. The larvae of ctenophores, sessile rotifers, brachiopods, sipunculids, echiurids, chaetognathes, and hemichordates all metamorphose by differential growth of the parts of the body and can be compared with the hemimetabolous development of many exopterygote insects, although since they lack the molting sequence of the insects, the transformations are rarely as marked as in the latter. Metamorphosis in the holothuroids and crinoids among the echinoderms can also be compared with that of the hemimetabolous insects, although echinoid, asteroid, and ophiuroid metamorphoses are considerably more dramatic, with development of the adult originating from particular parts of the larvae together with the loss of many larval structures, and consequently more nearly comparable with holometabolous insect metamorphosis. Similarly, metamorphosis in the aplacophoran and polyplacophoran molluscs can be compared with that of hemimetabolous insect metamorphosis, but in the remaining molluscs, metamorphosis is more drastic, culminating with that of the glochidium in the unionid lamellibranchs, which can be compared with the more extreme forms of holometabolous insect metamorphosis. Metamorphosis in the cestodes and monogenetic trematodes can also be compared with hemimetabolous insect development, particularly that of the stoneflies, mayflies, and damsel- and dragonflies, where purely larval structures are lost, the adults developing from the remaining larval tissues; but metamorphosis in the digenetic trematodes is more like that of holometabolous insects. The gordioid larva of the nematomorphans also metamorphoses with the loss of its purely larval structures. In the archiannelids and polychaetes generally, most of the trochophore larva becomes part of the adult, although some purely larval structures are cast off, but metamorphosis of the mitraria larva is more drastic. In the crustaceans, xiphosurans, and pycnogonids among the arthropods, and in the kinorhynchs, metamorphosis involves the addition of body segments, and in the crustaceans in particular, the modification or loss of larval features belonging to earlier stages; such progressive adult differentiation is akin to hemimetabolous insect metamorphosis. Although the cypris larva of cirripede crustaceans has been called a pupa, it is perhaps better considered an example of hypermetamorphosis (Section 3.4).

Metamorphosis of the phoronid actinotrocha larva, and the larvae of entoprocts, involves considerable reorganization of the larval body to produce the adult and can be compared with holometabolous insect metamorphosis. The development of parts of the adult body from particular ectodermal cells in the turbellarian Müller's larva is reminiscent of metamorphosis in some holometabolous insect orders, as is the development of the adult acanthocephalan from cells set aside in the larva. But perhaps the closest analogs of metamorphosis in the holometabolous insects are to be found in the metamorphosis of the cyphonautes larva of the polyzoa, where a large part of the adult body is derived from an ectodermal invagination of the larva, and also in the metamorphosis of the pilidium and Desor's larvae in the nemertines, where a number of ectodermal invaginations of the larvae form the epidermis and associated structures in the adult: such invaginations closely resemble the development of the imaginal discs of the holometabolous insects.

Care has been taken in this account not to ascribe any phylogenetic significance to the various larval types found in the invertebrates, or to their metamorphoses. But the fact that comparisons can be made between insect metamorphoses and those of other groups suggests that in general terms, the biological problems associated with the interpolation of larval stages into life histories with the consequent necessity for metamorphic transformations are capable of a limited number of solutions, and that these solutions have been exploited many times in many different organisms. Moreover, the concept of sequential polymorphisms as applied to insect development is equally applicable to the larval and adult stages of other invertebrates. Although only a few instances have been given here, the independent adaptations and evolutions of larval and adult stages point as strongly to the underlying genetic control of these stages in other invertebrates as well as in the insects.

The amphiblastula and parenchymula larvae of sponges, and the planula larvae of the cnidaria metamorphose with a usually abrupt change in form, and without any loss or destruction of larval tissue: all the cells of the larvae transform into those of the juveniles. This is a pattern unlike those found in other invertebrates, and it has been suggested that such larvae are best considered free-living embryonic stages, not to be compared with other larval forms. But the larvae of both sponges and cnidaria exhibit the functions of aggregation and site selection found in many other larvae (Section 4.1), and the planulae of different species show individual adaptations, such as the production of mucous strands to aid transport by water currents. The kind of metamorphosis found in these groups must be considered a consequence of their lowly grade of organization, the lack of the more complex tissue and organ systems found in other invertebrates enabling the rapid and entire transformation to take place.

3.4. Hypermetamorphosis

Various insects in a number of hemimetabolous and holometabolous orders show an abrupt change in larval form prior to the adult transformation: they are said to exhibit hypermetamorphosis. Examples are *Margarodes* and its allies in the Hemiptera, *Mantispa* and its relatives in the Neuroptera, various staphylinid, meloid, and rhipiphorid beetles, all the Strepsiptera, and Hymenoptera parasitica. Hypermentamorphosis is associated with a marked change in habit. The most common is from an active campodeiform larva, which seeks out its prey, to a more or less degenerate eruciform type, parasitic upon eggs, larvae, or other stages of a variety of insects or spiders. The general concept of sequential polymorphisms in insect development can encompass hypermetamorphosis as an extreme adaptation to different habitats during the life of the individual (Wigglesworth, 1954).

It is difficult to find a comparable phenomenon in other invertebrate groups. It has already been suggested that the changes in form that occur in the life histories of digenetic trematodes (Section 2.4.2) should be considered as a form of polyembryony rather than hypermetamorphosis; unlike the situation in insects, the transformations do not occur in one individual, but a series of offspring possess the different forms. The sequence of larval forms found in many crustaceans rarely show abrupt changes associated with different habitats, and can be considered the mechanical solutions to the problems associated with the support and movement of increasingly large planktotrophic organisms. Perhaps the best example would be the presence of the cypris stage in cirripede life histories: the planktonic nauplius giving rise to the cypris, highly adapted for site selection for the ensuing sessile adult stage.

4. CONTROL OF METAMORPHOSIS

To increase the chances of adult survival, metamorphosis must occur in the appropriate place at the appropriate time. The larva often has to monitor more or less long-term environmental cues to determine *when* to metamorphose; in addition, and particularly in sessile, sedentary, or other bottom-living animals, *where* to metamorphose is of the utmost importance.

4.1. Larval Settlement and Metamorphosis

Behavioral changes often precede larval settlement in marine organisms: a planktonic larva can be positively phototropic and negatively geotropic for much of its existence, but the tropisms are reversed when the bottom habitat is sought. Behavioral changes associated with pupation site selection are well known in insect larvae, and it is known also that such changes are under endocrine control.

Where adults can move freely about their tidal range, the larvae may settle anywhere with no special affinity for any algal species or structure (Underwood, 1972; Wigham, 1975). Other adult species may move freely only upon a particular substrate and larvae settling there metamorphose quickly, delay occurring until the correct substrate is encountered (Wilson, 1978). Where adults have particular food sources, larvae will settle on or near the source (Thompson, 1958, 1962; Fretter and Manly, 1977). Some species can be very selective: the hydroid *Proboscidactyla flavicirrata* is symbiotic with sabellid worms, living on the rims of their tubes. The planulae initially settle on the tentacles of the worms and when the latter withdraw into their tubes, the larvae are rubbed against the exterior of the tube where they adhere and metamorphose—but only after a period of contact with the tentacles of the worm (Donaldson, 1974). The planulae of other species will settle and metamorphose on the undersurfaces of structures, to prevent excessive silting (Cargo, 1979).

Marine larvae are stimulated to settle and metamorphose by a combination of biological and physical factors (Meadows and Campbell, 1972; Scheltema, 1974; Crisp, 1974, 1976). The biological factors can include coatings of microbiological films on the particulate or solid substrates, or of traces of biological materials produced by adults of the species (Wilson, 1955, 1968, 1970). The actual presence of conspecific adults or organic products derived from them can be potent factors in inducing larval settlement and metamorphosis (Larman and Gabbott, 1975; Fretter and Manly, 1977). Planulae of *Nemertisia antennina* will settle on living hydroids of the parental generation, but not the remains of the grandparental generation (Hughes, 1977). Barnett and Crisp (1979) have shown that settlement of cypris larvae is reduced by the presence of previous colonizers, but is biased toward settled spat of their conspecifics, so that a balance between territoriality and species gregariousness is achieved. Contrary to previous findings, settled spat of *Balanus balanoides* and *Elminius modestus* are found in contact with their conspecific adults more frequently than in contact with the other species (Barnett *et al.*, 1979). Such gregariousness exploits the properties of what must be a favorable habitat since a previous generation is already established there. Moreover, gregariousness during larval settlement has the advantage, in gonochoristic species, of bringing about sexual proximity in the succeeding adults. Presumably as an adaptation to the parasitic life, the cypris larvae of rhizocephalan barnacles do not show gregarious settling behavior, and the growth and development of a later parasite can be inhibited if one parasite is already present in a crab (Rainbow *et al.*, 1979).

The physical features that can affect larval settlement in marine organisms include particularly the texture of the substrate—the rugosity of a rock surface, the grade and shapes of sand grains, the contour and permanence of the substrate, exposure to water flow together with salinity and temperature extremes, light and shade. Prior to settlement and metamorphosis, the larva must therefore be able, through its sense organs, to monitor a variety of stimuli emanating from the biological and physical features of the habitat. The behavior of many larvae prior to settlement falls into three major phases: broad exploration, close exploration, and finally inspection (Crisp, 1974). Only if the information gained from each of the three phases proves satisfactory will the larva attach and metamorphose.

When a suitable habitat has been found, events are set in motion that often terminate in a very rapid metamorphosis. This is to say that the appropriate sensory input derived from the habitat is the trigger that sets off the transformation process. But the converse is equally of interest: that with the receipt of inappropriate sensory inputs, attachment and metamorphosis can be delayed and the larvae will search elsewhere. Thus the nature of the substrate can induce or delay metamorphosis. When metamorphosis is delayed for too long, the ability of larvae to discriminate between habitats is reduced: they will settle on progressively less favorable substrates and eventually they will metamorphose under quite inappropriate conditions. The sabellid worm, *Lygdamus muratus*, kept continuously in clean dishes eventually becomes "desperate" to metamorphose (Wilson, 1977), and unless they die first, the larvae are forced into a metamorphosis that is abnormal to a greater or lesser degree. Many other examples of abnormal delayed metamorphoses are known (Crisp, 1974, 1976).

If metamorphosis is the change from a genetically determined larval program to a genetically determined adult program (Section 3.3.1), what may the mechanisms be that operate the switch? A variety of chemicals, pH changes, CO_2 concentration, and other factors can induce metamorphosis in many marine invertebrates (Berrill, 1961; Crisp, 1974, 1976); their function could be related to the "exhaustion" of the larval tissues, or their programmed death, which then "forces" adult development to be initiated. But a more likely explanation, particularly in the bottom-living marine invertebrates, when metamorphosis follows close sensory monitoring of appropriate substrates, that is, when a developmental event results from a "correct" sensory input, is that a hormonal mechanism is involved, and in particular a neurosecretory mechanism. Neurosecretory cells are ubiquitous in the animal kingdom, and provide the ideal transducers of nervous into chemical messages. Abnormal delayed metamorphoses could result from the presumed neurosecretory hormone(s) being released in inappropriate amounts, or at the inappropriate time. The rapidity of the release of some neurosecretory hormones upon receipt of the appropriate stimuli in insects and vertebrates would certainly allow sufficient time for even the most rapid metamorphosis in these invertebrates, if the same principles applied. But until techniques are evolved for experimental and surgical interventions upon the generally very small larvae of marine invertebrates, a neurosecretory basis for their metamorphoses must remain speculative.

4.2. The Timing of Metamorphosis

In invertebrates with a relative long planktonic larval existence, and particularly those planktotrophic larvae that have to build up sufficient reserves for the metamorphic transformation, the timing of metamorphosis becomes of considerable importance. Comparison with the timing of metamorphosis in insect larvae may illuminate some of the possible control mechanisms involved.

In the insects, metamorphosis can be accelerated or delayed by a number of environmental factors that may shorten or extend individual larval instars, or may even vary the number of larval instars (Wigglesworth, 1965). The factors concerned are temperature, the quality and/or quantity of food, the actual availability of food, for example in blood-sucking species, crowding or isolation, and so on. In bi- or polyvoltine species, the incidence of a postembryonic diapause (arrested state of development) can delay metamorphosis for many months or even longer, and the length of day with or without the synergistic effects of temperature is of crucial importance for the induction or prevention of diapause in a particular generation.

Many of the natural variations in the timing of insect metamorphosis can be duplicated experimentally in the laboratory, and there is little doubt that the cerebral neurosecretory system with its control over the production of molting and juvenile hormones (see Granger and Bollenbacher, this volume) plays a central part in modifying larval development according to changes in the environmental cues. It would not be surprising, therefore, if similar neurosecretory mechanisms were to be found to control the timing of metamorphosis in other invertebrates also.

It is possible to speculate that since growth and regeneration in the juvenile nereid polychaete require a neurosecretory hormone (Highnam and Hill, 1977), growth of the mesodermal bands and trunk during metamorphosis of the trochophore requires a similar hormone, since the growth zone is similarly placed in the larva and the juvenile. In the nematodes, a neurosecretory hormone is involved in molting and analogs of the insect hormones can affect both growth and molting (Highnam and Hill, 1977). But it is only in the crustaceans that experimental evidence is accumulating that metamorphosis is controlled hormonally. The effects of eyestalk ablation in different species are rather equivocal (Costlow, 1963, 1966a,b; Little, 1969), perhaps because the X-organ–sinus gland system may mature differentially (Bellon-Humbert et al., 1978). But molting hormone appears to affect the epidermal cells in other crustacean larvae (Rao et al., 1973; Cheung, 1974; McConaugha, 1979), so that larval molting may be controlled in a manner similar to that in the adult, in which case environmental cues could dictate the timing of metamorphosis in crustaceans as in insects. It is regretted that so little information exists even about the histology of neurosecretion in other invertebrate larvae.

5. CONCLUSIONS

This chapter has attempted to show that the range and patterns of metamorphosis in the invertebrates follow the same principles, and that the differences that occur are not so profound as to make comparisons between groups impossible. The major functions of dispersal, habitat selection, and development of invertebrate larvae are balanced in different ways in different groups, and this varying proportionality of functions can impose variety in the timing and the rate of the subsequent metamorphosis. It is hoped that the information about insect metamorphosis, together with the concepts derived therefrom, which half of this volume pays tribute to, has helped at least to clarify and to organize information about metamorphosis in other invertebrates. It is unfortunate that all too often the relevant details about the control of metamorphosis in the noninsect groups are lacking.

REFERENCES

Balinsky, B. I., 1970, *An Introduction to Embryology*, 3rd ed., Saunders, Philadelphia.
Barnett, B. E., and Crisp, D. J., 1979, *J. Mar. Biol. Assoc. U.K.* **59**:581.
Barnett, B. E., Edwards, S. C., and Crisp, D. J., 1979, *J. Mar. Biol. Assoc. U.K.* **59**:575.
Barrington, E. J. W., 1979, *Invertebrate Structure and Function*, 2nd ed., Nelson, England.
Bellon-Humbert, C., Thijssen, M. J. P., and van Herp, F., 1978, *J. Mar. Biol. Assoc. U.K.* **58**:851.
Berrill, N. J., 1961, *Growth, Development and Pattern*, Freeman, San Francisco.
Cargo, D. G., 1979, *Int. J. Invertebr. Reprod.* **1**:279.
Cheung, P. J., 1974, *J. Exp. Mar. Biol. Ecol.* **15**:223.
Costlow, J. D., 1963, *Gen. Comp. Endocrinol.* **3**:120.
Costlow, J. D., 1966a, *Gen. Comp. Endocrinol.* **7**:255.
Costlow, J. D., 1966b, in: *Some Contemporary Studies in Marine Science* (H. Barnes, ed.), pp. 209–224, Hafner, Riverside, New Jersey.
Crisp, D. J., 1974, in: *Chemoreception in Marine Organisms* (P. T. Grant and A. M. Mackie, eds.), pp. 177–265, Academic Press, New York.
Crisp, D. J., 1976, in: *Adaptations to Environment* (R. C. Newell, ed.), pp. 83–124 Butterworths, London.
Donaldson, S., 1974, *Biol. Bull.* (*Woods Hole*) **147**:573.
Ede, D. A., 1978, *An Introduction to Developmental Biology*, Blackie, London.
Fretter, V., and Manly, R., 1977, *J. Mar. Biol. Assoc. U.K.* **57**:999.
Grant, P., 1978, *Biology of Developing Systems*, Holt, Rinehart, & Winston, New York.
Gurney, R., 1942, *Larvae of Decapod Crustacea*, Ray Society, London.
Highnam, K. C., and Hill, L., 1977, *The Comparative Endocrinology of the Invertebrates*, 2nd ed., Arnold, London.
Hughes, R. G., 1977, *J. Mar. Biol. Assoc. U.K.* **57**:641.
Hyman, L. B., 1951, *The Invertebrates*, Volume II, McGraw–Hill, New York.
Hyman, L. B., 1959, *The Invertebrates*, Volume V, McGraw–Hill, New York.
Larman, V. N., and Gabbott, P. A., 1975, *J. Mar. Biol. Assoc. U.K.* **55**:183.
Little, G., 1969, *Crustaceana*, (*Leiden*) **17**:69.
McConaugha, J. R., 1979, *Gen. Comp. Endocrinol.* **37**:421.
Meadows, P. S., and Campbell, J. I., 1972, *Adv. Mar. Biol.* **10**:59.
Rainbow, P. S., Ford, M. P., and Hepplewhite, I., 1979, *J. Mar. Biol. Assoc. U.K.* **59**:591.
Rao, R. K., Fingerman, S. W., and Fingerman, M., 1973, *Comp. Biochem. Physiol.* **44A**:1105.
Scheltema, R. S., 1974, *Thalassia Jugosl.* **10**:263.
Thompson, T. E., 1958, *Philos. Trans. R. Soc. London* **242**:1.
Thompson, T. E., 1962, *Philos. Trans. R. Soc. London* **245**:171.
Thorson, G., 1950, *Biol. Rev.* **25**:1.
Underwood, A. J., 1972, *Mar. Biol.* **17**:341.

Wigglesworth, V. B. 1954, *The Physiology of Insect Metamorphosis*, Cambridge University Press, London.

Wigglesworth, V. B., 1959, *The Control of Growth and Form*, Cornell University Press, Ithaca, New York.

Wigglesworth, V. B., 1965, *The Principles of Insect Physiology*, Methuen, London.

Wigglesworth, V. B., 1967, *Symp. Soc. Exp. Biol.* **2**:1.

Wigglesworth, V. B., 1970, *Insect Hormones*, Oliver & Boyd, Edinburgh.

Wigham, G. D., 1975, *J. Mar. Biol. Assoc. U.K.* **55**:45.

Wilson, D. P., 1932, *Philos. Trans. R. Soc. London Ser B* **221**:231.

Wilson, D. P., 1955, *J. Mar. Biol. Assoc. U.K.* **34**:531.

Wilson, D. P., 1968, *J. Mar. Biol. Assoc. U.K.* **48**:387.

Wilson, D. P., 1970, *J. Mar. Biol. Assoc. U.K.* **50**:1.

Wilson, D. P., 1977, *J. Mar. Biol. Assoc. U.K.* **57**:761.

Wilson, D. P., 1978, *J. Mar. Biol. Assoc. U.K.* **58**:467.

CHAPTER 3

Cell Structure during Insect Metamorphosis

M. LOCKE

> My purpose is to speak of bodies which have been changed into new forms.
>
> —Ovid. Metamorphoses.

1. INTRODUCTION

1.1. The Nature of Metamorphosis

Metamorphosis is the transformation in structure and function that became necessary when some organisms adapted to the need for a sequential occupation of more than one ecological niche (see Highnam, this volume). One of the interests of such organisms for biologists is that they may display commonplace developmental events in an exaggerated way, enabling us to notice and study special cases that nevertheless illustrate developmental processes of general significance. The role of the anatomist in such studies is to describe what happens. He must know the form of what is there. He must give hard substance to biochemical findings, pointing to microbodies as the structural home for the concept of peroxisomes, placing DNA and genes in nucleosomes, showing the cytoskeleton where movement occurs and ATPase at sites of energy transduction. The morphologist must be more adventurous: he must interpret structural change in cell and molecular terms, suggesting explanations

---

M. LOCKE · The Cell Science Laboratories, Zoology Department, University of Western Ontario, London, Ontario N6A 5B7, Canada.

for experiments that may seem far removed from structure. He must become a cell biologist.

In considering insect metamorphosis it is essential to realize how it differs from that in vertebrates. In general, the body form of a vertebrate is the result of a dynamic equilibrium between loss and replacement. Cells differentiate, die, and are replaced. Metamorphosis is an exaggeration of this, differing only in that the replacing cells have a new form. This is not true for most insects. Except in the higher Diptera, cell death and replacement is not the rule. Most insects maintain most of their form by cell division and the addition of cells that may become different from their progenitors but rarely die or lose their capacity to divide as a consequence of their differentiation. Metamorphosis comes about by remodeling the progeny of differentiated cells or even by remodeling differentiated larval cells themselves. Their capacity for remodeling makes insect cells especially interesting.

Changes in form at metamorphosis involve several levels of organization. At the grossest level, tissues are reorganized primarily by the redistribution of cells. The problems are ones of individual cell movement and the direction given to movement by surrounding cells, both in shaping tissues and in changing the distribution of cells within them (Section 2). At a finer level of organization, individual cells are remarkably plastic in their structure and are capable of almost completely turning over their cytoplasmic components so that one cell type, "differentiated" by the criteria used by vertebrate morphologists, may become another "differentiated" cell type (Section 3). Tissue composition and shape may also change through the net difference between cell division and destruction. The higher Diptera show an extreme condition of this where the destruction of one population of tissues coincides with recolonization by entirely new ones (Section 4). The problems of control posed by these changes require that a morphologist look for the structures appropriate for mechanisms by which hormonally initiated signals may arrive in the nucleus and the way that the commands are returned to the cytoplasm and coordinated with neighboring cells (Section 5).

This chapter describes and interprets what we know of cell changes during metamorphosis, correlating cell structure and function. In describing the structure it uncovers the complexity of control mechanisms needed to account for the cellular events. It also goes further than anatomical description and attempts to predict the structures that we should search for to explain aspects of metamorphosis that are presently mysterious.

1.2. Definition of Terms

Ecdysis is the process by which an insect emerges from its exuvium or old skin or the moment in time at which it emerges.

An instar is a developmental stage characterized by an integument

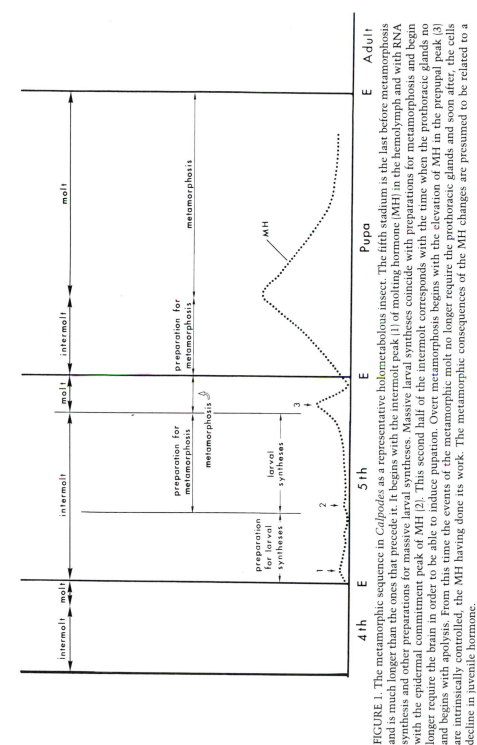

FIGURE 1. The metamorphic sequence in *Calpodes* as a representative holometabolous insect. The fifth stadium is the last before metamorphosis and is much longer than the ones that precede it. It begins with the intermolt peak (1) of molting hormone (MH) in the hemolymph and with RNA synthesis and other preparations for massive larval syntheses. Massive larval syntheses coincide with preparations for metamorphosis and begin with the epidermal commitment peak of MH (2). This second half of the intermolt corresponds with the time when the prothoracic glands no longer require the brain in order to be able to induce pupation. Overt metamorphosis begins with the elevation of MH in the prepupal peak (3) and begins with apolysis. From this time the events of the metamorphic molt no longer require the prothoracic glands and soon after, the cells are intrinsically controlled, the MH having done its work. The metamorphic consequences of the MH changes are presumed to be related to a decline in juvenile hormone.

that is outwardly unchanged between ecdyses, such as the fourth, fifth, or pupal instars.

A stadium is the interval between ecdyses, such as the fourth, fifth, or pupal stadia.

The intermolt/molt cycle refers to the sequence of development that repeats from stadium to stadium.

Molt events include the covert preparations for ecdysis beginning with the secretion of ecdysial droplets by the epidermis and ending when the form of the new instar is complete, usually soon after ecdysis. Molt events are for the most part intrinsically controlled, prothoracic glands no longer being required since the events are the result of earlier exposure to molting hormone (see Granger and Bollenbacher, this volume).

Intermolt events are those from ecdysis to the beginning of molting, and are for the most part extrinsically controlled, requiring the continued presence of hormones. It is the intermolt that is prolonged in a premetamorphic stadium.

Apolysis is the separation of the apical plasma membrane surface from the cuticle that takes place when the plasma membrane plaques involute as ecdysial droplets are secreted at the beginning of molting.

Metamorphosis occurs as a result of exaggerated changes at the last one (hemimetabola, exopterygota) or two molts (holometabola, endopterygota), and is usually recognized by characteristic alterations to the integument—the secretion of pupal or adult cuticle rather than larval cuticle. Overt metamorphosis corresponds to the beginning of molting but preparations for metamorphosis begin much earlier, making premetamorphic stadia much longer than previous ones.

The life cycle of a lepidopteran, *Calpodes ethlius*, is used in Fig. 1 to put these terms in perspective. The terminology is solely for convenience and must not be built upon like a legal edifice. It shows that although we may like to separate metamorphosis as a concept in our minds, it is only one of several concurrent activities in the organism. In particular, the covert preparation for metamorphosis may begin immediately after the previous ecdysis and may take place at the same time and in some of the very same cells that are engaged in exaggerated larval syntheses.

2. TISSUE REORGANIZATION

2.1. Metamorphic Changes in the Arrangement of Cells in Lepidopteran Fat Body

The fat body of *Calpodes* demonstrates the kinds of change in tissue morphology that can take place at metamorphosis with little or no cell division or cell death. The larval fat body is in the form of strips of cells making two concentric but incomplete cylinders suspended in the hemocoel around the gut and below the integument. In cross section the strips have the form shown in Fig. 2. From early larval life until the time of epidermal commitment for pupation (see O'Connor and Chang, this volume), the cells are in a single layer (Locke and Collins, 1968). At the time of pupal commitment by the epidermis (66 hr), the cells of the fat body reorganize into a two-layered structure. During the intermolt period devoted to larval syntheses, they maintain this form but increase in size with the deposition of lipid and glycogen and later with hemolymph protein until overt metamorphosis (defined as the time at 156 hr when the epidermis begins to deposit pupal cuticle). With overt metamorphosis, the fat body cells separate and seem held together only by the envelope of basal lamina around them (Locke and Collins, 1965; Dean, 1978). The cells may be tightly pressed together in the pupa but there is little coherent tissue structure. Even the most delicate dissection liberates free cells that are themselves easily damaged, giving an erroneous impression of cell destruction in all but ideally preserved tissues. In a second met-

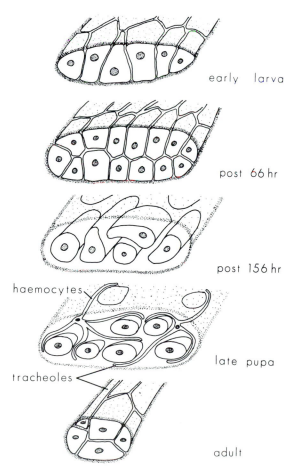

FIGURE 2. The sequence of tissue organization at metamorphosis in lepidopteran fat body. Longitudinally arranged strips of cells become bilayered as the fat body begins its premetamorphic massive larval syntheses at 66 hr in *Calpodes*. These cells separate at the beginning of larval/pupal metamorphosis (156 hr). They are rearranged into the adult pattern of lobular strings by hemocytes and tracheoblasts prior to the pupal/adult ecdysis. The changes are accompanied by alterations in ploidy. After Locke and Collins, 1968; Larsen, 1976; Walters, 1969, 1970; Dean, 1978.

amorphic rearrangement in the late pupa, these cells are pulled together by tracheoblasts (Larsen, 1976) or by hemocytes (Walters and Williams, 1966; Walters, 1969, 1970) into lobular strings around tracheae and tracheoles. The fat body thus undergoes at least three metamorphic changes in form corresponding in time to the epidermal activities of commitment to pupation, pupation itself, and pupal/adult metamorphosis.

2.2. Changes in the Location of Cells—Epidermal and Muscle Rearrangement

The relative proportions of segments and areas between recognizable landmarks on the integument often change markedly at a metamorphic molt. Some changes are the result of differential growth but others involve mainly the relocation of cells. For example, larval segmental muscles in the beetle, *Tenebrio*, may survive in the adult but with changed absolute and relative positions in the segment (Williams and Caveney, 1980a). Translocation occurs by two distinct mechanisms. Prior to pupation the muscles are carried to new positions by growth and perhaps by the movement of epidermal cells to which they are attached. After pupation certain muscles migrate over the hemocoel face of the epidermis to take up their new pupal positions in a way that suggests that the myoblasts sense their position from information in the epidermal gradient (Williams and Caveney, 1980b). It is well known that epidermal cells can exert a direct mechanical influence up to hundreds of micrometers and many cell bodies away from themselves. *Rhodnius* epidermal cells starved of oxygen extend processes to the nearest tracheole, drawing it closer to increase their oxygen supply (Wigglesworth, 1959). This could well be a model for the way that insect cells sense their position and rearrange themselves at metamorphosis. At the beginning of the prepupal molt of *Calpodes*, the apical face detaches from the cuticle by the loss of plasma membrane plaques in apolysis (Locke and Huie, 1979). It was believed that apolysis was related to a need for dividing cells to be separate from their superstratum, but mitosis in *Calpodes* occurs much earlier than apolysis, at the time that cuticle is secreted rather than separated (Locke, 1970). When the apical face detaches from the cuticle at metamorphosis, the basal surface develops long extensions and multiple hemidesmosomes as if the cells were crawling relative to the basal lamina. Plaque loss and cuticle separation in apolysis also occur at nonmetamorphic molts but over a much shorter time span (approximately 1–3 hr) than at metamorphosis (approximately 8–12 hr) as would be expected for the smaller extent of the rearrangement. There is thus a structure appropriate for cell relocation at the only time in the stadium that the epidermis is detached and free to move relative to the old cuticle (Fig. 3). The epidermis also detaches from the cuticle during the prepupal phase of epidermal rearrangement in *Tenebrio* (Williams and Caveney, 1980a), but here it detaches from the basal lamina as well. Complete

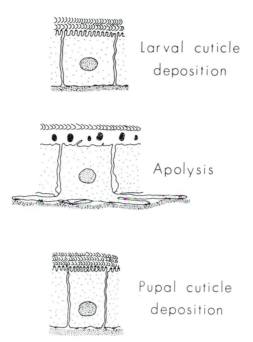

Larval cuticle
deposition

Apolysis

Pupal cuticle
deposition

FIGURE 3. When epidermal cells detach from the cuticle they have a morphology appropriate for sampling the surfaces of cells further away than their nearest neighbors in the epithelium and for moving relative to them. At the time of epidermal detachment from the cuticle (apolysis = ecdysial droplet secretion followed by plasma membrane plaque turnover in MVBs, Fig. 9), there are numerous basal extensions with a cytoskeleton and desmosomes.

detachment may therefore be necessary for movement to result from the cell multiplication that occurs at this time.

2.3. Changes in Composition and the Distribution of Reserves

A larval stadium before metamorphosis is always proportionally much longer than the preceding ones. For example, although the fourth stadium is only slightly longer than the third (14 days compared with 9 in *Rhodnius*, 4 days compared with 3 in *Calpodes*), the fifth premetamorphic stadium is twice as long as the fourth (28 days in *Rhodnius*, 8 days in *Calpodes*). The extra time needed for metamorphosis is not in the overt steps beginning with cuticle secretion; most insects only take about the same time to molt as to pupate ($1-1\frac{1}{2}$ days). The extra time is taken in the covert preparations for metamorphosis, which include massive accumulation and later redistribution of the nutritional reserves of protein, glycogen, and lipid needed subsequently for adult development.

The commitment of the epidermis to metamorphis, marking the beginning of epidermal preparations for the metamorphic molt, is coincident with a major shift in metabolism of many tissues toward the storage of nutritional reserves. This is shown in Fig. 4 for *Calpodes*. From the moment of pupal commitment the epidermis greatly increases its rate of lamellate cuticle deposition (number of lamellae increases 14-fold and thickness 6-fold, Locke, 1967). The fat body begins to secrete hemolymph proteins, the rate going from 0 to 1 mg/ml per hour (Collins, 1969), and increases its rate of deposition of lipid 18-fold (Locke, 1970). Although these activities all take place in earlier stages, their elevated rate is characteristic of preparation for metamorphosis and is associated with appropriate structural changes. In the epidermis the small size of the Golgi complexes (GCs) and the low frequency of secretory vesicles suggest that they play a minor role in lamellate cuticle deposition. The microvilli and plasma membrane plaques, on the other hand, have their highest density at this time of maximal secretion (Locke and Huie, 1979). In the fat body, the GCs enlarge with numerous secretory vesicles as the hemolymph protein increases from about 2 to 14% protein. Nutritive

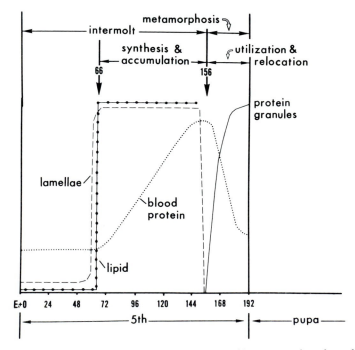

FIGURE 4. The premetamorphic stadium is characterized by massive larval syntheses, with first the accumulation of reserves and later their redeployment. The rate of accumulation of cuticular lamellae by the epidermis, of lipid by the fat body, and the accumulation of proteins, first in the hemolymph and later as fat body storage granules, during the fifth stadium of *Calpodes*. Accumulation of reserves begins at the time of epidermal commitment to pupation (66 hr). The switch from accumulation to utilization and relocation coincides with overt metamorphosis and independence from MH (156 hr). After Locke, 1967, 1970; Locke and Collins, 1968; Collins, 1969.

proteins stored temporarily in the hemolymph prior to metamorphosis are probably of general occurrence in insects (Loughton and West, 1965; Tobe and Loughton, 1969a,b). For example, there is a larval specific protein that disappears from the hemolymph at metamorphosis in the Dictyoptera (Kunkel and Lawler, 1974). Lipid droplets increase in size and deposits of glycogen enlarge. Peroxisomes containing urate oxidase attain their largest size and urate is not stored (Locke and McMahon, 1971). There is also one other major difference from nonpremetamorphic stadia. The fat body in fourth and earlier instars contains very large watery vacuoles. These do not develop in the fifth, premetamorphic stadium. Thus, preparation for metamorphosis involves a sharp shift in the utilization of reserves from cell growth to storage.

The switch to overt pupal syntheses involves the redistribution and/or reutilization of these reserves. The beginning of overt metamorphosis is signaled by the epidermis secreting ecdysial droplets. These are very much larger than those secreted at a larval molt in proportion to the extra thickness of lamellate cuticle that is to be digested. (The ecdysial droplets are presumed to be, or to have, some special relation to molting fluid proenzymes for cuticle digestion.) The larval cuticle may be several hundred micrometers thick and there is surely more than enough material to make the pupal cuticle. The fate of the excess is unknown. Fat body lipid declines, presumably as it is used up for some pupal activities. There is a major redistribution of hemolymph proteins (Locke and Collins, 1967). Wing imaginal discs and othe anlagen pinocytose proteins massively throughout their growth. Most other tissues pinocytose hemolymph proteins more at metamorphosis, presumably as a source of raw materials for their pupal syntheses. The biggest utilizer of hemolymph proteins is the fat body, which switches its function from synthesizing hemolymph proteins to storing them in large protein granules (Locke and Collins, 1968; Collins, 1969). There is also a switch in nitrogen metabolism as the Malpighian tubules shut down (Ryerse, 1978, 1979). The fat body loses its peroxisomes (Locke and McMahon, 1971) and gains urate granules (Tojo et al., 1978).

Thus, metamorphosis involves first the accumulation and later the redistribution of nutritional reserves in preparation for the change of form.

3. CELL REMODELING

The feature that distinguishes many gross structural changes in invertebrates, and particularly insects, is the degree to which their individual cells can be reprogrammed. This contrasts with metamorphoses like those in a caterpillar gut or the larval–pupal–adult transformation of a blowfly, which only appear to be sequential but in reality involve cells that have been differentiated in separate lines from much earlier in

their development. Sir Thomas Browne (1635) said of insects, "Who wonders not at the operation of two souls in those little bodies" (quoted in
Wigglesworth, 1954b). Cell and developmental biologists may paraphrase
this to "Who wonders not at the mechanism by which a genome equips
a cell for both its larval and adult lives." We can certainly find delight
and instruction in those insect cells that so readily change their function
while we watch.

The cell remodeling that accompanies the switch from larval to adult
function at metamorphosis proceeds in two stages. There is: (1) a loss of
the cell components that are presumably only appropriate for larval functions and (2) subsequent repopulation with organelles appropriate for an
adult. The synchrony with which insects remodel their cells makes them
particularly favorable for observing transient and rare events. Fragments
of the sequence proposed have been observed in numerous cell types,
supporting the idea that it has general relevance for eukaryotes.

3.1. The General Mechanism of Autophagy

The problem facing a cell that is intent upon destroying part of itself
is how to restrict the digestive environment to the particular organelles
that are scheduled for destruction. The answer, from studies on metamorphosing fat body and other tissues of *Calpodes*, is that components
to be digested are first isolated from the cytoplasm (Locke and Collins,
1965; Locke, 1967, 1969) and only then do primary lysosomes fuse with
the new, topologically external compartments, to digest their contents
(Locke and Sykes, 1975). This two-step principle is shown in Fig. 5.

In the fat body the first indication that an organelle is to be destroyed
is the presence of a tiny vesicle closely apposed to its surface. As more
vesicles fuse with it, the isolating envelope seems to creep over the
surface until investment is complete. The close apposition between the
envelope and what we may call its prey suggests that there may be a
special kind of adhesion between their surfaces. This would be required
to explain the specificity of destruction. The contents of autophagic vacuoles (AVs) are not a random sample of the cytoplasmic constituents of
the cell. Although most cell components are on the menu, they are eaten
separately and in a particular order—first peroxisomes, then mitochondria, followed by the rough endoplasmic reticulum (RER) some 12 hr
later. The RER is isolated in a particularly interesting way. The envelope
cuts off a fragment of about the same size as a mitochondrion, as though
there is a mechanism for making a vacuole of about that size independently of the shape and area of the RER surface to be covered. RER is also
carved up in this way in the epidermis, as is smooth endoplasmic reticulum (SER) in oenocytes (Locke, 1969) and prothoracic glands (Locke,
unpublished). In these examples also, the envelope is intimately associated with the object that it is enclosing, whether it is a membrane or
ribosomal surface. It follows from these observations that isolating en-

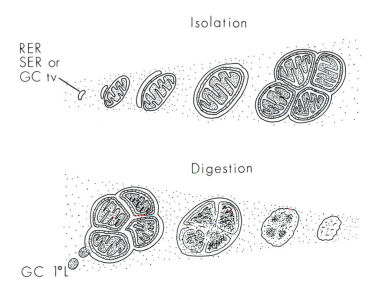

FIGURE 5. The general mechanism of autophagy, the first stage of remodeling. Isolation in an external compartment is followed by fusion with primary lysosomes (1° L) and digestion. Isolating envelopes are a distinct class of membranous cell component, being part of the vacuolar system. They may arise from the GC transition vesicle region (GC tv) or from nearby rough endoplasmic reticulum (RER) or smooth endoplasmic reticulum (SER) according to cell type. Isolation may be specific as in the fat body where peroxisomes are destroyed before mitochondria followed by RER, or relatively nonspecific as in metamorphosing oenocytes. Digestive enzymes from primary lysosomes are only added when investment is complete. After Locke and Collins, 1965, 1980; Locke and Sykes, 1975; Locke, 1969.

velopes and the vesicles from which they arise are distinct cell components that form in relation to the first phase of autophagy. Their membranes are unique in the way that they adhere to other cell components. (Confronting cisternae are somewhat similar in the way that a smooth face of RER approaches a plasma membrane or growing peroxisome.) The removal of cell components for digestion occurs with varying degrees of precision in different cell types. In oenocytes the AVs are mixed bags that may contain complete GCs at times. Even during RER removal in the fat body, mitochondria wrapped in RER may occasionally be lost. Presumably the isolating membranes only respond to objects at their surface and this is usually enough to give specific autophagy when needed.

The morphology suggests that the membrane of the isolating envelope arises from the forming face of the GC or from nearby endoplasmic reticulum (ER) (Locke and Collins, 1965, 1980). Hot osmium staining confirms this origin, for the only parts of the cell to stain are the contents of the envelopes, GC transition vesicles, and nearby ER and saccules (Locke and Sykes, 1975).

The chief characteristics of the isolating phase are:

1. Its specificity for certain organelles (in some cell types such as the fat body).
2. The intimate relation between the envelope and the object that it encloses.
3. The rather uniform size of the compartments formed independently from their contents.
4. The emptiness of the envelope lumen after most treatments except hot osmium, which reacts intensely. In particular, they do not contain acid phosphatase, the marker enzyme for lysosomes.

Once the cell components have been externalized in their own compartments in this way, they may either fuse directly with primary lysosomes to become AVs or first fuse with one another to make large (fat body) or even giant AVs (oenocytes). In this second step of autophagy, primary lysosomes carry digestive enzymes from the maturing face of the GC which may be equivalent to Novikoff's GERL (Locke and Sykes, 1975).

The chief characteristics of the digestive phase are:

1. Hydrolytic enzymes are not detectable in the isolation vacuole until after primary lysosomes have fused with it.
2. Digestion does not begin until this happens.
3. Isolation vacuoles may fuse with one another before and after fusion with primary lysosomes but not with any other membrane.

The general mechanism of autophagy outlined in Fig. 5 serves to remove and digest almost all cell components. The net result is a cell with reduced cytoplasm that is ready for redifferentiation by repopulation with new organelles, or one that is small enough to be phagocytosed in cell death.

3.2. Repopulation with New Organelles

Calpodes fat body cells begin the pupal stadium with few mitochondria, no peroxisomes, little RER, and few GCs but with much glycogen, lipid, stored hemolymph protein, and AVs containing RNA from partly digested RER. They have clearly switched from larval syntheses to storage. During the 36 hr before and the 36 hr after the pupal/adult ecdysis, the adult cells reduce their reserves and restore a full complement of organelles (Larsen, 1976). Figure 6 outlines the changes as the cells switch from storage to a structure appropriate for adult syntheses.

The first organelles to reappear are the peroxisomes, which lie adjacent to the RER cisternae from which they are presumed to arise as they do during the generation of larval peroxisomes in the fifth stadium.

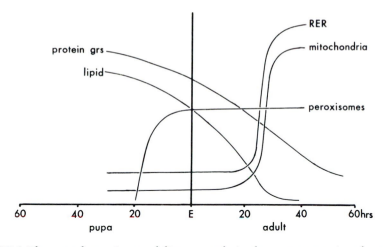

FIGURE 6. The second stage in remodeling, repopulation by a new generation of organelles. *Calpodes* pupal fat body cells contain few mitochondria, no peroxisomes, and little RER. At the pupal/adult metamorphosis new peroxisomes arise from the RER, new RER is formed, and a phase of mitochondrial division increases their abundance sixfold. Protein granules (grs) and lipid reserves decline as the new organelles arise. After Larsen, 1970, 1976.

For a very brief period after ecdysis, mitochondrial profiles show division figures, a large crista cuts the inner compartment into two and is followed by the outer membrane to form two daughter mitochondria (Larsen, 1970). At this time there is a sixfold increase in the number of mito-chondria per cell and a threefold increase in the surface area of RER membranes.

The end result is a cell presumably reequipped for adult activities. It would be especially interesting to know if the new peroxisomes are qualitatively different from the larval ones. Both have catalase, but is their complement of oxidases the same? The mitochondria are morpho-logically different. Larval fat body mitochondria are sausage shaped for most of the time (although they round up for division and after fusion during starvation). The adult mitochondria are spherical or oval. Do these differences reflect changed functions and enzyme compositions and if so are they the result of their mitochondrial lineage or of the new adult environment? The change in form at least suggests a kind of mitochon-drial differentiation. We may ask of the mitochondria the same questions that we ask of postmetamorphic adult cells. Are they the result of the proliferation of a stem line set aside to survive prepupal destruction, or is the mitochondrial autophagy random, with all mitochondria having the potential to assume the adult spherical form?

3.3. Membrane Changes and Membrane Turnover

An aspect of metamorphosis that has received little attention con-cerns changes in the nature of the cell surface. One may expect changes in surface receptors in response to changed components of the hemocoel, both hormonal and nutritional. Several observations suggest that such changes accompany metamorphosis in all or most cell types.

Hemolymph proteins and foreign tracer proteins are pinocytosed throughout larval development but the rate is higher in rapidly growing tissues such as wing discs and at metamorphosis (Locke and Collins, 1967). The selective uptake of proteins by the fat body has been described in several insects (Chippendale and Kilby, 1969; Loughton and West, 1965; Chippendale, 1970; Collins and Downe, 1970). The fat body presumably selects hemolymph proteins for uptake by specific binding to surface receptors at sites where they are later pinocytosed. At the larval/pupal metamorphosis the fat body of *Calpodes* takes up one of the main hemolymph proteins for storage (Collins, 1975). The hemolymph contains three main groups of antigens. Group I and II are present in both the fat body and the hemolymph throughout the last larval stage. Group III antigens are absent from the fat body but accumulate in the hemolymph until the initiation of pupation. They then appear in the fat body as it forms protein storage granules from sequestered hemolymph (Collins, 1974) in response to the prepupal ecdysteroid peak (Dean et al., 1980). The sequestration of Group III antigens and the switch to storage granule formation have been shown experimentally to be dependent upon the presence of active prothoracic glands and ecdysteroids (Collins, 1969, 1974). Thus, one aspect of metamorphosis in the fat body that is caused directly by molting hormone involves a change in response to a hemolymph protein. The structural changes involved in this shift in the rates of membrane turnover at metamorphosis are outlined in Fig. 7. Although the switch involves the complete uptake system [membrane, pinocytosis, fusion in storage granules rather than multivesicular bodies (MVBs), primary lysosomes, and the digestion of excess membrane], the initial step must involve the availability of specific protein binding sites and the uptake mechanism must involve the loss of much surface membrane. Assuming that there is membrane turnover rather than recycling, order of magnitude calculations show that the fat body plasma membrane would be replaced every 10 min during the uptake of protein. Even with some recycling and a margin of error in the calculations, it seems likely that fat body membranes are replaced rather rapidly at metamorphosis. An important question for the future will be to determine the relative roles of membrane turnover and membrane recycling during protein uptake.

Plasma membrane is also turned over in the fat body at the pupal/adult metamorphic molt. The frequency of MVBs is highest during the 40 hr centered upon ecdysis and they are absent 10 hr before and 10 hr after this time interval (Larsen, 1976) (Fig. 8). The minimal surface area of plasma membrane needed to form these MVBs is probably greater than the surface area of the whole cell. Thus it is probable that the plasma membrane turns over once again at the pupal/adult metamorphosis.

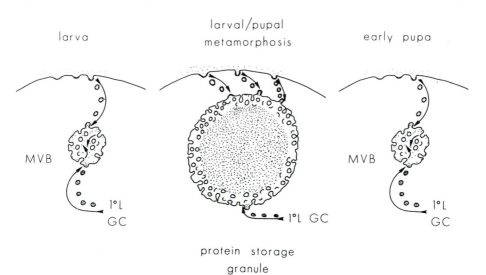

FIGURE 7. Membrane turnover in *Calpodes* fat body at metamorphosis. The switch from pinocytosis into multivesicular bodies (MVB) to massive uptake of hemolymph proteins into storage granules (Fig. 4) must involve increased plasma membrane turnover. The abundant microvesicles within the edge of forming granules suggest membrane digestion rather than recycling. Although other tissues do not store proteins at metamorphosis, they usually form more or larger MVBs.

The turnover of plasma membranes at metamorphosis is probably of general occurrence in all or most cell types. In the epidermis, the plasma membrane plaques are a morphological marker for the differentiated areas of membrane involved in cuticle secretion that lie at the tips of microvilli (Locke, 1976). The plaques involute by pinocytosis and are destroyed in MVBs as the level of the hemolymph ecdysteroid rises and

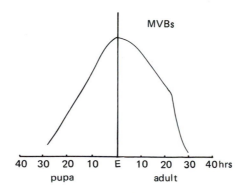

FIGURE 8. The increase in frequency of fat body MVBs at the pupal/adult ecdysis in *Calpodes*. MVBs increase from zero to become about 2% (~800 μm^3) of the cell volume at the time of emergence to the adult. The area of membrane needed to make the pinocytosis vesicles that could fill these MVBs suggests that the plasma membrane is turned over at this time. After Larsen, 1976.

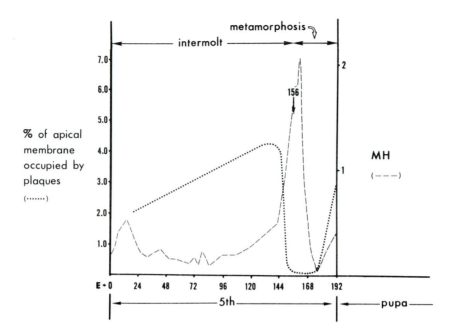

FIGURE 9. The turnover of the apical plasma membrane in *Calpodes* epidermis at metamorphosis. The plaques on the apical plasma membrane are pinocytosed and destroyed in MVBs at the beginning of overt metamorphosis. The absence of plaques leads to apolysis (cf. Fig. 3). MH, molting hormone. After Locke and Huie, 1979.

pupation begins (Locke and Huie, 1979) (Fig. 9). Plaques and apical membrane re-form in time for the secretion of pupal cuticle. At the beginning of pupation the Malpighian tubules cease to function in fluid transport (Ryerse, 1978, 1980). The loss of function is accompanied by the disappearance of basal infolds and a reduction in the dimensions of the apical microvilli. The membrane presumably disappears in the MVBs which increase in frequency at this time (Ryerse, 1979). At the pupal/adult ecdysis the frequency of MVBs rises again as the basal infolds and apical microvilli re-form. Thus, in Malpighian tubules the plasma membrane probably turns over at both larval/pupal and pupal/adult ecdyses.

The replacement of a cell surface at metamorphosis can be viewed as a special case of autophagy and organelle renewal. The end result is a cell equipped with a new plasma membrane for its changed existence.

3.4. The Cytoskeleton and Changes in Shape and Organelle Distribution

Most cell types change their shape at metamorphosis but the role of the cytoskeleton has been little studied. Epidermal cells, for example, become columnar and occupy a smaller area. As at molting, their apical surface is a mold for the cuticulin layer determining the new surface pattern. The shape of the cell surface itself depends upon the cytoskeleton (Filshie and Waterhouse, 1969). Fat body cells change from a cuboidal double layer to clusters of separate, almost spherical cells as they lose their junctions. It may be relevant that fat body cells change their cytoskeleton when placed in culture. Balls of filaments appear that are reminiscent of the filament skeins seen in BHK-21 cells treated with colchicine (Goldman, 1971). In other cells there are characteristic changes in the distribution of organelles at metamorphosis. In Malpighian tubules, mitochondria slide out of the microvilli and back into the cell body where they are destroyed in autophagy. This movement of mitochondria is one of the earliest structural correlates with the decline in fluid transport induced by 20-hydroxyecdysone at the larval/pupal metamorphosis (Ryerse, 1979). When fluid secretion resumes prior to the pupal/adult ecdysis, mitochondria are reinserted in new microvilli in a reversal of the shutdown appropriate for pupal life. This shows the overriding importance of the state of the system responding to the hormone since in this instance the peaks of ecdysteroid preceding the two metamorphic molts are correlated with diametrically opposite effects.

3.5. Cell Coupling and Junctions

The alterations in shape and cell distribution at metamorphosis involve extensive changes in junctions involved in coupling, adhesion, and the formation of permeability barriers (Lane, 1978). The coupling between cells changes at metamorphosis in the epidermis of *Tenebrio* larvae (Cavency and Podgorski, 1975; Caveney, 1978). Coupling first increases and then decreases transiently just prior to pupation. The initial increase can be stimulated *in vitro* by 20-hydroxyecdysone. The immediate cause of these fluctuations may lie in changes in the number of gap junctions or in their properties or both (Caveney, 1980). Gap junctions change their form dramatically during larval and pupal development in *Calliphora* (Lane and Swales, 1978a,b). At the end of larval life, gap junctional particles in perineurial and glial cells are arranged in maculae but move to disorganized arrays in the early pupa and back to maculae again as the adult develops. Insect cells are clearly model systems for studying the functions of junctions in development. We need many more such studies to understand the control of metamorphosis

3.6. The Control of Remodeling

The initial events in metamorphic remodeling coincide with the prepupal peak for hemolymph ecdysteroid (Dean *et al.*, 1980). Isolation for autophagy in the fat body (Locke and Collins, 1965), apical membrane involution in the epidermis (Locke and Huie, 1979), and mitochondrial withdrawal from microvilli in Malpighian tubules (Ryerse, 1979, 1980) all begin as the hormone reaches its peak. There is therefore a strong presumption that molting hormone (MH) (in the absence of juvenile hormone) induces autophagy. The induction of autophagy *in vitro* shows that this is a direct effect of the hormone on these cell types. Both fat body and Malpighian tubules initiate metamorphic structural changes when exposed to MH in culture (Dean, 1978; Ryerse, 1979, 1980). The cellular sequence itself is independent of hormone. The fat body in abdomens isolated without prothoracic glands after the hemolymph ecdysteroid peak has been reached, or such fat body explanted in tissue culture, complete their autophagic sequence. Initiation of the autophagic sequence is induced by MH but the sequence itself is intrinsically controlled. There is as yet no experimental work aiming to dissect the control of the various steps in the sequence. For example, is the initiation of the sequence dependent upon the earlier history of the cell that led it to become polyploid, or will any fat body cell without juvenile hormone respond to MH in this way? Is the formation of isolating envelopes dependent upon the presence of organelles of a particular kind to invest, or is it an obligatory response to a hormonal command to start the sequence? Is the presence of isolated organelles necessary for the formation of the appropriate kinds of primary lysosome or are they an obligatory consequence of the earlier formation of isolating envelopes? The precise timing of the steps in the phagic phase of remodeling gives hope that questions of this type may be answered.

The main phases of remodeling in the pupa do not correlate temporally with the hormonal peak that is presumed to initiate adult development. The peak for pupal ecdysteroids is early in the stadium (Dean *et al.*, 1980) but most of the morphological changes discussed are just pre- or post-pupal/adult ecdysis. Mitochondrial replication could be one of the last events set in train by ecdysteroids when adult development begins in the pupa or there may be fresh signals in the new adult controlling such development. We cannot assume that MH is itself sufficient for metamorphosis in all cell types. Some metamorphic responses require factors in addition to MH for their induction. For example, the evagination of wing discs requires a dialyzable factor from the fat body in addition to ecdysone (Benson *et al.*, 1974; Bergtrom and Oberlander, 1975).

4. CHANGES IN THE NUMBERS AND KINDS OF CELLS IN REMODELING TO MAKE NEW TISSUES

Metamorphosis involves the differentiation of completely new cell types (as in the formation of epidermal oenocytes), new tissues (for example, those involved with sex and reproduction), and new organs (the characteristic adult structures such as wings). Other new structures may also arise from rudiments or from undifferentiated precursors, but their further development depends upon the destruction of antecedents having functions limited for larval life. Whether new or replacements, the final form results from the number of cells left after local variations in division and death.

4.1. Cell Division

The phase in a premetamorphic stadium during which both larval syntheses and preparations for metamorphosis take place is also a time for mitosis and cell division. Although mitosis is probably not an obligatory step for metamorphosis, the timing of the decision to metamorphose rather than to molt coincides with the initiation of epidermal mitoses in *Calpodes* (Dean *et al.*, 1980) and *Manduca* (Wielgus and Gilbert, 1978; Wielgus *et al.*, 1979. These cell divisions coincide with the small but definite elevation of MH titer in the hemolymph forming the reprogramming peak that is known in several insects (see O'Connor and Chang, this volume). The frequency of divisions declines by the time of the major prepupal peak of MH, which coincides with the beginning of overt preparations for metamorphosis. These premetamorphic epidermal divisions differ from those in preceding molts because their local frequency must reflect the need for new areas and shapes related to metamorphosis rather than to molting. The pattern of frequency of mitosis must relate to the postmetamorphic form, and the future synthetic capabilities of these cells may also relate to this division. The way that a group of epidermal cells responds to MH by such and such a degree and perhaps kind of mitosis may be the result of an interaction between its state of determination and its appreciation of position in the epidermal gradient.

4.2. Cell Death

The use of cell death to accomplish metamorphic changes varies widely between different insects and between different tissues. It may involve anything from a minor adjustment in cell density to almost complete destruction of a larval organism prior to replacement by adult cells.

At its least obtrusive, cell death may occur to restore the cell density in relation to the determination of a new size as in *Rhodnius* tracheae (Locke, 1958, 1974) (Fig. 10). Tracheae are not committed to a certain cell density by mitosis alone. Cell death may be used quantitatively as a fine control for cell number. The extent of tracheal growth is signaled by the polarized transport of information, so that cell death may be dependent upon position in the epidermal gradient, one aspect of which is the polarized transport of growth factors.

Cell death is responsible for the complete loss of larval organs such as prolegs, silk glands, larval or pupal muscles (Lockshin and Williams, 1964, 1965a,b; Lockshin, 1969), and specializations of the integument. It is also used to cause minor pattern changes such as the fine sculpturing

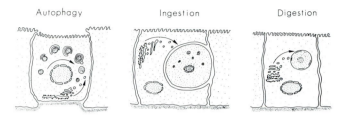

FIGURE 10. The steps involved in cell death. The morphology associated with cell death in *Rhodnius* epidermis. A cell scheduled for death first decreases in size by autophagy as outlined in Fig. 5 and by turnover within the neutral environment of the cell sap. The reduced cell is then ingested by its neighbors and digested in a phagic vacuole.

of the scalloped or tailed wings of Lepidoptera. The smooth outlines of pupal wings are carved by selective death into their adult outline.

In other tissues programmed destruction is coupled with replacement from reserve cells to give qualitatively different organs. Shortly before pupation the midgut of caterpillars is completely shed and replaced by a new epithelium derived from basal reserve cells. This principle of programmed death and replacement may apply to some other organs and tissues, but reports of the death of many fat body cells in Lepidoptera are probably misinterpretations due to the difficulty of fixing freely floating cells containing little but phagic vacuoles.

Cell death on a massive scale to allow the formation of a qualitatively different organism probably only occurs in the higher Diptera, where nearly all tissues of the adult are rebuilt from organ and tissue rudiments.

Although the morphogenetic use of cell death is widespread and variable, the basic cell biology involved is probably always the same. The problem is one of controlled digestion, how to degrade certain tissues without affecting others. Cell death probably always takes place in two steps: (1) autophagy, in which cytoplasmic components of cells about to die are segregated in autophagic vacuoles as already described, and (2) phagocytosis of the much reduced cell by its neighbors as in *Rhodnius* epidermis (Locke, 1967) or by hemocytes (Crossley, 1964, 1965, 1968). Both steps involve the release of primary lysosomes into confined compartments for controlled digestion. There is no evidence for the free release of lysosomes within a cell to cause its death and uncontained digestion. Lysosomes do not break and release their contents of digestive enzymes, they fuse and deliver them. The topological principle in the use of primary lysosomes involves their fusion with vacuoles that are part of an external compartment. The morphology of cell death in *Rhodnius* epidermis is interpreted in Fig. 10. Although most insects are neat and tidy in their cell death like this, others may leave much debris to be phagocytosed by hemocytes. Muscle proteins, for example, have been found to circulate in the hemolymph before their final digestion (Lockshin, 1975).

The fact that we can see the way that some cell components are digested after their isolation in autophagy should not blind us into ignoring other mechanisms for their loss. Many structures such as muscle protein filaments (Randall, 1970; Wigglesworth, 1956) and nuclei lose their substance within the neutral environment of the cell sap, and since ribosomes can turn over in bacteria without isolation, we may presume that they can do so in eukaryotes. The problem is especially complicated in muscle degeneration, where organelle autophagy involving lysosomes goes on side by side with muscle protein breakdown while within the cytoplasmic compartment (Lockshin and Beaulaton, 1974a,b,c; Lockshin et al., 1977; Schlichtig et al., 1977).

4.3. Polyploidy and Polyteny

It is usually supposed that polytene and polyploid cells, having resulted from nuclear replication without cell division, are incapable of further division. Although there are exceptions to this in that some polyploid cells can become diploid again (Berger, 1938; Grell, 1946; Wielgus et al., 1979; Nair and Locke, unpublished), such cells do not usually divide. Since metamorphosis results in an adult with a limited life span, they do not have to divide. It is perhaps for this reason that metamorphosis often involves increased ploidy, either with an acceleration of the process begun in earlier stadia as in *Calpodes* oenocytes, pericardial cells, and Malpighian tubules or with a new round of nuclear replication as in the fat body. Once a tissue is released from the necessity of having to divide again, it seems as though there are good functional reasons for having larger cells to make the kinds of tissues and organs that an insect finds useful in an adult. The higher Diptera, of course, have this release in the larva. Since the adult is formed from different cells, their larvae can afford to be made of large polyploid and polytene cells. Other insects follow this pattern to a lesser degree. Silk glands only used by the larvae of Lepidoptera also become greatly polyploid.

The fat body of *Calpodes* is a particular example where polyploidy occurs in the premetamorphic stadium but not earlier and is therefore an expression of metamorphic change. The number of cells is determined by divisions at the end of the fourth stadium. At the beginning of the fifth stadium there is a round of nuclear replication followed by polyploidy that lasts into the pupa (Locke, 1970) where there may be reduction before rearrangement to the adult form of fat body. Whereas the epidermal commitment to become pupal rather than larval occurs at about 66 hr into the fifth stadium, corresponding in time to the MH commitment peak (Dean et al., 1980), the fat body has done so at ecdysis. The fat body has already reached its full degree of ploidy (an overt commitment to metamorphosis) by the time that the epidermis begins the divisions that coincide with its pupal commitment.

Whatever the mechanism by which cells are induced to become polyploid rather than divide, it is one of the processes most characteristic of insects and must be considered in the morphogenesis of metamorphosis (see Sridhara, this volume).

5. THE MORPHOLOGY NEEDED TO EXPLAIN SOME ASPECTS OF METAMORPHOSIS

The aery spirit of biochemical and physiological hypothesis must be given the bones of morphology to articulate into credible developmental biology theory.

5.1. Hormonal Signals and the Response of the Nucleus

The stadium preceding the first metamorphic molt is longer and more complex than those preceding it, as noted previously. Although pupal epidermal characteristics may be determined at the end of the first third (e.g., at about 66 hr out of 192 hr in the fifth stadium of *Calpodes*), other events that are unique to this stadium (and therefore presumably related to metamorphosis) begin immediately after the fourth to fifth ecdysis. For example, the fat body becomes polyploid and epidermal nucleoli become multiple just after ecdysis, quite unlike the earlier stadia that precede ordinary molts. Metamorphosis begins with the fifth stadium. This is reflected in the levels of MH in the hemolymph, which also differ from earlier stages, with peaks corresponding to at least three characteristic stages of development, the intermolt, commitment, and prepupal peaks (Dean *et al.*, 1980). The MH is presumed to act in the nucleus after being taken up by cytosol receptors (Yund *et al.*, 1978; see O'Connor and Chang, Sridhara, and Fristrom, this volume). The principal steps in the hormonal stimulation of metamorphosis (or most other responses) are outlined in Fig. 11a, using an epidermal cell as an example. Can the morphologist point to an appropriate cell structure to confirm the plausibility of this scheme?

The first question concerns the way in which the MH gets from the hemolymph to its presumed site of action, the chromosomes of the nucleus. In mammalian cell types the first step is the binding of hormone to steroid receptors at the cell surface after which the plasma membrane is pinocytosed. Structurally appropriate vesicles increase in frequency in most insect tissues at molting and metamorphosis so that the first step in MH uptake could very well be binding to the plasma membrane and pinocytosis. This still leaves the hormone outside the cytoplasmic compartment until the vesicle membrane breaks or becomes permeable to MH. Once in the cell sap, the MH, with or without its receptor protein, is still two unit membrane barriers away from the nuclear compartment. The route that would allow the greatest specificity of uptake of molecules controlling nuclear processes would involve a repetition of membrane binding and pinocytosis but through the nuclear envelope. If the cytoplasmic face of the nuclear envelope were to have MH binding sites that pinocytosed when occupied, and if these later fused with the membrane of the nuclear face, there could be a specific pathway from the cytoplasmic to the nuclear compartment. Nuclear envelope binding sites have been described in some cells, and in *Calpodes*, there is an appropriate morphology at the time when MH is believed to be acting (Locke and Huie, 1980). Very tiny pinocytosis vesicles occur on both faces of the nuclear envelope and in its cisternae. The route for transport of MH from the hemolymph to the nucleus that is suggested by the morphology is outlined in Fig. 11b, which also shows the routes for the transport of other material. Messenger RNA packages can be traced as perichromatin granules from the edges of chromatin through the nuclear pores. Nuclear proteins and ribosomal precursor proteins may enter the nucleus by moving through the pores in the opposite direction. Ribosomal precursor granules move from the edges of the nucleoli through the pores and into the cytoplasm. There is thus a structure appropriate for the steps presumed to occur when MH induces metamorphosis.

5.2. Positional Signals and the Kind and Amount of Growth

The qualitative changes at a metamorphic molt and their differences from a nonmetamorphic molt tend to draw attention away from a feature common to both, that is, the increase in size, mere growth, the quantitative difference that occurs at all molts. The formation of a new organism within the old one in arthropods is basically a process for getting smaller—unless the new skin is inflated and expanded upon its release from the exuvium. The new size is limited by the area to which the integument can expand. Somehow the extent of growth is coordinated in different tissues and is related to previous nutrition. A clue to the mechanism for this coordination comes from studies on the growth of tracheae. Tracheae grow at each molt according to the respiratory demands that have been placed upon their tracheole terminations (Locke, 1958; Wigglesworth, 1954a). In this way the cross section of a main trachea grows in proportion to the new terminal branches that are added. Although MH causes cuticle deposition, the extent of growth requires something else (Ryerse and Locke, 1978). The information needed to specify the amount of growth is carried along the tracheal epithelium in a polarized way from termination to main branch. This intrinsic polarity cannot be reversed by pruning the branches. The polarity is probably one manifestation of the epidermal gradient in the surface integument (Caveney, 1973; Locke, 1959), since grafts that are removed from their normal gradient position also fail to grow (Locke, 1960). Grafting experiments of the integument of *Galleria* show that the position in the gradient can

Theory

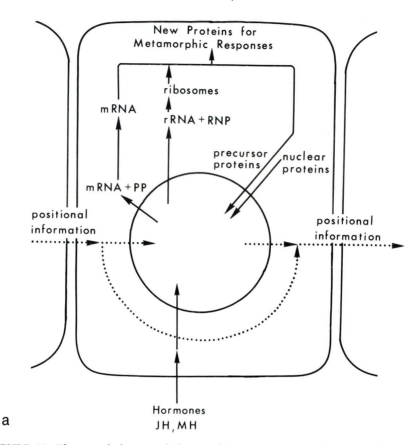

FIGURE 11. The morphology needed to explain some aspects of metamorphosis in an epidermal cell. (a) The principal steps in the hormonal stimulation of metamorphosis. (b) The morphology observed at the time of hormonal stimulation. After Locke and Huie, 1980.

also determine the kind of cuticle deposited at metamorphosis (Stumpf, 1968), suggesting that information passed from cell to cell determines qualitative as well as quantitative responses. The chief properties of the polarity/gradient are:

1. Information moves along an epithelium in one direction.
2. The information specifies both the extent and the kind of growth.
3. The information does not decay with distance moved.
4. The movement has a stable morphological base. After grafting experiments, gradients made visible as patterns only relax slowly.

Most workers have had difficulty coming to grips with the nature of the gradients. Simple diffusible morphogens (Crick, 1970), for example, do not account for these properties. It is relatively easy to think of structures

Observation

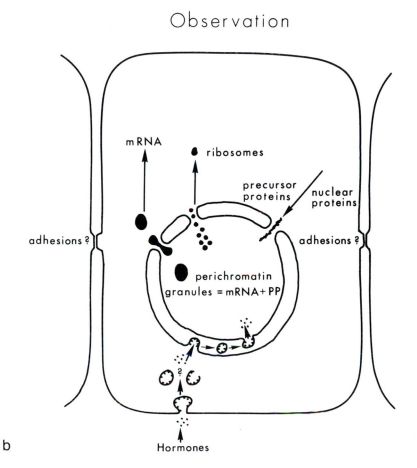

b

FIGURE 11. (*Continued*)

that could result in the polarized distribution of molecules. Can morphology suggest the kind of structure that should be looked for to confirm certain classes of explanation?

The relation between mitochondria (or chloroplasts) and the nucleus may be a model for the vectorial transport of informational molecules. The two compartments have separate but closely coordinated genomes, and many mitochondrial proteins are known to be specified by nuclear genes (Schatz and Mason, 1974). The two systems may influence one another through their protein products since there is no evidence that an RNA transcribed in one system is translated in the other. Since most mitochondrial proteins are made in the cytoplasm, how do they get to the mitochondria and is this transport a model for the way that cells influence one another in an epithelium?

The inner mitochondrial membrane contains stable patches of high density around sites that are regions of contact with the outer membrane

(Hackenbrock and Miller, 1975). Ribosomes associate with this outer membrane near the adhesions and could be the source of mitochondrial proteins. Some outer chloroplast membranes are also continuous with ribosome-bearing membranes (Crotty and Ledbetter, 1973). If epithelial cells have similar adhesions with a selective distribution that could function to transfer information in a gradient, they could easily have been overlooked. It would at least seem worthwhile to search for such special kinds of contact that might provide the morphological base for vectorial transport.

5.3. The Reception and Use of Nutritional and Precursor Molecules: Pinocytosis and Transepithelial Transport

In the premetamorphic stadium the cuticle may become a major carbohydrate and protein reserve in preparation for metamorphosis (as in *Calpodes* and *Rhodnius*). Although tracer amino acids injected into the hemocoel become cuticular proteins in minutes (Condoulis and Locke, 1966), it has not been demonstrated that all cuticular proteins are synthesized by the epidermis. Indeed, it has been claimed that some cuticular proteins are deposited from the hemocoel since proteins from the two sources are immunologically indistinguishable. The demonstration that vertebrate maternal antibodies can pass across the fetal gut (Abrahamson *et al.*, 1979; Rodewald, 1973) and that digestive enzymes can be pinocytosed from the gut and passed via the blood back through the gut wall for recycling without resynthesis (Diamond, 1978) should make us look more carefully at the insect epidermis. Does it have different categories of pinocytosis vesicles corresponding to different binding sites, some scheduling proteins for digestion in the apical MVBs (as has been demonstrated for foreign proteins, Locke and Krishnan, 1973) while others carry proteins across the epidermis to the cuticle?

6. DISCUSSION

6.1. The Onset of Metamorphosis

One of the generalizations about metamorphosis that emerges from this discussion is that it begins much earlier than a superficial examination of the premetamorphic stadium would lead us to expect. Events uniquely related to metamorphosis begin from the inception of the last larval stadium, before pupal cuticle is laid down and before the epidermal commitment to lay down pupal cuticle. Metamorphosis begins immediately after ecdysis when the intermolt activities differ from those in previous stadia. The epidermis has an elevated rate of RNA synthesis and the fat body has elevated RNA synthesis and polyploidy (Locke and Huie, 1980). Metamorphosis begins with the commitment to change the style of larval activities, particularly with the preparation to change the rate of synthesis of larval reserve molecules (Fig. 1), so that food reserves are accumulated (Fig. 4) rather than processed to become more larval cells.

6.2. Comparisons between Insects and Vertebrates

Insect development is characteristically conservative, cells and tissues being rearranged in a make-do and mend organization. Vertebrates change by replacement with less conservation. For example, the skin of a frog is derived from a germinal layer rather than from differentiated tadpole epidermal cells, whereas most adult insect cuticle is a new synthetic manifestation of cells that may have previously secreted larval structures.

Insect development is also characteristically cyclical and sequential whereas that of vertebrates is often the dynamic outcome of numerous concurrent and continuous activities. The contrast between cyclical activity and a dynamic equilibrium is well seen in comparisons between organelle development in insect fat body and in vertebrate liver. In rat liver the half-life of mitochondria is 5–10 days and of peroxisomes it is about $3\frac{1}{2}$ days. There are about 1200 mitochondria per cell, so that on average only $2\frac{1}{2}$ mitochondria are lost and replaced in each cell every hour. Even if mitochondrial division or envelopment were to take a whole hour for completion, only 1 in 500 might be observed to have a distinctive morphology in appropriate profiles. The synchrony of autophagy in the few hours prior to pupation and of mitochondrial replication in a few hours in the new adult, make it easy to see isolation for digestion and the formation of mitochondrial division figures. It is these synchronous sequences that make insect cells so favorable for observing transient events rarely seen in vertebrate cells.

ACKNOWLEDGMENTS This work was supported by Grant A6607 from the Natural Sciences and Engineering Research Council of Canada. Thanks are due to Phil Huie for technical assistance.

REFERENCES

Abrahamson, D. R., Powers, A., and Rodewald, R., 1979, *Science* **206**:567.

Benson, J., Oberlander, H., Koreeda, M., and Nakanishi, K., 1974, *Wilhelm Roux Arch. Entwicklungsmech. Org.* **175**:327.

Berger, C. A., 1938, *Carnegie Inst. Washington Publ.* **496**:209.

Bergtrom, G., and Oberlander, H., 1975, *J. Insect Physiol.* **21**:39.

Caveney, S., 1973, *Dev. Biol.* **30**:321.

Caveney, S., 1978, *Science* **199**:192.

Caveney, S., 1980, in: *VBW 80: Insect Biology in the Future* (M. Locke and D. S. Smith, eds.), Academic Press, New York.

Caveney, S., and Podgorski, C., 1975, *Tissue Cell* **7**:559.
Chippendale, G. M., 1970, *J. Insect Physiol.* **16**:1057.
Chippendale, G. M., and Kilby, B. A., 1969, *J. Insect Physiol.* **15**:905.
Collins, J. V., 1969, *J. Insect Physiol.* **15**:341.
Collins, J. V., 1974, *Can. J. Zool.* **52**:639.
Collins, J. V., 1975, *Can. J. Zool.* **53**:480.
Collins, J. V., and Downe, A. E. R., 1970, *J. Insect Physiol.* **16**:1697.
Condoulis, W. V., and Locke, M., 1966, *J. Insect Physiol.* **12**:311.
Crick, F., 1970, *Nature (London)* **225**:420.
Crossley, A. C. S., 1964, *J. Exp. Zool.* **157**:375.
Crossley, A. C. S., 1965, *J. Embryol. Morphol.* **14**:89.
Crossley, A. C. S., 1968, *J. Insect Physiol.* **14**:1389.
Crotty, W. J., and Ledbetter, M. C., 1973, *Science* **182**:839.
Dean, R. L., 1978, *J. Insect Physiol.* **24**:439.
Dean, R. L., Bollenbacher, W. E., Locke, M., Smith, S., and Gilbert, L. I., 1980, *J. Insect Physiol.* **26**:267.
Diamond, J. M., 1978, *Nature (London)* **271**:111.
Filshie, B. K., and Waterhouse, D. F., 1969, *Tissue Cell* **1**:367.
Goldman, R. D., 1971, *J. Cell Biol.* **51**:752.
Grell, M., 1946, *Genetics* **31**:60.
Hackenbrock, C. R., and Miller, K. J., 1975, *J. Cell Biol.* **65**:615.
Kunkel, J. G., and Lawler, D. M., 1974, *Comp. Biochem. Physiol.* **47B**:697.
Lane, N. J., 1978, in: *9th Int. Congr. on EM*, Volume II, p. 673.
Lane, N. J., and Swales, L. S., 1978a, *Dev. Biol.* **62**:389.
Lane, N. J., and Swales, L. S., 1978b, *Dev. Biol.* **62**:415.
Larsen, W. J., 1970, *J. Cell Biol.* **47**:373.
Larsen, W. J., 1976, *Tissue Cell* **8**:73.
Locke, M., 1958, *J. Microsc. Sci.* **99**:373.
Locke, M., 1959, *J. Exp. Biol.* **36**:459.
Locke, M., 1960, *J. Exp. Biol.* **37**:398.
Locke, M., 1967, in: *Insects and Physiology* (J. W. L. Beament and J. E. Treherne, eds.), pp. 69–82, Oliver & Boyd, Edinburgh.
Locke, M., 1969, *Tissue Cell* **1**:103.
Locke, M., 1970, *Tissue Cell* **2**:197.
Locke, M., 1974, in: *The Physiology of Insecta* (M. Rockstein, ed.), pp. 123–213, Academic Press, New York.
Locke, M., 1976, in: *The Insect Integument* (H. R Hepburn, ed.), pp. 237–258, Elsevier/North-Holland, Amsterdam.
Locke, M., and Collins, J. V., 1965, *J. Cell Biol.* **26**:857.
Locke, M., and Collins, J. V., 1967, *Science* **155**:467.
Locke, M., and Collins, J. V., 1968, *J. Cell Biol.* **36**:453.
Locke, M., and Collins, J. V., 1980, in: *Pathobiology of Cell Membranes* (A. U. Arstila and B. F. Trump, eds.), pp. 223–247, Academic Press, New York.
Locke, M., and Huie, P., 1979, *Tissue Cell* **11**:277.
Locke, M., and Huie, P., 1980, *Tissue Cell* **12**:175.
Locke, M., and Krishnan, N., 1973, *Tissue Cell* **5**:441.
Locke, M., and McMahon, J. T., 1971, *J. Cell Biol.* **48**:61.
Locke, M., and Sykes, A. K., 1975, *Tissue Cell* **7**:143.
Lockshin, R. A., 1969, *J. Insect Physiol.* **15**:1505.
Lockshin, R. A., 1975, *Dev. Biol.* **42**:28.
Lockshin, R. A., and Beaulaton, J., 1974a, *J. Ultrastruct. Res.* **46**:43.
Lockshin, R. A., and Beaulaton, J., 1974b, *J. Ultrastruct. Res.* **46**:63.
Lockshin, R. A., and Beaulaton, J., 1974c, *Life Sci.* **15**:1549.
Lockshin, R. A., and Williams, C. M., 1964, *J. Insect Physiol.* **10**:643.
Lockshin, R. A., and Williams, C. M., 1965a, *J. Insect Physiol.* **11**:123.

Lockshin, R. A., and Williams, C. M., 1965b, *J. Insect Physiol.* **11**:831.

Lockshin, R. A., Schlichtig, R., and Beaulaton, J., 1977, *J. Insect Physiol.* **23**:1117.

Loughton, B. G., and West, A. S., 1965, *J. Insect Physiol.* **11**:919.

Randall, W. C., 1970, *J. Insect Physiol.* **16**:1297.

Rodewald, R., 1973, *J. Cell Biol.* **58**:189.

Ryerse, J. S., 1978, *J. Insect Physiol.* **24**:315.

Ryerse, J. S., 1979, *Tissue Cell* **11**:533.

Ryerse, J. S., 1980, *J. Insect Physiol.* **26**:449.

Ryerse, J. S., and Locke, M., 1978, *J. Insect Physiol.* **24**:541.

Schatz, G., and Mason, T. L., 1974, *Annu. Rev. Biochem.* **43**:51.

Schlichtig, R., Lockshin, R. A., and Beaulaton, J., 1977, *Insect Biochem.* **7**:327.

Stumpf H. F., 1968, *J. Exp. Biol.* **49**:49.

Tobe, S. S., and Loughton, B. G., 1969a, *J. Insect Physiol.* **15**:1331.

Tobe, S. S., and Loughton, B. G., 1969b, *J. Insect Physiol.* **15**:1659.

Tojo, S., Betchaku, T., Ziccardi, V. J., and Wyatt, G. R., 1978, *J. Cell Biol.* **78**:823.

Walters, D. R., 1969, *Biol. Bull. (Woods Hole)* **137**:217.

Walters, D. R., 1970, *J. Exp. Biol.* **174**:441.

Walters, D. R., and Williams, C. M., 1966, *Science* **154**:516.

Wielgus, J. J., and Gilbert, L. I., 1978, *J. Insect Physiol.* **24**:629.

Wielgus, J. J., Bollenbacher, W. E., and Gilbert, L. I., 1979, *J. Insect Physiol.* **25**:9.

Wigglesworth, V. B., 1954a, *Q. J. Microsc. Sci.* **95**:115.

Wigglesworth, V. B., 1954b, *The Physiology of Insect Metamorphosis*, Cambridge University Press, London.

Wigglesworth, V. B., 1956, *Q. J. Microsc. Sci.* **97**:465.

Wigglesworth, V. B., 1959, *J. Exp. Biol.* **36**:632.

Williams, G. J. A., and Caveney, W., 1980a, *J. Embryol. Exp. Morphol.* **58**:13.

Williams, G. J. A., and Caveney, S., 1980b, *J. Embryol. Exp. Morphol.* **58**:35.

Yund, M. A., King, D. S., and Fristrom, J. W. 1978, *Proc. Natl. Acad. Sci. USA* **75**:6039.

CHAPTER 4

Hormonal Control of Insect Metamorphosis

Noelle A. Granger and Walter E. Bollenbacher

1. INTRODUCTION

In the swamps and forests of the Devonian era some 350 million years ago, there emerged a group of animals that today can be found in virtually every ecological niche on earth. With the number of their species variously estimated at 1–10 million, the insects display an evolutionary virtuosity unequaled by the rest of the animal kingdom combined. Within 100 million years of the first appearance of these arthropods, the adaptive features that permitted their ecological conquest had evolved, among them small size, functional wings, and a unique specialization of the life cycle known as metamorphosis.

The most primitive insects were wingless and underwent little change in their appearance during their maturation to adults other than the development of reproductive organs and an increase in biomass between periodic molts. With the evolution of the winged insects, however, different stages in the life cycle came to exhibit different forms, the process by which a change in form occurs being termed metamorphosis. The earliest winged insects underwent a type of metamorphosis in which the wings slowly develop as the insect molts into stages more and more closely resembling the adult in size and appearance. Today, these insects are represented by several orders, e.g., Orthoptera, Hemiptera, Homoptera, and their type of metamorphosis is termed gradual or hemimeta-

NOELLE A. GRANGER and WALTER E. BOLLENBACHER · Department of Anatomy and Department of Zoology, University of North Carolina, Chapel Hill, North Carolina 27514.

bolous. The immature stage of the hemimetabolous insect is referred to as a nymph (larva), which is distinguished from the adult by vestigial wings and incomplete genitalia. The most highly evolved winged insects undergo a dramatic type of postembryonic development in which there are three distinct stages (larval, pupal, and adult), separated by two metamorphic molts (Fig. 1). This type of metamorphosis is called complete or holometabolous, and it is best illustrated in the orders Lepidoptera, Diptera, Coleoptera, and Hymenoptera. The larval stage of the holometabolous insect is a wingless form specialized for feeding and growth. The pupal stage is a nonfeeding, quiescent stage within which adult structures are formed from pupal tissues or from anlagen present since embryogenesis. The adult is the reproductive stage and thus is identical to that of the hemimetabolous insect.

The dynamic developmental process of metamorphosis contributes in a unique way to the insect's success. As a result of metamorphosis, the different stages of the life cycle of holometabolous insects do not compete for survival. The larval form of an insect generally occupies an ecological niche that is optimally suited for feeding and growth and which

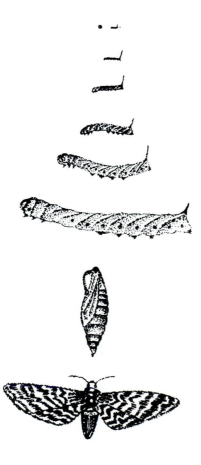

FIGURE 1. Life cycle of the tobacco hornworm, *Manduca sexta*. The cycle begins with the egg and newly hatched larva. Four larval molts separate periods of larval growth and development called instars, of which there are five. At the end of the fifth instar, there is the first metamorphic molt to the pupa. Depending on environmental conditions, the pupa may then enter a period of developmental arrest called diapause or immediately undergo further development resulting in a second metamorphic molt to the adult.

is different from the ecological niches occupied by the pupal and adult forms. The pupal form is designed to ensure the survival of the insect over periods of adverse seasonal changes and is generally found in a protected environment necessary for the development and subsequent emergence of the adult. The adult form is specialized for the dispersal of the species, and because of its mobility, can occupy a niche different from those of the larva and pupa, one suited for reproduction. Thus, metamorphosis, and particularly complete (holometabolous) metamorphosis, has enabled the insect to successfully exploit a variety of environments.

The events of insect metamorphosis are under the direct control of the insect's endocrine system and occur in a temporally precise and predictable manner. A series of early biological studies in insect physiology/endocrinology, by virtue of their simplicity of design and the profoundly intuitive interpretation of their results, have given rise to what is frequently termed the "classical scheme" for the hormonal control of insect metamorphosis. More contemporary studies, utilizing both biochemical and chemical methods, have verified the tenets of the classical scheme and have elaborated upon them. The results of these recent studies have emphasized the importance of the precise regulation of the endocrine glands involved in the hormonal control of metamorphosis. Thus, an examination of the hormonal control of the physiological events of metamorphosis becomes secondary to a study of the control of hormone production itself. It is this latter aspect of the endocrinology of insects with which this chapter will deal. To develop this topic, a critical evaluation will be made of those classical and contemporary studies that have contributed most significantly to our current understanding of the control of the endocrine glands involved in metamorphosis. On the basis of this evaluation, consideration will be given to those questions about the control of the endocrine glands with which future studies will be concerned as well as the methods best suited to answer them.

2. CLASSICAL METHODS FOR THE STUDY OF THE ENDOCRINOLOGY OF INSECT METAMORPHOSIS

Although the nature of insect metamorphosis stimulated conjecture by natural philosophers such as Aristotle and William Harvey, the search for the underlying mechanisms involved in this developmental process was not really begun until the extensive studies of H. E. Crampton in 1899. Until that time, insects had been investigated almost exclusively at a morphological level. By contrast, Crampton utilized a surgical manipulation termed parabiosis. In this procedure, immobilized insects are joined in such a way as to share a common circulatory system, thus enabling a study of the effects of the humoral agents of one insect on another. This study initiated a new era of investigation in which simple but effective surgical methods were developed to study for the first time, albeit indirectly, the fundamental aspects of metamorphosis (Schneider-man, 1967). In addition to parabiosis, these included ligation, extirpation, and implantation—transplantation techniques. Unlike parabiosis, the techniques for ligation and extirpation utilize a single insect in which different parts of the endocrine or nervous system are isolated by excision or ligation. These manipulations do not affect the immediate survival of the insect, and the hormonal function of the excised or ligated tissue can then be inferred from the effect of its removal. The implantation—trans-plantation of tissues and organs is technically easy with insect material and can involve a single insect or an exchange of tissue between insects. The normal function of the tissue is then implied by the overt effect(s) it exerts on the further development of the host. The ease with which these surgical procedures can be performed on insects and the dramatic nature of the results clearly were the causal factors in determining the structural components of the endocrine system and their basic interaction in the regulation of metamorphosis.

3. THE INSECT ENDOCRINE SYSTEM: STRUCTURAL COMPONENTS OF THE CLASSICAL SCHEME

In insects, as in all animals, both nervous and endocrine systems coordinate the functions of the tissues and organs of the body as well as interpolate the intrinsic and extrinsic stimuli that affect these functions. The result is a dynamically modulated response of the individual to its environment. In this instance, the response is the sequential occurrence of events comprising postembryonic development and metamorphosis. In insects, the components of the nervous and endocrine systems con-tributing to the neuroendocrine axis regulating metamorphosis are: neu-rosecretory cells (NSC) located in the brain and also in the ventral ganglia; neurohemal organs, serving as sites of release of NSC hormones; the corpora allata, the source of juvenile hormones; and the prothoracic glands, the source of ecdysone (Fig. 2).

The studies that led to the formulation of the classical scheme for the control of insect metamorphosis provided information that was both simplistic and indirect, yet which has withstood direct testing by more contemporary and sophisticated technologies. Before a discussion of the classical studies that elucidated the components of the neuroendocrine system responsible for metamorphosis, a brief description of the organs and tissues involved is necessary.

3.1. The Brain—Neurosecretory Cells

The insect brain is a bilaterally symmetrical structure composed of two lobes. These lobes are subdivisible into three embryologically dis-

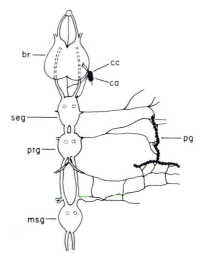

FIGURE 2. Anatomical arrangement of the brain, retrocerebral complex (corpus cardiacum and corpus allatum), thoracic nerve cord, prothoracic glands, and associated nerves in a last-instar larva of the wax moth, *Galleria mellonella*. br, brain; cc, corpus cardiacum; ca, corpus allatum; pg, prothoracic gland; seg, subesophageal ganglion; ptg, prothoracic ganglion; msg, mesothoracic ganglion. After Granger *et al.*, 1979.

tinct segments that later fuse: a protocerebral area, a deuterocerebral area, and a tritocerebral area. The importance of the brain in initiating metamorphosis derives from the function of specialized neurons called neurosecretory cells. Historically, NSC have been defined as neurons in which there is synthesis of material for release, a definition routinely based upon evidence obtained by cytological and/or histochemical methods. In actuality, this may not be an appropriate definition since all neurons synthesize material for release in the form of neurotransmitters for short-term synaptic conduction (Scharrer, 1977). Therefore, NSC have been redefined as nerve cells that release their product(s) into the circulatory system at a site some distance from their target cells, the products thus not involved in synaptic conduction. However, neurosecretory cells may also directly innervate the endocrine glands they are thought to affect, and it is known that neurotransmitters in some cases can act as hormones. Thus, an acceptable definition for an NSC has yet to be formulated, and this situation may not change until other definitive criteria are established. At present, these cells are defined mainly by their morphology, histology, and ultrastructure (Maddrell, 1974; Rowell, 1976; Berlind, 1977). In insects, the NSC are composed of a cell body, which is usually larger than that of a typical neuron and contains variable accumulations of stainable secretory material, and a long axon possessing several collateral branches close to the cell body. The termination site of the axon is characterized by repeated branching into many fine termini with swollen endings. These endings are considered the site of release (i.e., neurohemal site) and, as expected for such a structure, contain stainable material and granules. NSC have been described in both the brain and the ventral nerve cord of many insects and have been classified into different types depending on their size, location, staining properties and electron density, and number of vesicles. The most extensively investigated NSC in the insect brain are found in a region of the protocerebral cortex known as the pars intercerebralis. These NSC are organized in two large prominent clusters, one on each side of the median furrow and are collectively termed the medial NSC. More laterally in each hemisphere is another group called the lateral NSC. Both the medial and the lateral NSC are comprised of a heterogeneous population of cells whose axons contribute to the nerves running from the brain to the corpora cardiaca, bilaterally paired neurohemal organs, and from there to the corpora allata (CA) (Fig. 3). Until recently, the functions of these NSC were essentially unknown, but as shall be seen, their individual roles in postembryonic development are currently being defined. In addition to those found in the protocerebrum, NSC are located in the deuterocerebrum and tritocerebrum as well as in the optic lobes. The multiple sites of the NSC in the insect nervous system as well as their variable morphology suggest that each cell type, or possibly each cell, may have a specific function(s). The numerous physiological processes that have been linked to the presence of specific cells, particularly those in the pars intercerebralis, support this supposition.

NSC have also been identified in nervous tissue other than the brain. They are primarily located in ganglia, e.g., frontal, subesophageal, and ventral nerve cord ganglia, and have been observed in virtually every insect investigated. As with the cerebral NSC, there is as yet no known function for most of these cells although in several species, humoral factors involved in the control of diapause, diuresis, antidiuresis, and tanning have been associated with the ganglia (Maddrell, 1974).

3.2. Neurohemal Organs

The axons of a group or groups of NSC frequently terminate within a defined area. Such an area is termed a neurohemal organ, i.e., the site of release of neurosecretory material. The major neurohemal organ for the cerebral NSC in the insect is considered to be the corpora cardiaca (CC). The CC are transformed nerve ganglia that are located behind the brain and are connected to it by three main nerve bundles composed of nervous and neurosecretory axons. The first bundle, the nervis corporis

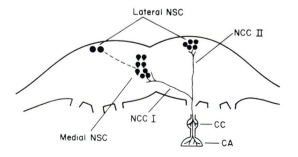

FIGURE 3. Topographical diagram of NSC in the Day 0 pupal brain of *Manduca sexta*. Shown are the medial and lateral NSC groups whose axons form the NCC I and NCC II, which innervate the corpora cardiaca and corpora allata. From Nijhout, 1975, and Agui *et al.*, 1979.

cardiaci I (NCC I), contains the axons of the protocerebral medial NSC on the contralateral side. This nerve bundle is frequently fused with the nervis corporis cardiaci II (NCC II), which contains axons from the protocerebral lateral NSC on the ipsolateral side. The last bundle, the nervis corporis cardiaci III (NCC III), contains axons from NSC located in the tritocerebrum. The CC contains the cell bodies of intrinsic NSC as well as the axons of NSC and of nonneurosecretory neurons traversing the organ to reach the CA. The intrinsic NSC apparently synthesize and release a variety of neurohormones, among them hyperglycemic and cardio-accelerating factors and the adipokinetic hormone (Maddrell, 1974).

The sites of release of the many other NSC, particularly those in the ventral nerve cord, were unknown until quite recently. It is now apparent that segmentally repeating structures termed perisympathetic organs occur in association with the unpaired ventral nerves of the sympathetic (visceral) nervous system (Raabe et al., 1974). NSC axons terminate in the perisympathetic organs, branching like those that end in the CC, and it is from these branches that neurosecretion is released into the hemolymph.

3.3. Prothoracic Glands

The "granulated vessels" described in the thorax of the goat moth caterpillar, *Cossus cossus*, in 1762 (Lyonet) are now termed the prothoracic glands (ventral glands, ecdysial glands). The prothoracic glands (PG) are epidermally derived structures and are usually paired organs lying along the dorsolateral body line in the head, prothorax, or thorax of the insect. These glands are generally found only in those stages of the insect life cycle that undergo molting and metamorphosis, e.g., larvae and pupae. Depending on the insect, they vary in form from compact to diffuse, but are composed of a relatively large, homogeneous secretory cell type. In some insect species, the PG receive direct nervous innervation from the ventral ganglia. The innervation appears to be predominantly nervous, but at least some of the innervating axons are neurosecretory.

3.4. Corpora Allata

The CA are retrocerebral organs of epidermal origin that were first identified as endocrine glands in 1913 (Nabert, 1913). They are typically paired structures and are composed of a single glandular cell type, surrounded by a layer of connective tissue and a basal lamella. The CA are commonly innervated from two sources, their primary innervation deriving from the NCC I and NCC II. Axons from these two nerves traverse the CC and exit as the nervis corporis allati I (NCA I), which contains both nervous and NSC axons terminating in the CA (Scharrer, 1964). Secondary innervation is provided by a nerve bundle from the subesophageal ganglion, the nervis corporis allati II (NCA II), which, like the NCA I, possesses both nervous and NSC axons.

4. STUDIES LEADING TO THE ELABORATION OF THE CLASSICAL SCHEME

In 1917, Stefan Kopec suggested that the larval brain of the gypsy moth, *Lymantria dispar*, released a factor that induced pupation. In a simple and straightforward experiment, he demonstrated that when a last-instar larva was ligated into anterior and posterior halves at a critical time during the instar, the anterior half of the animal pupated normally while the posterior half remained larval. Kopec subsequently demonstrated by extirpation that the absence of the brain before this critical time prevented pupation, while pupation occurred normally when extirpation took place after this time. Thus, the head critical period is the time during which the presence of the brain is necessary for pupation. Kopec's finding that the brain was an organ of internal secretion was the first demonstration of an endocrine function for nervous tissue in any animal. Paradoxically, Kopec's conclusion that the brain played the role of an endocrine gland in insect metamorphosis was not generally accepted until the 1930s. By that time, Ernst Scharrer (1928) had demonstrated an endocrine function for certain hypothalamic neurons in the vertebrate brain and Wigglesworth (1934) had successfully repeated Kopec's experiments using basically the same methods on a different insect, the blood-sucking bug, *Rhodnius prolixus*. Unlike the gypsy moth larvae, whose development proceeds according to a predictable, temporal program, *Rhodnius* larvae molt in response to the intake of a single large blood meal, the time between feeding and the ensuing molt increasing with each of the five larval instars. When a penultimate (fourth)-instar larva is decapitated within 4 days after feeding, the larva does not molt although it may survive for months. When decapitated 5 days or more after the blood meal, the animal molts normally except for the fact that it is unable to shed its old cuticle. Thus, the critical period for the brain in *Rhodnius* is the first 4 days of the instar. When Wigglesworth excised the region of the *Rhodnius* larval brain containing the NSC and implanted it into the abdomen of a larva decapitated before the end of the critical period, the larva molted. Since other regions of the brain and other cephalic structures failed to elicit this response, Kopec's original findings were corroborated, and for the first time the NSC were demonstrated to be the source of the brain's endocrine function. Wigglesworth termed the molt-eliciting factor produced by the cerebral NSC "brain hormone."

At approximately the same time, Hachlow (1931) conducted a series of experiments that suggested that the brain did not act alone in eliciting a molt. By transecting the bodies of lepidopteran pupae at different levels and sealing the cut ends, he was able to observe the development of the resulting body fragments and from the results infer the potential contribution of specific regions of the body to the development of the pupa as a whole. This study revealed that only fragments including the thorax were able to develop to the adult, which suggested the existence of an organ in the thorax also necessary for molting and metamorphosis.

Later, in 1938, Plagge repeated the experiments of Hachlow using a different surgical approach. He found that the implantation of brains would cause the pupation of a *Celerio lineata* larva from which the *in situ* brain had been removed, but would not elicit the pupation of an isolated abdomen of this larva. This result corroborated Hachlow's conclusion that, in addition to the brain, another factor was involved in the molting process. The simultaneous discovery that excision of the ring glands from the thoraces of the larval blowfly, *Calliphora erythrocephala* (Burtt, 1938), and the fruit fly, *Drosophila melanogaster* (Hadorn and Neal, 1938), prevented pupation confirmed that this factor originated in the thorax. However, since the ring glands of these cyclorrhaphous dipterans are complex structures, containing both the CC and the CA, the precise source of the factor could not be resolved. Finally, in 1940, Fukuda clearly demonstrated that the organ in the thorax responsible for molting was the PG. He found that when a larva of the commercial silkworm, *Bombyx mori*, was ligated just posterior to the PG, only the portion of the larva anterior to the ligature pupated. When a PG was implanted into the portion posterior to the ligature, pupation would then occur. Transection of a pupa behind the PG produced the same effects, and when either larval or pupal PG were implanted into the resulting isolated pupal abdomen, an adult molt ensued.

Although it was now established that both the brain and the PG released humoral factors necessary for the molt, the nature of the relationship of the brain and the PG in this process remained to be resolved. The unraveling of this mystery resulted from studies (Williams, 1952) of the adult development of the American silkmoth, *Hyalophora cecropia*. To investigate the relationship of the brain and the PG, Williams utilized diapausing pupae, i.e., pupae in a physiological state of developmental arrest, which is an overwintering phenomenon normally terminated by prolonged chilling. When brains excised from diapausing pupae or chilled pupae were implanted into isolated abdomens prepared from diapausing pupae, the abdomens did not molt. However, when these brains were implanted back into the pupae from which they came, the adult molt was elicited. The transplantation of chilled brains and PG from either diapausing or chilled pupae into the isolated abdomens of diapausing pupae resulted in the metamorphic molt of the pupal abdomens to adult abdomens. Thus, the interaction of the brain and PG was demonstrated conclusively to elicit insect molting. NSC present in the brain possess a trophic hormone that activates the PG to produce a second hormone that subsequently initiates a molt.

The early studies presented so far were directed at elucidating the

endocrine system involved in eliciting a molt. It might have been assumed that the interaction of the brain hormone and the molting hormone (MH) produced by the PG determined the form of the insect after
the molt as well, i.e., metamorphosis. However, it was soon apparent
that a third hormone was involved, one that was expected to be a "metamorphosis hormone," activating pupal or adult development. In actuality, this hormone is an "antimetamorphic hormone" that sustains
each stage of postembryonic development until the precise time for metamorphosis to the next stage.

The first clue that a hormone existed which determined the character
of the molt came from the 1934 study by Wigglesworth in which the
critical period for the brain hormone in *Rhodnius* was determined. The
observation initially suggesting that a humoral factor from the head determines the nature of the molt derived from experiments in which larvae
from early instars were decapitated at different times during the head
critical period. Some of the larvae that succeeded in molting showed
signs of precocious metamorphosis, e.g., partially differentiated adult genitalia and wings. As the head critical period for each instar came to an
end, the number of molting larvae that displayed adult development
decreased. Wigglesworth concluded that the head of the younger larva
contained a factor that would prevent the expression of adult characteristics.

The hormonal nature of this factor was then demonstrated in an
elegant parabiosis study (Wigglesworth, 1934, 1936). In this study, fifth
(last)-instar larvae decapitated early in the head critical period were joined
with fourth-instar larvae in the post critical period. As expected, the
fourth-instar larvae underwent a larval molt. The fifth-instar larvae
molted as well under the influence of endocrine system of their parabiotic
partners. However, rather than undergoing an adult molt, they developed
instead into sixth-instar larvae or larval–adult intermediates. This extra
(supernumerary) larval molt of the last-instar larvae confirmed the existence of a humoral agent that determined the character of the molt, or
in simpler terms, inhibited metamorphosis by maintaining the status
quo. Wigglesworth called this agent the "inhibitory hormone," "status
quo hormone," or "juvenile hormone."

Next, Wigglesworth identified the source of juvenile hormone (JH).
The brain was eliminated as the source since its removal at a critical
period did not result in precocious metamorphosis. By contrast, extirpation of the CA did lead to the formation of precocious adults, and
transplantation of CA from early instar larvae to last-instar larvae resulted in extra larval molts. Therefore, he concluded that in *Rhodnius*,
the CA were the source of JH. Similar experiments by Bounhiol (1938)
and Piepho (1943) utilized larvae of *Bombyx* and the wax moth, *Galleria
mellonella*, respectively, to demonstrate that the source and the action
of JH were the same in holometabolous insects. As with *Rhodnius*, the
extirpation of CA from early instar larvae resulted in precocious meta-

morphosis, e.g., tiny pupae and adults, and implantation of CA into last-instar larvae elicited extra larval molts.

By the early 1950s, there had emerged from these and many other studies the outline of our present understanding of molting and metamorphosis in insects, the so-called classical scheme of the endocrine control of insect postembryonic development and metamorphosis. According to this scheme, NSC of the pars intercerebralis of the brain release a "brain hormone" that stimulates the PG to release MH, which then initiates the molt. The character of the molt, i.e., larval–larval, larval–pupal, or pupal–adult, is determined by the titer of JH at or before the molt. It was hypothesized that MH in the presence of a high titer of JH elicits a larval molt and the presence of a reduced titer of JH evokes a pupal molt. Metamorphosis to the adult occurs in the absence of JH. Since its formulation, the classical scheme has served as the model for all investigations into the hormonal control of insect postembryonic development. Although far more technologically sophisticated studies have since been conducted, they have not succeeded in changing this scheme, but have only elaborated upon it. This is indeed a tribute to the men on whose work it is based. Appropriately, Williams (1958) once stated, "The test of scientific genius is the ability to reason to correct conclusions from inadequate evidence."

5. CHEMISTRY AND ENDOCRINOLOGY OF THE METAMORPHIC HORMONES

Even though the field of insect endocrinology began a period of exponential growth in the late 1950s, significant progress in refining the classical scheme was not made until the chemical structures of ecdysone and JH were elucidated in the 1960s (see Gilbert and Goodman, this volume). With this knowledge, critical investigations into the synthesis, titers, and modes of action of these hormones could be initiated.

A great many of the most significant studies have been conducted on holometabolous insects, whose metamorphosis, as previously stated, is the most highly evolved and probably the most complex. For this reason, further discussion of the endocrinological mechanisms involved in this process will deal almost exclusively with holometabolous development.

5.1. Brain Hormone

It is ironic that of the three hormones involved in the control of metamorphosis, brain hormone, which was the first to be discovered, has not yet been purified and chemically identified. In fact, until just recently, the specific cerebral NSC—the sites of synthesis of this hormone—were not known, and as yet its mode of action and even its titer during development have not been determined.

The primary reason for the dearth of information on brain hormone, or prothoracicotropic hormone (PTTH) as it is now called, is that an adequate biological assay for this hormone has not been available. Until recently, PTTH activity has been assessed exclusively with *in situ* bioassays. In general, bioassays to assess PTTH activity have utilized lepidopterans as test animals, the most popular being pupae of *Bombyx* or *Samia cynthia* which have been placed in a permanent state of diapause, so-called dauer pupae (Kobayashi and Yamazaki, 1974). A dauer pupa is created when the brain is extirpated from a freshly ecdysed pupa. The implantation of an active brain or the injection of a brain extract containing PTTH into these pupae can then initiate metamorphosis to the adult, which is scored as a positive response. Recently, a more rapid and possibly more sensitive *in situ* bioassay has been developed using larvae of the tobacco hornworm, *Manduca sexta* (Gibbs and Riddiford, 1977). For this assay, penultimate-instar larvae are neck-ligated just prior to PTTH release and a positive response to an injection of a brain extract containing PTTH can be either a larval- or a pupal-type molt (Bollenbacher *et al.*, 1979). It is obvious from the nature of these two bioassays that such procedures are typically lengthy and rather cumbersome, requiring large quantities of test material as well as large test populations.

Bioassays are also subject to interpretative variability due to their subjective nature. However, the most obvious drawback to these, as well as to other *in situ* assays for PTTH, is that PTTH activity is assessed indirectly. This situation can become problematic since nonspecific factors in a brain or brain extract may also evoke a positive response (Bollenbacher *et al.*, 1980). In addition, the PG often undergo spontaneous activation with a resulting molt, which would then be scored as a false-positive response. It is the lack of specificity of the bioassays that may be the single most critical factor responsible for the inconsistent data on PTTH.

Recently, a specific and direct means of monitoring the activation of the PG by PTTH has been achieved with the development of an *in vitro* assay for this neurohormone, which measures the synthesis of ecdysone by the PG of *M. sexta* (Bollenbacher *et al.*, 1979). This approach circumvents the problems inherent with the bioassays. In this assay, the basal rate of ecdysone synthesis by the PG is quantified by an ecdysone radioimmunoassay (RIA) and activation of a PG by PTTH is noted as an increase in the rate of ecdysone synthesis. A dose response of activation of the PG by brain extracts demonstrated conclusively that PTTH activates the PG directly. Furthermore, it was shown with this assay that only the brain or brain extracts activate the PG; other tissues or regions of the nervous system fail to elicit the response. Most importantly, the greater sensitivity and reproducibility of the *in vitro* assay in comparison to the bioassays suggest its applicability to a critical investigation of the chemistry and basic endocrinology of this neurohormone.

Studies of the cerebral NSC of insects historically have been descriptive in nature and have failed to identify a particular NSC as the source of a specific neurosecretory product (Maddrell, 1974; Berlind, 1977). Previous efforts to identify the PTTH NSC utilized classical extirpation and implantation methods, in which brain fragments were either removed or implanted into a decapitated assay animal. Unfortunately, this approach has yielded equivocal results since either the medial or the lateral NSC of the protocerebrum have been suggested as the source of the hormone depending on the study and/or type of insect used. These inconsistent findings may reflect species variation and/or positional variation of the NSC during development, but they may also be a result of the indirect nature of the *in situ* bioassay.

With the use of the *in vitro* assay for PTTH, the site of synthesis of this hormone in the brain of *Manduca* has been demonstrated in a direct and unequivocal manner (Agui *et al.*, 1979). In this study, the distribution of PTTH activity was determined for different regions of a Day 0 pupal brain with a dose–response protocol. Activity was found to be localized primarily in the region of the protocerebrum containing the lateral NSC, which had been previously implicated as the source of PTTH by the *Manduca* bioassay (Gibbs and Riddiford, 1977). Quantitative analysis of PTTH activity in groups of the different lateral NSC types and ultimately in the individual NSC revealed that PTTH activity could be localized to a single NSC. The success of this study was clearly a function of the *in vitro* assay in that it provided the necessary sensitivity to enable the assay of just one NSC. Although the identification of a single NSC as the source of a specific neurohormone in the insect brain is in itself novel, the major significance of the study is its demonstration of the potential of this approach for investigating the regulation of the synthesis and release of a defined neurosecretory product.

In addition to its use in the identification of the site of synthesis of PTTH, the *in vitro* PTTH assay has been used to identify the neurohemal organ for this hormone in *M. sexta* (Agui *et al.*, 1979, 1980). The CC were historically assumed to be the neurohemal organ for PTTH on the basis of very little critical evidence; e.g., the CC is morphologically a classic neurohemal structure, neurosecretory material is always observed in the CC, and exocytosis of neurosecretory material from the CC has been demonstrated. However, some older studies using the classical methods have suggested that the CA might also be the neurohemal organ. When the *in vitro* PTTH assay was employed to directly measure PTTH activity in the CC, the significant level of activity expected for a neurohemal organ was not found. Quantification of PTTH activity in the pupal brain, the CC, and also the CA revealed that the activity in the CA was consistently far greater than that in the CC and that fluctuations in PTTH activity in the CA during development paralleled that for the brain. These results suggest that the CA rather than the CC are the neurohemal sites for PTTH in *Manduca*. Support for this idea is provided by two additional studies. In the first, the axonal distribution patterns of the cerebral NSC to the CC and the CA in *Manduca* were traced by cobalt chloride back-filling. The axons from the lateral NSC were observed to terminate in the CA (Nijhout, 1975). In the second study, previously described, *in vitro* PTTH assay was used to demonstrate that one of the lateral NSC was the source of PTTH in the brain of *Manduca* (Agui *et al.*, 1980). The identification of the PTTH NSC and their axonal distribution in conjunction with the titer of PTTH activity in the CA provide compelling evidence that the CA is the neurohemal organ for PTTH. Ultrastructural observations of the CA of *Manduca* have revealed that axons containing significant quantities of neurosecretory granules are present at developmentally specific times, thus corroborating this conclusion (Sedlak, personal communication).

The chemical nature of PTTH has also proven elusive. Early studies employing *in situ* bioassays suggested that this neurohormone was one of several different types of molecules, e.g., a sterol, a mucopolysaccharide, or a protein (see Gilbert *et al.*, 1980). Recent investigations have essentially established that PTTH is a protein, and studies utilizing the *in vitro* PTTH assay suggest that it may exist in different molecular weight forms (Bollenbacher, unpublished observations). With the use of this new assay, the elucidation of the chemistry of PTTH should be forthcoming.

The fact that PTTH is chemically undefined has prevented a direct titer of this hormone during insect development. The only available PTTH titer was determined by bioassay of the brain, head, and thorax plus abdomen of *B. mori* (Ishizaki, 1969). However, because of the insensitive nature of the assay and the imprecise staging of the animals, a relationship between titer and development could not be established.

Hemolymph titers of PTTH have usually been inferred from ligation studies designed to determine when it is released and how its release is controlled. For example, the determination of a head critical period indirectly yields a hemolymph titer for PTTH since when sufficient PTTH has been released into the hemolymph, the insect will molt independently of the brain. In *Rhodnius*, distension of the abdomen by a large blood meal provides the nervous stimulus for the brain to secrete PTTH (Wigglesworth, 1934, 1936). The length of time the brain is required reflects the necessary period of release of the hormone, which in this insect increases with each larval instar. Presumably then, the PTTH titer in both the brain and the hemolymph of *Rhodnius* rises dramatically after a blood meal and remains high throughout the critical period. By contrast, in *Manduca*, PTTH release is governed by a circadian clock in conjunction with some form of allometry, i.e., release occurs at a certain time of the day and only after the animal reaches a critical weight (see

Section 6.1). During the last larval instar there are two head critical periods, implying that PTTH is released twice (Truman, 1972; Truman and Riddiford, 1974). The first release of the neurohormone at approximately the fourth day of the instar initiates wandering; the second release occurs approximately 2 days later and initiates the pupal molt. The apparent release of PTTH at two times during the instar is supported by the fact that there are two peaks in the hemolymph ecdysteroid titer at approximately the same time (Bollenbacher et al., 1975) (see Section 5.2). Nevertheless, the only information these types of studies can provide is an indirect indication of increases in the hemolymph PTTH titer because they do not directly monitor the hormone level. The in vitro assay for PTTH should permit a direct and simultaneous determination of the release and titers of this neurohormone at its site of synthesis, site of release, and site of action during insect metamorphosis. The information that can be derived from the titer data is critical to an understanding of the basic endocrinology of PTTH: whether synthesis and/or release are limiting factors; whether either or both of these processes determine the head critical period defined by ligation; whether PTTH is released continuously or in the pulsatile manner characteristic of some vertebrate neuropeptide hormones; whether different forms of the hormone operate at developmentally specific times and have different half-lives.

5.2. Molting Hormones (Ecdysteroids)

Although the studies of Wigglesworth, Fukuda, and Williams clearly demonstrated that the insect PG were the source of a "molting hormone," the methods employed did not provide any clues to the chemical nature of this material. To investigate the chemistry of this hormone, a method had to be developed with which MH activity could be assessed in extracts of tissues. The assay that contributed most significantly to the chemical identification of ecdysone was the Calliphora bioassay. In this assay, an abdomen from a late last-instar fly larva is isolated by ligation just prior to the release from the ring glands of MH, which is necessary for pupariation. The isolated larval abdomen thus will not molt into a pupal abdomen unless exposed to MH. In the bioassay, the frequency of molting of these isolated abdomens is a function of the amount of active material administered. This response permitted a quantification of MH activity, which is measured in Calliphora units. With this bioassay, Butenandt and Karlson were able to partially purify two compounds with MH activity from Bombyx pupae (see Bollenbacher et al., 1980). They termed these moieties α-ecdysone and β-ecdysone. The full structural identification of these compounds as polyhydroxylated keto steroids took another 10 years. Many subsequent studies have confirmed that these ecdysteroids, now termed ecdysone and 20-hydroxyecdysone, are the two principal ecdysteroids in insects during metamorphosis (see Gilbert and Goodman, this volume). Although seven additional ecdysteroids have since been identified in various insects and from various stages of the life cycle, their functions are at present not known (Gilbert et al., 1980).

Although both ecdysone and 20-hydroxyecdysone exhibited MH activity and thus could be the "molting hormone" produced by the PG, a causal relationship between the PG and these ecdysteroids remained to be established. It was not until 1974 that this question was finally resolved. In the past, attempts to extract MH activity from PG had been unsuccessful, suggesting that either the glands did not synthesize an ecdysteroid or that the organs simply did not store it. The breakthrough finally came with the concurrent development of in vitro culture methods for lepidopteran PG and microanalytical methods for ecdysteroid detection, i.e., radioimmunoassay, high-pressure liquid chromatography, gas–liquid chromatography, and mass spectrometry. The integration of these different techniques for the analysis of the in vitro product of the PG from Bombyx and Manduca resulted in the identification of ecdysone rather than 20-hydroxyecdysone as the product of these glands (see Bollenbacher et al., 1980). The PG were subsequently determined to be the only physiological source of ecdysone during larval–pupal development and probably early pupal life as well. The discovery that ecdysone was the only ecdysteroid synthesized by the PG raised the question of its actual endocrine function since it was found in both in situ and in vitro assays that 20-hydroxyecdysone was 100- to 200-fold more active than ecdysone in eliciting a molt (see Gilbert and King, 1973). Based on these data, it was suggested that ecdysone acts as a prohormone that undergoes hydroxylation at C-20 of the side chain to yield 20-hydroxyecdysone, the actual "molting hormone." This hypothesis was supported by the finding that under both in vivo and in vitro conditions, labeled ecdysone is readily oxygenated to 20-hydroxyecdysone. Subsequent chemical characterizations of the ecdysteroids present in various insect species at critical times during postembryonic life, such as the period preceding each molt, have revealed that 20-hydroxyecdysone is generally the predominant, if not the only, MH present. As a logical extension, it has been suggested that the hydroxylation of ecdysone to 20-hydroxyecdysone, which is carried out by ecdysone 20-monooxygenase, a cytochrome P-450 mixed-function oxidase, may be a critical point in the control of metamorphosis (Smith et al., 1980). This possibility is currently being extensively investigated using both in vivo and in vitro approaches (see Gilbert and Goodman, this volume).

The establishment of a titer of ecdysone and 20-hydroxyecdysone during insect postembryonic development would provide both support for the prohormone hypothesis and also the information on the temporal, quantitative, and qualitative relationships of these ecdysteroids necessary for future research on the action of these hormones and the control of their syntheses. Until as recently as 5 years ago, ecdysteroid titers were determined using the standard Calliphora bioassays. In general, such studies have corroborated the classical scheme in that there is a sharp increase in the ecdysteroid titer just preceding pupation and in the middle of the pupal stage. The more recent utilization of microanalytical techniques, especially radioimmunoassays, to determine the ecdysteroid titers for whole animals and in hemolymph has confirmed the bioassay data in considerable detail. The ecdysteroid titer has been found to undergo precisely modulated temporal and quantitative fluctuations in agreement with those predicted by bioassay. However, subtle but critical increases have also been found in the titer at developmentally specific times at which no increases in the ecdysteroid level were previously detected by bioassay (Bollenbacher *et al.,* 1975).

For example, during the last larval instar of insects in which the ecdysteroid titer has been determined at short time intervals, two ecdysteroid peaks have been found instead of one. At the present time, it appears that this phenomenon may be ubiquitous among all holometabolous insects and that the first, smaller peak causes commitment (reprogramming) of the epidermis for pupal syntheses, while the second elicits apolysis and synthesis of the pupal cuticle (see Chapter 8). During the previous larval instars, there still appears to be only one temporally acute increase in the titer, quantitatively similar to the second peak of the last instar. As previously mentioned, these increases in the ecdysteroid titer closely follow the head critical periods for PTTH release, indicative of a causal relationship between PTTH release and subsequent activation of the PG with the resulting increase in the ecdysteroid titer. Qualitative determinations of the ecdysteroids that contribute to the titer suggest that the ecdysone 20-monooxygenase could be of critical importance in regulating the 20-hydroxyecdysone titer, and thus could determine the course of postembryonic development (Smith *et al.,* 1980). For example, small peaks of ecdysone precede 20-hydroxyecdysone peaks and dramatic shifts in the ratio of ecdysone to 20-hydroxyecdysone occur in conjunction with these peaks.

This apparent modulation of the levels of ecdysone and 20-hydroxyecdysone is only one of several factors involved in regulating the qualitative and quantitative aspects of the ecdysteroid titer. To varying degrees, synthesis, degradation, excretion, and even sequestration undoubtedly contribute to the precise and critical fluctuations in the levels of the hormones (see Gilbert and Goodman, this volume). Of these, the control of synthesis and degradation are the most critical, and the regulation of the gland, i.e., hormone synthesis, will be considered in detail later in this chapter.

5.3. Juvenile Hormones

At the same time that the chemistry of ecdysone was being unraveled, the structure of JH was also identified. A serendipitous discovery by Williams (1956) provided a unique and plentiful source of JH for this purification, which had not been possible previously because the amount of material present in the insect during metamorphosis was insufficient for characterization. In the early 1950s, Williams attempted to prolong the lives of the nonfeeding adult *H. cecropia* moths by joining them parabiotically to diapausing *H. cecropia* pupae. He observed that when an adult male was joined to a chilled pupa, a "status quo" effect occurred when the pupa molted; i.e., the pupa did not undergo metamorphosis to the adult but instead underwent a second pupal molt. When an adult female was used, however, the pupa underwent normal adult development. These results suggested that JH was being supplied by the adult male. A comparison of the effect of implanting CA from either male or female adults into chilled pupae corroborated the parabiosis data, and Williams concluded that JH produced by the male CA was responsible for the status quo effect. The extraction of adult male abdomens resulted in a "golden oil," which, when injected into a pupa, induced a second pupal molt. The purification and chemical characterization of JH from this oil took approximately a decade to complete. The successful characterization by Röller and Dahm of JH I, a C_{18} aliphatic sesquiterpene, was in part due to the development of sensitive and efficient bioassays for the hormone. One was the *Galleria* wax test, which scores the effect of topically applied JH on the differentiation of pupal epidermis to adult epidermis, a positive response being the secretion of pupal rather than adult cuticle in the area where the hormone was applied.

Shortly after the elucidation of the structure of JH I, a second moiety with JH activity was identified in the *H. cecropia* golden oil. This C_{17} homolog of JH I was termed JH II. A third JH homolog was subsequently identified in the media of *in vitro* cultures of adult female *Manduca* CA (see Gilbert and Goodman, this volume). This molecule, JH III, is characterized by a 16-carbon skeleton. Since the identification of the three JHs, their occurrence has been studied in many species of Lepidoptera, Hymenoptera, Orthoptera, and Hemiptera. With the single exception of the lepidopterans, JH III has been the only homolog found. In the Lepidoptera, JH I and JH II are the predominant homologs during the larval stages, while JH III is the hormone present in the adult. This developmental peculiarity suggested that JH I and JH II are morphogenetic hormones and that JH III is a gonadotropic hormone. The hypothesis that JH III is not a morphogenetic hormone is supported indirectly by the fact that JH III has considerably less morphogenetic activity than JH I or JH II in bioassays. However, the converse does not appear to be the case since both JH I and JH II have gonadotropic activity equivalent to or even greater than JH III. Unfortunately, the present dearth of information about the JH from immature stages or orders other than Lepidoptera precludes any generalizations about the possible unique functions of the different homologs. However, in two instances where JH has been identified in larval stages of a hymenopteran (honeybee) and an orthopteran (grasshopper), JH III was the only homolog present. Thus, despite the bioassay data, there may be instances in which JH III acts as a morphogenetic hormone as well (see Gilbert *et al.*, 1980).

Although the structure of JH has been known for over a decade and microanalytical methods, e.g., gas–liquid chromatography and mass spectrometry, have been successfully employed for the analysis of the different homologs in a single sample, a complete quantitative titer of JH during postembryonic development has not been published thus far. There have been a number of titers determined by bioassays, e.g., the *Galleria* wax test or *Manduca* black larval assay, but as with all such assays, the results are qualitative, subject to interpretation, and possibly nonspecific. In part, the difficulty in obtaining a critical titer for JH has been due to the unique lipoidal nature of this molecule. That JH is present in very small amounts in the insect, that its chromatographic properties are similar to nonpolar lipids, and that it tends to be labile under most conditions have hindered development of a rapid and direct biochemical assay for this hormone.

Specific and sensitive radioimmunoassays for JH I, JH II, and JH III have been developed similar to those for ecdysteroids; however, there are problems in the use of these for the titer of JH in extracts of hemolymph, possibly because of the physical properties of this molecule. With the generation of JH antisera with greater specificity and affinity, and of better extraction procedures, this assay may become a viable approach to titering. Nevertheless, the JH radioimmunoassay currently available apparently will provide the means for investigating *in vitro* the regulation of JH synthesis by the CA, possibly along the lines of the *in vitro* PTTH system (Granger *et al.*, 1979).

Based on bioassay data, the JH titer appears to follow that predicted by the classical premise. In general, the hormone is present at precise times throughout larval life, increasing around the time of each molt, and then decreasing toward the end of the ensuing instar. With each succeeding instar, the peak level of JH in the hemolymph decreases; by the last larval instar, it is undetectable, and thus permits the onset of metamorphosis. In certain insects, e.g., *Manduca*, there appears to be a second increase in the JH titer, which occurs during late pharate pupal development and disappears abruptly just before the pupal ecdysis. This JH peak is apparently required to prevent the precocious development of adult imaginal structures because allatectomy of lepidopteran larvae during the last larval instar results in pupae with some adult characteristics (Kiguchi and Riddiford, 1978). After this last small peak in the titer, the JH levels do not increase again until after adult eclosion. In general then, these titer data tend to corroborate the classical concept of the way in which JH determines the character of the molt.

The regulation of the JH titer probably occurs at several different levels, i.e., synthesis, transport, sequestration, catabolism, and excretion (see Gilbert *et al.*, 1980). At the level of the corpora allata, synthesis rather than release appears to be the focus of control, since the CA immediately release JH following its synthesis. Mechanisms for the transport and metabolism of JH appear to vary among different species of insects, and in certain insects, e.g., Lepidoptera, binding proteins and specific esterases appear to play a major role in modulating the JH titer (see Gilbert and Goodman, this volume). By contrast, very little is known about the significance of sequestration and excretion in the regulation of the titer. In light of the obvious importance of the control of JH synthesis in this overall process, this aspect of the regulation of the titer will be discussed in detail later in this chapter.

6. CONTROL OF ENDOCRINE GLANDS

6.1. Cerebral NSC

Control of the endocrine function of a gland, a single NSC, or a group of NSC can be exerted by a variety of factors, the predominant ones being environmental, neural, and humoral. The interaction of these factors with the nervous system is the basis of a complex, dynamic system of endocrine regulation that can occur at several different levels. In the case of the PTTH NSC, potential levels of regulation exist with synthesis of the hormone, axonal transport of synthesized hormone to the neurohemal organ, and release of the hormone from the neurohemal site.

The exact mechanisms of regulating the activity of the PTTH NSC during postembryonic development have not yet been defined, primarily because the hormone is not chemically identified nor was its site of synthesis known until just recently. Nevertheless, existing information suggests that the endocrine activity of these NSC is controlled by a complex scheme and that all insects possess certain common regulatory mechanisms for the PTTH NSC. However, it is also evident that nuances of this basic scheme in any given insect have apparently evolved to meet the particular behavioral, physiological, and/or environmental characteristics of that insect's life cycle. With this in mind, the mechanisms for the regulation of the PTTH NSC will be discussed primarily for three insects, *Manduca*, *Rhodnius*, and *H. cecropia*, to illustrate clearly both the aspects of control fundamental to all three species as well as the modifications of this control specific to each one.

In *Rhodnius*, the regulation of the PTTH titer is apparently a result of the integration by the brain of both neural and humoral stimuli. Following a blood meal, stimulated abdominal stretch receptors are thought to promote the synthesis and release of PTTH by acting either directly on the PTTH NSC or indirectly via a circuit presynaptic to these NSC. Although there exists some neurophysiological data to support this ideal (Steel, 1978), it is mainly on the basis of cytological evidence that a neural control of release is proposed. The classical studies of Wigglesworth indicated that the medial NSC are the source of PTTH in *Rhodnius*, and more recent studies have demonstrated characteristic changes in the histological and ultrastructural appearance of these cells following a blood meal (Steel, 1978). Interpretative analysis of the successive morphological changes in the medial NSC as well as changes in the density of stainable material in their axons and estimates of the rates of their transport suggest that neural stimuli coordinate two phases of release of neurosecretory material during an instar. The first phase is comprised of the 6–8 days after feeding determined to be the head critical period; this phase coincides with increases in the electrical activity in medial NSC axons, the synthetic activity of the medial NSC, and accumulations of NSC material in the axon tract innervating the CC. Following the head critical period, both the electrical activity and the neurosecretory material in these axons decline to prefeeding levels, suggesting that a phase of release has ended. During the following 2 weeks, stainable material gradually accumulates in the medial NSC. At the time of ecdysis, 24 days after feeding, cytological changes indicative of accelerated synthesis and a second phase of release are evident.

The prolonged influence of feeding on the release of PTTH and possibly on its synthesis as well may be due to the adapted discharge rate maintained by the abdominal stretch receptors over a period of several weeks, although the electrical activity of the NSC axons persists in the isolated head of a fed larva (Steel, 1978). The release of PTTH in *Rhodnius* is apparently terminated at the end of the head critical period by a humoral mechanism. It has been suggested that once the 20-hydroxyecdysone titer reaches a critical level after the first phase of release, this ecdysteroid affects the medial NSC via a negative feedback mechanism to inhibit the release of PTTH (Steel, 1978). This idea is based on two observations: (1) following parabiosis of larvae decapitated at the end of the head critical period, i.e., when the 20-hydroxyecdysone titer is high, the medial NSC in the young larvae undergo morphological changes similar to those in late-instar larvae, and (2) the administration of 20-hydroxyecdysone to larvae 1–3 days after feeding elicits the same changes in medial NSC morphology as parabiosis with late-instar larvae. Thus, the regulation of the PTTH titer in *Rhodnius* at the time of the head critical period is accomplished by the brain through its integration of both neural and humoral stimuli. The factors governing the apparent second release of PTTH at ecdysis have not been investigated thus far, but it might be reasonable to assume that similar mechanisms are involved.

Although these and a number of other studies have revealed dramatic morphological, histological, cytochemical, and autoradiographic changes in various cerebral NSC as well as their axons and sites of release (see Gilbert *et al.*, 1980), they have all failed to establish a causal relationship between the changing appearance of NSC and the synthesis, storage, transport, and/or release of an identified neurosecretory product. Only when the structure of PTTH is known and methods subsequently developed for detecting and tracing it, will the roles of these various processes in the regulation of this hormone be resolved.

In *Manduca*, the modulation of the PTTH titer also appears to be governed by both environmental stimuli and humoral factors, which are integrated by the brain to affect the PTTH NSC. The release of PTTH in each stage of postembryonic development occurs at a precise time of day established by a circadian clock (Truman, 1972; Truman and Riddiford, 1974). The clock establishes a "gate" or period of time in each 24-hr cycle during which PTTH can be released. During larval development, the actual release of PTTH during the gate is dependent upon the reception by the brain of specific stimuli just before or while the gate is open. For the third and fourth (penultimate) larval instars, the gate is approximately 10 hr, beginning just after the initiation of the scotophase (lights off). The stimulus triggering PTTH release appears to be the attainment of a critical weight before the gate closes. Larvae reaching the critical weight after the closing of the gate on a given day continue to feed and grow until the next gate on the following day. In the last larval instar, there are apparently two gates for PTTH release (see Section 5.1). Similar to the situation in earlier instars, the first release of PTTH in the last instar is associated with the attainment of a certain weight while the gate is open (Nijhout and Williams, 1974a). Although there is reason to believe that the competency of the larval brain to release PTTH is thus established through some form of allometry, it should be pointed out that the data suggesting this are as yet only correlative and not causal. If indeed allometry is the causal factor in eliciting PTTH release, then the type of neural input regulating PTTH release in *Manduca* would be similar to that in *Rhodnius*, and thus the same fundamental mechanism in different insects may control the PTTH titer.

Humoral regulation of PTTH synthesis and/or release in *Manduca* is also believed to occur and apparently involves JH (Nijhout and Williams, 1974b). During the last larval instar, the JH titer decreases to undetectable levels during the 24 hr before the gate for PTTH release. It has been speculated that this decrease in the JH titer, which begins when the larva weighs approximately 5 g (approximately Day 3), establishes competency for the brain to release PTTH. Two lines of experimental evidence support this hypothesis: (1) injections of JH just before PTTH release on Day 4 delay by 1–2 days the overt indicants of the first increase in the ecdysteroid titer (e.g., dorsal vessel exposure and wandering); and (2) allatectomy of larvae weighing 5 g or less elicits wandering 1–2 days earlier than normal. Thus, JH appears to exert a negative control on the synthesis and/or release of PTTH by the NSC either directly or indirectly via an afferent neural pathway. This concept is not without precedent since larval diapause delays the normal larval–pupal molt in the rice stem borer, *Chilo suppressalis*, and the southwestern corn borer, *Diatraea grandiosella*, and is maintained in the presence of a high JH titer (see Chippendale, 1977).

Although there are a variety of stress factors (nutritional state, crowding, wounding) that can affect either directly or indirectly the synthesis and/or release of PTTH during metamorphosis, the most important factors contributing to the regulation of PTTH under normal conditions are environmental, in particular, photoperiod and temperature. The phenomenon of pupal diapause best illustrates their effects because both the initiation and the termination of this state of developmental arrest are regulated by the action of these two types of stimuli on the brain–prothoracic gland axis.

In *H. cecropia*, as well as other species of silkworms, pupal diapause results from a failure of the brain to secrete PTTH (Williams, 1952, 1956). The major stimulus for the initiation of diapause appears to be photoperiod, based on the observation that larvae raised in the laboratory under short-day conditions (<12-hr photophase) enter diapause after the larval–pupal molt just as they would under natural conditions in preparation for overwintering. The termination of diapause and subsequent metamorphosis to the adult will occur only following prolonged exposure to cold. Chilling seems to affect the brain directly since unchilled diapausing pupae will undergo adult development when implanted with brains from chilled pupae. Long-day photoperiod, rather than chilling, terminates pupal diapause in the Chinese oak silkmoth, *Antheraea pernyi* (Williams and Adkisson, 1964). However, as with temperature regulation, the center for photoperiod control is found in the brain. Several lines of evidence support this idea. For example, when the head of a diapausing *Antheraea* pupa is exposed to short-day illumination, diapause is maintained even though the abdomen may be exposed to long-day conditions. Brainless pupae, however, are insensitive to photoperiod. In addition, transplantation of the pupal brain to the abdomen shifts the location of photoperiodic sensitivity to the location of the transplanted brain. The mechanisms by which temperature and photoperiodic stimuli are transduced by the brain to affect the PTTH NSC are not known at the present time. However, extraretinal photoreception, which appears to be a general capacity of insects, is obviously involved in the case of photoperiodic stimulation (see Gilbert *et al.*, 1980). Extraretinal receptors are usually localized in the cerebral lobes of the brain, and in two insects, the oak silkmoth and the aphid, they are apparently found in the lateral area of the lobes. The failure of the potent synaptic inhibitor, tetrodotoxin, to inhibit PTTH release in the oak silkmoth suggests that the photoperiodic mechanism governing PTTH release involves cells with a nonneuronal function, i.e., the lateral NSC. If, as in *Manduca*, a single lateral NSC is the source of PTTH in *Antheraea*, then the extraretinal photoperiodic receptor may be a component of the PTTH NSC itself.

6.2. The PG

The regulation of the PG appears to be exerted by several of the same basic mechanisms involved in the control of the PTTH NSC, specifically by neural and humoral factors. Of the two, humoral control of gland activity has been most extensively studied and involves control by PTTH and possibly by JH and 20-hydroxyecdysone as well, acting either directly on the PG or indirectly via the brain.

The principal humoral factor involved in the control of the PG is, of course, PTTH. Thus, those stimuli that modulate synthesis and release of this neurohormone are in actuality the regulators of the PG. Although little is known about how PTTH synthesis and release are controlled (see Section 6.1), there is some information on the biochemical effect of PTTH on the PG. In general, this effect involves the modulation of gland biosynthetic activity, i.e., PTTH elicits an increase in the basal rate of ecdysone biosynthesis by the glands. The overall effect of PTTH on the PG is similar to that of ACTH on the adrenal cortex; therefore, it was assumed that the interaction of PTTH with the PG would similarly involve cyclic nucleotides.

The potential involvement of cyclic nucleotides as second messengers in the PTTH activation of the PG has been investigated in *Manduca* by an examination of adenylyl cyclase activity in the PG as well as by a titer of cyclic nucleotides in the glands during larval–pupal development (see Bollenbacher *et al.*, 1980). With the utilization of a cell-free system, adenylyl cyclase activity was demonstrated in the PG and was observed to increase at critical times during the last larval instar. The significance of this observation was underscored by the discovery that active PG from Day 5 of the last larval instar could be stimulated by either cyclic AMP (cAMP) or aminophylline (a phosphodiesterase inhibitor) to synthesize ecdysone at approximately twice the rate of a normal Day 5 gland. By contrast, inactive PG from Day 3 last-instar larvae could not be stimulated by cAMP, but were stimulated by phosphodiesterase inhibitors. Thus, both phosphodiesterases and adenylyl cyclase appear to be involved in differentially modulating cAMP levels and consequently gland activity. A comparison of cAMP levels in the PG during larval–pupal development to the ecdysone biosynthetic activity of the PG during this same period revealed that a temporally acute, 15-fold increase in the cAMP level immediately precedes the first ecdysteroid peak on Day 4, suggesting a causal relationship between cAMP levels and gland activation. Despite the timely occurrence of this cAMP increase, a second increase was not observed preceding the larger, second ecdysteroid peak in the last instar. Since ligation studies have indicated the existence of a second head critical period, i.e., a second period of PTTH release necessary for the activation of the PG and the subsequent large increase in the ecdysteroid titer on Day 7, it would appar that both cAMP-dependent and cAMP-independent activation of the PG exist. Alternatively, it is possible that more than one brain hormone exist, each activating the PG at precise times during development but via distinctly different mechanisms. With the establishment of the *in vitro* assay for PTTH, such possibilities can now be investigated directly.

In addition to the activation of the PG by PTTH, it has been proposed that JH exerts a stimulatory effect on the glands, either directly or indirectly (see Gilbert and King, 1973). The mimetic effect of JH was suggested by the observation that brainless lepidopteran pupae, when implanted with CA, undergo adult development. Since the CA is probably the neurohemal organ for PTTH, the putative effect of JH may actually represent PTTH activation of the glands. However, the application of JH or compounds with JH activity also elicits adult development in diapausing or brainless lepidopteran pupae. The actual effect of JH, if any, cannot be discerned from these results since the nature of the interaction of JH with the pupal neuroendocrine system, i.e., direct or indirect, is not evident and because such an effect has not been demonstrated under normal physiological conditions. It also must be recognized that the potential regulation of the PG by JH may be an endocrine event specific to larval diapause.

Recently, it has been suggested that in the cabbage armyworm, *Mamestra brassicae*, activation of the PG by JH late in the head critical period of the last instar elicits pupation (Hiruma *et al.*, 1978). It was found that JH analog treatment induced the pupation of larvae neckligated late in the head critical period and of isolated larval abdomens implanted with larval PG from late in the head critical period. Similar treatment of larvae neck-ligated at earlier stages and of abdomens implanted with precritical period PG did not elicit pupation. Thus, it was concluded that the effect of JH on the PG was a temporally restricted event, i.e., occurring only in the second half of the instar. Unfortunately, as with similar previous studies, a direct effect of JH on the PG was not demonstrated and, more importantly, high doses of JH analogs, which are not physiological molecules, have been used to evoke the biological response. If JH does directly activate the PG, this should be demonstrable with the utilization of the *Manduca in vitro* assay for PG activity. In fact, preliminary results from a study investigating the direct effect of JH I on *Manduca* PG from throughout the last larval instar and early pupal period have failed to demonstrate either activation or inhibition of the glands (Gruetzmacher, personal communication). This could mean that JH affects the ecdysteroid titer indirectly rather than directly, i.e., occurring at a level other than that of synthesis. JH might indirectly elicit a molt by affecting the PTTH activity of the brain, the catabolism of ecdysteroids, or the sensitivity of target tissue to the ecdysteroid titer.

Ecdysteroids themselves may also regulate the PG. Although it has not been shown that ecdysone, the product of the PG, has any feedback effect on the glands, it does appear that 20-hydroxyecdysone may exert such an effect. As previously stated, 20-hydroxyecdysone is thought to be indirectly involved in an indirect feedback on the PG by affecting the PTTH NSC (see Section 6.1). There is also some evidence for a direct positive effect on the PG by 20-hydroxyecdysone, but since this evidence is derived primarily from *in vivo* studies that utilized large doses of 20-hydroxyecdysone and ecdysone analogs, the evidence is at best equivocal.

Only with the development and utilization of *in vitro* systems that mimic normal physiological conditions can the question of feedback regulation in the control of the PG be resolved.

Although some progress has been made in understanding the humoral regulation of the PG, the question of neural regulation of the gland is completely unresolved. Innervation of the PG was first observed when the glands were initially described (Lyonet, 1762), and subsequent morphological studies have shown that larval and pupal lepidopteran PG are innervated from the prothoracic and mesothoracic ganglia and in some instances from the subesophageal ganglion. Recent ultrastructural analyses of this innervation have revealed the presence of neurosecretory axons within the glands. The source and structural composition of the nerve supply to the PG support the idea of some kind of nervous control of PG activity, and on this basis, studies have been directed at establishing a physiological role for the innervation. These studies have frequently involved nerve transections or destruction of the ventral ganglia. Unfortunately, such an approach makes it difficult to differentiate between the interruption of nervous and/or neurosecretory input directly inhibiting and/or stimulating gland function and the interruption of nervous input affecting PG activity via the PTTH NSC. The wax moth, *G. mellonella*, has been the focus of most of these studies, on the basis of which it has been concluded that in this insect propioceptive stimuli transmitted via the ventral nerve cord to the cerebral PTTH NSC are the major avenue of nervous regulation of the PG (Malá and Sehnal, 1978). The contribution of the direct innervation of the PG to the overall control of the gland appears to be only modulatory, with neurosecretory input from the ventral ganglia providing a subtle but perhaps important component. The existence in the ventral ganglia of inhibitory or stimulatory factors that modulate PG activity remains to be confirmed although the abundance of NSC in these ganglia and the neurosecretory axons that terminate in the PG support this possibility.

6.3. The CA

The synthesis of JH by the CA appears to be controlled by both neurosecretory and nervous mechanisms and under certain circumstances by environmental factors as well (Gilbert *et al.*, 1978, 1980). Although these same mechanisms were implicated in the control of the PTTH NSC and PG, their integration into a regulatory system for the CA may be far more complex. This complex nature of CA regulation is suggested in part by the possibility that separate and distinct mechanisms may be involved in modulating the synthesis of each of the three JH homologs during critical developmental periods. Furthermore, although neurosecretory control of these syntheses is likely, this control could be exerted either in a normal humoral manner via the hemolymph or by the cerebral NSC axons that terminate in the CA. Complicating any attempt to elucidate the mechanism of neurosecretory control is the fact that the nerves that innervate the CA in the NCA I and NCA II contain not only axons of NSC, but axons of ordinary neurons as well. Thus, the composition of the NCA I and NCA II suggests the existence of a nervous component in the control of the CA as well, the effects of which could not be easily distinguished from direct neurosecretory control.

The differentiation between nervous and direct neurosecretory control is probably the single most difficult aspect of resolving the regulation of the CA; the possible existence of either or both of these phenomena complicates the interpretation of the many *in vivo* studies that have investigated the effect of nerve severance on CA activity. Either type of control could be excitatory and/or inhibitory, further complicating the system. There is evidence for nervous inhibition of the CA in *Leucophaea maderae*, in which severance of nervous connectives to the CA results in an extra larval molt (Engelmann, 1959). By contrast, in the locust, *Schistocerca gregaria*, nervous stimulation of the CA may occur since severance of the nerves to the CA results in a rapid decline in the ability of the CA to synthesize JH (Tobe *et al.*, 1977). In this case, CA activity was assessed by an *in vitro* radioenzymological assay for JH synthesis. Alternatively, however, this increase and decrease in CA activity may reflect the absence of excitatory and inhibitory nonhumoral neurosecretory factors released directly from NSC axons terminating in the CA. Because of the difficulty in distinguishing a nervous component from a neurosecretory component in the control of the CA, it becomes questionable whether nervous control of the CA can be demonstrated at this time. The best approach to this problem may be one that first establishes the existence of neurosecretory control, and then, by process of elimination and redefinition, determines the existence of a strictly nervous contribution to the regulation of the gland.

Neurosecretory control of the CA has been proposed to involve both activating and inhibiting factors, termed allatotropins and allatohibins, respectively. Although the existence of neither has been directly demonstrated, there is some experimental precedent for an allatotropin, while the existence of an allatohibin is as yet primarily speculation. Several studies of larval insects, e.g., the wax moth, *G. mellonella*; the earwig, *Anisolabis maritima*; the locust, *Locusta migratoria*; and the silkmoth, *B. mori*, have indicated that the CA are activated by a neurohormone (see Gilbert *et al.*, 1978, 1980). For example, it was demonstrated in *Galleria* that the implantation of Day 0 last-instar larval brains into young last-instar larvae elicited an extra larval molt, but after a certain time in the last instar, larvae became refractory to implanted brains and underwent normal pupal metamorphosis. In preventing pupal development, the implanted brains exerted a juvenilizing effect, which presumably occurred through the release of an allatotropin. This allatotropic activity of the brain appeared to fluctuate developmentally, with the activity being greatest in brains early in each instar, a finding consistent with JH titer data demonstrating high levels of JH early in the instar (Granger and Sehnal, 1974). Localized cauterization of implanted brains indicated that the source of this allototropic activity was the medial neurosecretory cells. Although these data suggest the existence of an allatotropin, the indirect nature of these studies permits an alternative interpretation of the results: that the implanted brains activated the pro-thoracic glands to synthesize enough ecdysone to initiate a molt at a time when the JH titer in the larva was still high. Although a subsequent examination of the PTTH activity of the brain during larval development in *Galleria* tends to support the original interpretation of the implanta-tion studies (Malá *et al.*, 1977), the ultimately inconclusive nature of results obtained with the *in vivo* approach underscores the need for a system with which the question of CA regulation can be directly ap-proached.

An initial effort to investigate *in vitro* the neuroendocrine control of the CA during postembryonic development focused on a demonstration of an allatohibin as well as an allatotropin. The JH titer in last-instar larvae of *Manduca*, as measured by the *Manduca* black larval assay for JH, had previously been shown to decrease dramatically at approximately Day 2 (5 g) of the instar (Nijhout and Williams, 1974b). When JH bio-synthesis by brain–CC–CA complexes taken from early last-instar larvae and maintained *in vitro* was measured in the presence of hemolymph or a hemolymph fraction from Days 2 to 4, biosynthesis was substantially reduced (Williams, 1976). It was concluded that at a critical time during the last larval instar (post 5 g), an inhibitory factor was released into the hemolymph and acted on the brain to inhibit the release of an allatotropin and simultaneously stimulate the release of allatohibin. The existence of these regulatory factors has yet to be demonstrated *in vitro*, but the results of this first study argue convincingly for the implementation of such an approach in the investigation of putative allatotropins and al-latohibins.

Currently, two *in vitro* systems have been established to study CA regulation and have incorporated contemporary methods for directly quantifying rates of JH synthesis. The first of these approaches utilizes a JH radioenzymological assay in which the rate of methylation of JH at C-1, a terminal step in the JH biosynthetic pathway, is quantified *in vitro* (Pratt and Tobe, 1974). This method has been employed mainly to measure the biosynthetic activity of CA from adult female orthopterans during reproduction and after various *in vivo* surgical manipulations. Thus far, the data generated indicate that the rate of methylation by the adult CA *in vitro* parallels the JH titer, and also suggest that nervous, rather than neurohormonal, activation and inhibition of the CA occur in the adult female of several species. It must be pointed out that this method of JH quantification measures the rate of a terminal enzymatic step in JH biosynthesis, which may not reflect effects on the actual rate step(s) in the pathway and thus may not be an ideal point from which to study activation or inhibition. In addition, the technical difficulty of simultaneously monitoring the synthesis of more than one JH homolog with this assay precludes its routine use for the simultaneous measurement of the synthesis of the different homologs. The second *in vitro* approach for investigating the control of the CA has employed a radioimmunoassay (RIA) for JH I as the means of assessing CA activity (Granger *et al.*, 1979). Unlike the radioenzymological assay, the RIA quantifies *de novo* JH synthesis and thus would be more likely to detect any effect on regulatory steps in the JH biosynthetic pathway. In addition, with the development of specific RIAs for JH II and JH III, it should be a simple matter to monitor the synthesis of any or all of the homologs concurrently.

Thus far, the JH I RIA has been adapted to determine the *in vitro* biosynthetic activity of CA from *Manduca* larvae throughout the penultimate and last larval instars (Granger *et al.*, 1979, 1981). The results of this study have revealed that CA activity fluctuates dramatically during development in a manner temporally similar to fluctuations in the JH titer during this same period. The correlation between *in vitro* CA activity and the *in vivo* JH titer argues for the validity of this approach, but more importantly, the fluctuations in CA activity indicate the existence of regulation at the level of synthesis. Since this approach is basically the same as that developed to investigate the PTTH–PG system, it will undoubtedly facilitate investigations of the putative neurohormonal control of the CA during metamorphosis.

In addition to the possible neurohormonal and/or nervous control of CA activity, there is some evidence to suggest that JH itself can affect gland activity in a positive or negative manner, either directly or indirectly. Positive feedback has been suggested from an autoradiographic study in which an increase in nuclear RNA synthesis in CA of diapausing saturniid pupae occurred in response to the injection of large doses of JH (Siew and Gilbert, 1971). In other insects, e.g., a cockroach (*Diploptera punctata*) and the potato beetle (*Leptinotarsa decemlieata*), studies utilizing the radioenzymological assay for JH synthesis to measure CA activity *in vitro* after various treatments *in vivo* have suggested that the feedback effect of the hormone on the CA may be both positive and negative, depending on the endogenous titer of the hormone (see deKort and Granger, 1980). For example, in unilaterally allatectomized adults, which have artificially induced low JH titers, a compensatory increase in the activity of the remaining CA occurs whether or not the gland is innervated. This suggests that low concentrations of JH exert positive feedback on CA activity. By contrast, implantation of additional CA into an animal or treatment with high doses of JH or JH analogs results in a decrease in the *in vitro* activity of the host glands, implying that a high concentration of JH exerts negative feedback on the CA. Because the manipulations performed in these studies were carried out *in vivo*, it is not known if their effects on CA activity are direct or indirect. Evidence suggesting an indirect effect of JH on the CA comes from a study with *Galleria*, the results of which suggested that JH could inhibit the synthesis or release of the cerebral allatotropin (Granger and Borg, 1976). The implementation of a completely *in vitro* approach for investigating the question of feedback effects of JH on CA activity will circumvent the interpretative limitations of the previous studies and will permit an evaluation of these putative effects at physiological concentrations of the hormone.

In addition to the humoral and neural control of the CA, environmental stimuli such as raised temperature, chilling, humidity, crowding, starvation, and injury apparently affect the CA as well (see Gilbert *et al.*, 1978). In response to stressful environmental conditions, many larvae undergo extra larval molts, presumably in response to an increased level of JH synthesis by the CA. Any understanding of how these stimuli are transduced to affect the CA will probably have to await the elucidation of the primary neural mechanisms involved in the regulation of this gland.

7. SUMMARY

In the last two decades, there have been several significant additions to our knowledge of the endocrinology of insect metamorphosis. These include the elucidation of the chemistry of two of the three major hormones, the identification of the site of synthesis of all three, the establishment of the CA as the neurohemal organ for PTTH, and the demonstration of a direct activation of the PG by PTTH (brain hormone). In addition to these discoveries directly affecting our understanding of the regulation of the glands, significant advances have been made in the areas of hormone transport, biosynthesis, catabolism, and mode of action (see Chapters 5 and 8, this volume). The technologies and methods that have been or are currently being developed to provide this information suggest that we are at the beginning of an era of exponential growth of the field.

If only one conclusion is to be drawn from this chapter, it is that despite the minimal change in our basic understanding of the hormonal control of insect metamorphosis, the insect endocrine system has proven to be far from simple and primitive, involving hormonal and neural interactions and feedback relationships on many different levels. The beginning of our understanding of this complexity is illustrated in Fig. 4. Thus, there is every reason to believe that the complexity of the insect endocrine system may approach that of the vertebrates and that the same basic mechanisms of hormone regulation and action exist for both. This observation, coupled with the obvious and decided advantages of insects for endocrinological studies, makes it apparent that research on the control of insect endocrine glands may contribute significantly to our basic knowledge on the function of endocrine systems in general.

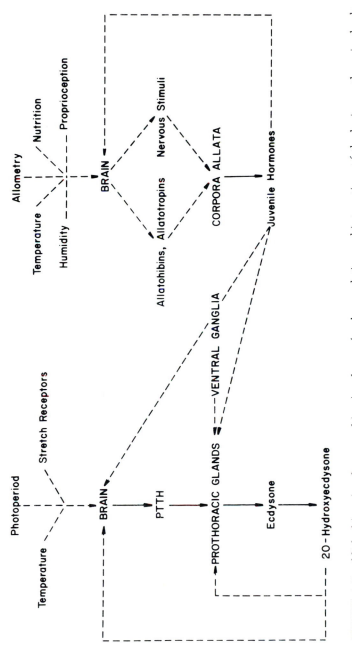

FIGURE 4. Established (——) and proposed (----) pathways for the regulation and interactions of the brain, prothoracic glands, and corpora allata in the endocrine control of insect metamorphosis.

ACKNOWLEDGMENTS. Work from the authors' laboratory was supported by Grants AM-02818, NS 15387-01, and NS 14816 from the National Institutes of Health and PCM-76-23291-A01 from the National Science Foundation. We thank Dr. Stan Lee Smith for his editorial comments, Nancy Grousnick for her excellent secretarial assistance, and Eva Katahira and Sheryl Niemiec for their artistic and technical contributions.

REFERENCES

Agui, N., Granger, N. A., Gilbert, L. I., and Bollenbacher, W. E., 1979, *Proc. Natl. Acad. Sci. USA* **11**:5694.

Agui, N., Bollenbacher, W. E., Granger, N. A., and Gilbert, L. I., 1980, *Nature (London)* **285**:669.

Berlind, A., 1977, *Int. Rev. Cytol.* **49**:171.

Bollenbacher, W. E., Vedeckis, W., Gilbert, L. I., and O'Connor, J. D. 1975, *Dev. Biol.* **44**:46.

Bollenbacher, W. E., Agui, N., Granger, N. A., and Gilbert, L. I., 1979, *Proc. Natl. Acad. Sci. USA* **10**:5148.

Bollenbacher, W. E., Agui, N., Granger, N. A., and Gilbert, L. I., 1980, in: *Invertebrate Systems in Vitro* (E. Kurstak, K. Maramorosch, and A. Dübendorfer, eds.), pp. 253–271, Elsevier/North-Holland, Amsterdam.

Bounhiol, J. J., 1938, *Bull. Biol. Suppl.* **24**:1.

Burtt, E. T., 1938, *Proc. R. Soc. London Ser. B* **126**:210.

Chippendale, G. M., 1977, *Annu. Rev. Entomol.* **22**:121.

Crampton, H. E., 1899, *Wilhelm Roux Arch. Entwicklungsmech. Org.* **9**:293.

deKort, C. A. D., and Granger, N. A., 1981, *Annu. Rev. Entomol.* **26**:1.

Engelmann, F., 1959, *Biol. Bull. (Woods Hole)* **116**:406.

Fukuda, S., 1940, *Proc. Imp. Acad. Jpn.* **16**:417.

Gibbs, D., and Riddiford, L. M., 1977, *J. Exp. Biol.* **66**:87.

Gilbert, L. I., and King, D. S., 1973 in: *The Physiology of Insecta,* Volume 1 (M. Rockstein, ed.), pp. 249–370, Academic Press, New York.

Gilbert, L. I., Goodman, W., and Granger, N., 1978, in: *Comparative Endocrinology* (P. J. Gaillard and H. H. Boer, eds.), pp. 471–486, Elsevier/North-Holland, Amsterdam.

Gilbert, L. I., Bollenbacher, W. E., and Granger, N. A., 1980, *Annu. Rev. Physiol.* **42**:493.

Granger, N. A., and Borg, T. K., 1976, *Gen. Comp. Endocrinol.* **29**:349.

Granger, N. A., and Sehnal, F., 1974, *Nature (London)* **251**:415.

Granger, N. A., Bollenbacher, W. E., Vince, R., Gilbert, L. I., Baehr, J. C., and Dray, F., 1979, *Mol. Cell. Endocrinol.* **16**:1.

Granger, N. A., Bollenbacher, W. E., and Gilbert, L. I., 1981, in: *Current Topics in Insect Endocrinology and Nutrition* (G. Bhaskaran, S. Friedman, and J. G. Rodriguez, eds.), pp. 83–105, Plenum Press, New York.

Hachlow, V., 1931, *Wilhelm Roux Arch. Entwicklungsmech. Org.* **125**:46.

Hadorn, E., and Neel, J., 1938, *Wilhelm Roux Arch. Entwicklungsmech.* **138**:281.

Hiruma, K., Shimada, H., and Yagi, S., 1978, *J. Insect Physiol.* **24**:215.

Ishizaki, H., 1969, *Dev. Growth Differ.* **11**:1.

Kiguchi, K., and Riddiford, L. M., 1978, *J. Insect Physiol.* **24**:673.

Kobayashi, M., and Yamazaki, M., 1974, in: *Invertebrate Endocrinology and Hormonal Heterophylly* (W. J. Burdette, ed.), pp. 29–42, Springer-Verlag, New York.

Kopec, S., 1917, *Bull. Int. Acad. Scie. Cracovie* **BB**:57.

Lyonet, P., 1762, *Traite Anatomique de la Chenille qui Ronge le Bois de Saule,* La Haye.

Maddrell, S. H. P., 1974, in: *Insect Neurobiology* (J. Treherne, ed.), pp. 307–358, Elsevier/North-Holland, Amsterdam.

Malá, J., and Sehnal, F., 1978, *Experientia* **34**:1233.

Malá, J., Granger, N. A., and Sehnal, F., 1977, *J. Insect Physiol.* **23:**309.

Nabert, A., 1913, *Z. Wiss. Zool.* **104:**181.

Nijhout, H. F., 1975, *Int. J. Insect Morphol. Embryol.* **4:**529.

Nijhout, H. F., and Williams, C. M., 1974a, *J. Exp. Biol.* **61:**481.

Nijhout, H. F., and Williams, C. M., 1974b, *J. Exp. Biol.* **61:**493.

Piepho, H., 1943, *Naturwissenschaften* **31:**329.

Plagge, E., 1938, *Biol. Zentralbl.* **59:**1.

Pratt, G. E., and Tobe, S. S., 1974, *Life Sci.* **14:**575.

Raabe, M., Baudry, N., Grillot, J. P., and Provensal, A., 1974, in: *Neurosecretion: The Final Common Neuroendocrine Pathway* (F. Knowles and L. Vollrath, eds.), pp. 60–71, Springer-Verlag, New York.

Rowell, H. F., 1976, *Adv. Insect Physiol.* **12:**63.

Scharrer, B., 1964, *Z. Zellforsch. Mikrosk. Anat.* **62:**125.

Scharrer, B., 1977, in: *Peptides in Neurobiology* (H. Gainer, ed.), pp. 1–8, Plenum Press, New York.

Scharrer, E., 1928, *Z. Vgl. Physiol.* **7:**1.

Schneiderman, H. A., 1967, in: *Methods in Developmental Biology* (F. W. Wilt and N. K. Wessells, eds.), pp. 753–766, T. Y. Crowell, New York.

Siew, Y. C., and Gilbert, L. I., 1971, *J. Insect Physiol.* **17:**2095.

Smith, S. L., Bollenbacher, W. E., and Gilbert, L. I., 1980, in: *Progress in Ecdysone Research* (J. Hoffmann, ed.), pp. 139–162, Elsevier/North-Holland, Amsterdam.

Steel, C. G. H., 1978, in: *Comparative Endocrinology* (P. J. Gaillard and H. H. Boer, eds.), pp. 327–330, Elsevier/North-Holland, Amsterdam.

Tobe, S. S., Chapman, C. S., and Pratt, G. E., 1977, *Nature (London)* **268:**728.

Truman, J., 1972, *J. Exp. Biol.* **57:**805.

Truman, J. W., and Riddiford, L. M., 1974, *J. Exp. Biol.* **60:**371.

Wigglesworth, V. B., 1934, *Q. J. Microsc. Sci.* **77:**191.

Wigglesworth, V. B., 1936, *Q. J. Microsc. Sci.* **79:**91.

Williams, C. M., 1952, *Biol. Bull. (Woods Hole)* **103:**120.

Williams, C. M., 1956, *Nature (London)* **178:**212.

Williams, C. M., 1958, *Sci. Am.* **198:**67.

Williams, C. M., 1976, in: *The Juvenile Hormones* (L. I. Gilbert, ed.), pp. 1–14, Plenum Press, New York.

Williams, C. M., and Adkisson, P. L. 1964, *Biol. Bull. (Woods Hole)* **127:**511.

CHAPTER 5

Chemistry, Metabolism, and Transport of Hormones Controlling Insect Metamorphosis

Lawrence I. Gilbert and Walter Goodman

1. INTRODUCTION

Since the changes in hormone titer must be precise if insect growth, development, and metamorphosis are to proceed normally, knowledge of the biosynthesis and degradation of the growth hormones as well as their possible control is of utmost importance if we are to understand metamorphosis at the biochemical level. Indeed, progress in this area of insect hormone metabolism and titer regulation has been remarkable during the past decade, particularly when compared to the less precise data accumulated on gland regulation (see Granger and Bollenbacher, this volume). In this chapter we will be concerned only with the three hormones known to control molting and metamorphosis, namely the prothoracicotropic hormone (PTTH) from the brain, the molting hormone (MH) (20-hydroxyecdysone) and the juvenile hormones (JH) secreted by the corpora allata (Fig. 1). However, we should not lose sight of the fact

LAWRENCE I. GILBERT · Department of Zoology, University of North Carolina, Chapel Hill, North Carolina 27514. WALTER GOODMAN · Department of Entomology, University of Wisconsin, Madison, Wisconsin 53706.

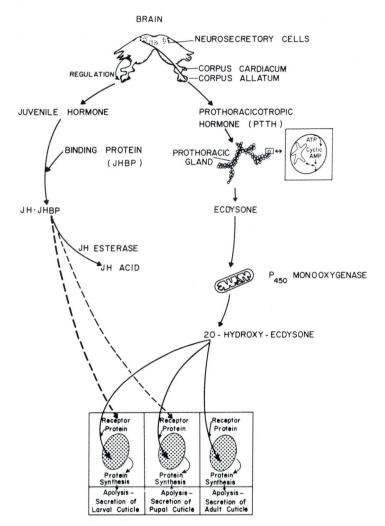

FIGURE 1. Scheme of the endocrine control of development for the tobacco hornworm, *Manduca sexta*. From Gilbert *et al.*, 1980.

that many more hormones exist in insects, although most are concerned with the everyday ability of the insect to modify its metabolism in response to changing environmental conditions (e.g., adipokinetic hormone, hyperglycemic hormone, diuretic hormone, hypolipemic hormone, etc.) rather than directing growth processes.

2. PROTHORACICOTROPIC HORMONE

Of the three hormones to be considered here, least is known about PTTH. This neurohormone is synthesized by specific neurosecretory cells in the insect brain and released from neurohemal organs distal to the brain. The control of synthesis and/or release is quite complex and for the most part unknown (Granger and Bollenbacher, this volume), although environmental and nutritional factors are known to play critical roles.

One cannot, of course, discuss the chemistry, transport, and metabolism of a hormone such as PTTH that has yet to be purified and characterized. We will discuss the little that is known about the chemical nature of PTTH and point out some of the reasons that its structural elucidation has resisted some rather persistent efforts, particularly by groups of Japanese researchers (e.g., Ishizaki, Kobayashi, Nishiitsutsuji-Uwo).

In most cases, the brains of the commercial silkmoth (*Bombyx mori*) have been utilized as the starting material for PTTH purification, presumably because of the abundance of these insects in Japan. Initial studies made use of larval and pupal brains, but the labor involved in extirpating brains from hundreds of thousands of pupae prompted developmental surveys that indicated that adult brains equaled or surpassed brains from other stages insofar as PTTH activity was concerned (Ishizaki, 1969). More recent attempts have therefore made use of acetone powders of frozen, male, adult heads as starting material. In the latest report from Ishizaki's laboratory (Nagasawa *et al.*, 1979), 96,000 *Bombyx* heads (~720 g) were extracted and 2.88 g of "crude" PTTH was obtained through a series of filtrations and precipitations. This was further purified by several column chromatography separations (Sephadex, DEAE–Sepharose, etc.) to yield 3.0 mg of "highly purified" PTTH, being effective in their bioassay (see below) at a dose of 6 ng.

Whereas proteolytic enzymes such as trypsin inactivated the highly purified PTTH, exopeptidases were ineffective. This substantiated the peptide nature of PTTH and indicated that the N and C termini of the molecule are blocked; further studies (e.g., oxidation by *N*-bromosuccinimide in urea) revealed that a tryptophan residue is required for activity, that sulfhydryl groups are not components of the active site, but that a disulfide bond is essential for activity. Size separation studies with known molecular markers indicated a molecular weight of between 4000 and 4800, approximately the same as mammalian ACTH, although molecular weights as high as 30,000 have been reported for PTTH in the past.

It is now about 20 years since the initial studies on PTTH purification were initiated and one could ask why this critical hormone has not yet been purified and characterized. Possibly, it is more of a problem with the bioassay than with the actual purification procedures. The usual bioassay animal is a debrained pupa of the saturniid silkmoth, *Philosamia cynthia ricini*, which is in a state of arrested development (diapause) because it no longer has the capacity to synthesize PTTH. It is of interest that in the case of the closely related *H. cecropia* silkmoth, such brainless pupae can live for some 2–3 years under the proper conditions, until the substrates supporting life processes are exhausted. When fractions containing PTTH are injected into these brainless *Philosamia* pupae, the animals initiate pupal–adult development within a few days as a result of PTTH activation of the prothoracic glands and subsequent synthesis of molting hormone (see Granger and Bollenbacher, this volume, and Fig. 1). Such *in vivo* assays are inherently plagued by a lack of rapidity, specificity, and reproducibility due in part to physiological variations between animals, even debrained pupae, and the presence of hydrolytic enzymes in the hemolymph that are capable of degrading peptides. In addition, there is the possible problem of species specificity; that is, PTTH from one species may not be entirely effective when assayed on pupae from another species. For example, it appears that the highly purified PTTH preparation derived from *Bombyx* brains is ineffective in another lepidopteran, *Manduca sexta* (C. M. Williams, personal communication).

Recently, a rapid, sensitive, and very reproducible assay for the PTTH of *Manduca* was developed using the amount of ecdysone synthesized by *Manduca* prothoracic glands *in vitro* as the endpoint (see Granger and Bollenbacher, this volume). Although utilization of this assay has not yet yielded the structure of the *Manduca* PTTH, some intriguing observations have been made (Bollenbacher, *et al.*, 1980). Although the pupal brain contains a molecule that exhibits PTTH activity and has a molecular weight similar to that noted above for *Bombyx*, a larger molecular weight form also was observed during kinetic analysis. This possibility of multimolecular forms (not subunits of a dimer or tetramer) of PTTH is currently being investigated, and the final story may be similar to that elucidated for the vertebrate ACTH. In the latter case, a 31,000-dalton glycosylated molecule is synthesized in the anterior pituitary and is comprised of both ACTH (amino acids 1–39) and β-lipotropin (amino acids 1–91); the ACTH can then act as a precursor for both α-MSH (amino acids 1–13) and corticotropin-like intermediate lobe peptide (amino acids 18–39) (see Krieger and Liotta, 1979). Both ACTH and the peptides derived from it exhibit ACTH biological activity, and there is some evidence that they can be synthesized in the brain as well as the pituitary. Whether PTTH may be processed in a similar manner is a matter now being pursued actively.

3. MOLTING HORMONE

3.1. Ecdysteroids—Chemistry and Occurrence

Figure 1 indicates that PTTH stimulates the prothoracic glands to increase their rate of ecdysone synthesis and that this ecdysone is subsequently converted by peripheral tissues (e.g., the fat body mitochondria of *Manduca* larvae) to 20-hydroxyecdysone, the insect MH. Both of these compounds are polyhydroxylated steroids (Fig. 2), and since insects do not possess the biochemical repertoire to form the cyclopentanoperhydrophenanthrene ring structure typical of steroids, they depend on a dietary source of steroids (cholesterol in carnivorous insects and β-sitosterol in phytophagous species). Steroids are therefore "vitamins" for insects. and the dietary steroids must be converted to ecdysone and 20-hydroxyecdysone as well as perform their normal physiological roles as constituents of cell membranes.

Studies on the chemistry of the MH were initiated in the early 1940s, but it was not until 1954 that the classic paper of Butenandt and Karlson appeared in which they described the crystallization of MH, having obtained 25 mg from 500 kg of *Bombyx* pupae (2×10^7-fold purification). The assay used was that devised by Gottfried Fraenkel in 1935 and consisted of ligating last-stage fly (*Calliphora*) larvae before MH was secreted. These isolated abdomens were devoid of any known growth hormones and underwent pupariation (contraction, rounding up, and tanning of the original cuticle to form a puparium) when injected with active fractions of MH. Karlson detected another substance with MH activity 2 years later and termed it β-ecdysone, the first substance being denoted as α-ecdysone. It was not until 1963, however, that an error in molecular weight was corrected to yield a formula of $C_{27}H_{44}O_6$ for α-ecdysone, a formula consistent with a four-ringed structure (Karlson *et al.*, 1963). This suggested that α-ecdysone was a steroid with an α,β-unsaturated ketone. Indeed, upon dehydrogenation of the crystalline material, the German group identified methylcyclopentenophenanthrene. Finally, Huber and Hoppe (1965) determined 3400 structural parameters by X-ray scattering, calculated a complete Fourier synthesis of the molecule, and determined the structure as 2β,3β,14α,22R,25-pentahydroxy-5β-cholest-7-en-6-one (Fig. 2). This was rapidly confirmed by several organic syntheses.

The second substance, β-ecdysone, was later shown to be identical to α-ecdysone except for an additional hydroxyl group at C-20, that is, 20-hydroxyecdysone (2β,3β,14α,20R,22R,25-hexahydroxy-5β-cholest-7-en-6-one; Fig. 2), and indeed was identical to the material with MH activity that was isolated and characterized from crustaceans (Hampshire and Horn, 1966). This latter material was termed crustecdysone and its presence in noninsect arthropods was not surprising considering the fact that these animals molt and possess a molting gland, the Y-organ, analogous to the prothoracic gland of insects.

To further complicate the nomenclature issue were the findings from the laboratories of Nakanishi and Takemoto (see Nakanishi, 1971) that a variety of polyhydroxylated steroids could be extracted from plants that had potent MH activity when assayed on ligated fly larvae and debrained moth pupae. Indeed, more than 30 of these "phytoecdysones" have been isolated and characterized. In some cases they constitute up to 1% of the dry weight of the plant and several have greater MH activity in bioassay

Cholesterol

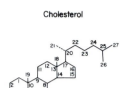

β – Sitosterol

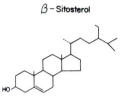

Ecdysone

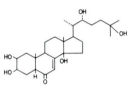

20 – Hydroxyecdysone

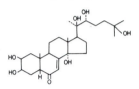

26 – Hydroxyecdysone

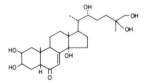

2 – Deoxyecdysone

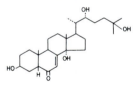

22 – Deoxyecdysone

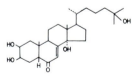

3 – Epi – 20 – Hydroxyecdysone

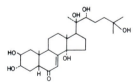

Ketol

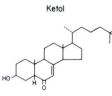

Ketodiol

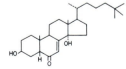

FIGURE 2. Representative ecdysteroids. Cholesterol is not an ecdysteroid but is the ultimate precursor for many insect steroids.

than α- or β-ecdysone. Finally, both α- and β-ecdysone were definitively identified in plant material, and plants have become the commercial source of these compounds for the insect physiologist over the past decade even though their function(s) in plants remains conjectural. As each phytoecdysone was characterized, it was named, usually, but not always, with the prefix indicating the genus or species of plant yielding the material; e.g., ponasterone A, viticosterone E, stachysterone D, polypodine B, podecdysone B, makisterone, etc. None of these names gave the uninitiated the slightest hint of the compound's structure and ultimately led to some confusion; e.g., β-ecdysone, crustecdysone, ecdysterone, and 20-hydroxyecdysone were terms describing the same molecule.

In 1978 it was proposed by leading investigators in the field that all of these substances from insects and plants be given the generic term *ecdysteroid*, that α-ecdysone be termed *ecdysone*, and that β-ecdysone and its synonyms now be called *20-hydroxyecdysone*. Although not perfect, it is the nomenclature now agreed to and it will be used in this volume.

3.2. Biosynthesis of Ecdysone and 20-Hydroxyecdysone

3.2.1. Ecdysone

The starting material for the biosynthesis of ecdysone is cholesterol or plant sterols that are efficiently converted to cholesterol by deethylation (e.g., removal of the 24-ethyl group of β-sitosterol) (see Thompson *et al.*, 1973). Early *in vivo* studies by Karlson's laboratory revealed a low (~0.0001%) but significant incorporation of injected [³H]cholesterol into ecdysone in fly larvae, while similar studies by Horn's group showed a 0.015% conversion of cholesterol to 20-hydroxyecdysone. With hindsight, this discrepancy in conversion rate can be explained as due to the efficient conversion of ecdysone to 20-hydroxyecdysone by fly tissues and the paucity of circulating ecdysone (Bollenbacher *et al.*, 1976). When administering a labeled compound such as cholesterol to an insect larva *in vivo*, it is theoretically possible to extract a labeled ecdysteroid, crystallize it repeatedly to constant specific activity, and calculate the percent conversion. However, this experimental strategy usually results in very low conversion rates since some of the cholesterol is utilized by dividing cells as membrane constituents, some is nonspecifically sequestered into lipophilic compartments of tissues such as the fat body, some is excreted, some is taken up by hemolymph lipoproteins, etc., and probably only a small component reaches the site of ecdysone biosynthesis. Further, if one is looking for unknown, short-lived intermediates in the biosynthetic pathway, it would be a most difficult task to purify and identify them from whole-body extracts. One *in vivo* approach that could yield significant information is the synthesis of putative, radiolabeled immediate precursors of ecdysone, the subsequent injection of these compounds into insects at critical developmental stages, and finally, isolation of labeled ecdysone and/or 20-hydroxyecdysone. The data generated from such studies by D. H. S. Horn's laboratory using fly (*Calliphora stygia*) larvae and William Robbins' group at Beltsville (Svoboda *et al.*, 1975; Kaplanis *et al.*, 1979) utilizing the tobacco hornworm, *M. sexta*, have been important and have laid the groundwork for the *in vitro* approach that will be discussed shortly. In addition, such studies have resulted in the identification of other ecdysteroids such as 26-hydroxyecdysone (Fig. 2), which may have a role in *Manduca* embryogenesis.

When *Calliphora* larvae were injected with a putative precursor, 2,22,25-trideoxyecdysone, label was found to be associated with 22-deoxyecdysone, ecdysone, 20-hydroxyecdysone, and possibly 2,22-dideoxyecdysone. It was therefore tentatively assumed that the sequence of hydroxylation was C-25, C-2, C-22, leading to ecdysone and then at C-20 to yield 20-hydroxyecdysone. Analogous studies with *Manduca* developing adults utilized [^{3}H]22,25-dideoxyecdysone as a precursor, and the results indicated that hydroxylation of the steroid nucleus (Fig. 2) precedes any side-chain hydroxylation, i.e., hydroxylation at C-2 must occur before hydroxylation at C-25. The obvious conclusion is that either one group erred or, more likely, that different insects, and perhaps the same insect at different developmental stages, use different biosynthetic pathways. It should be emphasized that these experiments are tedious and difficult, involving radiochemical synthesis of specific precursors and subsequent purification and characterization of the labeled products. Further, incorporation of any of these exogenously administered "precursors" into ecdysone is only suggestive since many are products of the imagination and talent of the organic chemist in that they have not yet to be identified as natural products in the insect. Thus, *in vivo* studies have yielded little concrete information and even that is equivocal. Indeed, if metabolites of putative precursors are identified, they could be catabolic products of ecdysone rather than intermediates on the biosynthetic pathway to ecdysone.

To circumvent at least some of these problems, e.g., nonspecific uptake of precursors, degradative mechanisms, isotope dilution, one could incubate the tissue responsible for synthesizing ecdysone in a medium to which is added the putative radiolabeled precursors. Up until 6 years ago, however, the site of synthesis had not been identified unequivocally, although the prothoracic glands were known to be required for molting. In addition, the physiological relationship between ecdysone and 20-hydroxyecdysone had not yet been fully elucidated, although it was shown that a variety of *Manduca* tissues could convert [^{3}H]ecdysone to [^{3}H]20-hydroxyecdysone *in vitro*, while a few, including the prothoracic glands, could not perform this hydroxylation (King, 1972). These observations suggested that the prothoracic glands synthesized and secreted ecdysone or a precursor or enzyme necessary for ecdysone biosynthesis elsewhere, and that 20-hydroxyecdysone was in any case a product of tissues peripheral to the prothoracic glands. To resolve the question, it was first necessary to determine the chemical nature of the product of the prothoracic glands.

During this time in the early 1970s, two groups began exploring the *in vitro* culture of the prothoracic glands with an eye toward identification of the product, one in Japan using the glands of larvae of the commercial silkworm (*B. mori*) (Chino *et al.*, 1974) and one in the United States utilizing *Manduca* larvae (King *et al.*, 1974). Almost simultaneously, microchemical analysis was being made available to the biologist with the development of high-pressure liquid chromatography (HPLC) and the perfection of electron-capture gas–liquid chromatography (EC-GLC). Of even greater importance for those interested in the function of the prothoracic gland was the development of a radioimmunoassay (RIA) for ecdysteroids by Borst and O'Connor (1972). Since that time a number of RIAs have been developed with varying affinities for ecdysone and 20-hydroxyecdysones (see Gilbert *et al.*, 1977; Reum and Koolman, 1979; Chang and O'Connor, 1979). This technique is sensitive (~25 pg), reproducible, and usually specific for a particular class of compounds, i.e., ecdysteroids. It is in essence a competitive-binding protein assay, and since RIA has become almost a routine technique in insect endocrinology, and will become an even more prevalent technique in the future, it is worth describing here.

In the case of ecdysteroids, a derivative of 20-hydroxyecdysone (e.g., carboxymethyloxine, *N*-hydroxysuccinamide, succinyltyrosine methyl ester, etc.) is conjugated to serum albumin, administered to rabbits, and an antiserum obtained. Depending on the type of derivative formed, the antibody will display varying affinity for ecdysone and 20-hydroxyecdysone, but will also react with other ecdysteroids and ecdysteroid metabolites. To derive a standard curve, one adds increasing amounts of unlabeled 20-hydroxyecdysone to a constant quantity of antibody and labeled 20-hydroxyecdysone. The higher the specific activity of the latter, the greater will be the sensitivity of the assay, and the preparation of [^{3}H]ecdysone and then [^{3}H]20-hydroxyecdysone at 60–70 Ci/mmole has been important in attaining enhanced resolution. As the amount of unlabeled 20-hydroxyecdysone is increased, less of the [^{3}H]ecdysteroid will bind to the antibody because of competition for the binding site between the unlabeled and the labeled material. One then separates bound ecdysteroid from unbound, usually by precipitation or absorption, and quantifies the amount of [^{3}H]20-hydroxyecdysone in the ecdysteroid–antibody complex or measures the amount not bound. The less [^{3}H]ecdysteroid bound, the greater is the amount of ecdysteroids in the sample being assayed, and the experimental data can be compared to a standard curve. If 20-hydroxyecdysone is used to establish the standard curve, then the experimental data are expressed in 20-hydroxyecdysone equivalents.

Each antiserum must be characterized as to specificity by examining the competitive ability (i.e., affinities for antiserum) of a variety of ec-

dysteroids and similar compounds. The sensitivity of the RIA is really
the smallest quantity of ecdysteroid that can be distinguished from a
blank within 95% confidence limits. In addition, the RIA must be precise,
reproducible, and accurate. In the case of the ecdysteroid RIA, two facts
must be kept in mind: (1) when the RIA is used to titer hemolymph and
tissue extracts, the data may be expressed in 20-hydroxyecdysone equivalents but actually are a composite of all ecdysteroids present that react
with the antibody; and (2) when the material to be assayed contains only
ecdysone and 20-hydroxyecdysone, it is possible to quantify the amount
of each ecdysteroid present by employing antibody preparations that have
differing affinities for each of these ecdysteroids.

By using an ecdysteroid RIA, it was soon demonstrated that the
medium in which *Manduca* larval prothoracic glands had been cultured
contained RIA-positive material, whereas when other tissues of the insect
were incubated *in vitro*, no such activity was noted in the medium. No
RIA-positive material was present in the prothoracic glands themselves,
while the medium ecdysteroid content increased with time. When bioassayed, the culture medium displayed molting hormone activity, i.e., elicited puparium formation in ligated fly abdomens. These data revealed
three important facts: (1) only the prothoracic glands synthesize and
secrete an ecdysteroid into the medium; (2) this ecdysteroid(s) possesses
MH activity; and (3) the glands do not store the ecdysteroid.

The culture medium was then subjected to a variety of chromatographic procedures (TLC, EC-GLC, HPLC) during which the RIA and
bioassay were used to monitor fractions, and the data indicated that
ecdysone was the secretory product of the prothoracic glands and that it
appeared to be the only ecdysteroid secreted by the glands. By culturing
about 350 pairs of prothoracic glands, 1 liter of medium was obtained
containing ~4 μg of the putative ecdysone; it was purified to homogeneity, and a mass spectrum was obtained that demonstrated conclusively
that ecdysone was the product of the prothoracic glands. This was the
same conclusion reached by our colleagues in Japan who studied the
prothoracic glands of *Bombyx* and identified the secretory product by
mass fragmentography. These conclusions have been corroborated in several other orders of insects since 1974. It should be noted that although
the prothoracic glands appear to be the sole source of ecdysone in lepidopteran larvae, other tissues appear to have the capacity to synthesize
ecdysone in other insects at specific developmental stages (e.g., ovaries,
oenocytes).

Aside from the general biological implications of the above, the prothoracic glands were established as an excellent *in vitro* system for the
study of ecdysone biosynthesis since: (1) ecdysone was the only ecdysteroid synthesized and secreted by the glands, and (2) the glands were
incapable of hydroxylation at the C-20 position to form 20-hydroxyecdy-
sone, the other major insect ecdysteroid. The studies to be discussed are
for the most part the result of a collaborative effort between this labo-

ratory and that of D. H. S. Horn at CSIRO. The latter group conducted radiochemical synthesis of putative precursors, while the ability of the prothoracic glands to incorporate the compound into ecdysone as well as the stoichiometry of conversion were determined by this lab (see Smith *et al.*, 1980). Using the data from *in vivo* studies as a base, the first studies attempted to demonstrate the incorporation of [^{3}H]- and [^{14}C]cholesterol into ecdysone, but significant amounts of incorporation were never observed, notwithstanding reports that glands from insects of other orders can accomplish the conversion *in vitro*. Indeed, little is known about the early steps in ecdysone biosynthesis except that these steps must involve the establishment of the 7,8-double bond characteristic of ecdysone, saturation of the 5,6-double bond of cholesterol, the addition of a keto group at C-6, and *cis* fusion of the A/B rings. One or more of these steps may take place outside of the prothoracic glands, perhaps explaining our inability to observe the conversion of cholesterol into ecdysone by the prothoracic glands *in vitro*. The demonstration of relatively large quantities of 7-dehydrocholesterol in the cockroach prothoracic glands by the Beltsville group suggests that if our premise is true, saturation of the 5,6-double bond of cholesterol follows formation of the 7,8-double bond and that 7-dehydrocholesterol is formed elsewhere and is specifically sequestered by the prothoracic glands. If so, it should be recoverable from the hemolymph where it is probably bound to a protein. The inherent instability of 7-dehydrocholesterol makes it difficult to work with although its conversion *in vivo* to 20-hydroxyecdysone has been demonstrated in *Calliphora* by Horn's laboratory. If 7-dehydrocholesterol is a precursor for ecdysone, one might expect that it would stimulate the rate of ecdysone synthesis by *Manduca* prothoracic glands *in vitro*, but it does not. It is of interest in this regard that at least two insects lack the ability to form the Δ^7 and thus require a Δ^7 sterol (or $\Delta^{5,7}$ diene) in their diet rather than cholesterol.

The first positive *in vitro* studies examining the sequence of hydroxylations used two labeled putative precursors: a 5β-ketol (a compound having the Δ^7 and basic side chain of ecdysone but only one hydroxyl group at C-3; 3β) and a 5β-ketodiol (same as the 5β-ketol but with an additional hydroxyl group at C-14; 3β, 14α). These compounds were individually incubated with very active, last-instar larval prothoracic glands, and the results revealed an efficient conversion of the ketodiol to ecdysone as was found in *Calliphora in vivo*. The ketol, however, was not hydroxylated to ecdysone but rather to 14-deoxyecdysone (identical to ecdysone except for the absence of the hydroxyl group at C-14). This suggests that the ketodiol is a natural intermediate although it has not yet been identified as an insect natural product. It also indicates that the failure of the prothoracic glands to hydroxylate the ketol at C-14 may be due to the lack of enzyme for this hydroxylation or that this hydroxylation occurs at an earlier step in the pathway, e.g., at the time of introduction of the 7,8-double bond or the formation of the C-6 ketone.

Analogous studies with 5α-ketodiol, a 5α-ketoriol, and a ketodiene-diol (Bollenbacher *et al.*, 1979) revealed a low level of metabolism of the 5α-ketosteroids by the prothoracic glands and suggested that only the 5β configuration is recognized by the enzyme complex responsible for introducing the hydroxyl group at C-2. This is in contrast to the enzymes that mediate the introduction of hydroxyl groups into the side chain, in which case the 5α and 5β epimers can serve as equal substrates. Indeed, if a 5α precursor exists in the prothoracic glands, it must function early in the biosynthetic scheme since the glands do not appear to possess an epimerase capable of converting a 5α-6-keto-Δ^7 sterol to the 5β form.

These and other investigations have not yielded as many conclusions as one would wish, but do indicate that there is a preferred order of side-chain hydroxylations, rather than the hydroxylations being random events. In essence, we can state that these *in vivo* and *in vitro* experiments conducted for the most part by three groups (USDA at Beltsville; CSIRO at Melbourne; University of North Carolina) are tedious and difficult, and that our knowledge is still fragmentary.

It has been demonstrated that the ovaries of a variety of insects have the capacity to synthesize ecdysone, which presumably plays a regulatory role in oogenesis and/or embryogenesis. Hoffmann's laboratory in Strasbourg has analyzed the ecdysteroids in the oocytes and ovarian tissue of the locust by chromatographic and mass spectrometrical analyses in an attempt to identify biosynthetic intermediates. Aside from the definitive identification of cholesterol and ecdysone, four ecdysteroids less polar than ecdysone were tentatively identified as: (1) 2,14,22,25-tetradeoxy-ecdysone; (2) 2,22,25-trideoxyecdysone; (3) 2,22-bis-deoxyecdysone; and (4) 2-deoxyecdysone. Based only on the order of increasing polarity, the authors suggest a biosynthetic sequence of cholesterol → (1) → (2) → (3) → (4) → ecdysone. This indicates that hydroxylation at C-2 in ring A occurs *after* hydroxylation is complete in the side chain, the sequence of hydroxylation being C-14 → C-25 → C-22 → C-2, in contrast to the sequence noted above for the studies involving *in vitro* biosynthesis of ecdysone from putative radiolabeled precursors by the prothoracic glands. Although the order of hydroxylations has not been definitively elucidated for either the locust ovary or the *Manduca* prothoracic glands, it would not be surprising if alternative pathways are available in different tissues, and particularly in species well separated by evolutionary time.

Ecdysone biosynthesis remains an important but unanswered question for those interested in insect metamorphosis. However, the use of new approaches such as those noted above will surely result in the establishment of one or more definitive routes in the next few years.

3.2.2. 20-Hydroxyecdysone–Ecdysone Monooxygenase

There is no question that ecdysone is the product of the prothoracic glands during larval–pupal and pupal–adult metamorphosis and that it is converted by hydroxylation to 20-hydroxyecdysone in other tissues such as the fat body. Before discussing this important conversion, the physiological relationship between these two ecdysteroids should be considered. Although ecdysone has numerous effects *in vivo*, it is probable that they are a result of hydroxylation of ecdysone to 20-hydroxyecdysone. Indeed, we now believe that ecdysone is primarily a prohormone and that 20-hydroxyecdysone is the principal MH of insects and probably all arthropods. This conclusion is based on the following observations: target tissues not capable of the conversion react *in vitro* to physiological concentrations of 20-hydroxyecdysone but require much greater than physiological amounts of ecdysone; in some insects the conversion is so rapid that the amount of ecdysone present is barely detectable; ecdysteroid receptors have more than a 1000-fold greater affinity for 20-hydroxyecdysone than for ecdysone (see O'Connor and Chang, Fristrom, this volume). This situation appears to be analogous to a variety of mammalian systems; e.g., vitamin D in the blood is converted to the hormonally active form, 1,25-dihydroxyvitamin D_3, by hydroxylations in the liver and kidney, A number of studies have demonstrated a changing ratio of ecdysone : 20-hydroxyecdysone in the hemolymph at particular developmental stages, and the possibility remains that a specific ratio of the two ecdysteroids is necessary to elicit a specific effect *in vivo*. In any case, the synthesis of 20-hydroxyecdysone from its substrate, ecdysone, is vital to the insect.

Since PTTH acts on the prothoracic glands to stimulate ecdysone biosynthesis in a manner analogous to the action of ACTH on the vertebrate adrenal glands, and since the adrenal cortex performs steroid hydroxylations by utilizing cytochrome P-450 oxidases located in both the mitochondria and the microsomes, it is not surprising that the adrenal steroid hydroxylation system was used as a model for investigating the ecdysone C-20 hydroxylase. This section will deal primarily with the work conducted by our laboratory on *Manduca* fat body (Smith *et al.*, 1979, 1980) although significant contributions have been made by other laboratories as well (Strasbourg, Beltsville, Liverpool).

With the synthesis of high-specific-activity [^{3}H]ecdysone, it became feasible to develop a radioenzymological assay in which [^{3}H]ecdysone is converted to [^{3}H]20-hydroxyecdysone, the two ecdysteroids separated by TLC, and the [^{3}H]20-hydroxyecdysone counted in a liquid scintillation spectrometer. Before such an assay could be used in a routine way, the ^{3}H-labeled product that migrated with standard 20-hydroxyecdysone during TLC was further separated by HPLC and unequivocally identified as 20-hydroxyecdysone by mass spectrometry and other procedures. The initial studies with various tissue homogenates utilized *in vitro* assay conditions developed for vertebrate adrenal steroid hydroxylases (monooxygenases) and included NADPH as the electron donor as well as [^{3}H] ecdysone. The results revealed that the larval fat body was the most potent source of 20-hydroxylase activity; because of its central role in the intermediary metabolism of insects, further studies utilized this tissue. Although the ultimate goal of this study was to examine the regulation of enzymatic activity and therefore to indirectly analyze how differing ratios of ecdysone:20-hydroxyecdysone are established, the two immediate aims were characterization of the enzyme system and subcellular localization.

Using fat body homogenates, the *in vitro* assay system was optimized (e.g., pH, buffer, incubation time) and the reaction was shown to be dependent on the addition of NADPH or a system that generates NADPH. The latter requirements indicated a monooxygenase (mixed-function oxidase) reaction (Fig. 3), and the fat body was then fractionated to determine the subcellular location of the enzyme system. By the use of differential centrifugation and sucrose gradient centrifugation, it appeared that most of the activity was localized in the mitochondrial fractions. Although mitochondrial monooxygenases had been described in a variety of vertebrate tissues including the adrenal cortex, none had been reported for any invertebrate and it was therefore necessary to corroborate this finding by ensuring that the mitochondrial fractions derived from the centrifugation procedures consisted primarily of mitochondria. This was accomplished in two ways: (1) electron microscopic examination of the various fractions, and (2) the use of marker enzymes, e.g., cytochrome c oxidase for mitochondria. The resulting data supported the supposition that the steroid hydroxylase system responsible for the conversion of ecdysone to 20-hydroxyecdysone resided in the mitochondria.

Some kinetic properties of this monooxygenase were determined (Table 1) and the apparent K_m calculated to be 1.6×10^{-7} M, which is in the range of the highest endogenous ecdysteroid concentration noted in last-instar *Manduca* larvae. In terms of regulation, it is of interest that 20-hydroxyecdysone is a competitive inhibitor of the reaction with an apparent K_i of about 2.7×10^{-5} M. Since this is well above the physiological range, it is unlikely that the C-20 hydroxylation is regulated by

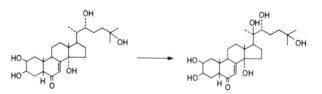

FIGURE 3. The conversion of ecdysone (left) to 20-hydroxyecdysone (right) by the mitochondrial, P-450 ecdysone 20-monooxygenase.

TABLE 1. SUMMARY OF THE ENZYMATIC
PROPERTIES OF *M. SEXTA* FAT BODY ECDYSONE
20-MONOOXYGENASE[a]

Optima
  Buffer: 0.05 M sodium phosphate
  Temperature: 30°C
  Cations: none stimulatory, Mg^{2+} inhibitory
  pH: 7.4–7.5
Cofactors
  NADPH or NADPH-generating system optimal
  NADH 10% as effective as NADPH
  Krebs cycle intermediates, ATP and ADP ineffective
Kinetics
  Hyperbolic saturation kinetics for ecdysone
    S_{90}/S_{10} = 77–83
    $K_m = 1.6 \times 10^{-7}$ M
    V_{max} = 2.5 ng/min/mg protein
  Competitive inhibition kinetics for 20-hydroxyecdysone
    $I_{50} = 1.3 \times 10^{-5}$ M
    $K_i = 2.7 \times 10^{-5}$ M

[a] From Smith *et al.*, 1980.

product inhibition unless localized accumulations occur, e.g., in the mitochondria, that exceed 10^{-5} M.

Since it has been demonstrated that certain vertebrate steroid hydroxylases are capable of oxygenating several substrates, i.e., one enzyme system may perform several physiological functions, it was of interest to probe this possibility for the ecdysone 20-monooxygenase. This was investigated in a series of analyses in which a variety of ecdysteroids were individually tested by adding increasing quantities to the standard radioenzymological assay. This results in a competition for hydroxylation by the enzyme between the added ecdysteroid and the [^{3}H]ecdysone. The results of this competition study indicated that the specificity of the insect enzyme depends on the number and position of the hydroxyl groups and the stereochemistry of the A/B rings and side chains of the competing ligand. 2,25-Dideoxyecdysone had a greater affinity for the active site of the enzyme than did ecdysone, although the latter compound was more effective than six other ecdysteroids. Polarity and isomerism appeared critical and since ecdysone was not the most effective competitor, it appears that the ecdysone 20-monooxygenase may function in other steroid hydroxylation reactions as well.

Having established the above, it was of interest to learn if this mitochondrial enzyme system included a cytochrome P-450 component, particularly since this is typical of vertebrate steroid hydroxylases and a mitochondrial cytochrome P-450 had not been described for any invertebrate. Kinetic studies with known inhibitors of vertebrate P-450 steroid hydroxylases (e.g., metyrapone, *p*-aminoglutethimide) revealed similar dose-dependent inhibition of the insect enzyme. It is well known that P-450 enzymes bind CO and that the complex exhibits a Soret band at 450 nm. When such CO difference spectra were obtained, the ecdysone 20-monooxygenase yielded a major Soret band at 450 nm. This indicated the presence of a heme P-450 molecule, albeit at a low concentration relative to what is noted for adrenal cortex mitochondrial preparations. In general, studies on the characterization of mitochondrial and microsomal cytochrome P-450 monooxygenases usually cease at this point and data such as those noted above are considered adequate to prove the involvement of a cytochrome P-450 system in the reaction under analysis. In reality, however, the above only indicates the presence of a P-450 in the assay mixture but does not prove that the P-450 is the terminal oxidase in the reaction under study, i.e., ecdysone monooxygenation. For such an unequivocal demonstration, one must obtain a photochemical action spectrum.

Although a rather complicated procedure conducted by only a few laboratories in the world, obtaining a photochemical action spectrum basically consists of measuring the reversal of CO inhibition of the monooxygenase reaction by bands of monochromatic light. This procedure cannot be conducted with a spectrophotometer since the wavelengths of light impinging on the reaction vessel at constant temperature must be of equal intensity. When a photochemical action spectrum was determined from 400 to 500 nm for the fat body ecdysone 20-monooxygenase, maximal reversal of CO inhibition of enzyme activity occurred with 450-nm light. This was the first photochemical action spectrum reported for any invertebrate steroid hydroxylase and clearly established the ecdysone 20-monooxygenase as a mitochondrial, cytochrome P-450 enzyme. This novel finding was of interest both because of the importance of the enzyme system in modulating the MH titer and because of the close relationship established between the mechanisms of steroid hydroxylation in vertebrates and invertebrates. The latter observation underlines the postulate that evolutionary forces preserve successful mechanisms, i.e., biochemical reactions in the mitochondria. Indeed, it has been proposed that the biochemical electron transport system involved in steroid biosynthesis (e.g., P-450 monooxygenases) may be of primordial origin (Sandor *et al.*, 1975), and therefore it is not surprising that the insect utilizes such a system.

It is of obvious importance for the insect to have the capacity to regulate its titer of MH, and the actual titer is influenced by: the rate of ecdysone biosynthesis by the prothoracic glands; hormone compartmentalization; the rate of hormone degradation; and, of course, by the activity of the ecdysone 20-monooxygenase. Whether the enzyme varies in activity during development is not known with certainty nor has its regulation been studied in detail. In the case of *Manduca*, the quality as well as quantity of the hemolymph ecdysteroids changes during development (see Smith *et al.*, 1980). For example, fourth-instar hemolymph contains

20-hydroxyecdysone almost exclusively, whereas fifth (last)-instar hemolymph contains substantial quantities of ecdysone, but mainly 20-hydroxyecdysone. The ecdysteroid peak responsible for initiating pupal–adult development is comprised predominantly of ecdysone. It therefore appears that the temporal sequence of ecdysone 20-monooxygenase activity is fourth-instar larva > fifth-instar larva > pupa. It remains to be investigated whether this activity is modulated by neurohormones or other regulatory molecules (e.g., JH) or indeed if it reflects actual enzyme synthesis rather than activation. This may become a major research area in itself, particularly since the activity of the enzyme appears to change dramatically within an instar, e.g., rising from Day 1 to Day 4 in *Manduca* fat body and then dropping precipitously at Day 7, the period of peak ecdysteroid titer.

3.3. Ecdysteroid Transport

Once ecdysone is synthesized by the prothoracic glands it is secreted into the hemolymph and transported to tissues such as the fat body to be converted to 20-hydroxyecdysone, which also must be transported in the hemolymph to target tissues. Although the relatively high solubility of ecdysteroids in aqueous media precludes the necessity for hemolymph carrier molecules (i.e., lipoproteins), it is possible that hemolymph macromolecules exist that bind ecdysteroids so as to prevent degradation and nonspecific uptake as in the case of JH (see Section 4.3). This possibility has been examined in several insects utilizing [³H]ecdysone or [³H]20-hydroxyecdysone, but an apparently saturable binding protein has only been demonstrated in the case of locust hemolymph after incubation with high-specific-activity [³H]20-hydroxyecdysone (Feyereisen, 1977). In the other instances, the fact that the binding was not saturable indicated a lack of specificity and/or low-affinity interactions. With the locust, however, the quantity of MH bound was inversely proportional to the endogenous ecdysteroid content; i.e., the most hormone bound in *in vitro* incubations was with hemolymph from a stage when the endogenous titer was lowest. The fact that the binding factor was a protein was demonstrated by incubation of the complex with ribonuclease or protease as well as heat treatment. In the latter two cases, binding was substantially impaired, indicating the existence of a 20-hydroxyecdysone–binding protein complex. Preliminary analysis suggests a molecular weight greater than 250,000, about 10 times that of the JH binding protein (see Section 4.3). Although the endogenous hormone titer of the stages studied exceeds the capacity of the binding protein by an order of magnitude (the reverse of the situation with the JH binding protein), the very low affinity of the molecule for other ecdysteroids, including ecdysone, indicates high specificity. Unlike the JH binding protein, no physiological role has yet been elucidated for this macromolecule.

3.4. Ecdysteroid Degradation

Although the conversion of ecdysone to 20-hydroxyecdysone may be an important regulatory point for alterations in the titer of MH, the data thus far accumulated indicate that inactivation mechanisms may be even more critical in establishing the titer of 20-hydroxyecdysone (Young, 1976). There have been a plethora of studies in which [³H]ecdysone was injected into a particular insect and the labeled metabolites isolated by thin-layer chromatography. Usually these metabolites were noted as more or less polar than the original injected substrate, were or were not conjugates, and have not been identified definitively. It has been shown that more metabolically active developmental stages appear to degrade and excrete the injected [³H]ecdysone at a more rapid rate. More precise information has been garnered in a few cases, notably by Koolman and Karlson (1978; Koolman, 1978) studying the fly, *Calliphora vicina*, and by the USDA Beltsville laboratory utilizing *Manduca* (Nigg *et al.*, 1974; Kaplanis *et al.*, 1980). These studies indicate that 20-hydroxyecdysone can be inactivated by hydroxylation; epimerization and conjugation with the products being 20,26-dihydroxyecdysone, 3-dehydro-20-hydroxyecdysone, 3-epi-20-hydroxyecdysone, and a variety of esters (glucuronides, glucosides, sulfates).

Ecdysone oxidase mediates the oxidation of ecdysone and 20-hydroxyecdysone to 3-dehydroecdysone (K_m = 9.8 × 10^{-5} M) and 3-dehydro-20-hydroxyecdysone (K_m = 3.1 × 10^{-5} M), respectively, plus H_2O_2 It is a soluble enzyme isolated from the fat body and gut of Diptera, Lepidoptera, and Dictyoptera, with most of the work conducted on *Calliphora*. Most organisms would utilize 3β-dehydrogenases to accomplish this oxidation, a reaction that would be reversible. Insects may be unique among eukaryotes in employing an oxidase that, for practical purposes, mediates an irreversible reaction.

When the entire development of this blowfly was analyzed for ecdysone oxidase activity, the highest specific activity was noted in fertile eggs. Enzymatic activity then declined to almost undetectable levels by the end of embryonic development, rose slightly during larval life, increased dramatically to about one-third the egg value at metamorphosis to the pupa, and remained at about that level through metamorphosis to the adult and in adult flies. It is of interest that when the ecdysteroid content of *Calliphora* was determined by RIA, the resulting developmental curve approximated that obtained for ecdysone oxidase activity, i.e., the largest peak in the egg and greatest increase at the time of pupariation. These results indicate that the enzyme may play a major role in MH inactivation, a premise supported by the observations that the 3-dehydro products are relatively inactive in bioassay and the reaction is essentially irreversible. The fact that the apparent K_m of the ecdysone oxidase is two or more orders of magnitude greater than that of the ecdysone monooxygenase suggests that in the presence of both enzymes, physiological amounts of ecdysone may be preferentially converted to 20-hydroxyecdysone rather than to 3-dehydroecdysone. The possible physiological significance of this "ecdysone" oxidase awaits further studies by the Marburg scientists.

The second degradative enzyme that has been studied in some detail is the ecdysone 3-epimerase (ecdysone dehydrogenase-isomerase; 3-dehydroecdysone reductase), which is a soluble enzyme derived from the midgut of *Manduca* larvae. Since the end product of the reaction mediated by this enzyme is 3-epiecdysone, an ecdysteroid with less than 20% of the MH activity of ecdysone, one can conclude that it is a degradative reaction. Preliminary substrate specificity studies indicate that the C-25 hydroxyl is requisite for substrate efficacy. The reaction requires oxygen and NADH and NADPH, similar to the requirements for the ecdysone 20-monooxygenase and ecdysone oxidase, and has an apparent K_m of 1.7 $\times$ 10^{-7} M for ecdysone and 4.7 $\times$ 10^{-7} M for 20-hydroxyecdysone. The oxygen is required for the conversion of ecdysone to 3-dehydroecdysone, and the reduced coenzyme is utilized for the reduction of the latter intermediate to 3-epiecdysone. When 20-hydroxyecdysone is the substrate, the intermediate would be 3-dehydro-20-hydroxyecdysone. The lack of success thus far in identifying either intermediate ecdysteroid suggests that they are short-lived. Like the ecdysone oxidase reaction, the formation of the epimer is basically irreversible.

The combined research of the Marburg and Beltsville laboratories suggests that the degradation of MH proceeds by oxidation to yield the 3-dehydro-20-hydroxyecdysone, which is then reduced to yield the epimer. It should be noted, however, that: (1) the real physiological significance of the oxidase remains conjectural; (2) the proposed intermediate in the epimerase reaction has not yet been identified as a natural product and this is also conjecture; and (3) the epimerase has only been studied in detail in *Manduca* and the oxidase only in *Calliphora*. Since only a handful of laboratories have explored the question in any detail and only two are dedicated to the problem, it is perhaps not surprising that progress in this area is so slow.

The identification of ecdysteroid conjugates in a variety of insects has led to the concept that conjugation is a valid means of degradation prior to excretion. However, it should be noted that these reactions are usually reversible and, therefore, that conjugation may be a means of sequestering ecdysone and 20-hydroxyecdysone for use as regulatory molecules at a subsequent stage of development. This may be particularly true in the case of conjugate sequestration in the oocyte during vitellogenesis and hydrolysis and utilization during embryonic development.

Heinrich and Hoffmeister (1970) identified the glucosides of 20-hydroxyecdysone and 25-deoxy-20-hydroxyecdysone in the fat body of *Calliphora* larvae, probably derived from the action of a glucosyltransferase that converts the ecdysteroid plus UDPglucose to the ecdysteroid 3β-glucoside plus UDP. The observation by the Beltsville laboratory that the meconium (excreted upon adult emergence) from adult *Manduca* moths contains significant quantities of nonecdysteroid sterol sulfates (e.g., cholesterol sulfate) suggested that sulfation may also be a means of ecdysteroid degradation. (It should be noted that sterol sulfates may be the precursors of vertebrate steroid hormones.) The larval gut tissue of the southern armyworm (Yang and Wilkinson, 1972) yielded an enzyme system requiring ATP that catalyzed the sulfation of ecdysone and other sterols and synthetic substrates. This is presumably a sulfotransferase that mediates the conversion of ecdysteroid + phosphoadenylsulfate to ecdysteroid sulfate + adenosine-3′, 5′-diphosphate. This soluble enzyme is similar in properties to a variety of analogous mammalian enzymes. Although the specificity of this enzyme to ecdysteroids is low compared to other steroids, the tentative identification of sulfate esters of ecdysone, 20-hydroxyecdysone, 3-dehydroecdysone, and 3-dehydro-20-hydroxyecdysone in locusts by the Marburg laboratory indicates that sulfation may be an important means of inactivating ecdysteroids prior to excretion. Finally, there is suggestive evidence for a role of the glucuronasyltransferases in ecdysteroid conjugation, but this enzyme is even less specific than the sulfotransferases.

It is important to note that in the entire sequence of synthetic reactions, beginning with cholesterol and ending with ecdysone and then 20-hydroxyecdysone, and the sequence of degradative steps giving rise to 3-dehydroecdysteroids, 3-epiecdysteroids, and conjugates thereof, almost nothing is known regarding possible control mechanisms. Although PTTH acts to stimulate ecdysone biosynthesis, presumably by acting at one or more levels between cholesterol and ecdysone, even its mode of action is really not understood. Because of the importance of maintaining and dissipating critical MH titers to the evolutionary and ecological success of arthropods, it is important that investigators begin to chemically dissect the systems noted in this section, as well as still unknown metabolic paths, so as to understand the fine control of metamorphosis at the biochemical level.

4. JUVENILE HORMONES

4.1. Chemistry and Occurrence

Although the initiation of molting can be ascribed to the MH, the character of the molt, i.e., the larval–larval or larval–pupal transformation, is influenced by another hemolymph-borne factor, JH (see Granger and Bollenbacher, this volume, and Fig. 1). Discovery of a rich source of JH in the abdomens of male *H. cecropia* moths by Williams (1956) coupled with the development of extremely sensitive bioassays made possible its chemical characterization. From 875 *H. cecropia* abdomens, Röller and Dahm (1968) isolated and purified 300 μg of the hormone with an overall purification factor of 125,000. Nuclear magnetic resonance and mass spectral analysis yielded the primary structure. The presence of double bonds prompted the study of the stereochemistry of the molecule, which was determined by comparison of the naturally occurring hormone with a synthetic series of all possible geometrical isomers. The fully characterized hormone was assigned the structure methyl (2E,6E,10cis)-(10R,11S)-10,11-epoxy-7-ethyl-3,11-dimethyl-2,6-tridecadienoate (Fig. 4).

Soon after the discovery of the first JH (JH I), the presence of a second hormone, JH II, was detected in lower concentrations (13–20%) in the *H. cecropia* extract. Another homolog, JH III, was isolated as the secretory product of the corpora allata of *M. sexta* cultured *in vitro*. Remarkably enough, the latter compound was first synthesized as a hormone analog a decade earlier, and was only later recognized as a naturally occurring homolog of the JH series. The homologs differ only in the length of the alkyl side chain (Fig. 4).

4.2. Structure–Activity Relationships

Current dogma holds that in order to elicit a biological response, a hormone must be recognized by specific receptor molecules within the target tissue. Such highly selective recognition implies that the size, conformation, and functional groups of the hormone are crucial in receptor–hormone interaction and subsequent biological activity. Insight into the structural requirements for hormonal activity can best be gained

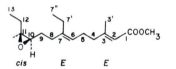

JH I

Methyl (2E,6E,10cis)-(10R,11S)-10,11-
epoxy-7-ethyl-3,11-dimethyl-2,6-trideca-
dienoate

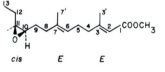

JH II

Methyl (2E,6E,10cis)-(10R,11S)-10,11-
epoxy-3,7,11-trimethyl-2,6-tridecadi-
enoate

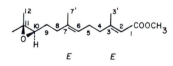

JH III

Methyl (2E,6E)-(10R)-10,11-epoxy-3,7,11-
trimethyl-2,6-dodecadienoate

FIGURE 4. The naturally occurring JH.

by examining hormone analogs in biological assays (Sláma *et al.*, 1974). Although a compound's activity may be influenced by factors such as rate of penetration, metabolism, mode of application, and species used, some broad generalizations have been developed. To exhibit morphogenetic activity, i.e., to inhibit metamorphic processes, the basic carbon chain must contain at least 10 $-CH_2-$ units, with an optimal length of 14–16 units. Although length is important, the total length of the molecule may be less significant than the intramolecular distances between important functional groups. Thus, addition of a methylene group between carbons 7 and 8 will have a different effect than the same modification between carbons 3 and 4, although the length of the chain will be the same. A large number of the active JH analogs, as well as the three naturally occurring hormones, have a branched carbon chain typical of the acyclic isoprenoid compounds. The characteristic side chain at C-7 and C-11 can be repositioned with only a modest decrease in biological activity; however, the loss of the side chain at C-3 will lead to nearly complete inactivation of the analog.

The naturally occurring hormones have three unsaturated positions, the $\Delta^{2,3}$, $\Delta^{6,7}$, and $\Delta^{10,11}$, which contribute significantly to biological activity. The geometrical configuration about the bonds is also important. The naturally occurring hormones have a 2*E*,6*E*,10*cis* (JH I and JH II) or a 2*E*,6*E* (JH III) configuration, both of which have been shown to be the most biologically active of the geometrical isomers. The other configurations display varying degrees of decreased activity, the 2*Z*,6*Z*,10*trans* isomer (the direct opposite of the natural hormone configuration) exhibiting the least activity. The chiral centers at the C-10 and C-11 positions in the naturally occurring hormones have been assigned a 10*R*,11*S* (JH I and JH II) or a 10*R* (JH III) configuration. It has been suggested that enantiomeric purity of the hormones is important for biological activity in some insects. Although biological studies have only partially resolved this question, recent evidence using the hemolymph JH binding protein from *M. sexta* indicates that in the case of JH III, the natural enantiomer, 10*R*, is the only enantiomer bound by the protein (see Section 4.3).

In addition to these requirements of size and shape, several functional groups are necessary to confer biological activity. In all three homologs, the C-1 position has been esterified with a methyl group, which significantly enhances the biological activity of the hormone. The loss of this substituent results in the formation of the apparently biologically inert JH acid (JHA). Conversely, expanding the terminal methyl to an ethyl or isopropyl enhances biological activity; however, the increased activity may be due not so much to enhanced interaction with the target cell as to less susceptibility to esterolytic cleavage by hemolymph esterases.

The epoxide moiety at the C-10,11 position makes the JH homologs unique among animal hormones. The loss of this functional group through hydration results in a biologically inactive metabolite, JH diol.

Biological studies have led to the conclusion that this functional group is necessary for JH activity throughout a wide range of insect species.

Structure–activity relationships within the homologous series are complicated by the hormone's dual functions, i.e., a reproductive as well as a morphogenetic hormone. The conspicuous absence of JH I and JH II during the reproductive phase of most insects has prompted speculation that JH I and JH II are morphogenetic hormones while JH III is a gonadotropic hormone (Lanzrein et al., 1975). Although it has been demonstrated that JH III has considerably less morphogenetic activity than the higher homologs, there is no evidence that JH III is more active as a gonadotropin. Indeed, exogenously applied JH I and JH II can replace and surpass JH III as a gonadotropin in allatectomized adult females. Moreover, early experiments involving transplantation of corpora allata from larva to adult and from adult to larva revealed that the products of the corpora allata during both the developmental and the reproductive periods were quite similar, at least in function if not in structure. Thus, differences in side-chain length do not appear to lead to a differential biological response.

4.3. Transport

Despite their evolutionary divergence better than 300 million years ago, insects have developed systems analogous to those of vertebrates that promote the conservation and maintenance of circulating hormone titers. Hemolymph and serum hormone binding proteins play important roles in the protection and transport of lipophilic hormones like the JH. Although soluble at relatively high levels (10^{-5} M) in an aqueous medium, JH are surface active and therefore readily partition into less polar environments such as basement membranes, fat body, etc. In order to move from the site of synthesis to target tissues through the hemolymph, JH are bound to specific macromolecules in the circulatory system.

The physicochemical interaction of JH with macromolecules in the hemolymph can be divided into two classes, based on affinity and binding capacity. The first is represented by the lipoproteins and is characterized by low affinity, low specificity, high molecular weight (>100,000), and high capacity for JH (more than one binding site for JH per protein molecule). The other category of proteins exhibits high affinity, high specificity, low molecular weight, and low binding capacity (one JH binding site per protein molecule).

4.3.1. Lipoproteins

Lipoproteins are the predominant proteins in the insect hemolymph (Gilbert and Chino, 1974) and may be considered functionally equivalent to the vertebrate serum albumins. Since these proteins bind a number of low-molecular-weight lipophilic compounds, it was this group of proteins that was first examined for protein–hormone interaction (Whitmore and Gilbert, 1972). Larval *Manduca* hemolymph contains at least six lipoproteins as determined by polyacrylamide gel electrophoresis, and one of these binds radiolabeled JH *in vivo* and *in vitro*. Since this protein contains a relatively high proportion of protein to lipid, it is classified as a high-density lipoprotein (HDL). Besides binding JH and JHA, the HDL binds a number of fatty acids as well, indicating that the protein–ligand interaction is nonspecific. Although the affinity, or binding strength, of this HDL for JH is relatively low ($K_a = 10^5$ M^{-1}), it can bind a number of JH molecules at the same time ($n \geq 1$) (Gilbert *et al.*, 1976). Thus, a high concentration of this protein in *Manduca* hemolymph may give rise to significant short-term interaction with JH, especially near the corpora allata. However, such interactions probably do not persist as the hormone circulates through the hemocoele since lipoproteins offer little protection from hemolymph esterases, which can inactivate the hormone (see Section 4.5). The low specificity and relatively low affinity of JH interactions with the HDL in *Manduca* further minimize the role of lipoproteins in overall JH titer regulation and transport in Lepidoptera, but lipoproteins may play important roles in other insects such as hemipterans.

4.3.2. JH-Specific Binding Proteins

The hemolymph of certain insects contains not only nonspecific, low-affinity lipoproteins that can bind JH, but a specific, relatively high-affinity JH binding protein (JHBP) as well (Fig. 1). This type of protein was first observed in the hemolymph of larval *Manduca* and has now been reported in a number of lepidopteran species (Gilbert *et al.*, 1980).

The initial *in vivo* studies on JHBP employed two strategies to determine the fate of JH in the hemolymph. First, radiolabeled JH I was injected into larvae and allowed to incubate *in vivo* before hemolymph was collected. The radiolabeled hemolymph was then fractionated by either gel filtration or polyacrylamide gel electrophoresis, and most of the label was found to be complexed to a single low-molecular-weight protein. When analyzed chemically, the radiolabeled compound bound to the protein proved to be JH, while unbound label consisted of metabolites of JH, e.g., JHA, JH diol, and JH acid–diol.

The second approach assumed that if this protein served a transport function, endogenous hormone would also be complexed to it. To test this hypothesis, the low-molecular-weight protein was rapidly purified from the hemolymph using preparative slab gel electrophoresis, and the area containing the protein was treated with organic solvent to extract the bound hormone from the protein. This material was bioassayed with the *Galleria* wax test and displayed considerable morphogenetic activity (Table 2). These studies revealed the existence *in vivo* of a specific binding protein for JH that is probably responsible for transporting the hormone in the hemolymph (Goodman *et al.*, 1978).

TABLE 2. JH BIOASSAY OF HEMOLYMPH
BINDING PROTEINS FROM *MANDUCA*
LARVAE[a,b]

Lipoprotein extract	1.30 ± 0.03
Lipoprotein extract (1:10 dilution)	0.80 ± 0.08
Lipoprotein extract (1:100 dilution)	0.14 ± 0.10
JHBP	2.50 ± 0.19
JHBP (1:10 dilution)	1.50 ± 0.37
JHBP (1:100 dilution)	0.50 ± 0.19
Acrylamide (no protein)	0.28 ± 0.14

[a] From Goodman *et al.*, 1978.
[b] The score range of the *Galleria* wax text is from 0 to 3, where 0 = no effect, 1 = minimal effect, 2 = intermediate effect, 3 = maximal effect.

4.3.3. Purification of JHBP

To better understand this unique molecule, methods were developed for its purification. JHBP was purified to homogeneity from 300 ml of late-fourth-instar hemolymph by a combination of gel filtration, ion-exchange and hydroxylapatite chromatography, and preparative gel electrophoresis with an overall purification of 240-fold (Goodman *et al.*, 1978). Homogeneity of the protein was established by analytical gel electrophoresis, sodium dodecyl sulfate–urea gel electrophoresis, and immunoelectrophoresis. Some of the characteristics of the protein are listed in Table 3. A comparison of vertebrate serum hormone binding proteins with the JHBP indicates that JHBP has a lower molecular weight than its vertebrate counterparts and a lower affinity for its specific ligand. Moreover, JHBP does not appear to be a glycoprotein as are many of the vertebrate serum transport proteins. Similarities between the proteins include one binding site and one polypeptide chain per molecule.

4.3.4. Specificity of JHBP

Since a number of compounds exhibit JH activity in bioassays, the specificity of JHBP was examined. Several techniques have been devel-

TABLE 3. SOME PROPERTIES OF THE
M. SEXTA JHBP[a]

Molecular weight	
SDS gel electrophoresis	28,000
Gel filtration	29,500
Sedimentation equilibrium	28,000
$S_{20,w}$	2.2
pI	5.0
Binding sites per molecule (n)	1.0
Subunits	1

[a] From Goodman *et al.*, 1978.

oped for the analysis of hormone binding by proteins, of which the most widely used is the competitive protein-binding assay. A brief outline of the protocol for this assay may be useful. The protein is stripped of hormone, then incubated with saturating amounts of [^{3}H]-JH I in the presence of either unlabeled JH I or competitor, and a slurry of hydroxylapatite is then added. The latter specifically binds and precipitates the protein, affecting a separation of bound and unbound hormone. The pellet of hydroxylapatite containing the protein is washed several times and then radioassayed to determine the amount of hormone bound. Low levels of radioactivity bound signify that the competitor is recognized by JHBP and has displaced JH I, while high levels of label indicate that the protein does not bind the competitor.

Using this technique, it was demonstrated that the protein is highly specific for the JH homologs; within the homologous series, there are differences in affinities based on the hydrophobicity of the hormones. The affinity for JH I, the least polar of the naturally occurring hormones, was greater than that for JH II and approximately 10-fold greater than that for JH III, the most polar of the series (Table 4). A wide range of other substances were also tested, including the morphogenetically inactive metabolites of JH I, JHA and JH diol, and several analogs possessing JH activity. Only those molecules with a carbon chain similar to JH and containing an epoxide and a methyl ester group in the proper positions displayed binding activity. Modification of the C-1 terminus with hydrophobic groups, e.g., ethyl or propyl, enhanced binding activity.

Equally important to binding to the JHBP is the geometry of the hormone. When the racemates of the seven geometrical isomers of JH I were compared, structural variation between the isomers was systematically reflected in their binding activity. Competition studies indicated that the configuration around the 2,3-double bond of JH is quite important; when the naturally occurring 2E isomer was replaced with the 2Z isomer, a marked reduction in binding was observed. Replacement of the 6E isomer with the 6Z configuration reduced binding, but to a lesser degree than the rearrangement at the 2 position. Changing the 10cis epoxide to the 10trans resulted in a modest decrease in binding. There is now evidence that stereochemistry about the epoxide also has an important role in binding. The affinity for the 10R (the naturally occurring enantiomer) is greater than that for the racemic mixture and far greater than for the 10S enantiomer (Table 4).

Based on the available evidence, the binding site can be envisioned as a hydrophobic groove or pocket on the surface of the protein with specific sites for the epoxide and methyl ester of JH. The primary binding forces are likely to be hydrophobic with the majority of interactions taking place along the alkyl side chains and with the terminal methyl group at C-1 of the JH molecule. Predictions based on this model suggest that any change in the hydrophobic face formed by the side chains and terminal methyl group would lead to decreased binding.

TABLE 4. SPECIFICITY OF THE *MANDUCA*
JHBP[a]

	Compound	Ratio of association constants
1.	Homologs, analogs, fatty acids[b]	
	JH I	1.00
	JH II	0.35
	JH III	0.08
	JH I acid	—
	ZR 515	—
	ZR 512	—
	R20428	—
	Precocene II	—
	Methyl epoxy palmitate	—
	Methyl epoxy stearate	—
	Palmitic acid	—
	Oleic acid	—
	Linolenic acid	—
2.	JH I geometrical isomers	
	2E,6E,10cis	1.00
	2E,6E,10trans	0.64
	2E,6Z,10trans	0.27
	2E,6Z,10cis	0.12
	2Z,6E,10cis	0.08
	2Z,6E,10trans	0.03
	2Z,6Z,10cis	0.02
	2Z,6Z,10trans	0.01
3.	JH III stereoisomers	
	JH III 10R	6.32
	JH III 10R,S	1.00
	JH III 10S	0.20

[a] From Gilbert *et al.*, 1980.
[b] Because each group had a different ratio (R = unbound/bound in the absence of unlabeled hormone) and separation of bound and unbound was performed differently, groups 1, 2, and 3 cannot be compared directly. Dashes signify no displacement at the highest concentration used.

4.3.5. Hemolymph Titer of JHBP

To determine the relative concentrations of binding sites, hence the concentration of binding protein, multiple equilibrium dialysis was performed (Goodman and Gilbert, 1978). Diluted hemolymph from different developmental periods was put into individual dialysis bags and the bags incubated together in a buffer containing [^{3}H]-JH. The concentration of unbound hormone inside and outside of the bags was allowed to reach equilibrium as measured by the number of counts in the buffer versus those in a bag of JHBP-free hemolymph and aliquots from each experimental bag were radioassayed. Thus, a hemolymph sample having twice the concentration of binding protein should show a corresponding increase in radiolabeled hormone.

By utilizing this technique, binding activity in the hemolymph of fourth-instar *Manduca* larvae was shown to be highest at the beginning of the instar, decreased about 24 hr after ecdysis, and then increased between 55 and 72 hr (time of ecdysis to the fifth instar). A correlation between hemolymph JH titers and the levels of JHBP was noted at the beginning of the fourth and fifth instars. During the rest of the instar, there was an inverse correlation between hormone and protein (Fig. 5). Since the JH titer fluctuates greatly during development, it was suggested that limiting concentrations of JHBP may be involved in regulating JH levels. Should the total number of binding sites per unit volume fall below the amount of JH available to fill those binding sites, the unbound hormone would be metabolized by hemolymph general esterases (see Section 4.5). Nevertheless, it should be noted that a high concentration of JHBP in the presence of a low concentration of JH will not necessarily bind all of the available hormone. Indeed, in the case of the lipoprotein–JH interaction, where the affinity is low, only a small percentage of the hormone is actually bound. Moreover, factors such as ionic strength, pH, etc. will affect the equilibrium of bound and unbound hormone. However, at no time does the combined concentration of JH I, JH II, and JH III exceed that of JHBP; in most instances, the JH titer is much lower.

According to the classical scheme, the JH titer must be relatively high at the end of each instar and the beginning of the following instar to assure the character of the larval–larval molt, and perhaps the increased levels of JHBP are necessary to maintain this increased JH titer. However,

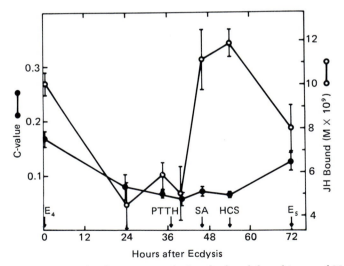

FIGURE 5. The relative JH binding capacity during the fourth larval instar of *M. sexta*. The hemolymph samples were dialyzed together in the presence of [³H]-JH I. The bound and unbound hormone were separated by immunoprecipitation and plotted on the basis of [³H]-JH bound and of the ratio of [bound hormone]/[unbound] × [protein concentration], which results in a C value. From Goodman and Gilbert, 1978.

since the general protein concentration increases concurrently with JHBP, the correlation of the latter with the JH titer may be misleading.

4.3.6. Site of Synthesis of JHBP

Since the JHBP plays a critical role in conserving JH levels (see Section 4.3.7), knowledge of its site of synthesis, rate of release, and possible regulation is important. To examine this problem, JHBP levels in various tissues were examined by immunoelectrophoresis, autoradiography, and immunoprecipitation of JHBP complexed with [^{3}H]-JH (Nowock et al., 1975). Tissues suspected of synthesis were also incubated in vitro with ^{14}C-labeled amino acids to permit incorporation of label into newly synthesized proteins, and then were screened for JHBP. The fat body was shown to be the primary site of synthesis for JHBP, and kinetic studies revealed that the newly synthesized protein was not stored; rather, it was rapidly released from the fat body. It should be noted that immunochemical examination detected no JHBP in corpora allata homogenates or medium in which corpora allata were cultured.

4.3.7. Role of JHBP

In analyzing the physiological role of the JHBP, an examination of the better characterized vertebrate serum hormone binding proteins may prove useful. The functions of these proteins include: (1) transfer of the hormone throughout the circulatory system; (2) protection of the hormone from enzymatic attack; (3) prevention of nonspecific adsorption to the walls of blood vessels; (4) protection from nonspecific excretion; (5) protection from uptake and metabolism by the liver; and (6) facilitation of hormone diffusion from its biosynthetic site to the circulatory system by establishing a gradient (Westphal, 1980).

To date, two of these functions have been ascribed to JHBP, i.e., transport of the hormone and protection from enzymatic attack. The role of the binding protein in protecting JH from certain hemolymph esterases has been recognized in several insects (see Section 4.5). The effect of JHBP on JH metabolism is evident in Fig. 6; i.e., when fat body was incubated with JHBP in vitro, hormone metabolism was significantly reduced.

Based on the dynamics of hormone binding and on an in vitro morphogenetic assay, it has been suggested that cellular recognition of the hormone is initially a function of the JH–JHBP complex. Kinetic studies have demonstrated that virtually all of the JH will be bound to JHBP under physiological conditions, indicating that the bound hormone is important to JH action. More cogent evidence for this hypothesis has been presented using a morphogenetic assay in which JH blocks MH-induced cuticle deposition. When wing disc epidermis was incubated with high levels of unbound JH (10^{-4} M), little inhibition was noted; however, when minute concentrations of JHBP (8×10^{-9} M) were added to the medium, a large synergistic effect was observed. Thus, it appears

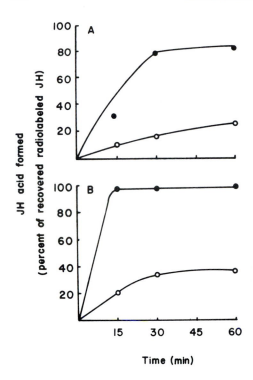

FIGURE 6. The effect of JHBP on the degradation of JH in *M. sexta* fat body *in vitro*. The open circles represent the presence of binding protein and JH, the closed circles represent JH in the presence of serum albumin. Graph A represents mid-fourth-instar relative metabolic rate and B represents mid-fifth-instar relative metabolic rate. The medium was assayed in each case. Data from Hammock *et al.*, 1975.

that recognition of the complex rather than the unbound hormone is important to biological activity (Sanburg *et al.*, 1975a).

4.4. Biosynthesis

Soon after the elucidation of the chemical structures of JH I and JH II, studies were initiated on the biosynthetic processes leading to the carbon skeleton. These early investigations using the common terpenoid precursors *in vivo* were inconclusive; although acetate was nominally incorporated into JH, it was suggested that the hormone was not synthesized via the normal isoprenoid pathway. Moreover, the origin of the side chains on JH I and JH II presented an enigma since these are the only known naturally occurring isoprenoid derivatives that display an ethyl side chain.

The importance of homomevalonate as an intermediate was first recognized by Schooley and his colleagues (1973, 1976) at Zoecon and has been extensively studied by that group. It was suggested that the extra carbon of JH I and JH II was derived from propionate via an intermediate, homomevalonate, and that biosynthesis of the hormone followed a normal terpene pathway. By incubating the corpora allata with radiolabeled precursors *in vitro*, extracting the medium, and subjecting the products to selective degradation, they showed that propionate did provide the extra carbon in JH II. Since homomevalonate is a necessary intermediate in this proposed scheme and had not been previously identified as a natural product in animals, attempts were made to isolate this compound from the corpora allata. Recently, this group has demonstrated the presence of homomevalonate in the glands, thus adding convincing evidence to the hypothesis that JH I and JH II are synthesized via the normal terpene pathway (Baker and Schooley, 1978).

According to the proposed scheme, the enzyme complex dimethylallyl transferase condenses one mevalonate and two homomevalonate molecules to form the carbon chain for JH I. Likewise, the JH II carbon chain is synthesized from one homomevalonate and two mevalonate molecules, and the JH III skeleton is formed from three mevalonate molecules. The condensation and subsequent elongation reaction is the C-4 to C-1 nucleophilic addition of homoisopentenylpyrophosphate to homodimethylallylpyrophosphate or dimethylallylpyrophosphate. The reactions in JH biosynthesis are apparently both sequential and specific; of the eight possible combinations of ethyl and methyl groups at the 3, 7, 11 positions, only three have been identified from natural sources (Fig. 7). How one homolog is preferentially synthesized by regulation of intermediates is still unclear.

The terminal steps in JH biosynthesis, the incorporation of the methyl group at C-1, and the epoxidation of the C-10,11 position have been characterized *in vivo* and *in vitro*. Epoxidation of the molecule is

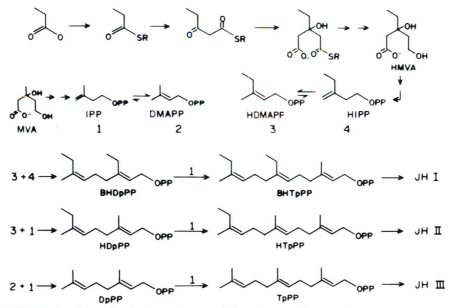

FIGURE 7. Scheme for the biosynthesis of the carbon skeletons of the three JH. MVA, mevalonic acid; HMVA, homomevalonic acid; IPP, isopentenylpyrophosphate; DMAPP, dimethylallylpyrophosphate; HDMAPP, homodimethylallylpyrophosphate; HIPP, homoisopentenylpyrophosphate; BHDpPP, bishomodiprenylpyrophosphate; BHTpPP, bishomotriprenylpyrophosphate; HDpPP, homodiprenylpyrophosphate; HTpPP, homotriprenylpyrophosphate; DpPP, diprenylpyrophosphate; and TpPP, triprenylpyrophosphate. From Schooley *et al.*, 1976.

most probably performed by a mixed-function oxidase requiring molecular oxygen and NADPH as with the ecdysone 20-monooxygenase. As in the case of the methyl ester synthetase (see below), the oxidase is associated with the microsomal fraction of homogenates of the corpora allata (Reibstein *et al.*, 1976). Biosynthetic studies have demonstrated clearly that the ubiquitous methyl donor, *S*-adenosylmethionine (SAM), is the source of the methyl group at the C-1 terminus. When either farnesenate or methionine was incubated with homogenates of the corpora allata, no methyl farnesenate was formed. Conversely, when farnesenate, *S*-adenosylmethionine, and the corpora allata fraction (12,000g) were incubated together, the terminal acid was alkylated with high efficiency. Since this reaction is highly specific and efficient, it is now being exploited in the laboratories of Pratt and Tobe to monitor the synthesis of the homologs *in vitro*.

The sequence in which the epoxide and methyl ester may be introduced is outlined in Fig. 8, and evidence has been presented for both pathways in different insects. It is possible either that the pathways are species specific or that different pathways are utilized under different conditions. This problem is compounded by the relative impermeability of the corpora allata to various precursors, and despite the use of gland homogenates, this difficulty has not been entirely overcome since homogenization leads to release of degradative enzymes.

4.5. Degradation

The clearance of JH from hemolymph and peripheral tissues at critical periods during the insect life cycle is essential for growth, development (Fig. 1), and reproduction. The reduction in JH titer can be brought about by cessation of glandular activity and/or by hemolymph metabolism of the hormone. Since the regulation of the corpora allata is discussed elsewhere (Granger and Bollenbacher, this volume), the discussion that follows will focus on degradation of JH by hydrolysis of the epoxide or methyl ester groups (Fig. 9).

In *M. sexta*, larval hemolymph contains two types of esterases. The

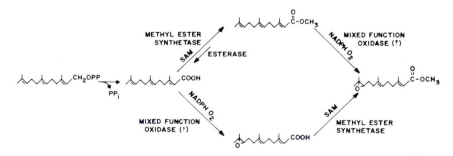

FIGURE 8. Possible biosynthetic pathways for the terminal steps in JH biosynthesis.

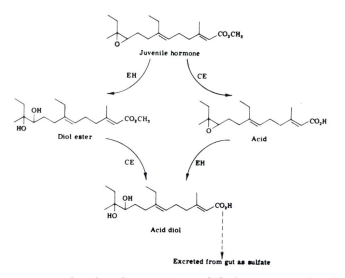

FIGURE 9. Routes of JH degradation. EH, epoxide hydratase; CE, carboxylesterase.

first type, the general esterases, mediates esterolytic cleavage in a wide range of compounds and can be identified by their ability to hydrolyze the synthetic substrate, 1-naphthyl acetate. They are readily inhibited by the serine esterase inhibitor diisopropylfluorophosphate and other organophosphates. These esterases are apparent throughout later larval life and can effectively hydrolyze unbound JH. When subjected to gel filtration, this group of proteins is usually associated with the relatively high-molecular-weight fraction (>100,000).

A second population of lower-molecular-weight esterases, termed JH-specific esterases, arise during the last larval instar (and possibly at other times). These enzymes are characterized by their insensitivity to diisopropylfluorophosphate, their substrate specificity, and their ability to metabolize JH bound to JHBP. At present, no JH-specific esterases have been purified to homogeneity although some physical parameters have been established. According to Sanburg *et al.* (1975b), there are three specific esterase populations, as demonstrated by isoelectric focusing, with isoelectric points ranging from pH 5.3 to 5.5. The JH-specific esterases have molecular weights estimated at about 67,000 and an apparent K_m of 1×10^{-6} M for JH III.

The levels of JH-specific esterase during development vary independently from the general esterase levels, lending support to the hypothesis that these enzymes play an important role in JH titer regulation. In fifth-instar *Manduca*, the titer of JH-specific esterase increases at Day 2, peaks at Day 4, and then drops to undetectable levels thereafter until Day 9, at which time the titer increases only to drop soon after pupation (Vince and Gilbert, 1977) (Fig. 10). The peaks of esterase activity occur at those times when JH titer must be reduced to ensure a normal molt.

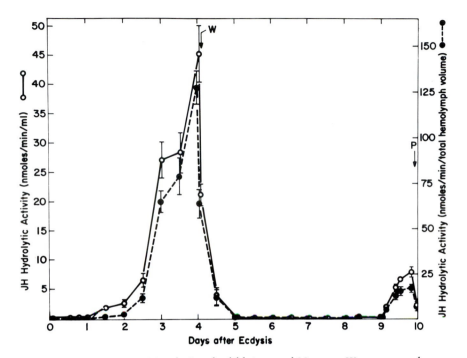

FIGURE 10. JH esterase activity during the fifth instar of *M. sexta*. W represents the wandering or burrowing stage when animals stop feeding and burrow underground. P represents larval–pupal ecdysis. From Vince and Gilbert, 1977.

Since the peak of activity may occur after the decline in JH hemolymph titer is noted, the JH-specific esterase may act primarily to "clean up" peripheral tissues so that pupal development can begin. The specific esterases also may have a central role in other physiological events such as larval or adult diapause where the presence of JH either delays or prevents developmental processes.

In contrast to the extracellular degradation of JH by the specific esterases of the hemolymph, the metabolism of the epoxide moiety is performed primarily within the cell. The primary source of this enzyme is the microsomal fraction of cell homogenates of the fat body, imaginal discs, and midgut. This enzyme, epoxide hydratase, catalyzes the hydration of the epoxide (Fig. 9); however, little more is known about its characteristics and specificity.

The final step in metabolism of JH is the conjugation of the acid or acid–diol to either sulfate, glucuronide, or glucoside compounds. Conjugation to these water-soluble molecules promotes the excretion of the inactivated hormone, and it is unlikely that these conjugation reactions are reversible as is the case with ecdysteroids (see Section 3.4).

In conclusion, the regulation of JH titer can be viewed as a relatively complex system. As noted by Granger and Bollenbacher (this volume), the regulation of the synthetic activity of the corpora allata by neural mechanisms, allatotropins and allatohibins, has not yet really been analyzed in a critical manner. However, once JH is secreted into the larval hemolymph of *M. sexta*, we do know that a specific binding protein (JHBP) with relatively high affinity for the hormone forms a complex with JH (JH:JHBP; Fig. 1), and that this complex may play a variety of roles, including protection of JH from esterolytic cleavage by the ubiquitous general esterases of the hemolymph. When the JH titer must be decreased to allow normal metamorphic events to occur, JH-specific esterases are released from the fat body into the hemolymph leading to the hydrolysis of bound JH (Fig. 1). It is therefore critical for the insect to always have a "saturating level" of JHBP and to possess regulatory mechanisms for allowing the temporally precise synthesis and secretion of the JH-specific esterases. These enzymes are probably controlled both by substrate "induction" and by as yet unidentified neurohormones, and this area is fertile for further research.

Control of epoxide hydratase activity has not yet been approached, nor has the mechanism of conjugation and excretion of the JH acid, JH diol, and JH acid–diol. All of these processes will contribute to the JH titer of the insect, which itself is of critical importance. The work on JHBP and JH-specific esterases is a beginning, but as is also the case for MH, the important questions regarding insect hormone metabolism and transport have certainly not yet been answered.

ACKNOWLEDGMENTS. Work from the authors' laboratories was supported by Grant AM-30118 from the National Institutes of Health. This is in part a contribution from the College of Agricultural and Life Science, University of Wisconsin. We thank Dr. S. Smith for his helpful comments and Nancy Grousnick for excellent secretarial assistance.

REFERENCES

Baker, F. C., and Schooley, D., 1978, *Chem. Commun.* **1978**:292.

Bollenbacher, W. E., Goodman, W., Vedeckis, W. V., and Gilbert, L. I., 1976, *Steroids* **27**:309.

Bollenbacher, W. E., Faux, A. F., Galbraith, M. N., Gilbert, L. I., Horn, D. H. S., and Wilkie, J. J., 1979, *Steroids* **34**:509.

Bollenbacher, W. E., Agui, N., Granger, N. A., and Gilbert, L. I., 1980, in: *Invertebrate Systems in Vitro* (E. Kurstak, K. Maramorosch, and A. Dubendorfer, eds.), pp. 253–271, Elsevier/North Holland, Amsterdam.

Borst, D. W., and O'Connor, J. D., 1972, *Science* **178**:418.

Butenandt, A., and Karlson, P., 1954, *Z. Naturforsch. Teil B* **9**:389.

Chang, E. S., and O'Connor, J. D., 1979, in: *Methods of Hormone Radioimmunoassay* (B. M. Jaffe and H. R. Behrman, eds.), pp. 797–814, Academic Press, New York.

Chino, H., Sakurai, S., Ohtaki, T., Ikekawa, H., Miyazaki, N., Ishibashi, M., and Abuki, H., 1974, *Science* **183**:529.

Feyereisen, R., 1977, *Experientia* **33**:1111.

Fraenkel, G., 1935, *Proc. R. Soc. London Ser. B* **118**:1.

Gilbert, L. I., and Chino, H., 1974, *J. Lipid Res.* **15**:439.

Gilbert, L. I., Goodman, W., and Nowock, J., 1976, in: *Actualité sur les Hormones d'Invertébrés* (M. Durchon and P. Joly, eds.), pp. 413–434, CNRS, Paris.

Gilbert, L. I., Goodman, W., and Bollenbacher, W. E., 1977, in: *Biochemistry of Lipids II* (T. Goodwin, ed.), pp. 1–50, University of Maryland Press, Baltimore.

Gilbert, L. I., Bollenbacher, W. E., Goodman, W., Smith, S. L., Agui, N., Granger, N., and Sedlak, B. J., 1980, *Recent Prog. Horm. Res.* **36**:401.

Goodman, W., and Gilbert, L. I., 1978, *Gen. Comp. Endocrinol.* **35**:27.

Goodman, W., O'Hern, P., Zaugg, R. H., and Gilbert, L. I., 1978, *Mol. Cell. Endocrinol.* **11**:225.

Hammock, B., Nowock, J., Goodman, W., Stamoudis, V., and Gilbert, L. I., 1975, *Mol. Cell. Endocrinol.* **3**:167.

Hampshire, F., and Horn, D. H. S., 1966, *Chem. Commun.* **1966**:37.

Heinrich, G., and Hoffmeister, H., 1970, *Z. Naturforsch.* **25**:358.

Huber, R., and Hoppe, W., 1965, *Chem. Ber.* **98**:240.

Ishizaki, H., 1969, *Dev. Growth Differ.* **11**:1.

Kaplanis, J. N., Thompson, M. J., Dutky, S. R., and Robbins, W. E., 1979, *Steroids* **34**:333.

Kaplanis, J. N., Weirich, G. F., Svoboda, J. A., Thompson, M. J., and Robbins, W. E., 1980, in: *Progress in Ecdysone Research* (J. Hoffmann, ed.), pp. 163–186, Elsevier/North-Holland, Amsterdam.

Karlson, P., Hoffmeister, H., Hoppe, W., and Huber, R., 1963, *Liebigs Ann. Chem.* **662**:1.

King, D. S., 1972, *Gen. Comp. Endocrinol.* **3**:221.

King, D. S., Bollenbacher, W. E., Borst, D. W., Vedeckis, W. V., O'Connor, J. D., Ittycheriah, P. I., and Gilbert, L. I., 1974, *Proc. Natl. Acad. Sci. USA* **71**:793.

Koolman, J., 1978, *Hoppe-Seyler's Z. Physiol. Chem.* **359**:1315.

Koolman, J., and Karlson, P., 1978, *Eur. J. Biochem.* **89**:453.

Krieger, D. T., and Liotta, A. S., 1979, *Science* **205**:366.

Lanzrein, B., Hashimoto, M., Paramakovitch, V., Nakanishi, K., Wilhelm, R., and Lüscher, M., 1975, *Life Sci.* **16**:1271.

Nagasawa, H., Isogai, A., Suzuki, A., Tamura, S., and Ishizaki, H., 1979, *Dev. Growth Differ.* **21**:29.

Nakanishi, K., 1971, *Pure Appl. Chem.* **25**:167.

Nigg, H. N., Svoboda, J. A., Thompson, M. J., Kaplanis, J. N., Dutky, S. R., and Robbins, W. E., 1974, *Lipids* **9**:971.

Nowock, J., Goodman, W., Bollenbacher, W. E., and Gilbert, L. I., 1975, *Gen. Comp. Endocrinol.* **27**:230.

Reibstein, D., Law, J. H., Bowlus, S. B., and Katzenellenbogen, J. A., 1976, in: *The Juvenile Hormones* (L. I. Gilbert, ed.), pp. 131–146, Plenum Press, New York.

Reum, L., and Koolman, J., 1979, *Insect Biochem.* **9**:135.

Röller, H., and Dahm, K. H., 1968, *Recent Prog. Horm. Res.* **24**:651.

Sanburg, L., Kramer, K., Kézdy, F., Law, J. H., and Oberlander, H., 1975a, *Nature (London)* **253**:266.

Sanburg, L., Kramer, K., Kézdy, F., and Law, J. H., 1975b, *J. Insect Physiol.* **21**:873.

Sandor, T., Sonea, S., and Mehdi, A. Z., 1975, *Am. Zool.* **15**(suppl. 1):227.

Schooley, D., Judy, K., Bergot, J., Hall, M. S., and Siddall, J. B., 1973, *Proc. Natl. Acad. Sci. USA* **70**:2921.

Schooley, D., Judy, K., Bergot, J., Hall, M. S., and Jennings, R. C., 1976, in: *The Juvenile Hormones* (L. I. Gilbert, ed.), pp. 101–117, Plenum Press, New York.

Sláma, K., Romaňuk, M., and Šorm, F., 1974, *Insect Hormones and Bioanalogues*, Springer-Verlag, New York.

Smith, S. L., Bollenbacher, W. E., Cooper, D., Schleyer, H., Wielgus, J. J., and Gilbert, L. I., 1979, *Mol. Cell. Endocrinol.* **15**:111.

Smith, S. L., Bollenbacher, W. E., and Gilbert, L. I., 1980, in: *Progress in Ecdysone Research* (J. Hoffmann, ed.), pp. 139–162, Elsevier/North-Holland, Amsterdam.

Svoboda, J. A., Kaplanis, J. N., Robbins, W. E., and Thompson, M. J., 1975, *Annu. Rev. Entomol.* **20**:205.

Thompson, M. J., Kaplanis, J. N., Robbins, W. E., and Svoboda, J. A., 1973, *Adv. Lipid Res.* **11:**219.

Vince, R., and Gilbert, L. I., 1977, *Insect Biochem.* **7:**115.

Westphal, U., 1980, in: *Steroid Receptors and Hormone Dependent Neoplasia* (J. L. Wittliff and O. Dapunt, eds.), pp. 1–17, Masson, New York.

Whitmore, E., and Gilbert, L. I., 1972, *J. Insect Physiol.* **18:**1153.

Williams, C. M., 1956, *Nature (London)* **178:**212.

Yang, R. S. H., and Wilkinson, C. F., 1972, *Biochem. J.* **130:**487.

Young, N. L., 1976, *Insect Biochem.* **6:**1.

CHAPTER 6

Macromolecular Changes during Insect Metamorphosis

S. Sridhara

1. INTRODUCTION

Insects have been used extensively as the experimental organisms of choice to study biochemical changes associated with the developmental patterns of higher eukaryotes. One has only to recall the contributions of *Drosophila* to biochemical genetics to realize its potential for probing the biochemistry of development. An in-depth discussion of macromolecular changes and control of gene activities during insect metamorphosis is limited by the format of this volume. First, much of the biochemical work is in relation to hormonal changes and has been reviewed recently (Sridhara *et al.*, 1978). Second, much of the interesting recent progress in this field has occurred with two systems, both of which are discussed individually in this volume (Chapters 7 and 8). Third, due to the diversity of organisms and tissues studied as a function of a specific developmental age or hormonal milieu, few generalizations can be drawn from published data on macromolecular changes, particularly those pertaining to protein and RNA.

In view of these restrictions and to avoid redundancy, only those systems that have contributed significantly over the past 10 years or have the potential to generate new concepts will be considered in detail. The older experiments have been summarized succinctly in the first edition of this volume (Wyatt, 1968) and will not be discussed here.

A brief discussion of the general pattern of insect development is necessary to understand and appreciate the questions raised and to cor-

S. SRIDHARA · Department of Biochemistry, University of Mississippi Medical Center, Jackson, Mississippi 39216.

relate the data presented with the biology of the organisms. The developmental stages of insects may be divided into two categories: embryonic and postembryonic. Postembryonic development is quite different in hemimetabolous and holometabolous insects. In the former, the newly hatched larva resembles the adult very closely so that after a succession of molts, only a few changes are necessary at metamorphosis to produce the characteristics of the adult, i.e., the young and the adult have similar morphological characteristics. In contrast, drastic developmental changes occur during metamorphosis of larva to pupa to adult in holometabolous insects. It is thought, without much experimental evidence, that the differences between holo- and hemimetabolous insects are more quantitative than qualitative.

Larvae of holometabolous insects secrete a larval cuticle at each larval molt, a term that includes detachment of the old cuticle from the epidermis, apolysis, replication of epidermal cells, secretion of a new cuticle, digestion and resorption of the old cuticle, and finally shedding or ecdysis of the remnants of the old cuticle. At metamorphosis a morphologically distinct pupal or adult cuticle is secreted by the epidermis. The molting process and the nature of the cuticle secreted, whether larval, pupal, or adult, depend upon at least three hormones: prothoracicotropic hormone (PTTH), molting hormone (MH; 20-hydroxyecdysone), and juvenile hormone (JH) (see Granger and Bollenbacher, this volume). PTTH is released under the influence of neural or environmental cues and activates the prothoracic glands to synthesize and release ecdysone, which is converted to 20-hydroxyecdysone in other tissues (see Gilbert and Goodman, this volume). The latter acts on a variety of tissues and, in the case of epidermal cells, causes them to deposit a cuticle. JH, which also acts on various tissues including epidermis, influences the cell's capacity to synthesize a particular kind of cuticle.

The insect's life depends on this process of proper cuticle deposition. Thus, MH stimulates the synthetic activities necessary for growth and molting, while JH determines which of several possible synthetic activities take place. Since holometabolous insects exhibit profound morphological changes at different developmental stages determined by the interactions of these hormones, and the respective patterns are suited for that specific period of life involving different functions, one can assume extensive changes in the pattern of proteins synthesized. Both qualitative and quantitative changes in messenger RNAs (and probably other RNAs) can be expected to underwrite these alterations in protein synthesis. Consequently, the general hypothesis has been that different gene sets specific for larval, pupal, and adult stages exist and that these sets are turned "on" or "off" in sequence by the interaction of MH and JH. A corollary to this would be that both hormones function at the transcriptional level (Sridhara et al., 1978).

Metamorphosis in Lepidoptera and higher Diptera has long attracted the attention of biologists and biochemists interested in the biochemical basis of the consequences of such metamorphosis. In the conversion of an actively feeding larva or maggot into the volant adult moth or fly, two intimately related but distinct processes are involved: the breakdown of some larval tissues accompanied and/or followed by development and differentiation of the adult tissues (see Locke, this volume). Both of these phenomena occur within the hardened outer cuticle of the pupa or puparium. Simplistically, the larval stage can be considered as a period of growth and storage of reserves that will be used subsequently for the synthesis of adult structures within a closed system. The breakdown of macromolecules accumulated during larval growth and the concomitant macromolecular synthesis occurring within this cleidoic system have made the period just prior to larval–pupal metamorphosis and the period of adult development particularly challenging for biochemical analysis.

2. CHANGES IN PROTEIN AND RNA DURING LARVAL–PUPAL–ADULT METAMORPHOSIS

Available evidence, principally cytological, points to the histolysis of larval tissues, e.g., components of the alimentary canal, fat body, muscle, etc., which form about 80–90% of the mass of the larva. The extent of this breakdown is much greater in Diptera than in Lepidoptera. Since adult tissues begin to develop while histolysis progresses, histolysis and histogenesis may occur simultaneously. As both occur in a "closed system," the materials necessary for histogenesis should be derived from the products of histolysis. The logical extension of this concept is that macromolecules such as nucleic acids and proteins present in larval tissues are broken down and the resulting substrates are employed for the synthesis of imaginal tissues. If so, one can pose several questions, most of which remain unanswered: (1) Do the hormones controlling metamorphosis also directly control both cellular breakdown and synthesis, and if so, how are these phenomena integrated? (2) Is tissue breakdown due to autolysis and/or phagocytosis and how are they controlled? (3) What is the nature of the changes in lysosomal enzymes, how are their activities regulated, and what is the fate of the degradative products? These questions gain more importance as attempts are made to correlate cytological observations with quantitative analysis of the biochemical parameters involved.

Before analyzing the available data relating to the above questions, it is necessary to describe the life cycle of one insect that has been used consistently in these research areas, namely the blowfly, *Calliphora erythrocephala* (Meig) (= *vicina* R. D.). The larva hatches from the egg and becomes fully grown at about 4 days (it consumes meat or a synthetic diet) and leaves its food and crawls around actively for up to 3 additional days. This wandering stage, during which the larva purges its gut of ingested food, is followed by a quiescent stage of about 1 day in duration

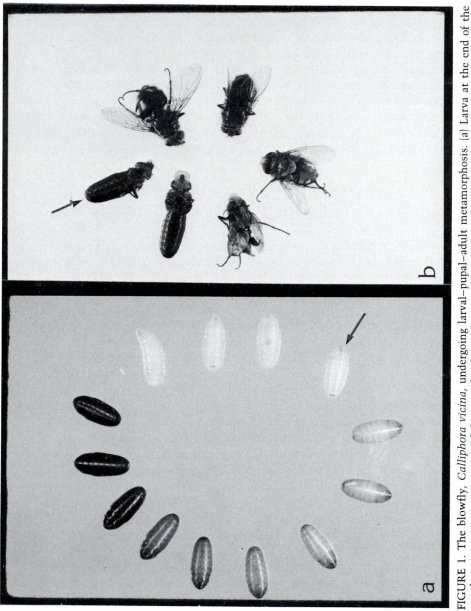

FIGURE 1. The blowfly, *Calliphora vicina*, undergoing larval–pupal–adult metamorphosis. (a) Larva at the end of the wandering stage stops moving, withdraws its head into the body, and gradually becomes a barrel-shaped pupa. This is the "white pupa" stage (arrow). Notice how the color of the outer cuticle changes during the next 6 hr (clockwise) as the puparium forms. (b) Adult has developed during 9 days in the puparium; the fully formed fly emerges in the course of a few hours from the puparium (counterclockwise from arrow).

(the seventh day!!). During this time, the larval head is withdrawn into the body and the insect gradually rounds up to become a white, barrel-shaped pupa, immobile because the larval muscles have become detached from the external cuticle. At this so-called "white pupa" stage, various internal changes characteristic of early metamorphosis have already commenced. During the ensuing 4–6 hr, the external cuticle becomes progressively sclerotized and darker (Fig. 1a). About 20 hr later, an internal molt occurs, at which time the insect becomes a pharate adult. This period of pharate adult development lasts about 9 days and terminates with the emergence of the fully formed, but immature, adult fly (Fig. 1b). During the first 4–5 days of pharate adult development, the internal milieu of the insect is a thick, creamy brei due to larval tissue breakdown.

The pattern of growth and development in other blowflies (*Lucilia cuprina*, sheep blowfly; *Phormia regina*, black blowfly) and other dipterans (*Drosophila melanogaster*, fruit fly; *Sarcophaga bullata*, flesh fly) is similar to that of *Calliphora*, except that the total time spent in any stage is variable, usually shorter. On the other hand, the time spent at these stages is much longer in Lepidoptera (*Manduca sexta*, tobacco hornworm; *Hyalophora cecropia*, silkmoth; *Bombyx mori*, commercial silkworm). The important stages involved in our discussion are the wandering stage, larval–pupal metamorphosis, and pupal–adult metamorphosis.

2.1. Breakdown of Macromolecular Components

2.1.1. Proteins

There is conclusive cytological evidence for the disappearance of some larval tissues during metamorphosis, although the causative factor(s) and extent of breakdown at the molecular level have not been examined extensively. Measurement of acid phosphatase activity points toward an increase of lysosomal activity during this regression of larval tissues, particularly in the fat body and salivary glands. Since acid phosphatases, along with other hydrolases with acidic pH optima, are involved in intracellular and extracellular digestion, it is assumed that the breakdown of larval tissues is due to the activity of lysosomal hydrolases liberated at the moment of cell death. These ideas draw support from the considerable evidence that exists in a wide variety of organisms associating cellular autolysis with lysosomal function. It is quite possible that the control of the level of hydrolases by MH may be a general phenomenon of histolysis in insect tissues (see Locke, this volume). The proliferation of lysosomes in response to 20-hydroxyecdysone may be due to the induction of synthesis of hydrolases. Another point of confusion is the fact that morphological breakdown does not necessarily mean either cell death or complete breakdown of macromolecules, which may occur at a later time. The only report claiming detection and measurement of adequate proteolytic activity necessary for the total breakdown of larval proteins (Smith and Birt, 1972) raises more questions than it answers (discussed below).

Two forms of proteolytic activity, distinguished by their pH optimum of 4.1 and 2.8, respectively, were found to be associated with particulate material from homogenates of larvae and pupae of *Lucilia*. The activity with a pH optimum of 2.8 was present at high levels throughout development, while the pH 4.1 activity showed two maxima of activity, one at pupariation and the other commencing midway during pharate adult development, reaching a peak just before adult eclosion. Such lysosomal associated activities (cathepsins) should normally be solubilized by detergents; however, the above activities were refractory to solubilization not only by Triton X-100, but also by sonication at acid pH. It is quite possible that these enzymes are located on the membrane or inside of protein-containing particles destined for hydrolysis. If so, these enzymes may not be responsible for the proteolytic activity occurring during larval–pupal–adult metamorphosis. Another problem with this study is that the activity was quite unstable and was lost within 2 hr at 0°C. This, along with the authors' conclusion that an earlier report of their inability to detect proteolytic activity was due to a large pool of amino acids which made it difficult to detect proteolysis, makes it likely that the proteolytic activity they purport to measure is not due to enzymatic activity at all, but is a measure of the release of nonspecifically bound amino acids and peptides from the particulate material during incubation. As there are no other studies relating to this phenomenon, it can be concluded that there is little evidence for protease involvement during metamorphosis. In fact, during the entire period of pupal and pharate adult development, only minor changes are observed in the concentration of total free amino acids. The fluctuation in both the amount of protein and the free amino acid concentrations is minor compared to the drastic morphological changes involved in histolysis. This raises the question of whether the degradation of larval proteins proceeds to completion, i.e., the production of amino acids, or whether peptides are utilized for the synthesis of adult proteins.

2.1.2. RNA and Ribosomes

There is a direct relationship between growth rate, protein synthesis, and ribosomal content in various tissues of a wide variety of organisms. Such a relationship would predict the maximum amount of ribosomes in mature larvae and also both quantitative and qualitative changes during metamorphic molts. Since, in most eukaryotic cells, ribosomal RNA accounts for more than 80% of total cellular RNA, it would be expected that the total RNA content of the insect or its tissues would be proportional to the number of ribosomes present. An analysis for RNA and ribosomes from the time of cessation of feeding to the emergence of the adult fly has been carried out in *Calliphora* (Sridhara and Levenbook,

TABLE 1. CHANGES IN WET WEIGHT, TOTAL RNA, AND NUMBER OF RIBOSOMES DURING GROWTH AND DEVELOPMENT OF THE BLOWFLY, *C. VICINA*[a]

Days after hatching	Wet weight (mg/insect)	Total RNA (µg/insect)	Ribosomes[b] (µg/insect)	rRNA as % of total RNA[c]
4 (larva)	80	390	540	76
5	72	340	470	76
6	67	340	400	65
7 (white pupa)	63	330	300	50
8 (pharate adult)	59	335	290	47.5
9	55	330	280	47
16	54	340	280	45
18 (young emerged adult)	42	220	270	67.5
22	54	225	260	63.5

[a] From Sridhara and Levenbook, 1974.
[b] Ribosome quantity calculated assuming that 10 A_{260} units are equivalent to 1 mg.
[c] Based on an empirical finding that RNA comprises 55% by weight of blowfly ribosomes.

1974). The values presented in Table 1 do show that ribosomal and RNA content are highest in the mature larva. However, throughout pharate adult development not only is the number of ribosomes constant, but it is some 45% lower than that in the fully grown larva. The absolute number of ribosomes does not correlate with the anticipated and measured rates of protein synthesis (see below). Most surprising is the fact that the fat body, which accounts for 10–12% of larval fresh weight, contains almost half of the total ribosomes of the larva, the concentration being approximately 25 mg/g fat body, a value about eight times that found in vertebrate liver. This probably accounts for the enormous quantity of hemolymph proteins synthesized by the fat body during the last larval instar (see Locke, this volume). Most of these ribosomes are degraded during the last 3 days of larval life (i.e., during the wandering stage leading to pupariation), while the number of ribosomes in the remainder of the insect's body remains essentially constant thereafter. These quantitative values confirm cytological observations on the fat body, as well as other reports showing little change in RNA content from about 2 days prior to pupariation until adult emergence.

The RNA from the ribosomes of the fat body are not degraded completely, but are converted to a partially degraded (4–7 S) relatively insoluble product. This accounts for the constancy of total RNA content, notwithstanding the fall in ribosome content to about 50% during pharate adult development. This insoluble RNA probably is stored in the protein–RNA granules observed under the microscope. That this pellet RNA is in fact derived from fat body ribosomes has been proven both by labeling experiments and by base analysis. The adult flies do not contain this pellet RNA, as it is eliminated after degradation to nucleosides and bases (Protzel *et al.*, 1976). This accounts for the observation that a direct correlation between ribosomes and RNA content, which did not exist during pharate adult development, occurs in the newly emerged adult.

The constancy of ribosomes throughout pharate adult development suggests either metabolic stability of the ribosomal population (probably derived from larval muscle) or essentially equal rates of breakdown and *de novo* synthesis. Absolute metabolic stability is incompatible with various observations where administered isotopic precursors are incorporated into rRNA, and a breakdown of some prelabeled RNA occurs. Quantitative measurements of the rates of ribosomal breakdown have been carried out by following the specific activity of rRNA labeled with radioactive uridine, either by injection or by feeding. It can be safely assumed that the maximum specific activity reached by larval ribosomes at cessation of feeding would be similar in a large batch of animals and that the change in this specific activity can be followed as a function of development. Such an analysis yields values of half-lives of 8.8 days for 26 S rRNA and 7.8 days for 18 S RNA (Fig. 2), and the decay follows first-order kinetics. As cytidine is derived from uridine, both bases are labeled in RNA and the rate of decay of rRNA can be followed by measuring the decay rates of the specific activity of uridine and cytidine nucleotides derived by hydrolysis of rRNA. These two nucleotides decay at different rates (UMP half-life 8.5 days; CMP half-life 10.7 days). The half-life of rRNA (9.6 days) based on these data is very close to that calculated above (Protzel *et al.*, 1976). This means that less than 50% of the ribosomes present at pupariation are degraded at emergence or that less than 50% of the ribosomes present in the adult are newly synthesized.

The computations involved in the above calculations assume the presence of a steady-state homogeneous pool of rRNA (ribosomes in the

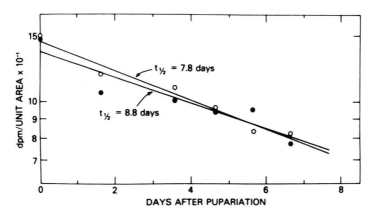

FIGURE 2. Time course of decrease in specific radioactivity of RNAs (18 and 26 S) during pharate adult development of the blowfly, *C. vicina*. Larvae were raised on a synthetic diet containing [³H]uridine. RNA was extracted and analyzed by polyacrylamide gel electrophoresis. 18 S RNA (○); 28 S RNA (●). Specific radioactivity is expressed in relative units. From Protzel *et al.*, 1976.

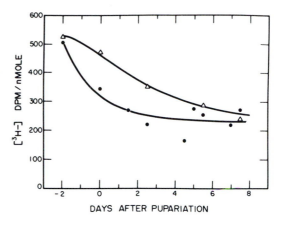

FIGURE 3. Time course of decay of specific radioactivity of UMP in rRNA and in soluble pools of UTP during the late larva, white pupa, and pharate adult stages of *C. vicina*. The line passing through the experimentally determined UTP values (●) is an exponential of the form $U(t) = Ae^{-\beta t} + C$. The line passing through the experimental points for RNA-UMP (△) is a least-squares best fit. From Protzel *et al.*, 1976.

mature larva) decaying exponentially with unlabeled input, but radioactive output. Unfortunately, in a closed system like that of *Calliphora* to which labeled precursor has been administered during growth, the UTP precursor pool (probably CTP pool as well) is not unlabeled as expected, but is labeled throughout pharate adult development. The contribution of such a labeled pool to the calculated decay rates can be taken into account and the rRNA turnover rate computed by employing the differential equation, $dR/dt = -K(R - U)$, where at any time t, K is the fraction of the rRNA pool replaced per unit time, and R and U are the specific activities of rRNA-UMP and the soluble UTP pool, respectively. The time course of change in the soluble UTP pool can itself be specified by fitting an exponential curve of the type $U = Ae^{-\beta t} + C$ to the determined values of U. Such an exponential curve appears to fit the experimental points very well (Fig. 3). The parameters of A, β, and C can be obtained by a least-squares reiterative curve-fitting program. From these data, K can be calculated and the best theoretical curve for the time course of R when fitted to the actual experimental values can be computed. Such an analysis now leads to a half-life of 2.2 days, in marked contrast to 8.5 days when the radioactivity of the precursor pool is ignored. A half-life of 2.2 days means that 31% of the rRNA from the pool turns over per day, or that at least 93% of the ribosomal pool that existed during the wandering stage has been degraded during pharate adult development. In other words, ribosomes found in the adults have all been synthesized *de novo*.

2.2. Synthesis of Macromolecular Components

2.2.1. Proteins

Convincing evidence exists demonstrating that the mechanisms of protein synthesis in insects are typical of other eukaryotic systems. During larval growth and adult development, changes in the ribosomal profiles from heavy polysomes to monosomes are observed, but no correlation is observed between these patterns and the extent of protein synthesis. Such data should be interpreted with caution since all the lysosomal and other nucleases present in the various histolyzing tissues will be released when whole larvae or pupae are homogenized, a condition adding to the difficulty of isolating polysomes. The gross level of protein synthesis during pupal and early adult development has been a subject of controversy since much of the work revolves around administration of labeled amino acids and following their incorporation into total acid-insoluble material, and, in addition, the use of protein synthesis inhibitors such as cycloheximide and puromycin to block incorporation. These studies, not surprisingly, indicate that protein synthesis accompanies adult development.

Attempts have also been made to quantify these incorporation data by including the dilution of the label that occurs within the system by the large amino acid pools present and to account for any turnover. This can be achieved by employing the equation $K_p = K_a \Delta p / \Delta q$, where K_p is the total rate of incorporation of amino acids into protein, K_a is the total rate at which the amino acid leaves the pool via all the pathways, and Δp and Δq are the changes in successive values of the amount of label in total protein and the amount of label in the amino acid pool during several time intervals, respectively. Based on certain assumptions, one can make accurate measurements of K_p (Dinamarca and Levenbook, 1966). This has been done with alanine and lysine in the black blowfly and with lysine in the sheep blowfly. Calculations show that between 25% and 60% of the total protein present in the adult is derived from *de novo* synthesis from amino acids, which also means that an equal amount of larval protein has to be degraded to amino acids. More important is the shape of the curve for the rates of protein synthesis as a function of development (Fig. 4). Compared to the rates of synthesis during larval growth, the synthetic rate follows a U-shaped curve where during the first third of pharate adult development it falls to very low values, stays at that low level during the next third, and then rises again during the last third of that period. The reason why the patterns in the two studies are shifted is not clear. However, this pattern of protein synthesis is analogous to the best known U-shaped pattern in the rate of respiration during metamorphosis. Wyatt (1968) theorized that, since during this period there is little muscular activity, one would expect protein synthesis to be a major energy user and that the rates of respiration would show some correlation with those of protein synthesis. The results of Fig. 4 not only lend support to this hypothesis, but permit the conclusion that protein synthesis occurs mainly during the last third of pharate adult development.

A good part of the new synthesis, both observed and calculated, is probably due to the development of thoracic flight muscles. *De novo* synthesis of muscle proteins and mitochondrial proteins occurs during

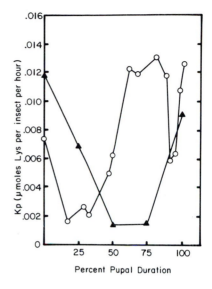

FIGURE 4. Rates of incorporation of lysine into protein (K_p) during metamorphosis of the black blowfly, *P. regina* (▲), and sheep blowfly, *L. cuprina* (○). Redrawn from Dinamarca and Levenbook, 1966, and Williams and Birt, 1972.

the period immediately before and after emergence, i.e., during the upturn in the U curve. Cytochrome c, an essential component of respiration and a good barometer of respiratory activity, is synthesized extramitochondrially and subsequently transferred into the mitochondria. Williams *et al.*, (1972) quantified the amount of cytochrome c by direct analysis and also by following the incorporation of labeled lysine in the blowfly, *Lucilia*. The rate of synthesis calculated from the pool size and incorporation of [^{14}C]lysine was sufficient to account for the amount of cytochrome c found in the newly emerged adult. The level of cytochrome c remained constant at about 25 μg/g during most of the developmental period within the puparium, started to increase 1 day prior to emergence, and increased to about 250 μg/g after emergence. Very similar results have been obtained for cytoplasmic α-glycerophosphate dehydrogenase activity, which, along with mitochondrial α-glycerophosphate oxidase, is an essential component of adult flight muscle energetics (Campbell and Birt, 1972). Administration of cycloheximide at the time of maximal synthesis blocks accumulation of both components by about 98% for a short time. These results leave no doubt that the proteins studied, and probably all the other proteins in the flight muscles, are synthesized *de novo* at about the time of adult emergence. It may be pointed out that in all holometabolous insects, the increase that occurs in soluble α-glycerophosphate dehydrogenase activity, and probably of other components required for flight activity, does not occur at the same period. The variability appears to be related to the time when the adult acquires the ability to fly. The honeybee and Colorado beetle, for example, are unable to fly for 4–8 days after eclosion, and only 10–20% of the maximal enzyme activity is found at emergence. On the other hand, the housefly, fruit fly, and sheep blowfly all can fly within a few hours after eclosion; in these insects, 30–70%

of maximal activity is already present at eclosion and reaches a maximum within a day or two. Similar correlations are found even in hemimetabolous insects such as locusts and grasshoppers. Consequently, the results obtained from studies of thoracic flight muscle development cannot be extended to other tissues that are fully developed by the time of emergence.

The studies on thoracic muscle development also have led to other interesting results. In *Calliphora*, the flight muscle fibers increase rapidly in length as early as the third or fourth day of development; subsequently, longitudinal growth becomes more gradual, but diameter growth occurs rapidly after the sixth day of development. The early elongation can occur in th absence of oxygen or when protein synthesis is blocked by cycloheximide (Houlihan and Newton, 1978). In both cases, the increase in diameter is blocked. It is believed that the longitudinal increase is due to the incorporation of preformed actomyosin into existing muscle fibers, while the increase in diameter is the result of *de novo* synthesis of actomyosin. Studies on the synthesis and accumulation of actomyosin itself reveal an increased rate of synthesis toward the end of pharate adult development. This peaks just after emergence and declines within 2 days. This actomyosin is most likely employed for the growth in muscle diameter. The normally found delay in the growth of myofilaments compared to the peak in the accumulation of actomyosin is interpreted as indicating that the developing muscle accumulates actomyosin for some time before the final assembly of myofilaments (Campbell and Birt, 1975).

Studies of the development of adult muscle offer an opportunity to study the contribution of larval macromolecules to the adult. Thomson (1975), while discussing the reconstruction of muscle tissues in flies, points out two phases in the histolysis of larval muscle. Some of the larval muscles vacuolate and are then phagocytosed by granular hemocytes, which become engorged by cell debris. However, a certain group of larval muscles do not break up completely in spite of the appearance of striations and vacuoles because phagocytosis does not occur. Instead, myoblast nuclei penetrate the cells while larval nuclei degenerate. The adult nuclei enlarge and striations begin to appear. The subdivision of muscle fibers occurs before the myofibrils become clearly differentiated. This early establishment of the template of the larval contractile machinery by the adult genome, followed by growth both longitudinally and in thickness, promises to be an important experimental model for the study of the utilization of larval macromolecules by adult tissues without complete breakdown of some of the larval components.

2.2.2. RNA and Ribosomes

We have already discussed how either 50% or 95% of the larval ribosomes are turned over, depending on the method of computation against a constant rRNA pool. This means that for every ribosome that is degraded, a new one is made at the same time. Instead of depending on the decay rates in the specific activity of rRNA, it is much easier to obtain a direct estimate of the rates of rRNA synthesis during metamorphosis by administering a labeled precursor and following its appearance in rRNA. This has been carried out under carefully controlled experiments, taking into consideration both the rate of incorporation of UMP into rRNA and the average specific activity of the UTP pools. The rate of rRNA biosynthesis, which actually means ribosome synthesis since the calculations are based on the specific activity of both 18 and 26 S RNAs, is maximal in the late-third-instar larva, as expected. It then drops markedly at the white puparium stage, as for protein synthesis and respiration. The main difference is that 2 days later, a somewhat lower value is observed, this value being maintained for the duration of adult development (Protzel and Levenbook, 1976). This indicates that there is no rRNA synthesis component corresponding to the upswing of the U-shaped curve seen in Fig. 4. The average rate of RNA synthesis during this time is 1.7 nmole UMP incorporated per insect per hour. When integrated over the period from pupariation to adult emergence, this leads to a value corresponding to 87% replacement of the ribosomal pool, the corresponding value from degradation studies being 93% (Protzel *et al.*, 1976). Both values indicate major *de novo* synthesis of adult fly ribosomes and little carry-over of larval ribosomes.

The relative constancy in the rate of rRNA synthesis is quite puzzling in view of the variable rates of protein synthesis seen during pharate adult development. If ribosomes were synthesized in the newly developing tissues for the support of protein synthesis, one would expect ribosome synthesis and content to reflect this U-shaped curve. In fact, in the thorax of the sheep blowfly not only does rRNA increase from about 10 μg per thorax during the last half of pharate adult development, but there is also a close correspondence between the maximum number of ribosomes and the maximal activity of these ribosomes in protein synthesis, as well as the appearance of polysomes (De Kort, 1975). In similar studies with the thoracic muscles of the tobacco hornworm, the synthesis of myosin was greater in polysomes than in ribosomes. Pharate adult muscle 3–4 days before adult emergence contained a greater ratio of polysomes to monosomes than did the muscle of newly emerged adults (Chan and Reibling, 1973). This agrees with the observed synthesis of actomyosin in *Calliphora* muscle, also prior to adult emergence. Assuming for the moment that, in fact, much of the thoracic muscle ribosomes appear, or at least are functional, late in pharate adult development, two hypotheses can be proposed. First, all of the muscle ribosomes are newly synthesized *de novo* from precursors reclaimed from larval ribosomes. Second, the ribosomes of larval muscles that develop into adult flight muscle are not liberated at all, but the newly made mRNA in the flight muscles recruits the ribosomes to form polysomes. No systematic study comparing the synthesis of ribosomes (or rRNA) in thoracic flight muscles with that of the whole body has been carried out to decide between these possibilities.

A few general comments and a discussion of the results presented above are now in order. All quantitative analyses of protein or RNA synthesis are based on certain assumptions. The basic assumption is that the acid-soluble pool of the precursor employed is, in fact, the source from which that amino acid or UTP is withdrawn for protein and RNA synthesis, respectively. Second, for the calculations, it is assumed that recycling or return of the label to the pool is negligible. Unfortunately, neither of these assumptions is true even in the simplest of experimental systems (e.g., He La cells). They are therefore probably not true in the complicated, closed insect system where cytolysis is occurring concomitantly with histogenesis and the tissues are bathed by hemolymph containing profoundly high levels of free amino acids.

The concept that a general intracellular pool serves as the sole precursor of amino acids for protein synthesis is a hotly debated point, even in simple systems such as cell cultures. Indeed, there is some evidence that amino acids from an extracellular pool are funneled directly to the protein synthetic machinery, bypassing the intracellular pool. The controversy as to which pool represents the best source of amino acid for protein synthesis continues and alternative interpretations of data are possible (Airhart et al., 1974). This argues against any general hypothesis concerning the precursor supply of amino acids for protein synthesis. The mere possibility of the existence of different metabolic pools makes it much more difficult to account for recycling of label (which is assumed not to occur in the calculations) and also to determine the specific radioactivity of an amino acid at the site of protein synthesis. Pool exchanges present serious problems if compartmentalization occurs and the amino acid derived from degradation is either trapped in a larger nonutilizable pool or tapped immediately for new synthesis. In either case, the amount of protein synthesis determined by incorporation studies will be incorrect, i.e., overestimated or underestimated.

Similar difficulties are encountered with the measurement of RNA synthetic rates, particularly in view of the evidence showing that nuclear and nucleolar RNA polymerases draw on separate pools of UTP in He La cells. The second major objection is the assumption that no reutilization of the radioactive precursor occurs, while this in fact occurs and probably to a considerable extent. The different decay rates of UMP and CMP within the same rRNA, although only labeled uridine was employed as the precursor, attest to this point (Protzel et al., 1976). Furthermore, reutilization has to occur because massive elimination of cytoplasmic and other elements and degradation of RNA and DNA are observed during the later part of development when selective cytolysis, which plays an important role in reconstruction, occurs. The above indicates that the chances of a true representation of the kinetics of either protein or RNA turnover are remote at this time. However, in the case of protein synthesis, these problems could be avoided. Direct quantitative measurements of protein synthesis can be made on whole insects or individual tissues by following the specific activity of the amino acids attached to the tRNAs rather than the acid-soluble pool. This is not more technically difficult than other procedures, but results in data not impaired by the assumptions and problems mentioned above.

Finally, it is clear that most adult tissues of Diptera are formed by the proliferation and differentiation of imaginal discs (at least 11 in *Drosophila*; see Fristrom, this volume). The progressive formation of adult structures from these nests of cells is asynchronous; e.g., wings are fully developed 75% through adult development while flight muscles mature just before or after adult eclosion, depending upon the insect. This suggests the presence of a progressively expanding pool of ribosomes and protein during the latter half of adult development. Conversely, in most larval cells undergoing cytolysis, the cell nuclei become pycnotic, ribosomal synthesis is impaired, and preexisting ribosomes and protein are either degraded or expelled during the first third of the period. The most intriguing question is, in view of the constancy of the total rRNA and protein pools and the known changes in larval cells undergoing degradation and growing adult cells, how is the process of degradation and synthesis coordinated? That coordination must occur cannot be questioned since blocking either protein or RNA synthesis does not lead either to a decrease in macromolecule content or to a large increase in the precursor pools. It is difficult to imagine how this is carried out, particularly since many conclusions derived from biological observations contradict biochemical studies. Histolysis commences in the larval tissues during the wandering stage, and soon after pupation the internal contents consist of a thick, milky brei with little recognizable structural organization. The criteria for cell degeneration and death at this stage of metamorphosis vary from species to species and even from tissue to tissue within a species. This variability in biological definition is due to observations indicating that degeneration in some tissues begins with the condensation of chromatin and the accumulation of lysosome-like structures and, finally, attack by phagocytes. In others, phagocytosis occurs at an earlier stage when cytological changes and lysosomal structures are rarely seen (see Locke, this volume).

These facts, when considered together with biochemical data, require a redefinition of histolysis in order to account for the loss of larval structures at a time when the rates of protein synthesis, energy requirements, and protein synthetic capacity of ribosomes all reach their lowest levels. It is possible that a large proportion of the macromolecules present in larval tissues undergoing histolysis are not completely broken down, but rather reorganized, distributed, and stored so that they may be degraded and reutilized at a later date, i.e., during the last third of adult development when the upswing in the U-shaped curve is observed. New approaches are required to determine and quantify the temporal and spatial relationships between partially broken, larval cellular organelles and growing adult tissues. Perhaps a better understanding of the processes involved can be achieved by repeating some of the experiments conducted on flies with Lepidoptera. Due to the large size of the latter, it should be possible to isolate pieces of specific tissues in which macromolecular components can be labeled to high specific radioactivity and implanted in other insects at various stages of development. One would then follow both cytologically and biochemically the fate of these implants so as to arrive at qualitative and quantitative correlations between biological and biochemical changes.

2.3. Cuticular Proteins

The central role of epidermal cells in insect development is due to their capacity to secrete cuticle. The adult cuticle with colored scales is far different from the thick, brown, and brittle pupal cuticle and the thin, translucent, scaleless larval cuticle. It is reasonable to assume that such differences may result from differences in the cuticular proteins. Numerous papers on cuticular proteins have been published in which amino acid compositions are given or gel electrophoresis analyses are presented (Hackman and Goldberg, 1976; Andersen, 1979). The conflicting results obtained are due in part to differential solubility of the cuticular proteins at different developmental stages and to the use of different solvents for extraction. A few generalizations that can be made are that cuticular proteins have a high proline content and the amount of sulfur-containing amino acids is quite low. Little information can be derived from the gross amino acid composition of a complex mixture of proteins although even such a crude analysis demonstrated differences between larval, pupal, and adult cuticular proteins of the mealworm, *Tenebrio molitor*. Another generalization is that the qualitative differences between larval and pupal cuticular proteins are much less pronounced than between adult cuticular proteins and those of earlier stages.

While many adult structures arise from specialized groups of cells (imaginal discs) that have no corresponding larval differentiation products, it appears that the abdominal epidermal cells of some insects persist throughout life so that the same cells or their descendants are responsible for the production of the different forms of cuticle. Based on this information, Roberts and Willis (1980) have extracted soluble proteins from the untanned abdominal cuticle of larval, pupal, and adult beetles soon after ecdysis (early) and 24 hr after ecdysis (late). The extracted protein mixtures were analyzed by electrophoresis in acrylamide gels in the presence of sodium dodecyl sulfate (SDS), a procedure that separates polypeptides based on their molecular weights. Two facts appear to be rather clear from the results (Fig. 5): (1) numerous, but variable numbers of bands are seen at all stages, indicating the complexity of the cuticular proteins; and (2) the patterns of bands obtained and their relative quan-

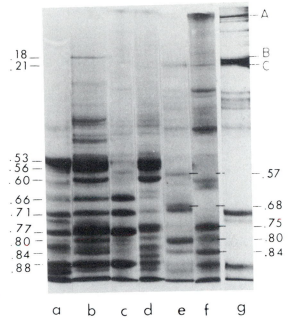

FIGURE 5. Comparison of the banding patterns of hemolymph and extracts from abdominal cuticles from larval, pupal, and adult mealworms (*Tenebrio molitor*). The insects were maintained for 1 hr (early) and 24 hr (late) after ecdysis prior to cuticle removal. The numbers indicate R_f values relative to the dye. (a) Early larva, (b) late larva, (c) early pupa, (d) late pupa, (e) early adult, (f) late adult, and (g) hemolymph from an early larva. From Roberts and Willis, 1980.

tities are unique to each stage, supporting the general assumption that cuticle at different stages of development is comprised in part of unique proteins. A comparison of polypeptide patterns indicates that the larval and pupal cuticular proteins are more closely related to one another than either group is to the adult cuticular proteins. The majority have molecular weights between 5,000 and 24,000. Their studies have led to another important conclusion (Roberts, 1976), i.e., the second pupal cuticle that the epidermal cells secrete after treatment with JH has a polypeptide profile identical to that of normal pupal cuticle. In contrast, a second pupal cuticle, by morphological criteria, produced by treatment with compounds like actinomycin D and mitomycin C is comprised of proteins characteristic of both pupa and adult.

3. PUPAL–ADULT METAMORPHOSIS: WING DEVELOPMENT

The development of wings from imaginal discs has been a favorite system for the study of genetics and pattern formation (fruit flies) and biochemical studies on hormone action (silkmoths). Only the latter will be considered. A number of species of insects adapt to hostile environments by entering a state of arrested development (diapause) either as embryos, larvae, pupae, or adults. Diapausing silkmoth pupae characterized by a very low level of metabolism and no growth contain a single

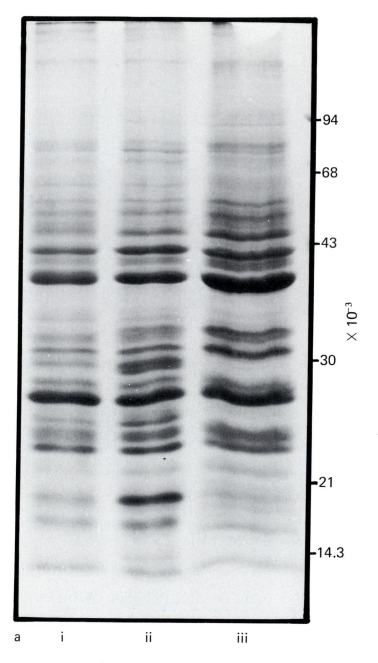

a   i   ii   iii

FIGURE 6. SDS–polyacrylamide electrophoresis of cuticular proteins from silkmoth (*A. polyphemus*) wing tissue (numbers indicate molecular weight standards). (a) Adult cuticular proteins (scales) extracted with (i) urea, (ii) guanidine hydrochloride, and (iii) SDS. (b) A comparison of cuticular proteins extracted with (i) urea and (ii) guanidine from (c) adult wings and (d) second pupal cuticle obtained by treatment with JH.

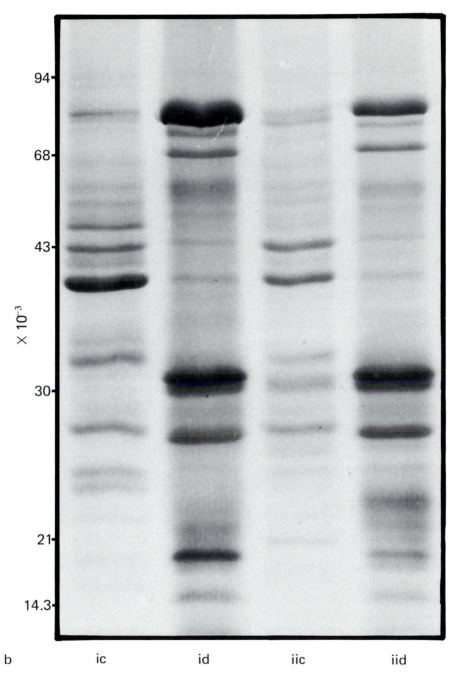

FIGURE 6 (*Continued*)

layer of wing epidermis which can be easily removed. These epidermal cells respond to MH by growth and then differentiation to the adult, i.e., secretion of adult cuticle, formation of scales, etc. Consequently, one either can elicit adult wing development *in vivo* or *in vitro* by the addition of 20-hydroxyecdysone or can prevent development by providing a hormone-free milieu. Initiation of wing development is characterized by enhanced synthesis of all classes of RNA within about 24 hr, and one notes a rapid formation of polysomes within a few hours where only monosomes existed previously. This indicates recruitment of previously present ribosomes by either newly made or activated mRNAs and the initiation of protein synthesis (Wyatt, 1972). It is also possible to prevent the development of the adult by administration of JH to the pupa. Such treatment results in a pupal–pupal molt, i.e., a second pupal cuticle is deposited by the epidermal cells in about a week. Whether pupal and adult silkmoth cuticular proteins differ has been analyzed by extracting soluble proteins from untanned, second pupal cuticle and adult wing (scales) cuticle with various solvents and analyzing these mixtures by SDS–acrylamide electrophoresis. It can be concluded from many such analyses as represented in Fig. 6 that: (1) both cuticles yield at least 25–30 polypeptides, among which are at least 10 that are unique to each stage and these cover a range of molecular weights from 10,000 to 90,000; (2) the pupal cuticle contains a larger proportion of low-molecular-weight ($<25,000$) polypeptides compared to adult cuticular proteins; and (3) while the total amount of proteins extractable by aqueous buffer is less than 1/10th of that soluble in urea, guanidine, or SDS (all containing 1 mM β-mercaptoethanol), all the extracts give similar electrophoretic patterns except for some minor variations (Fig. 6). These results provide the basis for analyzing macromolecular changes in the wing epidermal cells.

The developmental changes that occur in wing epidermal DNA, RNA, and protein have been measured in the oak silkmoth, *Antheraea pernyi*, and related to MH titer (Nowock *et al.*, 1978). The total DNA content of the wing epidermis increases beginning at Days 2 and 3 of adult development, presumably as a result of endomitosis in the scale-forming complexes. Subsequently, there is another surge of DNA replication, which is probably followed by cell division (Fig. 7). These cellular processes related to DNA synthesis are essentially over by Day 7, but there is a final increase in DNA content at about Day 10, which is followed by a cytolytic phase. Total RNA content, on the other hand, increases in three phases with maxima at 2, 7, and 9 days of adult development. The decline in both DNA and RNA content after 9 days of adult development is coincident with the time of selective cell death leading to formation of the adult wing structure. Immediately after the initiation of development, the wing protein content increases slightly, but remains at a constant level for the first 5 days, at which time there is a dramatic increase attributed to the beginning of the deposition of adult cuticle. The patterns of change in the content of these macromol-

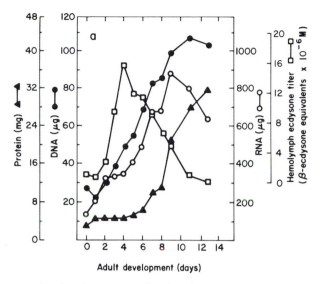

FIGURE 7. Macromolecular changes correlated with hormonal changes during adult wing development in the oak silkmoth, *A. pernyi*. Wings of pupae (Day 0) or developing adults were analyzed for the components by standard procedures. All data are expressed as values per set of wings of one animal. Ecdysteroid concentration was determined in the hemolymph by a radioimmunoassay procedure and data expressed as 20-hydroxyecdysone equivalents. From Nowock *et al.*, 1978.

ecules are similar during adult development of another lepidopteran, *Pieris brassicae* (Lafont *et al.*, 1976).

A. *pernyi* contains a low but physiologically significant level of MH, even in pupae stored at low temperature. Soon after transfer to room temperature, there is a reduction in hemolymph ecdysteroid concentration, indicating uptake by the tissues and/or degradation. Within 24–48 hr the prothoracic glands produce more ecdysone needed for further developmental changes (Fig. 7).

Since RNA is an essential component of protein synthesis and various RNAs have to be transcribed to account for the observed changes, attempts were made to correlate these changes with RNA polymerase activity. The enzymatic activity develops in a bimodal fashion, which closely follows the ecdysteroid titer. During the first day after transfer from low temperature to room temperature, the activity of both RNA polymerase I and II increases, probably as a result of the presence of endogenous MH. Class I is specific for transcription of RNA, class II for hnRNAs, and class III for 5 and 4 S RNAs. The dramatic increase in the hemolymph ecdysteroid titer up to 5 days correlates well with a further enhancement of the activity of both class I and class II (Nowock *et al.*, 1978). These results are in agreement with other data showing stimulation of uridine incorporation into RNA at these time periods (Wyatt, 1972). Analysis of RNA polymerase activity in isolated nuclei confirmed the results obtained by estimation of the enzymes in whole wing epidermis. It is not possible from these results to determine whether the increased activity is due to an actual increase in enzymatic activity or to increased template capacity, or even activation of preexisting enzymes. However, it is clear that the RNA polymerase activity undergoes much more pronounced changes than either RNA or MH. A subtle, direct control of RNA polymerase activity by MH is highly unlikely.

A major drawback with these studies is the variation in time of adult development among the various well-studied silkmoths. This period is influenced both by the history of the animals before and after diapause and by the endogenous MH titer of pupae, which is negligible in silkmoths such as *H. cecropia*. To avoid some of these problems, diapausing pupae of *A. polyphemus* or *H. cecropia* can be induced to develop fairly synchronously by administration of 20-hydroxyecdysone. The content and rates of synthesis of RNA and protein have been followed in *A. Polyphemus* pupae induced either to produce an adult cuticle by injection of MH (20-hydroxyecdysone) or to produce a second pupal cuticle by

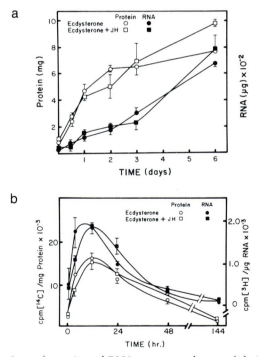

FIGURE 8. A comparison of protein and RNA content and rates of their synthesis in wing epidermis of *A. polyphemus* treated with 20-hydroxyecdysone (O) or 20-hydroxyecdysone and JH (□). (a) RNA (filled) and protein (open) content of wings determined at various periods after hormone treatment. (b) Incorporation of [14C]leucine (open) or [3H]uridine (filled) into acid-precipitable material. Wings were removed from pupae at the times indicated and cultured in Grace's insect medium with either 1 μCi [3H]leucine or 20 μCi [3H]uridine + 250 μmole cold uridine for 3 hr. The label in acid-precipitable material was determined by standard procedures. Redrawn from Katula, 1978.

TABLE 2. POLY-A-CONTAINING RNA IN WING EPIDERMIS OF
A. POLYPHEMUS FOLLOWING HORMONE TREATMENT[a,b]

Treatment	Time after treatment (hr)	% recovery of poly-A	% poly-A-containing mRNA	Poly-A-containing RNA (μg/pair of wings)
—	0	90.0	1.63	0.51
MH[c]	24	85.7	1.24	1.7
	48	98.1	1.47	2.5
	72	96.4	2.10	3.8
	144	111.4	1.03	8.9
MH + JH	24	109.1	1.67	2.4
	48	106.5	2.54	3.1
	72	110.2	1.56	4.2
	144	94.8	1.77	14.2

[a] From Katula, 1978.
[b] One to three milligrams of total RNA was fractionated on oligo dT cellulose. Total RNA before fractionation and bound RNA were hybridized to [³H]poly-U to measure the recovery and percent poly-A.
[c] MH, 20-hydroxyecdysone.

injection of MH and JH. Again, as with *A. pernyi* (Nowock *et al.*, 1978), *H. cecropia* (Wyatt, 1972), and *P. brassicae* (Lafont *et al.*, 1976), the RNA and protein content of the wing epidermis begin to increase rapidly within 4–6 hr after hormone administration (Fig. 8a). It is of interest that by the sixth day, when the MH–JH-treated pupae have already deposited a second pupal cuticle, the RNA content in the two treatment groups is similar, while the protein content of the MH–JH-treated animals is much higher than in the MH-treated pupae. This may be due to the secretion of the second pupal cuticle rather than an adult cuticle. Precursor incorporation studies, which take into consideration the pool sizes and kinetics of incorporation, demonstrate that the rates of increase correspond to the observed increase in RNA polymerase activity and RNA content. However, there are no dramatic differences in the rates of synthesis of either RNA or protein between cells that are synthesizing a second pupal cuticle and cells that will synthesize adult cuticular proteins (Fig. 8b) (Katula, 1978).

One mechanism by which JH acts to change the pattern of synthesis of cuticular proteins could be by eliciting qualitative and/or quantitative changes in specific gene products (mRNAs). Since mRNAs comprise about 2% of the total cellular RNA, one would not expect to discern these changes by measuring total RNA. This problem can be investigated by isolating mRNAs from wing epidermis at various stages of development and comparing the products translated by each group *in vitro*. This has been carried out and preliminary results show that the percentage of mRNA (isolated as poly-A-containing RNA) does not reflect any gross differences during normal development or between pupae receiving the two hormone treatments (Table 2). These different mRNA populations have been translated utilizing the wheat germ cell-free system employing either [³⁵S]methionine or [¹⁴C]leucine and the products analyzed by

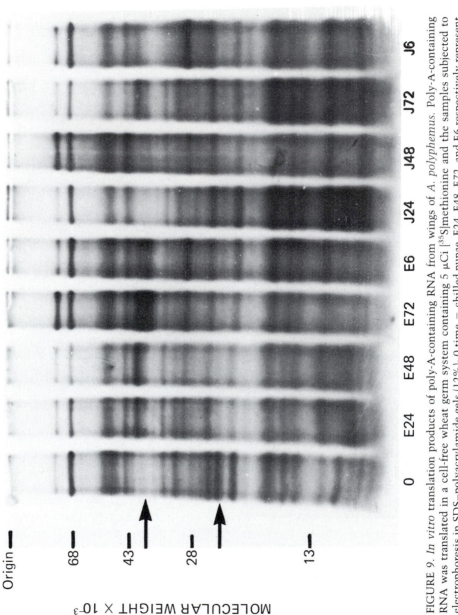

FIGURE 9. *In vitro* translation products of poly-A-containing RNA from wings of *A. polyphemus*. Poly-A-containing RNA was translated in a cell-free wheat germ system containing 5 μCi [^{35}S]methionine and the samples subjected to electrophoresis in SDS–polyacrylamide gels (12%); 0 time = chilled pupae. E24, E48, E72, and E6 respectively represent pupae 24 hr, 48 hr, 72 hr, and 6 days after 20-hydroxyecdysone administration. J24, J48, J72, and J6 respectively represent pupae 24 hr, 48 hr, 72 hr, and 6 days after administration of 20-hydroxyecdysone and JH. From Katula, 1978.

SDS–acrylamide electrophoresis and autoradiography (fluorography). A whole range of products from 13,000 to 68,000 daltons are observed with all mRNA populations (Fig. 9). Assuming that the intensity of labeling of any particular polypeptide band corresponds to the quantity of the corresponding mRNA within the mRNA mixture, an assumption that has been shown to be valid in other instances, two mRNAs that are barely detectable in 0-hr samples become prominent very early in MH-treated pupae, while the same mRNAs can be detected at low levels only 3–6 days after MH–JH treatment. On the other hand, an mRNA coding for a polypeptide of approximately 29,000 daltons is found in large amounts in 0-hr samples, but rapidly disappears in hormone-treated animals. The preponderance of low-molecular-weight components is not due to the presence of imcomplete translation products since analogous *in vivo* studies result in similar low-molecular-weight products. These results are in keeping with observations revealing a preponderance of low-molecular-weight polypeptides in the pupal cuticle.

A more detailed analysis of these translation products has been carried out by two-dimensional analysis; polypeptides are separated based on their isoelectric points in the first dimension (isoelectric focusing) and then are further separated on the basis of molecular weight (SDS–acrylamide electrophoresis). The resulting fluorograms (autoradiograms) are analyzed by comparing the intensity of the label in a particular spot with that of its surrounding neighbors. This yields an approximation of the proportion of a particular mRNA within a particular mRNA population. The most interesting periods for comparison are 0 hr, MH 6, MH–JH 6, and MH 10. The first one corresponds to chilled pupae, MH 6 to the period in which no adult cuticle is being made, MH–JH 6 to the time when the second pupal cuticle has been deposited, and MH 10 to when adult cuticle is beginning to be actively deposited (Fig. 10). A comparison of many such fluorograms with those of other developmental periods results in the following picture: (1) wing epidermis from chilled pupae contains many mRNAs in common with those from hormone-stimulated tissue; (2) spot 40, which is not present at 0 hr, and a group of spots (21–24), which are barely detectable at 0 hr, all increase considerably after MH treatment; (3) JH treatment does not prevent the increase in spot 40, but does inhibit spots 21–24; (4) some spots (33–35, 38, 39) appear to be either unique or in much higher quantities at 0 hr; (5) certain spots (1, 3, 10, 30, 31) in the low-molecular-weight range are more intense in MH–JH 6 than in MH 6. Comparison of MH 6 or MH–JH 6 with MH 10 indicates that a number of spots in the middle- and high-molecular-weight ranges increase in intensity, indicating that the corresponding mRNAs might code for adult cuticular proteins. Some mRNAs from wing epidermis actively engaged in cuticular secretion are not present in the 0- or 24-hr mRNAs. These results provide strong evidence that new mRNAs appear in wing tissue, perhaps under hormonal control, at times when the epidermis is secreting pupal or adult cuticle, but do not allow correlations between specific cuticular proteins and their corresponding mRNAs.

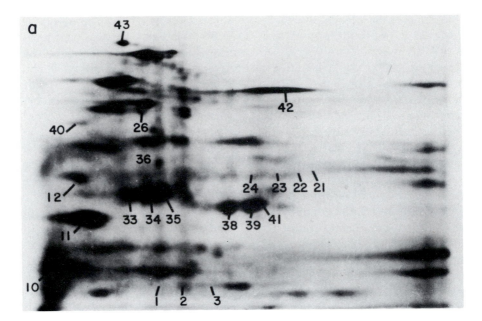

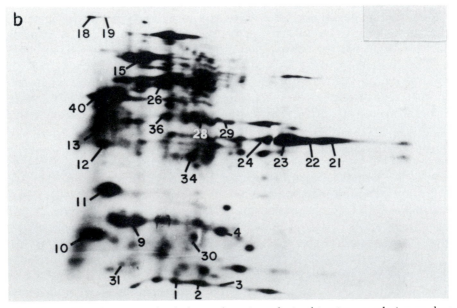

FIGURE 10. Two-dimensional gel electrophoretic analysis of *in vitro* translation products of poly-A-containing RNA from wing epidermis. (a) chilled pupae; (b) 6 days after MH treatment; (c) 6 days after MH–JH treatment; and (d) 10 days after MH treatment.

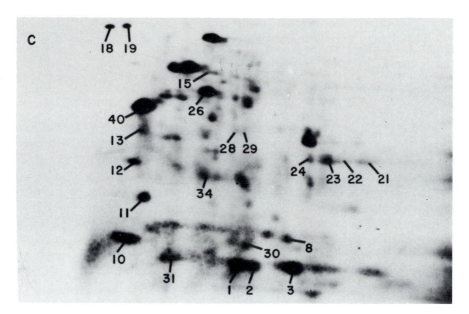

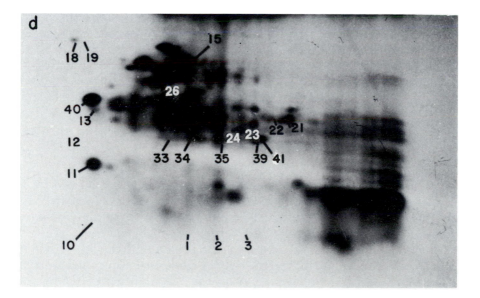

FIGURE 10 (*Continued*)

However, further studies with the wing epidermis system at the molecular level may help in elucidating the mechanisms of action of JH and MH.

4. LARVAL–PUPAL METAMORPHOSIS

Many biochemical studies on insect metamorphosis have utilized the last larval instar of Lepidoptera and Diptera. The tissues of choice for these studies have been the epidermis, which is the primary target of the hormones (see O'Connor and Chang, this volume), the fat body, which plays a central role in intermediary metabolism and performs the functions of the vertebrate liver and adipose tissue, and the hemolymph, which transports macromolecules, metabolic substrates, and hormones to other tissues. Much recent work is related to hormones (see Chapters 5, 7, and 8 of this volume), and only those areas that fall outside of endocrinology will be considered here. Notwithstanding, it is necessary to present the basic hormonal changes that occur during larval–pupal metamorphosis in order to correlate them with macromolecular alterations when necessary. The best studied insect in this respect is the tobacco hornworm, *M. sexta*, and Fig. 11 shows that during the last instar, JH falls to very low values by about 4 days, concomitant with a low, but significant, surge of MH. This small ecdysteroid peak in the absence of JH initiates all the behavioral changes that occur subsequently: stoppage of feeding, wandering, emptying of the gut, burrowing into the earth, etc. In addition, this small surge determines the capacity of the epidermis to respond to the larger peak of MH about 4 days later. Although no such detailed studies exist for any dipterans at this time, the preponderance of data suggest that hormonal changes similar to those observed in *Manduca*, *Pieris*, and other lepidopterans also occur in Diptera.

4.1. Epidermis

The abdominal epidermis of the fifth-instar tobacco hornworm is rapidly displacing *Calliphora* as the most popular object of study for probing epidermal biochemistry. The pioneering work on the tanning of *Calliphora* cuticle by Karlson and Sekeris has been reviewed recently (Sridhara *et al.*, 1978). The following is a brief summary of the salient features of the *Manduca* system: (1) the epidermal cells that are synthesizing larval cuticle undergo a change in "commitment" (reprogramming) at about Day $3\frac{1}{2}$, when the JH titer falls to a low level and a small burst of MH is noted (Fig. 11); (2) this change in commitment is covert since its actual expression, secretion of pupal cuticle, occurs several days later when the large ecdysteroid peak occurs; (3) JH, if present at the time of the first small ecdysteroid peak, prevents this reprogramming; (4) once the change in commitment has occurred, JH has no effect on these cells. Considerable effort is being put forth to understand in biochemical or molecular terms the meaning of commitment, reprogramming of genetic

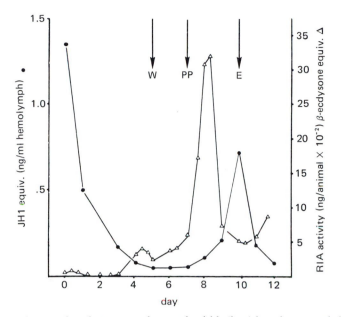

FIGURE 11. Ecdysteroid and JH titers during the fifth (last) larval instar of the tobacco hornworm, *M. sexta*. W, PP, and E represent the onset of the wandering stage, pharate pupal development, and ecdysis, respectively. RIA, radioimmunoassay. From Sridhara *et al.*, 1978.

information, differences between "covert" and "overt" differentiation, etc. The interesting aspect is that the same epidermal cells continue to secrete larval cuticle while all these events are in progress. In fact, larval cuticle deposition (i.e., elaboration of cuticular protein) follows a sigmoidal curve, the greatest rate being observed between 2 and 4 days, exactly at the time at which reprogramming takes place (Fig. 12; Wielgus and Gilbert, 1978). The thickness of the cuticle increases about 10-fold between 1 and 7 days of the last instar, at which time apolysis takes

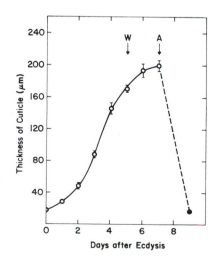

FIGURE 12. Kinetics of cuticle deposition during the fifth larval instar of the tobacco hornworm, *M. sexta*. Each point represents the average cuticle thickness of 8–14 animals ± S.E. Larval cuticle (○); pupal cuticle (●); W, wandering; A, apolysis. From Wielgus and Gilbert, 1978.

place. The pupal cuticle that probably begins to be deposited on Day 8 can be seen on Day 9. These results imply that either the mRNAs for larval cuticular proteins are stable and continue to be translated while the internal program is being modified to that of secretory pupal cuticle, or that that pupal program must be activated before the larval program is turned "off."

One critical question is whether a specific and critical replication of genomic DNA occurs at the time when the epidermal cells (or any other cells) undergo reprogramming and are committed to synthesize pupal cuticle several days later. Autoradiographic studies on the incorporation of [^{3}H]thymidine in both *Galleria* and *Manduca* epidermis showed that no DNA synthesis occurred at this time, meaning that DNA synthesis is not required for this cellular reprogramming (Krishna Kumaran, 1978; Riddiford, 1976). However, Wielgus *et al.* (1979) carried out more direct quantitative analyses with *Manduca*, not only of the amount of DNA per cell and the incorporation of [^{3}H]thymidine into DNA, but also of the MH titer in the experimental animals. Two rounds of DNA synthesis without an intervening mitosis occur exactly at the time of the small ecdysteroid peak (Fig. 13). Mitosis occurs approximately 6 days later, when apolysis and secretion of the second pupal cuticle commence (Figs. 12 and 14). These results are taken to support the "quantal mitosis" theory of cytodifferentiation, which postulates that covert differentiation occurs during a period of DNA synthesis, but the actual expression occurs in the daughter cells after mitosis. As the authors caution, these results only point to a temporal relationship between DNA synthesis and the change in commitment, but do not prove that one is required for the other. The direct way to prove such a requirement would be to block DNA synthesis and determine whether reprogramming occurs. This has apparently been done by inhibiting DNA synthesis in *Manduca* epidermis *in vitro* with cytosine arabinoside and showing that the change in commitment still occurred (Riddiford, personal communication). It is necessary that this observation be confirmed not only by autoradiographic analysis of [^{3}H]thymidine incorporation, but by a quantitative analysis of both incorporation and DNA content per cell. It is imperative that this matter be resolved since it is a central question in all of developmental biology and can be approached with the use of insect epidermis.

It is necessary to emphasize, however, that the molecular differences between larval and pupal cuticle are really not clear. If the *Tenebrio* system is any indication (Section 2.3), the similarities between larval and pupal cuticular proteins are more significant than the differences. If the proteins that comprise larval and pupal cuticles are the same or very similar, any minor differences observed could be due to posttranslational or other modifications (glycosylation, attachment of lipids, limited proteolysis, phosphorylation, etc.). The nature of the cross-linking of these proteins might account for the differences in texture, elasticity, and color of the cuticle, criteria that have been employed in all the above studies.

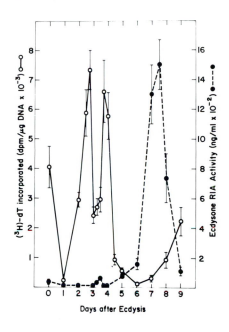

FIGURE 13. Relationship between [³H]thymidine uptake into epidermal DNA and hemolymph ecdysteroid titer during the fifth instar of the tobacco hornworm, *M. sexta*. The ecdysteroid peak at Day 3 + 12 hr was found to be significantly different from Day 3 and Day 3 + 18 hr (P < 0.001 for each). These same larvae served as donors, both for epidermis and for hemolymph. From Wielgus *et al.*, 1979.

It is known that cuticular proteins of different stages can be cross-linked through the same compound, *N*-acetyl dopamine, either at the β position of the aliphatic side chain or in the aromatic ring (Sridhara *et al.*, 1978; Andersen, 1979). The former type of cross-linking results in a colorless cuticle, while the latter type yields a dark brown cuticle. Enzymes having different specificities could control the type of cross-linking; therefore, the hormonal regulation of cuticular texture and color may be mediated through such enzymes. Only a detailed biochemical analysis of larval and pupal *Manduca* epidermal cuticular proteins can prove or disprove this hypothesis.

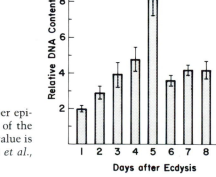

FIGURE 14. The relative DNA content per epidermal cell during the fifth larval instar of the tobacco hornworm, *M. sexta*. The Day 1 value is arbitrarily fixed as diploid. From Wielgus *et al.*, 1979.

4.2. Fat Body

As its name implies, the fat body contains a large amount of lipid, but it also contains considerable quantities of other macromolecules (protein, glycogen, rRNA, etc.). As noted previously for *Calliphora*, the fat body constitutes only 10–15% of the larval weight but contains almost 50% of the total ribosomes. An important feature of this tissue in last-instar holometabolous insects is that it synthesizes and secretes large amounts of protein into the hemolymph (see Locke, this volume). At the time of wandering, it becomes an important storage organ that actively accumulates certain proteins from the hemolymph and stores them in organelles variously called protein granules, albuminoid bodies, spherules, protein bodies, etc. A perusal of various publications on the fat body leaves the impression that it is a homogeneous, uniform tissue, but it is not. It does consist of a sheet of cells spread throughout the body, but the distribution is not uniform, and some studies indicate that the functions of the cells in various regions may be different. During metamorphosis in higher dipterans, the cells comprising the fat body are released by the breakdown of the basement membrane and these cells degenerate in time as the adult fat body cells differentiate from the growth of either precursor cells or imaginal fat body cells. This destruction of the fat body cells begins at the anterior end of the larva and progresses posteriorly (Thomson, 1975). Some larval fat body cells do persist in the newly eclosed adult, but disappear within a few days. In lepidopterans, many of the fat body cells remain intact and contribute to the organization of the pupal and the adult fat body.

It is difficult to compare biochemical data on the fat body reported by different laboratories, sometimes utilizing the same species, due to differences in timing specific developmental stages. The proper frame of reference for expression of biochemical data for developmental studies may be the weight of the larva during the feeding stages and the time elapsed from the cessation of feeding during the wandering stage rather than "days after ecdysis" since the latter does show variation. When individual tissues are employed, fresh tissue weight might be preferable. This conclusion is supported by the observation that somatic size plays a decisive role in controlling metamorphosis in *Manduca* (Nijhout, 1975), as it may in *Calliphora* and *Galleria*. The relationship between larval weight, tissue weight, and macromolecular content in *Manduca* fat body is exemplified in Fig. 15. Since many reviews have appeared on fat body metabolism in recent years (e.g., Price, 1973; Thomson, 1975; Keeley, 1978), this discussion will be restricted to only a few specific aspects.

The fat body is the site of synthesis of many hemolymph proteins including the JH binding proteins and JH-specific esterases involved in the control of JH titer (see Gilbert and Goodman, this volume). This biosynthetic activity takes place during the feeding stage; at about the time of cessation of feeding, hemolymph protein again returns to the fat body (see Locke, this volume), where it is stored in protein granules.

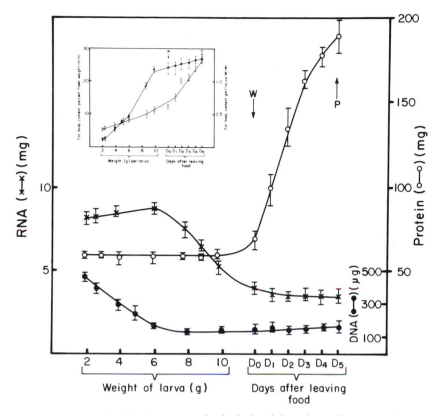

FIGURE 15. Macromolecular changes in the fat body of the tobacco hornworm, *M. sexta*, during the fifth larval instar. All results are expressed per gram tissue weight. Inset indicates changes in the fat body content in relation to larval growth. W, Wandering; P, pupation.

These storage bodies occupy most of the cell volume at this time. The process of reorganization and breakdown of the various granules and organelles occurs at different times and in different locations as required and is referred to as cell restructuring (Larsen, 1976). Unfortunately, very little biochemical data are available to explain and support these studies, which utilized cytological techniques for the most part.

The best studied hemolymph protein sequestered by the fat body is calliphorin, isolated from the blowfly, *Calliphora*; lucilin is an identical or similar protein isolated from the sheep blowfly, *Lucilia*. It now appears that drosophilin, from the fruit fly, *Drosophila*, falls into the same category. Very briefly, all three proteins are rich in aromatic amino acids and methionine, comprise between 50 and 80% of the hemolymph proteins in the mature larva, are large molecules of approximately 500,000 daltons, and consist of six identical subunits. The hemolymph of other flies and even Lepidoptera contains particles 100–150 Å in diameter that resemble calliphorin under the electron microscope. This has led to the assumption that a large-molecular-weight protein corresponding to calliphorin may be ubiquitous in insects. The fly protein exists as a hexamer *in vivo*, while on electrophoresis in SDS, a single polypeptide of 80,000–85,000 daltons is observed. Invariably, at least two closely associated bands differing by a few thousand daltons are observed. Extensive electrophoretically detectable polymorphism of lucilin subunit patterns (4–9 bands) is seen in wild and laboratory populations of *Lucilia*. Genetic, electrophoretic, immunological, and structural analysis of these populations point to the existence of 12 or more closely related structural loci, all probably located on chromosome 2 (Thomson *et al.*, 1976). A similar, but less extensive analysis of drosophilin shows it to exist in at least three forms, all derived probably from gene duplication and mutation.

These proteins form a major proportion of the total hemolymph protein in mature larvae (40% drosophilin, 70–85% calliphorin) and are taken up for storage in the fat body protein granules. The concentration of calliphorin is 7 mg per animal in the mature 90-mg larva, falls to 3 mg at pupariation, and to 0.03 mg at adult emergence. The decrease in the hemolymph protein concentration during larval–pupal transformation is due in large measure to removal of this protein from the hemolymph into the fat body protein granules, while the decrease during pupal–adult transformation is due to degradation. It must be emphasized that only a small percentage of this hemolymph protein is sequestered by the fat body.

The mRNAs from the fat body of early-third-instar larvae of *Calliphora* and *Drosophila* have been translated *in vitro* and the major product is calliphorin or drosophilin (Sekeris *et al.*, 1977; Kemp *et al.*, 1978). Sekeris and his collaborators have shown that the peak of calliphorin synthesis almost coincides with the peak of calliphorin in the hemolymph. They have also found that the mRNA for calliphorin persists beyond the time when calliphorin biosynthesis ceases, indicating that the shut-off of calliphorin and other protein synthesis in the fat body at cessation of feeding is not due to lack of mRNAs (Sekeris and Scheller, 1977; Sekeris *et al.*, 1977).

One major problem exists in the above studies of calliphorin and analogous proteins, particularly those experiments involving quantification of these proteins by the standard Lowry procedure, which actually measures the tyrosine content of the protein. Calliphorin and other fly storage proteins have a high tyrosine content (12–15%) compared to bovine serum albumin (3.36%), which is usually employed as the standard in the Lowry precedure. Consequently, the calculated amount of these proteins is amplified three- to fourfold over the actual amount present. When a correction factor is applied, the high value of 75–80% of the total hemolymph protein is lowered to 20–30%. A second procedure that is employed to quantify calliphorin and its subunits in a mixture of proteins is measurement of the relative intensity of the bands after separation of the mixture by SDS–acrylamide electrophoresis. This procedure is at best

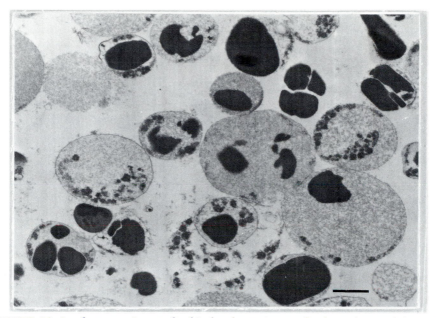

FIGURE 16. An electron micrograph of isolated protein granules from the pupal fat body of the silkmoth (*H. cecropia*). Some granules appear to have swollen and/or ruptured outer membranes. ×9400. Bar = 1 μm. From Tojo *et al.*, 1978.

semiquantitative since it is doubtful that all the polypeptides take up the dye as a direct function of their concentration. These comments are not meant to minimize the contributions of the above studies to our understanding of fat body function, but are aimed at future studies in which genetic and biochemical analyses will be carried out in order to understand the relationships between mRNA content, synthetic and degradation rates, etc., all based on the specific activity of the protein under study. In such cases, the purified protein should be used to develop the standard curve.

The above studies have used the storage proteins of Diptera. Although a large amount of cytological work exists for Lepidoptera, little biochemical analyses have been reported until recently. The work of Tojo *et al.* (1978) should be an impetus to the study of storage proteins in Lepidoptera larvae which yield large quantities of individual tissues. This group has purified the protein granules of the larval fat body of the American silkmoth, *H. cecropia*, and determined that the granules are membrane bound and consist of 98% protein. Within the membrane, one notes an amorphous outer zone and a dense core showing a periodic structure and layers of crystallization (Fig. 16). Two proteins, 1 and 2, which could be named hyalophorin 1 and 2 in keeping with precedence, can be extracted from the granules or from the hemolymph depending on the stage of development. These two proteins are distinct by electrophoretic analysis and antigenic properties, but resemble other storage

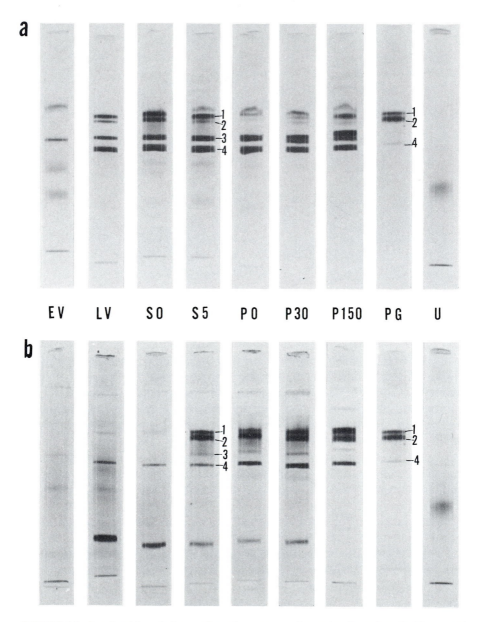

FIGURE 17. Acrylamide gel electrophoretic patterns of proteins from female *H. cecropia* of different stages. (a) Hemolymph; (b) fat body; EV, early fifth-instar larva; LV, late-fifth-instar larva; SO, larva on the day of spinning; S5, 5 days after spinning; PO, freshly molted pupa; P30, 30-day pupa; and P150, 150-day pupa. Also shown are proteins solubilized from isolated protein granules (PG) and urate granules (U). From Tojo *et al.*, 1978.

TABLE 3. AMINO ACID COMPOSITION OF FAT BODY STORAGE
PROTEINS

| Amino acid | Mole percent amino acids | | | | |
| | | | | Hyalophorin[c] | |
	Calliphorin[a]	Lucilin[a]	Drosophilin[b]	1	2
Ala	3.18	3.32	4.11	4.6	3.6
Arg	3.15	3.44	4.24	5.5	5.1
Asp	11.32	11.88	12.67	13.3	13.6
Glu	9.82	10.59	10.30	9.5	9.6
Gly	5.25	5.14	7.04	5.4	3.9
His	3.22	2.92	2.96	1.4	0.9
Ile	3.9	3.8	3.71	4.5	5.2
Leu	5.68	6.93	6.91	8.6	10.6
Lys	8.0	8.16	6.91	8.5	10.0
Met	4.42	3.52	5.38	7.0	4.9
Phe	10.9	11.11	9.35	6.6	6.2
Pro	2.99	2.96	3.21	2.3	1.9
Ser	4.2	3.72	3.59	4.9	4.9
Thr	4.7	4.29	5.2	5.9	5.5
Trp	1.74	—	—	—	—
Tyr	11.16	11.8	9.26	5.1	6.8
Val	5.8	6.35	4.69	7.0	7.5
Cys (half + cysteic acid)	0.5	Tr	0.47	Tr	Tr

[a] Data from Thomson et al., 1976.
[b] Data from Wolfe et al., 1977.
[c] Data from Tojo et al., 1978.

proteins in being hexamers of subunits with molecular weights of 85,000
and 89,000, respectively. The probable reasons for the existence of two
isomeric subunits are the same as for lucilin or drosophilin, namely
duplication and mutation of a parent gene. Table 3 compares the amino
acid composition of dipteran storage proteins with that of hyalophorins.
It is clear that the major difference is in the content of tyrosine and
phenylalanine, which are much lower in the hyalophorins. The difference
between the proteins of these two orders may be more profound, as antibodies to calliphorin cross-react with proteins in the hemolymph of
several species of Diptera but not those of Lepidoptera.

The appearance and distribution of hyalophorins follow the same
pattern as that reported for calliphorin. These proteins are absent from
the hemolymph of early last-instar larvae, appear at midinstar, and reach
maximum titer at the time of spinning (analogous to the wandering stage).
Subsequently, there is a rapid decline in the hemolymph content with
a concomitant increase in the fat body (Fig. 17). Notice that bands 3 and
4 appear to be the major hemolymph constituents, in contrast to dipteran
hemolymph where calliphorins or drosophilins are the major bands. They
appear in the fat body only in minor quantities when compared to bands

1 and 2, the hyalophorins. This implies that the uptake of proteins and the formation of granules have specificity with respect to both the protein and the fat body and are not due to a general shift in fat body metabolism. The control mechanisms that convert the fat body from a protein-synthesizing machine to one of accumulation and storage are unknown at this time.

Despite the accumulation of much information about these storage proteins, little is known about their fate and function. It was thought for some time that calliphorin could be a tyrosine carrier, the solubility of tyrosine being low, for tanning purposes during pupariation. This now seems highly unlikely for the following reasons. (1) The change that occurs in calliphorin content during larval–pupal metamorphosis is slight. (2) It is difficult to imagine what biological purpose would be served by converting a soluble protein into granules within the fat body if it must be hydrolyzed almost immediately to obtain tyrosine. (3) If this occurred to any major extent, the pool sizes of amino acids other than tyrosine should increase and that does not happen. (4) Much of the calliphorin is lost during pharate adult development when no tanning takes place. (5) In several flies, other components are employed as tyrosine precursors, e.g., tyrosine-O-phosphate in *Drosophila*, β-alanyltyrosine in *Sarcophaga*, γ-glutamylphenylalanine in *Musca* (Sridhara *et al.*, 1978). The important point is that the majority of the granules disappear during the period just before to just after adult emergence, coinciding with the upward swing of the U-shaped curve for protein synthesis. In fact, in Diptera one observes a regular arrangement of these granules between the developing muscle fibers and they disappear as the flight muscle matures. One can assume that proteolysis in the granules and protein synthesis in muscles occur simultaneously. Similarly, in Lepidoptera, at or shortly after adult emergence, the periphery of the granules, which have slowly crystallized during the pharate adult stage, become granular and by 20 hr postemergence one can detect acid phosphatase activity in that region. It can be surmised that the macromolecules in the granules are solubilized and released as degradation products to supply substrates necessary for the development of other structures. Now that the granules and their contained proteins can be isolated in pure form, one should be able to label the proteins *in vivo* or *in vitro* and follow their fate.

5. CONCLUSIONS

Insects in general, and holometabolous insects in particular, appear to be ideal model systems for the study of development, differentiation, aging, etc. Unfortunately, data on molecular interactions are sparse because of a paucity of investigators willing to work with small organisms as well as the difficulty in dissociating biochemical changes per se from those related to hormone action. Conversely, the finest biochemical analysis will be of no avail if the results and conclusions cannot correlate to, or explain, observed biological facts as exemplified by the studies on macromolecular changes in *Calliphora* (Section 2). Taking stock of progress in the field during the last decade, it is clear that no substantiative solutions have been found to any of the major problems. There has been real progress in the sense that a better appreciation of the problems is possible, due to progress in insect endocrinology and in the molecular biology of eukaryotic cells. During the next decade, resolution of many of the complex questions can be expected by the application of a multitude of approaches and techniques to some of the excellent systems now available. Just 2 years ago we predicted that genetic analysis of *D. melanogaster* at the endocrinological level would add a new dimension to the study of mechanisms of hormone action at the molecular level. This has already come to pass (see O'Connor and Chang, this volume) and one can predict great progress in the fields of hormone action and biochemistry during insect metamorphosis.

ACKNOWLEDGMENTS. The preparation of this article and the author's research were supported by Grant GM 26549 from the National Institutes of Health, Grant PCM 7903276 from the National Science Foundation, and NIH Biomedical Research Support Grant 5S07-RR05386. It is a pleasure to acknowledge the contribution of Dr. K. Katula to work on wing epidermis. I thank Ms. Thelma Carter for her expert technical assistance and Ms. Romie Brown for secretarial assistance.

REFERENCES

Airhart, J., Vidrich, A., and Khairallah, E. A., 1974, *Biochem. J.* **140**:539.
Andersen, S., 1979, *Annu. Rev. Entomol.* **24**:29.
Campbell, A. J., and Birt, L. M., 1972, *Insect Biochem.* **2**:279.
Campbell, A. J., and Birt, L. M., 1975, *Insect Biochem.* **5**:43.
Chan, S. K., and Reibling, A., 1973, *Insect Biochem.* **3**:317.
De Kort, C. A. D., 1975, *Dev. Biol.* **42**:274.
Dinamarca, M. L., and Levenbook, L., 1966, *Arch. Biochem. Biophys.* **117**:110.
Hackman, R. H., and Goldberg, M., 1976, *Comp. Biochem. Physiol.* **55B**:201.
Houlihan, D. F., and Newton, J. R. L., 1978, *J. Insect Physiol.* **24**:757.
Katula, K., 1978, Ph.D. thesis, Northwestern University, Evanston, Ill.
Keeley, L. L., 1978, *Annu. Rev. Entomol.* **23**:329.
Kemp, D. J., Thomson, J. A., Peacock, W. J., and Higgins, T. J. V., 1978, *Biochem. Genet.* **16**:355.
Krishna Kumaran, A., 1978, *Differentiation* **12**:121.
Lafont, R., Mauchamp, B., Pennetier, J. L., Tarroux, P., and Blais, C., 1976, *Insect Biochem.* **6**:97.
Larsen, W. J., 1976, *Tissue Cell* **8**:73.
Nijhout, H. F., 1975, *Biol. Bull. (Woods Hole)* **149**:214.
Nowock, J., Sridhara, S., and Gilbert, L. I., 1978, *Mol. Cell. Endocrinol.* **11**:325.
Price, G. M., 1973, *Biol. Rev.* **48**:333.
Protzel, A., and Levenbook, L., 1976, *Insect Biochem.* **6**:631.
Protzel, A., Sridhara, S., and Levenbook, L., 1976, *Insect Biochem.* **6**:571.

Riddiford, L. M., 1976, in: *The Juvenile Hormones* (L. I. Gilbert, ed.), pp. 198–219, Plenum Press, New York.

Roberts, P. E., 1976, Ph.D. thesis, University of Illinois, Urbana.

Roberts, P. E., and Willis, J. H., 1980, *Dev. Biol.* **74:**59.

Sekeris, C. E., and Scheller, K., 1977, *Dev. Biol.* **59:**12.

Sekeris, C. E., Perassi, R., Arnemann, J., Ullrich, A., and Scheller, K., 1977, *J. Insect Physiol.* **7:**5.

Smith, E., and Birt, L. M., 1972, *Insect Biochem.* **2:**218.

Sridhara, S., and Levenbook, L., 1974, *Dev. Biol.* **38:**64.

Sridhara, S., Nowock, J., and Gilbert, L. I., 1978, in: *Biochemistry and Mode of Action of Hormones*, Volume 20 (H. V. Rickenberg, ed.), pp. 133–188, University Park Press, Baltimore.

Thomson, J. A., 1975, *Adv. Insect Physiol.* **11:**321.

Thomson J. A., Radok, K. R., Shaw, D. C., Whitten, M. J., Foster, G. G., and Birt, L. M., 1976, *Biochem. Genet.* **14:**145.

Tojo, S., Betchaku, T., Ziccardi, V. J., and Wyatt, G. R., 1978, *J. Cell Biol.* **78:**823.

Wielgus, J. J., and Gilbert, L. I., 1978, *J. Insect Physiol.* **24:**629.

Wielgus, J. J., Bollenbacher, W. E., and Gilbert, L. I., 1979, *J. Insect Physiol.* **25:**9.

Williams, K. L., and Birt, L. M., 1972, *Insect Biochem.* **2:**305.

Williams, K. L., Smith, E., Shaw, D. C., and Birt, L. M., 1972, *J. Biol. Chem.* **247:**6024.

Wolfe, J., Akam, M. E., and Roberts, D. B., 1977, *Eur. J. Biochem.* **79:**47.

Wyatt, G. R., 1968, in: *Metamorphosis: A Problem in Developmental Biology* (W. Etkin and L. I. Gilbert, eds.), Appleton–Century–Crofts, New York.

Wyatt, G. R., 1972, in: *Biochemical Action of Hormones*, Volume II (G. Litwack, ed.), pp. 385–490, Academic Press, New York.

CHAPTER 7

Drosophila Imaginal Discs as a Model System for the Study of Metamorphosis

James W. Fristrom

1. INTRODUCTION AND BACKGROUND

The essence of metamorphosis in insects comprises those developmental programs that dispose of or modify larval tissues and produce the form and function of adult tissues. In *Drosophila* these metamorphic programs are executed during two developmental periods, the so-called prepupal period, immediately following puparium formation, and the pupal period. During the prepupal period, the imaginal discs, the embryonic precursors of many adult cuticular structures, undergo morphogenesis to create the external form of the adult, including the head, legs, and wings. During the pupal period, internal adult structures develop—viscera, muscles, nerves—and the detailed external differentiation occurs to produce bristles, hairs, sensillae, and other components. Fortunately, the metamorphosis of imaginal discs occurs *in vitro* under defined culture conditions (Fristrom *et al.*, 1977; Mandaron *et al.*, 1977; Milner, 1977) to produce not only basic adult structures, e.g., legs and wings, but under optimal conditions even musculature and neural tissue and the fine sculpturing of bristles and hairs in their characteristic arrays (Mandaron *et al.*, 1977; Milner, 1977).

JAMES W. FRISTROM · Department of Genetics, University of California, Berkeley, California 94720.

The studies by the imaginal disc group at Berkeley center on *in vitro*, ecdysteroid-induced metamorphic changes that are equivalent to those that occur *in vivo* during the prepupal period, namely the morphogenesis, called evagination, that produces appendages and the differentiation that produces a typical insect cuticle, the pupal cuticle, composed of an epicuticle and a chitin-containing procuticle that has a characteristic laminar appearance when viewed by transmission electron microscopy. This pupal cuticle is apolysed *in situ* during pupal development and gives way to the adult cuticle with its characteristic bristles and hairs. Virtually all of our work utilizes imaginal discs isolated *en masse* from late-third-instar larvae. The procedure for isolating discs is a continually evolving one that in its latest incarnation yields about 2.5 million discs per day from 1 to 1.5 liters of settled third-instar larvae (Eugene *et al.*, 1979). Such a yield is equivalent to about 5×10^{10} cells or 4 g wet weight of tissue. Discs isolated using these procedures are biologically competent, forming adult structures when transplanted to a developing host (Fristrom and Mitchell, 1965) and responding to ecdysteroids *in vitro* in a manner and on a timetable highly similar to that that occurs *in situ* during the late-larval and prepupal periods. The *in vitro* metamorphosis of mass-isolated discs provides the opportunity to study conveniently the molecular and cellular mechanisms by which ecdysteroids elicit disc metamorphosis. I shall describe here the general properties of mass-isolated discs and the characteristics of their response to ecdysteroids, concentrating on those events occurring *in vitro* that are equivalent to those taking place *in situ* in late larvae and prepupae. As implied from the title I shall attempt to set these considerations in a framework dealing with metamorphosis as a universal biological phenomenon. In particular, befitting the overall scope of this volume, I will compare and contrast the results we have obtained with *Drosophila* discs with metamorphic processes in amphibians. There are numerous similarities, suggesting underlying "rules" for metamorphosis.

1.1. General Characteristics of Mass-Isolated Imaginal Discs

Imaginal discs arise as invaginations of the embryonic epidermis. The number of embryonic cells that gives rise to a disc has been estimated to be around 50 (reviewed by Nothinger, 1972). During larval life discs increase by cell division, with a substantial number of those divisions occurring during the last, the third, larval instar. Based on the review by Nothiger (1972) we estimate that an early third-instar disc has, on the average, only 10–20% of the number of cells in a late-third-instar disc. Late-third-instar discs have distinctive shapes, allowing the identification of their developmental futures. Five major types of discs are present: eye–antenna, leg, wing, haltere, and genital. In mass-isolated discs, the antennal portion of the eye–antenna disc is typically absent, remaining attached to the brain, the connection between the two discs breaking during grinding of the larvae. The wing discs are the largest, containing an estimated 60,000 cells, the leg and haltere discs the smallest, containing around 10,000 cells (Fristrom 1972). Mass-isolated discs are composed of 17% wing, 71% leg and haltere, 11% eye, and about 1% genital discs. I estimate that 44% of the cells present are in wing discs, 30% in legs and halteres, 25% in eyes, and less than 1% in genital discs. Thus, the majority of cells are found in discs that evaginate to form appendages (wings, legs, halteres), but a substantial number of eye cells are present.

Discs themselves are composed of both epidermal and nonepidermal cells. The nonepidermal or adepithelial cells give rise developmentally to internal structures such as muscles and nerves. Adepithelial cells migrate into the disc during early development. The number of adepithelial cells in a disc has not been determined and may vary with differing disc types, but they are assumed to constitute 10% or less of the total. The adepithelial cells undergo little development during the prepupal period. Development of these cells occurs primarily in the pupal period when they divide and then form neural and muscle tissue. The disc epithelium itself is divided into two components, a columnar to cuboidal epithelium and the disc proper. The disc proper is the portion of the epithelium that gives rise to adult structures. Also present is the so-called peripodial membrane, a squamous epithelium that constitutes in part the tube by which the disc remains attached to the larval epidermis during larval life. The developmental distinction between the disc proper and the peripodial membrane may in part be artificial. Mandaron (1978) reports that parts of the peripodial membrane give rise to proximal structures in the adult cuticle.

In summary, mass-isolated discs constitute a heterogeneous population of discs, composed primarily of epithelial tissue. Insofar as all discs respond to the hormonal stimulus by manufacturing both an epi- and a procuticle, the heterogeneity may be less on a molecular level than would appear from the variety of disc structures and tissue types. However, when considering molecular events involved in evagination of the appendages *per se*, the contribution of eye cells must be taken into consideration as a possible source of confusion.

2. CHARACTERISTICS OF DISC METAMORPHOSIS

2.1. Late-Larval and Prepupal Development

The essential characteristics of imaginal disc development are summarized in Table 1. Larval discs are competent from at least the middle of the third instar to respond to an ecdysteroid by undergoing morphogenesis. The hormone titer, as judged by both direct and indirect meas-

TABLE 1. THE *IN VIVO* TIMETABLE OF
EARLY IMAGINAL DISC
METAMORPHOSIS

Developmental characteristic	Time (hr)[a]
Competence to respond to 20-hydroxyecdysone	−48 to −24
Increase in titer of 20-hydroxyecdysone to metamorphic levels	−6
Partial evagination first detected	0
Epicuticle deposition detected	0
Evagination complete	+4 to +6
Procuticle deposition detected	+4
Pupation (head emergence)	+12
Apolysis of first (pupal) cuticle	+24

[a] At 25°C, relative to puparium formation.

urements, increases about 6 hr prior to puparium formation. At the time of puparium formation discs are partially evaginated, and the epicuticle is partially deposited. About 4 hr after puparium formation discs are well evaginated and the first stages of deposition of the chitin-containing procuticle can be detected by transmission electron microscopy. Thus, evagination and epicuticle deposition are first detected *in situ* 6 hr and procuticle deposition 10 hr after exposure to increased hormone concentration.

The morphological differences between an unevaginated larval and a fully evaginated leg disc are shown in Fig. 1. Intermediate levels of evagination can be identified and expressed on a scale from 0 (no evagination) to 10 (full evagination), called the evagination index (Chihara *et al.*, 1971). The degree of evagination as a function of time or as a function of the concentration of a particular ecdysteroid is easily determined.

2.2. Specificity of Ecdysteroid Induction of Disc Metamorphosis

In insects, metamorphosis is induced when the prothoracic gland secretes a prohormone, ecdysone, which is then converted by peripheral tissues to the hormone, 20-hydroxyecdysone (see Gilbert and Goodman, this volume). Not all ecdysteroid target tissues in *Drosophila* are capable of converting the prohormone to the hormone. For example, imaginal discs in culture metabolize neither ecdysone nor 20-hydroxyecdysone (Chihara *et al.*, 1971), nor the phytoecdysteroid ponasterone A (an ecdysteroid found in plant material) (Yund *et al.*, 1978). This is in apparent contrast to amphibians, where target tissues characteristically are capable

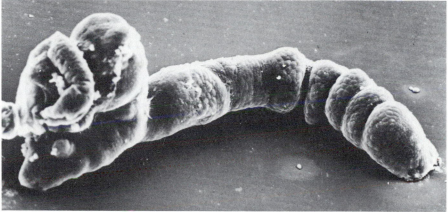

FIGURE 1. Scanning electron micrograph of an unevaginated leg disc (top) with the peripodial membrane removed to reveal the "disc proper" and a fully evaginated leg disc (bottom) showing the segments characteristic of an insect leg. From Fristrom *et al.*, 1977, with permission.

of converting the presumed prohormone thyroxine (T_4) to the hormone triiodothyronine (T_3). (Because of the extensive coverage of the subject elsewhere in this volume, I will not cite papers dealing with amphibian metamorphosis.) The physiological significance of which target tissues metabolize the prohormone and which do not in insects is not within the scope of this chapter. The absence of metabolism of these three ecdysteroids and presumably of all other ecdysteroids in imaginal discs, coupled with the conveniently assayed activity of ecdysteroids *in vitro* using the evagination index, provides a simple, biologically significant means for determining the inherent activity of ecdysteroids (i.e., evagination is a normal developmental process occurring during metamorphosis). We have previously reported on the assay of 25 ecdysteroids (Fristrom and Yund, 1976) and have tested others as well. Three major points derive from these studies. First, 20-hydroxyecdysone induces "half-evagination" (i.e., a score of 5 on the evagination index) at a concentration of about 5×10^{-8} M. Complete evagination is induced at about 6.5×10^{-8} M 20-hydroxyecdysone, a concentration of this ecdysteroid similar to that found *in vivo* at the time of puparium formation (Borst *et al.*, 1974; Hodgetts *et al.*, 1977). Second, approximately 500 times as much of the apparent prohormone, ecdysone (half-evagination at 2×10^{-5} M), is required to induce *in vitro* evagination, a concentration far in excess of the level of any ecdysteroid found *in vivo* at puparium formation, and a result entirely consistent with the proposed prohormone–hormone relationship between ecdysone and 20-hydroxyecdysone (see Gilbert and Goodman, Granger and Bollenbacher, this volume). Third, the phytoecdysteroid ponasterone A (PNA) is the most active hormone in the group we have surveyed and is required in about 1/30th the concentration of 20-hydroxyecdysone (half-evagination for PNA at 1.7×10^{-9} M). The differential activities of the ecdysteroids and the absence of activity of other steroids underscore the physiological specificity of the induction of disc metamorphosis by ecdysteroids. All three ecdysteroids, 20-hydroxyecdysone, ecdysone, and PNA, induce the entire set of disc metamorphic responses at the same concentration at which each induces evagination. This result is consistent with the existence of but one metamorphic hormone detection and response system in discs (or more precisely, consistent with the existence of a detection and response system composed of one or more receptors each with indistinguishable affinities for each of the three analogs).

2.3. Ecdysteroid Receptors in Imaginal Discs

The actions of hormones on target tissues are mediated by macromolecules, called receptors, that interact specifically, usually with high affinity and low capacity, with a hormone. The responses of discs to differing ecdysteroids provide criteria for identifying ecdysteroid receptors in disc cells. First, the equilibrium dissociation constant K_d, the concentration of hormone at which half of the receptor molecules are occupied by hormone, should be similar to the concentration of hormone needed to induce complete evagination. Second, the number of receptor molecules that bind hormone with an affinity characterized by this dissociation constant should be limited. This property is one measure by which specific physiologically significant binding can be distinguished from nonspecific binding to a variety of cellular components that are present in essentially unlimited amounts. Third, the affinities of the receptor for differing ecdysteroids should reflect the differing physiological activities of the varying receptors. Namely, because PNA is more potent than 20-hydroxyecdysone in inducing evagination, we expect the K_d for PNA to be proportionally smaller than the K_d for 20-hydroxyecdysone.

My colleague, Mary Alice Yund, has been chiefly responsible for the investigation of ecdysteroid receptors using both whole discs and cell-free extracts of discs. The results of her work have been published in a series of papers (Yund and Fristrom, 1975; Yund *et al.*, 1978; Yund, 1979) and are reviewed here.

2.3.1. Characteristics of Binding of [³H]-PNA and [³H]20-Hydroxyecdysone to Imaginal Disc Receptors

The investigation of receptors for ecdysteroids in imaginal discs has utilized binding of [³H]20-hydroxyecdysone to receptors in intact discs and the binding of [³H]-PNA to receptors in cell-free extracts. Studies on the binding of [³H]20-hydroxyecdysone to receptors in cell-free extracts have not proven possible because of the instability of the complex between the receptor and 20-hydroxyecdysone and because of a high level of nonspecific binding. The properties of the interaction between the receptor and these two ecdysteroids are summarized in Table 2. Notice that the estimated binding constants for the two hormones are very similar to the concentrations at which these two hormones induce complete evagination. Thus, approximately half of the receptor sites are occupied by hormone at concentrations of hormone that induce complete evagination. Furthermore, the K_d for PNA is about 1/60th of that for 20-hydroxyecdysone. We can conclude that the increased activity of PNA results from increased affinity for the ecdysteroid receptor (not, for example, from increased transport into disc cells). Also of note is the high instability of the hormone–receptor complex, with a half-life of binding of [³H]20-hydroxyecdysone to the receptor of about 30 min at 0°C. Thus, when exogenous hormone is removed, the bound hormone rapidly dissociates from the receptor. The affinity of the receptor for the natural hormone, 20-hydroxyecdysone, is low compared with that of steroid receptors in avian and mammalian tissues (K_d values approximately 10^{-10} to 10^{-12} M). The basis for the high K_d in insect tissues appears to stem mainly from the instability of the 20-hydroxyecdysone–receptor complex, i.e., the rate constant for binding to the receptors is similar in both insect and vertebrate tissues, but the rate constant for dissociation is much higher in insect tissues. I will note parenthetically here, and return to this point later, that the instability of the receptor–hormone complex in *Drosophila* in particular, and perhaps in insects in general, may reflect a requirement during development to switch rapidly from physiological states of high hormone levels to states of low hormone levels (see Granger and Bollenbacher, this volume).

Studies on the binding of [³H]-PNA to receptors in cell-free extracts of discs have allowed further characterization of ecdysteroid receptors.

TABLE 2. PROPERTIES OF ECDYSTEROID RECEPTORS IN IMAGINAL DISCS[a]

Parameter	20-Hydroxyecdysone	Ponasterone A	$\dfrac{\text{Ponasterone A}}{\text{20-hydroxyecdysone}}$
Association rate constant (K_1)	1.5×10^5 M^{-1} min^{-1}	1.2×10^7 M^{-1} min^{-1}	80
Dissociation rate constant (K_{-1})	3×10^{-2} min^{-1}	3.6×10^{-2} min^{-1}	0.83
Half-life of binding	23 min	19 min	0.83
Equilibrium binding constant (K_{-1}/K_1)	2×10^{-7} M	3×10^{-9} M	0.015
Equilibrium binding constant (Scatchard)	—	3.3×10^{-9} M	0.0165[b]
Number of receptors per cell	500–1000	1000 ± 300	ca. 1

[a] Data from Yund and Fristom, 1975 and Yund et al., 1978.
[b] Compared to equilibrium binding constant derived from ratio of rate constants.

As binding of [³H]-PNA to receptor is sensitive to proteases and to *N*-ethylmaleimide (an agent that disrupts disulfide bridges) but is not sensitive to RNase or DNase, the receptor is presumed to be a protein. Sephacryl S-200 gel filtration of the receptors released from disc cells indicates that the extracted receptor has a molecular weight equivalent to that of a globular protein, of about 400,000. In avian and mammalian tissues the receptor is a dimer, each subunit binding 1 molecule of hormone, the molecular weight of the dimer being about 200,000. If receptors in imaginal discs are similar to those in vertebrate tissues, the above results suggest that, under our conditions, the receptor is extracted as a tetramer. This may seem speculative, but more evidence will be presented suggesting that disc receptor subunits have properties similar to those of vertebrate steroid receptors.

There is, to my knowledge, but a single other published study on ecdysteroid receptors (see O'Connor and Chang, this volume) in which the criteria necessary to demonstrate specific binding of a ligand have been met (Maroy *et al.*, 1978). Using *Drosophila* cell cultures these investigators found receptors for ecdysteroids with binding properties indistinguishable from those described for imaginal discs. These receptors also appear to be proteins, but may exist in cells as monomers and dimers rather than tetramers (again a matter of speculation, but see O'Connor and Chang, this volume).

2.3.2. Number of Receptors in Disc Cells

The number of receptors present can be estimated from a Scatchard plot (Scatchard, 1949), where the ratio of bound to free ligand is plotted as a function of bound ligand (see Gilbert and Goodman, this volume). Alternatively, the number of receptors can be estimated by determining the saturation level for specific binding of a ligand. Using the former approach with [³H]-PNA and the latter with [³H]20-hydroxyecdysone, the number of receptors in disc cells has been estimated to be about 1000 per cell (500–1300). A similar estimate has been made for the number of ecdysteroid receptors present in *Drosophila* tissue culture cells (see O'Connor and Chang, this volume). Interestingly, for hormone receptors that act, or putatively act, on the genome, the ratio of the number of receptors to the amount of DNA in a cell is relatively constant (ca. 500–2000/pg DNA) for a variety of hormones (receptors for ecdysteroids, vertebrate steroid hormones, T_4 in mammalian liver) and also for bacterial repressors, e.g., the *lactose* repressor in *Escherichia coli* (10 per cell or 2000/pg DNA). (The action of bacterial repressors involves the binding of low-molecular-weight ligands.) The significance of the constancy of the ratio of receptors to DNA is moot. Bacterial repressors are proteins that bind to any double-helical DNA with low affinity and to specific DNA sequences with high affinity (Lin and Riggs, 1972; von Hippel *et al.*, 1974). It has been proposed (Lin and Riggs, 1975; Yamamota and

Alberts, 1975) that the number of receptor or repressor molecules present in a cell is a function of the "need" for these proteins to keep the DNA under continuous surveillance and to find the appropriate sites on which they operate, i.e., to which they bind with high affinity. Thus, given assumptions about the relative affinity of repressors for DNA in general, specific sites in particular, and time constraints in responses, it follows that proteins that "act" on the DNA in a manner similar to that of repressors will be present in cells in similar amounts relative to the DNA content. Thus, the similarity in numbers of repressor molecules and hormone–receptor molecules as a function of DNA content is a necessary, but insufficient, piece of information to suggest that the basic mechanism of action of these two groups of proteins is similar, i.e., they act to modify transcription by binding to specific sites along the DNA.

2.3.3. Specificity of Ecdysteroid Binding to Receptors

An expected property of receptors mediating the action of ecdysteroids on discs is that the receptor's affinity for different ecdysteroids should parallel the physiological activity of the ecdysteroids. Because the molar proportions of PNA:20-hydroxyecdysone:ecdysone in inducing evagination are about 1:60:25,000, we expect the ability of these ecdysteroids to displace bound radiolabeled hormone from the receptors to occur with similar differences in effectiveness and affinity. More precisely, the K_d values of the receptors for the three ecdysteroids should also parallel the physiological activity of the ecdysteroids. The ability of unlabeled hormones to displace bound [^{3}H]-PNA is summarized in Table 3. The data therein reveal that PNA is the most effective competitor and ecdysone the least effective. The relative effectiveness of the competitors in displacing bound labeled ligand is similar to their relative physiological activities with proportions of 1:26:36,000 for PNA:20-hydroxyecdysone:ecdysone. From Table 2, where the K_d values for binding of PNA and 20-hydroxyecdysone are presented, we see that the K_d for 20-hydroxyecdysone is about 67 times that of the K_d for PNA. Thus, the relative

TABLE 3. A COMPARISON OF THE EFFICIENCY OF INDUCTION OF EVAGINATION AND THE DISPLACEMENT OF BOUND [^{3}H]PONASTERONE A BY ECDYSTEROIDS

Ecdysteroid	Concentration to displace 50% of bound [^{3}H]-PNA (μM)	Ratio to PNA	Concentration to induce half-evagination (μM)	Ratio to PNA
Ponasterone A	0.002	1	0.0008	1
20-Hydroxyecdysone	0.052	26	0.05	62.5
Ecdysone	72	36,000	20	25,000

affinities of the three ecdysteroids for the receptor approximately parallel the physiological activity of the three steroids, and we can conclude with assurance that the receptors that have been identified in disc cells are the ones that mediate the effects of the hormone on discs.

2.3.4. Subcellular Location of Receptors

The argument that the number of receptors in a cell is dictated by the need to keep a comparatively constant amount of DNA under surveillance presumes that the receptors act on DNA, or in eukaryotic cells, perhaps on chromatin. The site of action of ecdysteroids has long been controversial. The observations in the late 1950s and the early 1960s that ecdysteroids modulated puffing patterns in polytene chromosomes (see Kroeger, first edition of this volume) led to the early proposal by Karlson (1962) that molting hormone acted directly on the chromatin to alter transcriptional activity, i.e., to change puffing patterns. An alternative view, championed particularly by Kroeger, held that the hormone acted on the cell and/or nuclear membrane to alter ionic ratios (particularly $K^+: Na^+$) that were in turn responsible for the alteration in transcriptional activity. Kroeger's view has been criticized (Cherbas and Ashburner, 1976) and defended (Kroeger, 1977). In discs we have reported that substantial perturbations in the $Na^+ : K^+$ ratio in disc cells as a result of incubation with ouabain, a specific inhibitor of Na^+/K^+ ATPase (the so-called sodium pump), do not affect the ability of 20-hydroxyecdysone to induce evagination. Furthermore, 20-hydroxyecdysone does not alter the binding of [^{3}H]ouabain to specific receptors in disc cells, nor does it alter the activity of Na^+/K^+ ATPase. Thus, in *Drosophila* discs we find no evidence supporting a change in ionic ratios as a general mediator for ecdysteroid action.

The cellular location of specific binding of [^{3}H]-PNA and 20-hydroxyecdysone has been investigated (Yund *et al.*, 1978; Yund, 1979). Basically, Yund has found that within 1 hr after addition of the hormone, specifically bound ecdysteroid is predominantly found in the nuclei (i.e., 95% or more) and that there is little specific binding in the cytoplasm. Yund has also demonstrated that the movement of ecdysteroid into the nuclei occurs at 0°C as well as at 25°C, although the rate of movement is slower at 0°C than at 25°C (8 hr vs. 30–60 min). In this regard, discs contrast with vertebrate tissues, where a temperature-dependent step is required for translocation of a steroid–receptor complex from the cytoplasm to the nucleus. The absence of a temperature-dependent transition turns out, for another reason, not to be surprising.

In a separate study, we reported that in discs mass-isolated from third-instar larvae and not exposed to the hormone by incubation *in vitro*, 95% of the receptor was found in the nuclear fraction. We interpret this result to indicate that in discs isolated from third-instar larvae, the unbound receptor is predominantly located in the nuclei. We recognize that discs are not hormonally naive. They have been exposed to high hormone titers during the molts between the first and the second instars and between the second and the third instars. The discs are also exposed to low ecdysteroid titers (ca. 0.02 μg/g wet wt) during intermolt periods. If the nuclear location of the receptors results from prior hormone exposure, it is not a result of high titer exposure occurring during the previous molt. As noted above, the number of cells in discs increases 5- to 10-fold during the final instar. Thus, in terms of the numbers of cells present in mature discs, only 10–20% of the cells could have been exposed to a high hormone titer. We conclude that the nuclear location of the receptor results either from exposure to the low hormone levels present during intermolt periods, or from other, yet unidentified, factors. Discs appear to differ from *Drosophila* tissue culture cells, where, according to Maroy *et al.* (1978), the ecdysteroid receptors are predominantly located in the cytoplasm and then translocate to a nuclear site following binding with an ecdysteroid in the manner of avian and mammalian steroid receptors (see O'Connor and Chang, this volume).

2.3.5. Hormonal Effects on Chromatin

The nuclear location of the ecdysteroid receptor in discs is not sufficient evidence to demonstrate that the receptor actually acts on the chromatin to elicit its effects. The hormone–receptor complex could, for example, act on the nuclear membrane. There is suggestive evidence that the hormone–receptor complex does indeed interact with the chromatin and is or becomes, in essence, a nonhistone chromatin protein. The receptor is released from nuclei by 0.3 M KCl, a condition known to displace nonhistone chromatin proteins from chromatin. The receptor is also displaced from nuclei by aurin tricarboxylic acid, a compound that interferes with protein–DNA interactions. Other evidence suggesting chromatin as a site of action of the hormone–receptor complex comes from a very different kind of experiment (Hill, Watt, and Fristrom, unpublished).

One of the striking effects of ecdysteroids is the induction of puffs in polytene chromosomes. Such a dramatic effect on transcription suggested to us that substantial changes in the repertoire of nonhistone chromatin proteins—proteins presumed to regulate transcription—might occur in response to an ecdysteroid. To test the possibility that incubation with an ecdysteroid alters the nonhistone chromatin proteins in discs, we performed experiments in which nonhistone chromatin proteins, labeled as a result of *in vitro* incubation of discs with [^{35}S]methionine, were extracted from discs that were incubated with or without 20-hydroxyecdysone. We also considered the possibility that such an approach might provide a means of identifying the ecdysteroid receptor without having to depend on the binding of labeled hormone. Such an identification might result from either of two possibilities, the first being that in the presence of an ecdysteroid the receptor would be isolated in the nonhistone chromatin protein fraction, but would not be in this fraction in the absence of an ecdysteroid. Such an hypothesis seemed reasonable in view of the situation in vertebrates, where the hormone–receptor complex translocates to the nucleus following steroid hormone binding. Second, there was the possibility of ecdysteroid-induced synthesis of the ecdysteroid receptor, with its increased synthesis allowing its identification. There are examples in vertebrates in which steroid hormones apparently stimulate the synthesis of their receptors.

A series of experiments were conducted. In one experiment two populations of discs were incubated in Robb's medium for 2 hr, one with and one without 20-hydroxyecdysone, and then incubated with [^{35}S]methionine for 5 hr in the presence of hormone. The nonhistone proteins were isolated following a modification of the procedure of Hill and Watt (1977) and were then subjected to two-dimensional electrophoresis followed by fluorography (autoradiograms of the gel). The two-dimensional patterns of nonhistone proteins synthesized during the incubation period reveal the presence of "families" of proteins with constant molecular weight but varying isoelectric points that we presume represent charge modifications of a single protein. Although the total number of individual electrophoretic entities is high, if the families really are modifications of single polypeptides, then the number of polypeptides being synthesized is comparatively low. The patterns of "spots" in the two-dimensional gels (see Sridhara, this volume) run with nonhistone proteins extracted from discs incubated with or without 20-hydroxyecdysone are very similar. Three spots are seen in which the relative level of synthesis in the presence of hormone is markedly increased relative to synthesis in the absence of hormone. Two of these are particularly interesting. They have isoelectric points around 4.5 and apparent molecular weights of about 88,000 and 108,000. In addition, the 110,000 spot resolves into two components with slightly differing mobilities during SDS gel electrophoresis. Provocatively, steroid receptors in vertebrate tissues have similar isoelectric points and molecular weights (Kuhn et al., 1975). In addition, the higher-molecular-weight component is resolved into two components of slightly differing electrophoretic migration in SDS gel electrophoresis (Kuhn et al., 1975). I have already noted that the receptor, as identified by binding of [^{3}H]-PNA during sieving on Sephacryl S-200, has a molecular weight around 400,000. This result, assuming a native receptor composed of two 88,000- and two 108,000-dalton polypeptides, is compatible with the results from two-dimensional electrophoresis of disc nonhistone chromatin proteins.

All this, of course, is circumstantial and speculative. Nevertheless, I believe the circumstantial evidence is compelling and points to the evolutionary conservation of steroid receptors. Thus, it appears, as Karlson (1962) suggested over 15 years ago, that the site of ecdysteroid action is on the genome itself.

2.3.6. Thyroid Hormone Receptors in Amphibians

The properties of ecdysteroid receptors in discs can be compared with the properties of T_4 and T_3 receptors in vertebrates. Receptors that bind T_4 and/or T_3 have been identified in both invertebrates (e.g., jellyfish) and vertebrates (e.g., mammals, amphibians). In mammals, receptors have been found in nuclei of brain, liver, and in a cell line derived from a pituitary tumor. These receptors have binding constants of about 10^{-11} to 10^{-10} M and bind T_3 with about 10 times the affinity of T_4, a property consistent with the proposed prohormone–hormone relationship for T_4 and T_3. Based on Scatchard analysis only one population of receptors is present. The nuclear receptors bind to DNA, and can be considered to belong to the nonhistone protein fraction. There are approximately 1000 receptors/pg DNA, a number similar to the number of ecdysteroid receptors in disc nuclei and to bacterial proteins that regulate transcription. There is no temperature-dependent step in the entry of ligand into the nuclei. In addition to the nuclear receptors, receptors for T_4 and T_3 have also been identified in the cytoplasm of mammalian cells.

Receptors for T_3 and T_4 have also been identified in tail and liver of premetamorphic and metamorphic tadpoles. As in mammals, there are nuclear receptors present. However, in contrast to mammals, two types of nuclear receptors are found, one type that binds both T_3 and T_4 with approximately equal affinity (K_d about 10^{-10} M) and another type that has a much greater (about 1000-fold) affinity for T_3 than for T_4. The numbers of the two classes of receptors differ, the T_3 and T_4 receptors being present in about 800 sites per tail cell and 2300 per liver cell, the T_3 receptors in about 1500 sites per tail cell and 12,300 per liver cell. These receptors are presumed to interact with chromatin. Entry of ligand into nuclei appears to involve a temperature-independent process, occurring at both 0 and 25°C. Receptors are also found in the cytoplasm. These receptors bind T_3 with far greater affinity than T_4 and are present in larger amounts than the nuclear receptors. Their affinity for T_3, however, is less than that of nuclear receptors.

Thus, we find in amphibians and mammals some (but not all) receptors that are, in distribution, number, and ability to bind ligand *in vivo*, similar to the ecdysteroid receptors in discs. We presume these receptors are instrumental in the action of thyroid hormones. However, the presence of multiple receptors, both within and without the nucleus, makes it difficult to identify which receptor population is critical for the metamorphic response. If we assume that both nuclear and cytoplasmic receptors, including both classes of nuclear receptors, are involved in hormone action, then it seems clear that the hormonal detection and response mechanisms in amphibians are more complex than those found in *Drosophila* imaginal discs.

2.4. Effects of Ecdysteroids on Discs

2.4.1. Effects on Cell-Membrane Properties

The above results from analysis of ligand binding indicate that there is but one population of ecdysteroid receptors in discs and that these receptors have a predominantly nuclear location, putatively bind to chromatin, and operate on the genome. These conclusions need some qualifications. First, receptors with differing structure but common affinities for ecdysteroids cannot be distinguished by simple binding assays. Second, low numbers of molecules with different binding properties, e.g., 5–10 per cell, would not have been identified by the assays. Third, receptors with greatly differing affinities for ecdysteroids, particularly if the affinities are much higher (i.e., characterized by far lower K_d values), might not be detected.

The existence of alternative sites of action, or alternative mechanisms of action, is implied by some effects that ecdysteroids have on disc cells. Several years ago, Jensen and De Sombre (1972) noted that a variety of steroid hormones commonly affect transport of some molecules into cells, including glucose, thymidine, and uridine. These effects are rapid, occurring within a few minutes after exposure of target cells to the steroid hormone, and are not sensitive to actinomycin D, indicating that the alterations in uptake do not depend on transcription. One can conclude that such effects on transport are general to steroid action and are an additional means by which steroids alter the physiology of target cells. We have investigated effects of ecdysteroids on transport of uridine, glucose, and thymidine into disc cells. 20-Hydroxyecdysone dramatically increases the uptake of uridine into discs within 30 min after addition of the hormone to the culture medium (Raikow and Fristrom, 1971). The hormone also affects the transport of glucose into disc cells, but the situation is womewhat more complex. Glucose appears to enter discs by two separate channels identifiable by the existence of transport with two K_m values, a low-K_m channel that would operate at low concentrations of glucose, and a high-K_m channel that would operate at high concentrations. Analysis of double-reciprocal plots of uptake data indicates that ecdysteroids affect only the low-K_m channel, decreasing the K_m and making transport more efficient at low concentrations of glucose (Chihara and Fristrom, 1973; Siegel and Fristrom, 1978). This effect is rapid, occurring within 5–10 min after addition of the steroid hormone, and we believe it to be independent of action on the genome (though insensitivity to inhibition by actinomycin D has not been convincingly demonstrated).

2.4.2. RNA Synthesis

2.4.2a. Ribosomal RNA. A major effect of ecdysteroids on discs is to increase both the synthesis and the processing of rRNA (see Sridhara, this volume). Synthesis was measured as incorporation of uridine into

RNA, corrected for alterations in transport, but not for specific activity of the acid-soluble counts (Raikow and Fristrom, 1971), and as the incorporation of UTP into RNA by isolated nuclei (Nishiura and Fristrom, 1975). Increased processing of the 38 S rRNA precursor into mature 28 and 18 S molecules was revealed by analysis of newly synthesized RNAs on sucrose gradients and by gel electrophoresis (Petri *et al.*, 1971).

We have proposed a model by which ecdysteroids increase rRNA synthesis in imaginal discs (Nishiura and Fristrom, 1975; Siegel and Fristrom, 1978). The major properties of the model are as follows.

1. A rapidly turning-over protein inhibits polymerase I activity in discs.
2. The ecdysteroid–receptor complex terminates the transcription of the mRNA for the polymerase-inhibitory protein.
3. Turnover of this mRNA in the absence of its synthesis leads to a reduction in the amount of the inhibitory protein in disc cells and increases RNA polymerase I activity.
4. Inhibition of transcription of this mRNA is maintained following dissociation of the receptor–ecdysteroid complex as a result of removal of exogenous hormone.

Evidence supporting the model comes in part from observations demonstrating that inhibiting protein synthesis increases RNA synthesis and polymerase I activity in discs to the same degree as incubation with 20-hydroxyecdysone. Inhibition of protein synthesis does not, however, augment the increased RNA synthesis stimulated by incubation with 20-hydroxyecdysone. Mixing experiments in which polymerase I activity is measured in combined extracts from discs previously incubated *in vitro* with and without an ecdysteroid indicate that an inhibitor of RNA polymerase I is present in discs not exposed to the hormone. Finally, the removal of hormone from the culture medium does not result in a return to control level of polymerase activity. In contrast, if reversible inhibitors of protein synthesis are used, polymerase activity returns to a control level following resumption of protein synthesis.

The increase in rRNA synthesis produced by ecdysteroids in imaginal discs is characteristic of the action of steroid hormones on target tissues. However, not all ecdysteroid target tissues in *Drosophila* larvae respond to ecdysteroids with increased rRNA synthesis. Recall that two major developmental programs are initiated by ecdysteroids: the development of adult precursor tissues and the degeneration of some larval tissues (see Locke, this volume). In salivary glands, which are programmed to degenerate during metamorphosis, ecdysteroids do not appear to increase rRNA synthesis, but instead may produce a reduction in rRNA synthesis. Thus, insofar as stimulation of overall RNA synthesis is concerned, there may be a dichotomy in responses of tissues programmed to develop into adult structures vs. those programmed to degenerate.

2.4.2b. Messenger RNA. For the purposes of discussion here I will consider mRNA or pre-mRNA to be anything other than rRNA or tRNA (a definition that would certainly offend many). mRNAs can be divided into two classes, those that contain poly-A tails at their 3' ends (poly-A$^+$ RNA) and poly-A-free (poly-A$^-$) mRNAs. Studies have been conducted on the effects of ecdysteroids on synthesis of poly-A-containing RNAs in discs. One such study (Silvert and Fristrom, unpublished) utilizes complementary DNA (cDNA) made against poly-A$^+$ RNA isolated from discs incubated with 20-hydroxyecdysone for 16 hr and cloned in *E. coli* using plasmid pBR322. This cDNA library was used to examine the mRNA sequences present after 0, 6, and 18 hr of incubation with 20-hydroxyecdysone. Three major classes of poly-A$^+$ RNAs were identified: (1) those present at all three times and complementary to a cDNA clone; (2) those present at 6 and 18 hr but absent at 0 hr; (3) those absent at 0 and 6 hr but present after 18 hr. We conclude from these preliminary results that incubation with 20-hydroxyecdysone changes the profile of the poly-A$^+$ RNA sequences present in discs and that the appearance of new sequences occurs over differing timetables (see also Sridhara, this volume).

An alternative approach to the effects of ecdysteroids on mRNA synthesis was taken by Bonner and Pardue (1976). These workers isolated poly-A$^-$ and poly-A$^+$ RNAs from discs incubated *in vitro* with and without 20-hydroxyecdysone and with [^{3}H]uridine. These [^{3}H]-RNAs were *in situ* hybridized to polytene chromosomes of salivary glands. Autoradiographs were made and the distribution of grains over the chromosomes was compared for the two populations of RNAs. Because the levels of labeling were low, it was necessary to perform a statistical analysis in which regions of the chromosomes were demarked into segments. Most regions formed hybrids with RNA from both control and hormone-stimulated discs. Several regions were labeled only by poly-A$^+$ RNA from ecdysteroid-stimulated discs, in particular region 67B (more precisely 67B11). One site, 47C, was found to form hybrids only with control poly-A$^+$ RNA. There are no genetic or other data regarding the nature of the products encoded at these loci. Using poly-A$^-$ RNA, most of the *in situ* hybrids were formed at the loci specifying rRNA, 5 S rRNA, and histones. Bonner and Pardue concluded from their results that 20-hydroxyecdysone modulates the synthesis of specific RNA sequences in discs.

In summary, from both of these studies, we conclude that the nature of RNA sequences changes in discs incubated with 20-hydroxyecdysone. Such a result is supportive of the possibility that 20-hydroxyecdysone alters transcription of disc mRNAs.

2.4.3. Protein Synthesis

I have described instances in which 20-hydroxyecdysone may affect synthesis of specific proteins in discs, including both stimulated synthesis of the ecdysteroid receptor and inhibited synthesis of a protein that possibly regulates RNA polymerase I activity. The hormone also affects the synthesis of other proteins in imaginal discs as revealed by double-label experiments utilizing one-dimensional SDS gel electrophoresis and causes an approximately 1.7-fold increase in overall protein synthesis (Siegel and Fristrom, 1974; Fristrom *et al.*, 1974). We expect that the altered synthesis of particular proteins should reflect the nature of the overall biological responses of a tissue to an effector (see Sridhara, this volume). Because ecdysteroids cause an increase in rRNA synthesis and processing, one can anticipate that, coordinated with these effects, there will be increased synthesis of ribosomal proteins. Also, imaginal disc evagination apparently results from organized short-distance movement of cells in the epidermis, a process called cell rearrangement (Fristrom and Fristrom, 1975; Fristrom, 1976; Fristrom and Chihara, 1978). For such rearrangement to occur there must be increased motility driven by the cell's motors, i.e., actin–myosin complexes, and there must be a map that guides the cells into their new locations. We believe this map is laid out in the language of proteins or glycoproteins on the cell surface. There is also increased secretory activity involved in the deposition of the extracellular cuticle (see Locke, this volume), a process that also involves the cell surface and for which an increased amount of Golgi is produced in response to 20-hydroxyecdysone. From these perspectives one can anticipate that there could be synthesis of new cell-surface proteins. Finally, the discs deposit a procuticle composed of both chitin, a polymer of *N*-acetylglucosamine, and cuticular proteins. One expects that an ecdysteroid will also stimulate the synthesis of proteins necessary for the formation of the procuticle, in particular, chitin synthetase and the cuticular proteins. Furthermore, we expect synthesis of proteins in these differing cell fractions to occur in separate time frames. Increased synthesis of ribosomal proteins should occur relatively soon after hormonal stimulation. Altered synthesis of cell-membrane proteins, both for cell rearrangement and as part of the increased secretory activity of discs, should follow. Finally, the synthesis of the proteins involved in the formation of the cuticle should occur. Furthermore, a means must exist to assure the correct temporal synthesis of the proteins in these celluar fractions.

1. Ribosomal proteins. Increased synthesis of ribosomal proteins was reported by Siegel and Fristrom (1974). Included in the increased synthesis were proteins not removed from ribosomal subunits by 0.3 M KCl, the core ribosomal proteins, as well as a putative initiation factor of about 70,000 daltons that was removed from the ribosomes by 0.3 M KCl. These effects on synthesis of ribosomal proteins are first detected 1–2 hr after addition of the hormone, and are substantial by 3 hr.

2. Cell-membrane proteins. To study the synthesis of cell-membrane

proteins, a vesicle fraction has been isolated by sucrose gradient centri-
riched in Na^+/K^+ ATPase, a membrane marker, and is presumed to be
enriched for other cell-membrane proteins as well. Analysis by two-dimensional gel electrophoresis and fluorography of ^{35}S-labeled proteins
synthesized *in vitro* in the presence and absence of 20-hydroxyecdysone
demonstrates substantial changes in the synthetic pattern of proteins
found in this membrane fraction, both increases and decreases in the
synthesis of specific polypeptides being detected. Such changes are first
detected 4–8 hr after the addition of 20-hydroxyecdysone to the culture
medium, and are prevalent 8–12 hr after the start of incubation.

3. Cuticular proteins. Cuticles isolated from prepupae and early
pupae contain about six major proteins and several minor proteins that
are released by detergents or 6 M urea and are resolved by electrophoresis
(Silvert and Fristrom, unpublished). These proteins are of varying molecular weights, ranging from about 15,000 to 85,000. The pupal cuticular
proteins differ in size and in one-dimensional "fingerprints" from the five
major proteins found in larval cuticles, which have molecular weights
of 10,000–16,000. The precise time after addition of the hormone when
these proteins are first synthesized has not been determined, but is presumed to be about 10 hr after the hormone is first added to the culture
medium. It is at this time that deposition of chitin is first detected.

There is an interesting and important aspect of the synthesis of pupal
cuticular proteins and of the formation of the cuticle itself that is worthy
of mention. In our hands, high levels of synthesis of these proteins depend
on removal of exogenous 20-hydroxyecdysone from the culture medium
5–7 hr after the start of incubation. The withdrawal of the hormone,
because of the instability of the hormone–receptor complex, should result
in rapid dissociation of bound ligand. In the absence of hormone withdrawal we find little or no synthesis of the procuticle as measured by the
deposition of chitin. According to the data published on the *in vivo* levels
of 20-hydroxyecdysone (Hodgetts *et al.*, 1977), there is also a drop in
hormone level *in vivo* at a corresponding time. Furthermore, chitin deposition is first detected *in vivo* by electron microscopy about 10 hr after
the increase in titer of 20-hydroxyecdysone that occurs just prior to puparium formation. Thus, the times of initial deposition of the procuticle
in vivo and *in vitro* are very similar.

In summary, the synthesis of proteins in several cell fractions in
response to hormonal stimulus occurs on different schedules. Increased
ribosomal protein synthesis appears to precede synthesis of membrane
proteins, which in turn, at least in part, precedes synthesis of cuticular
proteins. The overall mechanisms by which these temporal phases are
accomplished are not understood, but we believe they are regulated, to
some degree, by changes in the titer of 20-hydroxyecdysone.

2.4.4. Effects of Thyroid Hormones

The alterations in membrane properties and synthesis of macromolecules above are characteristic of effects induced by hormones in general and by steroid hormones in particular. In a similar fashion, T_4 (or T_3) induces changes in synthetic activities of target cells, producing overall increases in RNA and protein synthesis. The increases in RNA synthesis apparently result from changes in chromatin properties (template activity) and increases in polymerase activity. There is also evidence for increased processing of the 40 S rRNA precursor into mature 28 and 18 S forms. Accompanying the overall increase in protein synthesis are increases in synthesis of particular proteins characteristic of the developmental programs of specific tissues, e.g., increases in urea-cycle enzymes in the liver, increased net synthesis of collagen in the thigh bone, and collagenase and acid phosphatase in the resorbing gills of the tadpole. It also appears that as a generality the overall increases in macromolecular synthesis, while characteristic of those juvenile tissues giving rise to adult structures, are not characteristic of those tissues programmed to degenerate, e.g., tail tissue in most amphibians. In this regard, the juvenile tissues in amphibians are like those in *Drosophila*, the developmental program leading to degeneration not requiring overall increases in synthetic activity.

In contrast to imaginal discs where there is evidence that a reduction in the concentration of 20-hydroxyecdysone regulates the temporal response to hormonal stimulation, there appears to be no evidence suggesting that the drop in T_4 titer following metamorphosis is itself a signal for the reading of a late gene program. Nevertheless, the half-life of binding of T_3 to nuclear receptors is short (about 30 min) and is compatible with the possibility that changes in titer are critical for the temporal regulation occurring during amphibian metamorphosis.

3. DISCUSSION

As indicated in the Introduction, one focus of this chapter was to compare the action of hormones in both amphibian and insect metamorphosis using *Drosophila* imaginal discs as a model for one aspect of metamorphosis. The general characteristics of the hormonal response mechanisms in amphibians and in *Drosophila* imaginal discs are contrasted and compared in Table 4. Many striking parallels are evident. However, it does not follow that these similarities result from general laws governing hormonal response mechanisms in metamorphosis, rather than similarities in hormonal action independent of the function of the hormonal mechanism being considered. Still there are some striking aspects that are not characteristic of all hormonal response mechanisms.

TABLE 4. METAMORPHIC RESPONSES TO HORMONES IN *DROSOPHILA* IMAGINAL DISCS AND AMPHIBIANS

Characteristic	Imaginal discs	Amphibians
Prohormone	Ecdysone	Thyroxine
Hormone	20-Hydroxyecdysone	Triiodothyronine
Receptors		
Variety	One apparent type	Multiple types
Location	Nuclear	Nuclear and cytoplasmic
K_d	10^{-7} M	10^{-10} and 10^{-8} M
Half-life	~ 30 min	~ 30 min
Site of action	Chromatin	Chromatin
RNA synthesis		
rRNA	Increased synthesis, increased processing	Increased synthesis, increased processing
mRNA	New sequences appear	New sequences appear
Template activity	—	Increased
Polymerase	Polymerase I increased	Polymerase increased
Protein synthesis	Increased	Increased

3.1. Direct Nuclear Action of Hormones

Not all hormones act directly on the nucleus to elicit physiological changes in target cells. Polypeptide hormones, for example, act indirectly through "second messengers" to effect changes. Also the action of many vertebrate steroid hormones involves an interaction between hormone and receptor in the cytoplasm followed by a temperature (energy)-dependent transition by which the hormone–receptor complex translocates to the nucleus where it apparently interacts with the chromatin. In the case of both ecdysteroids in discs and thyroid hormones in amphibians, the interaction between receptor and hormone apparently occurs initially in the nucleus, and the formation of a hormone–receptor complex in the nucleus is independent of an energy-dependent transition. That both insects and amphibians are poikilothermic is not a determining factor because direct binding of T_3 to nuclear receptors in the absence of a temperature-dependent transition also occurs in rats.

3.2. Instability of Hormone–Receptor Complex

In addition to the nuclear location of the unoccupied hormone receptor, there appears to be one other similarity in the properties of the hormone receptors in amphibians and insects, namely the instability of the hormone–receptor complex. To wit, at 0°C, the half-lives of the hormone–receptor complexes in both imaginal discs and amphibians are about 30 min. (The estimate for amphibians is a gross approximation and may be smaller, but certainly is not much larger.) In *Drosophila* imaginal discs we believe that the dissociation of the hormone from its receptor is particularly important in the temporal program of events that occurs during the prepupal period, being necessary for the formation of the pro-cuticle and the synthesis of its components. To my knowledge no similar results have been obtained with T_4-induced synthetic events in amphib-ians, but the instability of the hormone–receptor complex suggests such a possibility.

3.3. Multiplicity of Target Tissues

In adult organisms T_4 plays a central role in regulating the level of general metabolism. As a general rule (adult amphibian liver is an ex-ception), every tissue responds to the hormone. In larvae of *Drosophila* it seems that virtually every tissue in the organism responds to ecdys-teroids. The responsiveness of most tissues to the hormonal stimulus is not surprising because virtually all tissues in an organism undergo some change during metamorphosis. Direct action of the hormone on the tis-sues would seem to be the most economic way for coordinating the multiple changes occurring in various tissues during metamorphosis.

3.4. Evolutionary Origin of Metamorphic Hormone Responsiveness

The mechanisms by which target tissues in insects respond to ec-dysteroids and in amphibians to thyroid hormone (see White and Nicoll, this volume) have clearly evolved independently, yet have many simi-larities. In addition to the shared physiological characteristics, both insect and amphibian metamorphic systems utilize hormones of ancient vin-tage. The use of T_4 and steroids to regulate physiological activities in lower invertebrates is well documented. Curiously, despite its ancestry, T_4 does not appear to be used in insects to regulate metabolic processes. Steroids are utilized in both insects and vertebrates. In both types of organisms these evolutionarily ancient hormone systems appear to have been "captured" for use during metamorphosis. T_4, as noted before, clearly plays an important role in regulating metabolism at stages other than metamorphosis (see Frieden, this volume). Ecdysteroids also have functions during the life cycle of insects separate from metamorphosis (e.g., vitellogenesis). I suspect, though the evidence is scarce, that ec-dysteroids, like T_4, may function in insects to regulate overall cellular metabolism. Further, I suspect that the role of these hormones as general regulators of metabolism is evolutionarily far older than their roles in metamorphosis. Thus, the use of the hormones for metamorphosis may be an evolutionary adaptation of an existing hormonal system.

One wonders about the mechanisms that allow the use of these hormone systems for differing purposes. What allows target tissues to distinguish a metamorphic signal from one regulating general metabolism? In imaginal discs, although we have direct evidence for the existence of only one hormone detection and response system functioning during metamorphosis, we suspect that a second system that responds to ecdysteroids at far lower concentrations of hormone also exists. Two results suggest the existence of a second response and detection system in discs. First, Logan *et al.* (1975) have reported that there is a transient increase in incorporation of [³H]thymidine into DNA in response to metamorphic concentrations of 20-hydroxyecdysone, but that an increase in incorporation into DNA occurring over a longer time scale also results from exposure of discs to lower concentrations of this ecdysteroid (ca. 10^{-9} M). Second, in the *ecd-1* mutant isolated by Garen *et al.* (1977), which at restrictive temperature fails to pupariate because of the absence of metamorphic levels of 20-hydroxyecdysone, we find that the discs remain in an immature state similar to that of discs isolated from early third-instar larvae (Doctor and Fristrom, unpublished), again suggesting responsiveness of discs to low levels of hormone. These kinds of observations suggest the existence of a receptor system responding to lower concentrations of 20-hydroxyecdysone than the receptors functioning in metamorphosis. Possibly these receptors were of older evolutionary origin and have given rise to the ones that are responsible for metamorphosis. In *Drosophila* I would argue that such an evolution has involved gene duplication and the evolution of one receptor gene to produce a polypeptide product that senses ecdysteroids at higher concentrations than the ancestral form. Thus, the metamorphic response is superimposed on an evolutionarily older response. Putatively, the reduced sensitivity to the hormone might be tied directly to an increase in the instability of the ecdysteroid–receptor complex. (Vertebrate steroid–receptor complexes are far more stable than *Drosophila* ecdysteroid–receptor complexes acting in metamorphosis.) If true, we might propose another explanation for the instability of the ecdysteroid–receptor complex, namely a result of the increase in the K_d, or a decrease in the affinity, of the receptor for 20-hydroxyecdysone. Noting that in amphibians there are multiple receptors in metamorphic tadpole tissue, I wonder which of these receptors is essential for metamorphic responses, and which, any or all, are evolutionarily ancient? Perhaps amphibian metamorphosis involves the use of evolutionarily new receptors? If true, this might be one more condition in which discs and amphibians are similar and in which discs serve as a general model for metamorphic target tissues.

ACKNOWLEDGMENTS. I am indebted to Dr. Mary Alice Yund for critically reading the manuscript. Previously unpublished work from the author's laboratory was supported in part by USPHS Grants GM19937 and AG02063.

REFERENCES

Bonner, J. J., and Pardue, M. L., 1976, *Chromosoma* **58**:87.
Borst, D., Bollenbacher, W., O'Connor, J. D., King, D., and Fristrom, J. W., 1974, *Dev. Biol.* **39**:308.
Cherbas, P., and Ashburner, M., 1976, *Mol. Cell. Endocrinol.* **5**:89.
Chihara, C., and Fristrom, J., 1973, *Dev. Biol.* **35**:36.
Chihara, C., Fristrom, J., Petri, W., and King, D., 1971, *J. Insect Physiol.* **18**:1115.
Eugene, O., Yund, M. A., and Fristrom, J., 1979, *Tissue Culture Manual* **5**:1055.
Fristrom, D., 1976, *Dev. Biol.* **54**:163.
Fristrom, D., and Chihara, C., 1978, *Dev. Biol.* **66**:564.
Fristrom, D., and Fristrom, J. W., 1975, *Dev. Biol.* **43**:1.
Fristrom, J., 1972, in: *The Biology of Imaginal Disks* (H. Ursprung and R. Nothinger, eds.), pp. 109–154, Springer, Berlin.
Fristrom, J., and Kelly, L., 1976, *J. Insect Physiol.* **22**:1697.
Fristrom, J., and Mitchell, H. K., 1965, *J. Cell Biol.* **27**:445.
Fristrom, J., and Yund, M. A., 1976, in: *Invertebrate Tissue Culture: Research Applications* (K. Maramorosch, ed.), p. 161, Academic Press, New York.
Fristrom, J., Gregg, T., and Siegel, J., 1974, *Dev. Biol.* **41**:301.
Fristrom, J., Fristrom, D., Fekete, E., and Kuniyuki, A., 1977, *Am. Zool.* **17**:671.
Garen, A., Kauvar, L., and Lepesant, J. A., 1977, *Proc. Natl. Acad. Sci. USA* **74**:5099.
Hill, R., and Watt, F., 1977, *Chromosoma* **63**:57.
Hodgetts, R., Sage, B., and O'Connor, J. D., 1977, *Dev. Biol.* **60**:310.
Jensen, E., and DeSombre, E., 1972, *Annu. Rev. Biochem.* **41**:203.
Karlson, P., 1962, *Prosp. Biol. Med.* **6**:203.
Kroeger, H., 1977, *Mol. Cell. Endocrinol.* **7**:105.
Kuhn, R., Schrader, W., Smith, R., and O'Malley, B., 1975, *J. Biol. Chem.* **250**:4220.
Lin, S., and Riggs, A., 1972, *J. Mol. Biol.* **72**:671.
Lin, S., and Riggs, A., 1975, *Cell* **4**:107.
Logan, W., Fristrom, D., and Fristrom, J., 1975, *J. Insect Physiol.* **21**:1343.
Mandaron, P., 1978, in: *Abstracts, International Conference on Molecular and Developmental Biology of Insects*, p. 96, Heraclion, Crete.
Mandaron, P., Guillermet, C., and Sengel, P., 1977, *Amer. Zool.* **17**:661.
Maroy, P., Dennis, R., Beckers, C., Sage, B., and O'Connor, J. D., 1978, *Proc. Natl. Acad. Sci. USA* **75**:6035.
Milner, M., 1977, *J. Embryol. Exp. Morphol.* **37**:105.
Nishiura, J. T., and Fristrom, J., 1975, *Proc. Natl. Acad. Sci. USA* **72**:2984.
Nothiger, R., 1972, in: *The Biology of Imaginal Discs* (H. Ursprung and R. Nothiger, eds.), *Results and Problems in Cell Differentiation*, Volume 5, pp. 1–34, Springer-Verlag, Berlin.
Petri, W., Fristrom, J., Stewart, D., and Hanly, E., 1971, *Mol. Gen. Genet.* **110**:245.
Raikow, R., and Fristrom, J., 1971, *J. Insect Physiol.* **17**:1599.
Scatchard, G., 1949, *Ann. N.Y. Acad. Sci.* **50**:660.
Siegel, J., and Fristrom, J., 1974, *Dev. Biol.* **41**:314.
Siegel, J., and Fristrom, J., 1978, in: *The Biology of Drosophila*, Volume IIa (M. Ashburner and T. Wright, eds.), pp. 317–394, Academic Press, London.
von Hippel. P., Rozvin, A., Gross, C., and Wang, A., 1974, *Proc. Natl. Acad. Sci. USA* **71**:4808.
Yamomoto, K., and Alberts, B., 1975, *Cell* **4**:301.
Yund, M. A., 1979, *Mol. Cell. Endocrinol.* **14**:19.
Yund, M. A., and Fristrom, J., 1975, *Dev. Biol.* **43**:287.
Yund, M. A., King, D., and Fristrom, J., 1978, *Proc. Natl. Acad. Sci. USA* **75**:6039.

CHAPTER 8

Cell Lines as a Model for the Study of Metamorphosis

JOHN D. O'CONNOR AND ERNEST S. CHANG

1. INTRODUCTION

Upon first thought it might appear unlikely that a coherent paradigm concerning the hormonal regulation of insect metamorphosis could be developed from a study of established insect cell lines. However, during the past 5 years the potential usefulness of such cell lines has been forcefully demonstrated in a number of laboratories (Courgeon, 1972, 1975; Cherbas *et al.*, 1977; Maroy *et al.*, 1978; Chang *et al.*, 1980). It seems appropriate prior to a detailed discussion of the recent data obtained from the use of insect cell lines that two obvious questions be answered. What advantages does the use of cell lines offer and what cell types are represented in the presently established cell populations?

1.1. Advantages of Cell Lines

The administration of a hormone to a normal or endocrine-deficient insect (obtained by either surgical ablation, ligation, or mutation) results in a number of tissue-specific responses (see Sridhara, this volume). However, it is difficult to resolve in such experimental systems whether the observed effects are the result of a direct action of the exogenous hormone

JOHN D. O'CONNOR · Department of Biology, University of California, Los Angeles, California 90024.   ERNEST S. CHANG · Department of Animal Science, University of California, Davis, California 95616, and Bodega Marine Laboratory, P.O. Box 247, Bodega Bay, California 94923.

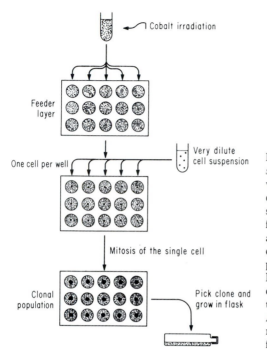

FIGURE 1. Cloning of cultured *Drosophila* cells. The technique illustrated was first applied to cultured *Drosophila* cells by Richard-Mollard and Ohanessian (1977). Irradiation prevents cells from dividing but permits metabolism and thus "conditioning" of the medium. One nonirradiated cell is then put into each well on top of the feeder layer. After a period of time the single cell has undergone sufficient division to appear as a small colony in the well. At this time the cloned cells can be removed from the wells and grown in flasks.

upon the tissue in question or the end product of a cascade initiated by the hormone at a site removed from the tissue being studied. Thus, in order to demonstrate that 20-hydroxyecdysone induced the evagination of imaginal discs directly without the intervening mediation of another tissue, imaginal discs were removed from insects, exposed to hormone *in vitro*, and were seen to evaginate (see Chapter 7 for a detailed description of this system).

In an analogous fashion single cell types can be obtained from heterogeneous populations of cells by cell-cloning techniques. In the case of cell lines from *Drosophila melanogaster*, specific and distinct populations of cells have been obtained by the technique of Richard-Mollard and Ohanessian (1977) illustrated in Fig. 1. When exposed to 20-hydroxyecdysone the cloned populations respond differentially. Several undergo spectacular shape changes while others clump or elevate specific enzyme concentrations (Cherbas *et al.*, 1989; O'Connor *et al.*, 1980). The use of such clonally derived cell lines offers several distinctive advantages to the investigation of metamorphosis. First, such cells can be grown inexpensively in copious quantities in the absence of serum. Thus, grams of hormonally responsive cells derived from a single precursor can be at hand as needed without the laborious and time-consuming necessity of dissection. Moreover, since the original cells were obtained from 8- to 12-hr embryos (Echalier and Ohanessian, 1970), it is unlikely that they have had a previous exposure to ecdysteroids or juvenile hormone. Therefore, one is permitted to examine the initial response of cells to hormone without the complications of previous exposure, which is not the case if tissues from ultimate or penultimate larval instars are used as the experimental material. Furthermore, the absence of endogenous hormone in the growth media and the inability of cell lines to metabolize ecdysteroids (Maroy *et al.*, 1978) permit experimental designs precluded in the whole organism or in organ culture. Lastly, in the case of cell lines obtained from *Drosophila*, it may well be possible to use the extensive genetic information of this insect to unravel several of the primary questions of metamorphosis.

1.2. Cell Types Present in Established Cell Lines

The first description of a cell line or continuously replicating cell population established from an insect source is generally acknowledged to have occurred in 1962 when Grace published his now classic manuscript (Grace, 1962) delineating the conditions for the continuous replication of cells from the ovaries of diapausing *Antheraea pernyi* pupa. By 1971 there were an additional 13 species of insects from which cell lines had been established (see Brooks and Kurtti, 1971, for reveiw of these early studies). In most of these cases either ovaries or whole embryos were used as the starting materials. Since these explanted tissues consisted of a heterogeneous population of cells, it was likely that the cell line once established would also contain two or more cell types. Indeed, several of the early descriptions of established insect cell lines cite the multiplicity of morphological types present in culture (Grace, 1962; Schneider, 1972). However, it should be noted that although cells may be described as "fibroblastlike," or "epitheliocytes," in most cases the tissue of origin or normal fate of cells in continuous culture is unknown. It has been suggested that continuous lines established from embryos of *D. melanogaster* consist of both imaginal and larval cells. Schneider (1972) implanted embryonic cells shortly after initial subculturing into third-instar larvae and obtained definite adult cuticular structures. However, it is not clear that upon further subculturing that such cuticular derivatives can be induced by implantation of cells into metamorphosing larvae. This would suggest either that imaginal cells are lost from the cultured population with time or that imaginal cells are retained in the population but lose their ability to synthesize cuticular derivatives.

Other less direct evidence has also suggested the presence of imaginal cells in *Drosophila* cell lines. Roberts (1975) and Moir and Roberts (1976) have demonstrated an antigenic similarity between imaginal line cells and the embryonic K_c cell line established by Echalier and Ohanessian (1969). Likewise, Debec (1974) has reported similarities in isozymes between imaginal discs and selected *Drosophila* cell lines. Although such data are equivocal, the accumulated evidence suggests that the existence of imaginal cells in cell lines is a distinct possibility.

Perhaps more important to the subject of metamorphosis than "What cells are present in culture?" is the question, "How do cell lines respond to the metamorphic hormones 20-hydroxyecdysone and juvenile hormone?" Although several cell lines have been exposed to these hormones in hopes of eliciting a response, to date the most dramatic effects have been observed in cell lines obtained from *Drosophila* embryos. Thus, in the following sections we will attempt to describe the mode of action of these hormones, primarily upon cultured *Drosophila* cells, and indicate the value of such an experimental system to the study of metamorphosis.

2. RESPONSES OF ESTABLISHED CELL LINES TO ECDYSTEROIDS

In 1971 Reineicke and Robbins observed that the addition of 20-hydroxyecdysone to an established insect cell line (RML-10, obtained from ovarian tissue of *Antheraea eucalypti*) resulted in both a decreased uptake of thymidine and a reduction in the cellular growth rate. After 5 days there were fewer cells in the media than at the start of the experiment. Up to this time insect cell lines were considered either nonresponsive toward ecdysteroids or inappropriate subjects for studying the mode of action of hormones. Subsequently, however, a wide range of responses toward ecdysteroids have been described in various cell lines, several of which will be discussed below.

2.1. Morphological Alterations

In 1972 Courgeon described the morphological alteration of *Drosophila* K_c cells in response to various ecdysteroids. In the presence of 10^{-8} M 20-hydroxyecdysone, the originally spherical, loosely adherent cells become flattened and spindle shaped, with an increased tendency toward aggregation or clumping (Fig. 2). This particular response toward 20-hydroxyecdysone has now been reported from a number of laboratories (Cherbas *et al.*, 1977; O'Connor *et al.*, 1980). Recently L. Cherbas *et al.* (1980) tested a large variety of ecdysteroids (Fig. 3) and found ponasterone A (III) and muristerone A (IV) to be almost 10-fold more active in inducing this morphological change than 20-hydroxyecdysone (II), the predominant naturally occurring ecdysteroid of *Drosophila*. Indeed, the relative effective doses of ponasterone A, 20-hydroxyecdysone, and ecdysone (I) on cultured *Drosophila* cells are not only similar to their effective doses in other ecdysteroid-responsive systems (e.g., evagination of imaginal discs; puffing of polytenic salivary gland chromosomes), but also identical to their relative affinities toward both the nuclear and the cytosolic ecdysteroid receptors found in these cells (Maroy *et al.*, 1978).

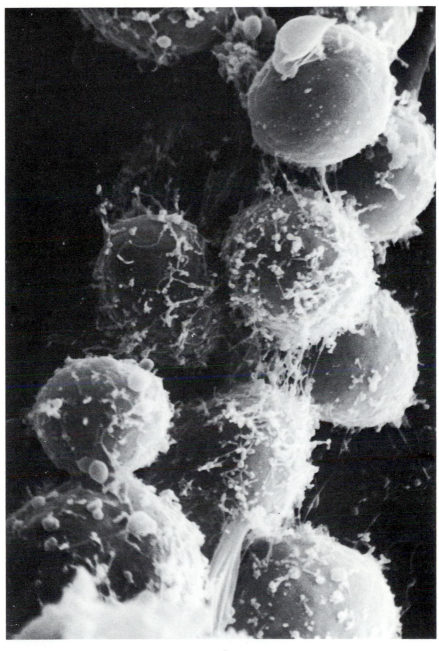

A

FIGURE 2. Scanning electron micrographs of cultured *Drosophila* cells. (A) Untreated cells grown in the absence of ecdysteroids; note the round and solitary appearance. (B, next page) Cells of clone that respond to ecdysteroids by extension of long cellular processes. (C, page 247) Cells of clone that respond by forming variously sized clumps of cells detached from the surface of the flask. Reprinted courtesy of Dr. C. M. Alvarez.

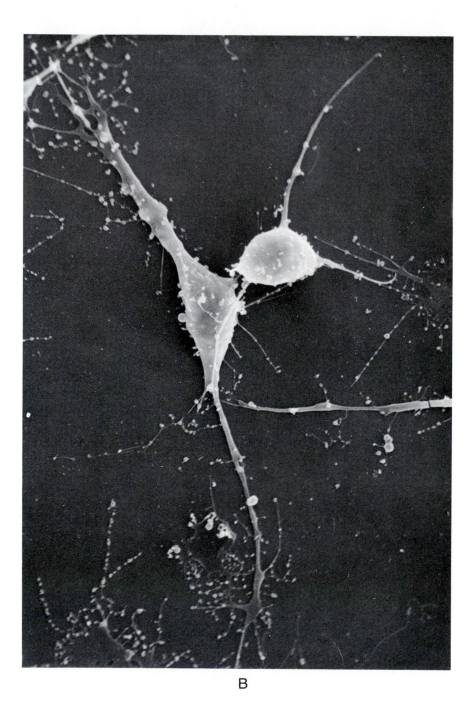

B

FIGURE 2 (*Continued*)

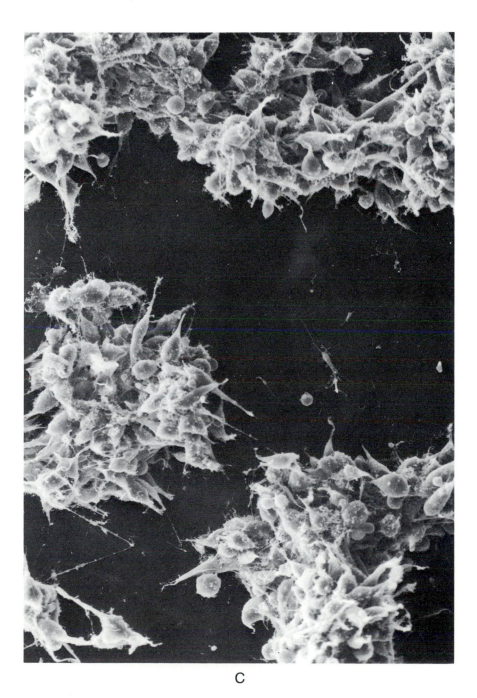

C

FIGURE 2 (*Continued*)

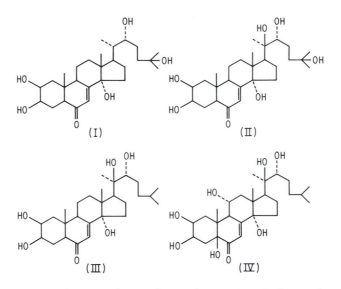

FIGURE 3. Structure of some ecdysteroids. I, ecdysone; II, 20-hydroxyecdysone; III, ponasterone A; IV, muristerone A.

This latter observation led to the suspicion that the morphological alterations induced by ecdysteroids are receptor mediated (see Section 2.5).

The induction of cellular process formation so clearly illustrated in Fig. 2 requires the continuous presence of hormone. The data in Table 1 indicates that only 18% of cells exposed to hormone for up to 32 hr and then washed and assayed for cellular extensions at 72 hr exhibit process formation, whereas 63% of cells maintained continuously in hormone for 72 hr exhibit processes. After 6 days in the presence of hormone, the extended and arborized cell surfaces are retracted and the

TABLE 1. TEMPORAL RESPONSE OF CELLS
EXPOSED TO 20-HYDROXYECDYSONE[a]

Exposure time[b] (hr)	Cells with processes[c]
0	3.8
2	7.8
8	11.3
12	4.5
24	10.9
32	17.5
72	62.8

[a] Data modified from Stevens et al., 1980.
[b] 2×10^{-7} M 20-hydroxyecdysone was present in the media for the indicated length of time, washed, and resuspended in hormone-free medium.
[c] Percent of cells with extensions as assayed 72 hr after the initial hormone addition.

cells revert to their originally smooth, spherical state. Subsequent administration of 20-hydroxyecdysone (up to 10^{-5} M—100 times the effective primary dose) at intervals of up to 2 months following hormone withdrawal fails to elicit the morphogenetic response.

The inability of cells to respond to a second exposure of hormone coincides with a loss of ecdysteroid receptor activity and is discussed in detail in Section 2.5.

A common hypothesis concerning the hormone-induced alteration in shape suggested that elongation was affected by the synthesis of microtubules in the same plane as the long axis of the cell. However, it has been rather convincingly demonstrated that although the assembly of microtubules may occur in the elongated cell processes, tubulin is neither synthesized nor turned over at a rate significantly different from that found in nonecdysteroid-treated cells (Berger *et al.*, 1978). Thus, while it is possible that disassembly of microtubules and reassembly in a different orientation may be hormonally mediated, there is no net increase in rate of tubulin accumulation. At present the underlying mechanism for the marked change in shape in unknown.

2.2. Enzyme Induction

In 1977 Cherbas *et al.* demonstrated that a subline of the K_c population, K_c-H, responded to 20-hydroxyecdysone not only with a striking change in morphology, but also a significant elevation of the enzyme acetylcholinesterase (AChE). Since AChE is restricted to the nervous system in *Drosophila*, its induction by ecdysteroids lends support to the suggestion that the original K_c line contained cells derived from lateral ectoderm some of which were destined to become neural cells.

Do ecdysteroids affect neural tissues *in vivo* during normal development? Clearly, during metamorphosis there is a rewiring of the organism as it changes from a crawling larva to a flying adult! In addition, the overall length of the neural tube is reduced during metamorphosis consistent with the smaller size of the adults in comparison to last-instar larvae. This shortening of the ventral nerve cord has been seen to occur *in vitro* as a direct action of 20-hydroxyecdysone (Robertson and Pipa, 1973). Thus, while it has not been claimed that the K_c-H line represents a determined neuroblast population of cells, they behave in many ways like neural cells and can be more expeditiously studied in their neurallike parameters than the authentic *in vitro* counterparts.

In addition to AChE, another subline obtained from *Drosophila* embryos responds to physiological doses of 20-hydroxyecdysone by an elevation of β-galactosidase (Best-Belpomme *et al.*, 1978). Although this enzyme has been extensively investigated in prokaryotes, its induction, subcellular location, and functional utility in *Drosophila* are poorly understood. At this time it simply stands as an additional example of a steroid-inducible enzyme system.

2.3. Cell-Surface Changes

In 1972 Courgeon remarked that following the administration of 20-hydroxyecdysone, K_c cells appeared to form variously sized aggregates, whereas prior to treatment these cells were generally solitary. A number of other laboratories have since confirmed this observation (Rosset, 1978; Cherbas *et al.*, 1980), and visual documentation of the phenomenon is presented in Fig. 2. In addition, cells exposed to ecdysteroids seemed to be less adherent to the glass or plastic surface of the culture vessel. Moreover, there is an indication that hormone-treated cells become less agglutinable by concanavalin A (Con A) (Metakovskii *et al.*, 1975). All of these results suggest a change in surface properties of the ecdysteroid-treated cells for which there is at least some evidence. Metakovskii *et al.* (1976) have demonstrated that following exposure to 20-hydroxyecdysone, *Drosophila* cells alter the electrophoretic distribution of glycoproteins labeled with galactose, N-acetylmannosamine, and N-acetylglucosamine. Under the influence of hormone a new protein of approximately 200,000 molecular weight appears labeled with galactose, along with three proteins ranging in molecular weight from 85,000 to 120,000. In contrast, the label from N-acetylglucosamine does not appear in the large protein following hormone treatment and is reduced in amount in the three smaller proteins. The label from N-acetylmannosamine is reduced in both the larger and the smaller proteins. It is not known whether the varied distribution of carbohydrates represents an altered glycosylation of constitutive membrane proteins or the *de novo* synthesis of a new glycoprotein. Since mannoside sugars are bound by the lectin Con A, these data offer an explanation for the reported decreased agglutinability of ecdysteroid-treated cells in the presence of Con A.

Perhaps the most dramatic effect of ecdysteroids on cell surfaces is the induction of cuticle synthesis. Riddiford (1976) has been able to regulate the type of cuticle synthesized by cultured epidermis of *Manduca sexta*. However, no such response has yet been evoked from a continuous cell line.

2.4. Cell-Cycle Changes

An event that appears to be common to all the above-described cellular responses to ecdysteroids is the cessation of division. Courgeon (1975) observed that in addition to changing their morphology, K_c cells stopped dividing following exposure to physiological concentrations of hormone. Wyss (1976) and Beckers *et al.* (1980) have suggested that lower concentrations of hormone stimulated population doubling, whereas at concentrations of 10^{-7} M and above, population doubling stopped. In addition, Rosset (1978) noticed that the incorporation of radioactive thymidine into DNA decreased markedly 10–12 hr following administration of 20-hydroxyecdysone to cultured *Drosophila* cells. If, in fact,

Rosset's observation could be extended to show that cell division stopped approximately 12 hr after exposure to ecdysteroids, then this would precede any morphological or enzymatic induction by at least 24 hr.

Recently, in a very detailed study of the cell cycle in a clonal population of *Drosophila* cells, Stevens *et al.* (1980) have demonstrated that 12 hr following treatment with either 10^{-9} M ponasterone A or 10^{-7} M 20-hydroxyecdysone, K_c cells are blocked in the G_2 stage of the cell cycle. By 48 hr the cells have begun to undergo either morphological or enzymatic induction, which is complete by 96 hr. However, in contrast to earlier reports (Courgeon, 1972; Cherbas *et al.*, 1980), these G_2-arrested cells do not lyse or die. Rather, by 120 hr after hormone exposure they begin to exit from G_2, divide, and reenter a normal cell cycle. Such a temporary cell cycle arrest in response to molting hormone raises generally interesting questions.

Can such an arrest in G_2 be demonstrated *in situ?* Of what functional significance is the arrest at G_2 as opposed to the more conventional G_1 arrest in differentiated cells?

In response to the first question, the epidermal cells of both *M. sexta* (see Sridhara, this volume) and *Tenebrio molitor* (Besson-Lavoignet *et al.*, 1981) duplicate their DNA prior to cuticle synthesis at ecdysis. Following ecdysis, the DNA level returns to the normal diploid level of epidermal cells. Clearly, the polytene nuclei of Diptera, most evident perhaps in salivary glands but present in a number of other tissues, could be in a state of G_2 arrest. Thus, in insects it may be the rule rather than the exception that cells responsive to ecdysteroids exhibit their phenotypic response in a state of G_2 arrest.

Of what functional significance a block at this stage of the cell cycle would be is not clear. However, since the block occurs following DNA replication but prior to the hypercondensation of the chromatin at metaphase, G_2 arrest may permit a more rapid transcription of RNA due to the tetraploid nature of the nuclei in G_2. Alternatively, such an arrest might permit a genomic reorganization such that the daughter cells respond differentially to hormonal signals. Such speculation simply indicates that a great deal of work still awaits those interested in the above phenomenon.

2.5. Ecdysteroid Receptors

At the present time the most popular concept of the mode of action of steroid hormones suggests that the steroid diffuses into the cell and combines with a specific protein "receptor" which as a complex then enters the nucleus of the responsive cell and initiates transcription at one or more specific loci (see Fristrom, this volume). That ecdysteroids function in specifying transcription was first suggested by Karlson (1963) after examining the data of Clever (1961), who had demonstrated that 20-hydroxyecdysone induced transcriptional activity at laterally extended

TABLE 2. CHARACTERISTICS OF CELLULAR
ECDYSTEROID RECEPTORS[a]

Parameter	Receptor	
	Cytoplasmic	Nuclear
K_D		
Ponasterone A	3×10^{-9} M	3×10^{-9} M
20-Hydroxyecdysone	1×10^{-7} M	1.5×10^{-7} M
Ecdysone	1×10^{-5} M	6×10^{-6} M
S value	4	6
$(NH)_2SO_4$ precipitation	33% saturated	50% saturated
Phosphocellulose binding	No	Yes

[a] From Maroy et al. (1978) and O'Connor et al. (unpublished observations).

loci of polytene chromosomes called puffs. An extensive series of experiments has confirmed that ecdysteroids are capable of inducing puffing and the associated transcriptional activity (Ashburner et al., 1973). However, only recently has conclusive evidence for the existence of ecdysteroid receptors been presented (Maroy et al., 1978; Yund et al., 1978).

Using the biologically active ecdysteroid, ponasterone A, we have been able to describe several kinetic as well as physical parameters of both the cytoplasmic and the nuclear receptors (Table 2). First, the binding affinity of several ecdysteroids is correlated directly with their biological activity. Thus, the K_d (equilibrium dissociation constant, a measure of binding affinity) for ponasterone A is approximately 20-fold lower than for 20-hydroxyecdysone and more than 500-fold lower than for ecdysone. The S value for the cytoplasmic receptor centrifuged in a salt-free sucrose gradient is approximately 4 S, while under similar circumstances the nuclear receptor migrates as a 6 S particle (Fig. 4). It is possible that the nuclear and cytoplasmic binding proteins are identical and the difference in S values simply represents a dimerization of the cytoplasmic receptor as it enters the nucleus. Alternatively, it is possible that upon entering the nucleus the cytoplasmic receptor becomes associated with a nonreceptor protein, thereby increasing the S value. Finally, it is quite possible that the cytoplasmic and nuclear receptors are two different proteins that because of steric demands have the same kinetic properties with various steroids but have two distinct amino acid sequences. Most of the data available to date would favor the latter interpretation. Not only are the S values different for the respective receptor preparations but there is also an obvious variation in their precipitation by ammonium sulfate and their binding to phosphocellulose. While such evidence is equivocal, it certainly suggests that the two receptors are distinct moieties.

In addition to the physical characterization of the receptor molecules, it is interesting to note their subcellular distribution. After examining

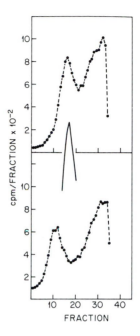

FIGURE 4. Velocity sedimentation of ecdysteroid binding proteins. The top figure illustrates the sedimentation of the cytosol receptor ligand complex in a 10–40% reorientating sucrose gradient; the bottom figure represents the sedimentation profile of the nuclear receptor ligand complex under similar centrifugal parameters. The solid line between figures represents the sedimentation of a hemoglobin standard. The ligand in this case was [³H]ponasterone A. Details have been published in O'Connor *et al.* (1980).

imaginal discs, Yund *et al.* (1978) were the first to suggest that there existed a resident population of nuclear receptors (see Fristrom, this volume). Thus, the nuclei of third-instar imaginal discs possessed approximately 95% of the receptor activity present in disc tissue. In contrast, there appears to be an almost equivalent distribution of receptors between the nucleus and the cytosol of cultured cells. While it is possible that in the cultured cells the cytosolic receptor may have a function separate and distinct from that of nuclear receptors, confirmation of such a supposition awaits further experimentation. Suffice it to say that all ecdysteroid-responsive tissues or cells thus far examined possess receptors, whereas those tissues or cells not exhibiting a response to ecdysteroids have no demonstrable receptor activity in either the nucleus or the cytoplasm. The means by which the receptor-mediated events occur will certainly be one of the most intensely investigated areas of insect endocrinology in the near future (see Sridhara, this volume).

3. RESPONSES OF ESTABLISHED CELL LINES TO JUVENILE HORMONES

3.1. Proliferation and Morphological Effects

As illustrated in Chapter 4, the actions of ecdysteroids are antagonized by the juvenile hormones (JH) *in vivo*. Similar antagonisms have not been as adequately demonstrated in established insect cell lines as

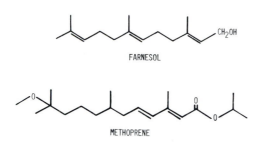

FARNESOL

METHOPRENE

FIGURE 5. Structure of two juvenile
hormone analogs.

in whole animal or organ culture systems (e.g., *Drosophila* imaginal discs). However, there are some observations suggesting that JH is an effective inhibitor of ecdysteroid-mediated processes at the cellular level.

The preceding sections have described the effects of ecdysteroids upon the cell cycle and morphology in the K_c *D. melanogaster* cell line. Preliminary observations (Courgeon, 1975) have suggested that a JH (the homolog was not identified) was able to prevent the 20-hydroxyecdysone-induced inhibition of multiplication and alteration in cell morphology. This effect was only seen at 10^{-4} M JH, a concentration greater than its limit of solubility (Kramer *et al.*, 1974).

Working with a clonal subline of the K_c cells, Wyss (1976) demonstrated a slight increase in cell proliferation with low concentrations of ecdysone (6.3×10^{-7} M) and 20-hydroxyecdysone (6.3×10^{-9} M). At high concentrations ecdysteroids inhibited cell proliferation. Depending upon the ecdysteroid concentration, low levels of a JH analog were able to inhibit the ecdysteroid effects on cell proliferation. This JH analog not only tended to make the ecdysteroid less effective (i.e., requiring a higher concentration for a similar effect), but was also inhibitory to growth in the absence of ecdysteroid (Lezzi and Wyss, 1976). The explanation for these results is apparently more complex than a simple antagonism.

An early report (Mitsuhashi and Grace, 1970) indicated that farnesol (Fig. 5) was able to antagonize the slight stimulation of proliferation of *A. eucalypti* cells by 20-hydroxyecdysone. However, this may have been a nonspecific effect, particularly since farnesol has been shown to have little JH activity in most systems examined (Bowers, 1971).

Working with the *Drosophila* K_c cell line, we have not been able to demonstrate these antagonistic effects of JH or analogs upon the ecdysteroid-mediated inhibition of proliferation. Indeed, because of the metabolism of JH (see below), very little native hormone remains in the medium after only a few hours. Until clones can be isolated that have lower rates of hormone metabolism or chemostats can be devised to maintain a constant level of JH in culture media, the rapid rate of metabolism of an authentic JH will make the demonstration of its effect on cell lines very difficult.

3.2. Macromolecular Synthesis

When JH I was added to cell cultures of the lepidopterans *A. eucalypti* and *Trichoplusia ni*, followed by radioactive labeling with uridine, it was seen that the amount of radiolabel incorporation into RNA was decreased (Cohen and Gilbert, 1972). However, it was observed that significant inhibition was also seen with other lipids that do not have JH activity, implying a nonspecific toxicity. In addition, the permeability of the RNA precursor may have been altered by the hormone treatment, thus affecting the amount of label available for incorporation and not actually decreasing the net amount of synthesis. It was suggested that JH may be affecting the plasma membrane of these cells in culture.

The effects of JH I and JH analogs were also investigated in a cell line derived from ovaries of the mosquito *Culex molestus* (Himeno *et al.*, 1979). It was seen that high concentrations (10^{-4} M) of JH I, the JH analog methoprene, or farnesol (Fig. 5) significantly decreased cell proliferation. Methoprene also affected [^{3}H]thymidine and [^{3}H]uridine incorporation. However, its effect upon precursor permeability again was not examined. Thus, while JH may affect macromolecular synthesis in cell lines, the data are as yet fragmentary and incomplete. In addition, JH appears to antagonize some actions of ecdysteroids on selected cell lines. Much more work, however, needs to be done to determine if JH is mediating its effects upon specific sites in the cells or whether the above observations are due to the nonspecific cytotoxicity of JH and its analogs as suggested by Cohen and Gilbert (1972). In support of the concept of hormonal heterophylly (the ability of invertebrate hormones to affect vertebrate cells; Burdette, 1974), JH has been shown to influence macromolecular synthesis in established cell lines of mammalian origin (Kensler and Mueller, 1978; Chmurzyńska *et al.*, 1979). The problems of nonspecific toxicity and radiolabel permeability remain to be fully addressed in these systems.

3.3. Juvenile Hormone Metabolism

Part of the problem in the elucidation of the effects of JH upon insect cell lines is the rapid metabolism of the hormone. We have examined various aspects of JH degradation by the *Drosophila* K_c cell line. When a concentrated suspension of 10^8 cells/ml was incubated in the presence of 10^{-8} M [^{3}H]-JH I, it was seen that only 18.0% of the recovered radioactivity from the medium was native hormone after 10 min. The major metabolite (72.9% of the total recovered radioactivity) was JH I acid (see Gilbert and Goodman, this volume). No labeled JH I could be detected after 35 min, while 90.1% of the radioactivity was the acid metabolite. The radioactivity associated with the cells from the above incubations was also examined. The cells were first separated from the medium by centrifugation and washed twice. After homogenization, thin-layer chromatography revealed that 70.3% of the cell-associated radioactivity was

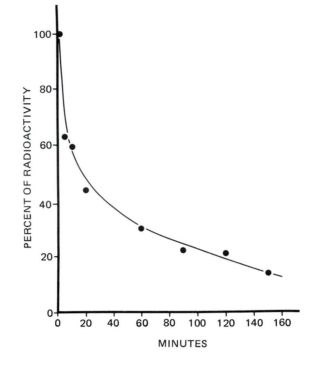

FIGURE 6. Time course of metabolism of [³H]-JH I by *Drosophila* K_c cells. One million cells per milliliter were incubated with 1×10^{-8} M hormone. At the times indicated the incubations were terminated and extracted with ethyl acetate. The extracts were analyzed by thin-layer chromatography and the amount of native hormone was determined. The JH acid comprised the majority of the radioactive metabolites although the diol and acid–diol were also present.

TABLE 3. METABOLISM OF [³H]-JH I BY
VARIOUS SUBCELLULAR FRACTIONS OF
DROSOPHILA K_c CELLS[a]

Fraction	Hormone metabolized (pmole/mg protein)
Whole cells	0.94
Whole homogenate	0.32
Nuclei	0.13
Mitochondria	0.32
Microsomes	1.03
Cytosol	0.16

[a] Approximately equal amounts of protein were incubated with 10 pmole [³H]-JH I for 15 min. The incubations were extracted with ethyl acetate and analyzed on thin-layer chromatography plates, the JH acid represented the majority of the radioative metabolites, although the diol and acid-diol were also present.

native hormone after 10 min and that 37.5% remained after 35 min. The remainder of the radioactivity was due to the acid, diol, and acid–diol metabolites (see Gilbert and Goodman, this volume).

In addition, after 35 min the total amount of radioactivity associated with the cells is as much as six times greater per unit of packed cell volume as compared to an equivalent amount of medium volume. The cells are therefore able not only to concentrate the hormone (presumably intracellularly) from the medium, but also to prevent its enzymatic degradation. This retention of the hormone is most likely due to macromolecular binding (see below).

Using a more dilute cell suspension (10^6 cells/ml, a concentration representative of a proliferating culture), the time course of the metabolism of [^{3}H]-JH I was obtained (Fig. 6). At this concentration of cells, there appears to be an initial rapid disappearance of the hormone with a half-life of approximately 10 min. At later times, however, the metabolism of the hormone appears to be slower. The loss of the native hormone from the medium is mirrored at each time point by the appearance of the JH metabolites.

Following homogenization, various subcellular fractions were examined for JH metabolic activity (Table 3). It was seen that the majority of this activity was located in the microsomal fraction (100,000g pellet).

3.4. Juvenile Hormone Receptors

Although several laboratories have documented the existence of an extracellular JH binding protein in the hemolymph of various insect species (see Gilbert and Goodman, this volume), the demonstration of an intracellular binding macromolecule (receptor) for JH has only recently been shown. An early description of a putative receptor for JH from T. molitor epidermis (Schmialek et al., 1973) seems to have been complicated by the formation of detergent micelles (see Akamatsu et al., 1975, for a discussion). More recently, Riddiford and Mitsui (1978) have provided some evidence for a high-affinity JH receptor in the nuclei of epidermis from M. sexta.

Because of the absence of endogenous hormones, the K_c cell line appears to be an ideal system for the examination of a JH receptor. JH I binding was examined in the high-speed supernatant fraction of K_c cell homogenates. Using gel permeation chromatography, velocity sedimentation centrifugation through sucrose gradients, and dextran-coated charcoal to separate bound from unbound hormone, it was demonstrated that a high-affinity receptor for JH I was present in the cytosol. This binding was saturable and specific, as evidenced by competition with excess unlabeled hormone (Chang et al., 1980). Determinations of the kinetic rate constants of binding yielded approximate values of $K_a = 1 \times 10^{-6} \, M^{-1}$ min^{-1} (Fig. 7) and $K_d = 1 \times 10^{-2}$ min^{-1} (Fig. 8). These values result in an estimated equilibrium dissociation constant of 1×10^{-8} M. These

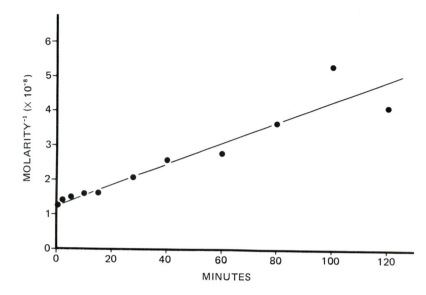

FIGURE 7. Rate of association of JH I with the cytoplasmic receptor. 1×10^{-8} M [^{3}H]-JH I was incubated with a 50% solution of cytosol and buffer (10 mM Tris, 5 mM MgCl$_2$, 150 mM KCl, pH 6.9 at 22°C) containing 10^{-5} M diisopropylphosphorofluoridate. Parallel incubations were also carried out in which excess unlabeled JH I (1×10^{-5} M) was added. Binding was determined by means of dextran-coated charcoal and represents specific binding (the difference between tubes incubated without and with excess JH I).

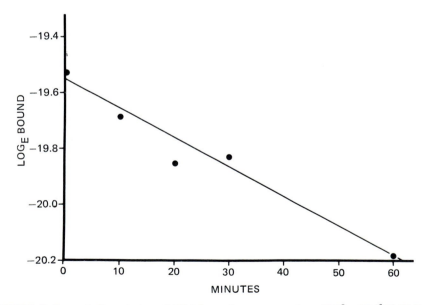

FIGURE 8. Rate of dissociation of JH I from the receptor. 1×10^{-8} mM [^{3}H]-JH I was incubated with a 50% solution of cytosol in buffer for 3 hr and then excess unlabeled JH I was added. At the designated times, aliquots were removed and assayed for bound hormone by means of dextran-coated charcoal.

TABLE 4. SPECIFIC BINDING OF [³H]-JH I
BY VARIOUS SUBCELLULAR FRACTIONS
OF *DROSOPHILA* K_c CELLS[a]

Fraction	Hormone bound (pmole/mg protein)
Whole cells	0.15
Whole homogenate	0.013
Nuclei	0.14
Mitochondria	0.01
Microsomes	0.0
Cytosol	0.71

[a] Approximately equal amounts of protein were incubated with 1.0 pmole [³H]-JH I without and with excess unlabeled JH. The unbound hormone was separated from the bound by means of dextran-coated charcoal. The difference in the amount of binding between tubes without and with excess JH was the specifically bound hormone.

data are consistent with the value obtained for the equilibrium dissociation constant as determined by Scatchard analysis (see Gilbert and Goodman, this volume).

The receptor has a molecular weight of approximately 80,000 and does not bind the acid or diol metabolites of JH I. It appears to bind the three naturally occurring JH with approximately equal affinity.

Although the majority of the specific binding appears to be concentrated in the cytosol, small but significant binding can also be demonstrated in washed nuclei (Table 4).

4. CONCLUSIONS

The use of established cell lines to study the effects of both the ecdysteroids and the juvenile hormone(s) has increased dramatically during the past few years. Several cell lines respond to ecdysteroids in a manner consistent with their putative embryonic origin. These responses together with the ease of handling cell lines and their abundant availability make them very attractive systems in which to study particular metamorphic mechanisms. At present, cell lines have been used extensively to study the mechanism by which ecdysteroids affect cellular differentiation. In the future it is clear that the modulation of the ecdysteroid effect by JH or its analogs will be studied in great detail using selected continuous cell lines.

While the responses of cultured cells to metamorphic hormones may not be identical to those seen *in situ*, the underlying mechanism is probably very similar. Thus, the receptor–hormone interactions and their subcellular distribution, the induction of new transcripts, and the alteration of cellular architecture are all parameters of metamorphosis that can be studied to great effect in continuous cell lines. However, the cultured cell cannot replace the intact organism. Indeed, the cultured cell is a *model* but the detailed examination of such cells may well reveal a critical clue in our understanding of metamorphosis.

REFERENCES

Akamatsu, Y., Dunn, P. E., Kézdy, F. J., Kramer, K. J., Law, J. H., Reibstein, O., and Sanburg, L. L., 1975, *Control Mechanisms in Development* (R. H. Meints and G. Davies, eds.), pp. 123–149, Plenum Press, New York.

Ashburner, M., Chihara, C., Meltzer, P., and Richards, G., 1973, *Cold Spring Harbor Symp. Quant. Biol.* **38**:655.

Beckers, C., Maroy, P., Dennis, R., O'Connor, J. D., and Emmerich, H., 1980, *Mol. Cell. Endocrinol.* **17**:51.

Berger, E., Ringler, R., Alahiotis, S., and Frank, M., 1978, *Dev. Biol.* **62**:458.

Besson-Lavoignet, M. T., and Delachambre, J., 1981, *Dev. Biol.* **83**(2):255.

Best-Belpomme, M., Courgeon, A.-M., and Rambach, A., 1978, *Proc. Natl. Acad. Sci USA* **95**:6102.

Bowers, W., 1971, in: *Naturally Occurring Insecticides* (M. Jacobson and O. G. Crosby, eds.), pp. 307–332, Dekker, New York.

Brooks, M. A., and Kurtti, T. J. 1971, *Annu. Rev. Entomol.* **16**:27.

Burdette, W. J., 1974, *Invertebrate Endocrinology and Hormonal Heterophylly*, Springer-Verlag, New York.

Chang, E. S., Coudron, T. A., Bruce, M. J., Sage, B. A., O'Connor, J. D., and Law, J. H., '980, *Proc. Natl. Acad. Sci. USA* **77**:4657.

Cherbas, P., Cherbas, L., and Williams, C. M., 1977, *Science* **197**:275.

Cherbas, P., Cherbas, L., Demetri, G., Manteuffel-Cymborowska, M., Savaklis, C., Yonge, C. D., and Williams, C. M., 1980, *Gene Regulation by Steroid Hormones* (A. K. Roy and J. H. Clark, eds.), pp. 278–308, Springer-Verlag, New York.

Cherbas, L., Yonge, C., Cherbas, P., and Williams, C., 1980, *Wilhelm Roux Arch. Entwicklungsmech. Org.* **189**(1):1.

Chmurzyńska, W., Grzelakowska-Sztabert, B., and Zielińska, Z. M. 1979, *Toxicol. Appl. Pharmacol.* **49**:517.

Clever, U., 1961, *Chromosoma* **12**:607.

Cohen, E., and Gilbert, L. I., 1972, *J. Insect Physiol.* **18**:1061.

Courgeon, A. M., 1972, *Exp. Cell Res.* **74**:327.

Courgeon, A. M., 1975, *C. R. Acad. Sci.* **280**:2563.

Debec, A., 1974, *Wilhelm Roux Arch. Entwicklungsmech. Org.* **174**:1.

Echalier, G., and Ohanessian, A. M., 1969, *C. R. Acad. Sci.* **268**:1771.

Echalier, G., and Ohanessian, A. M., 1970, *In Vitro* **6**:162.

Grace, T. D. C., 1962, *Nature (London)* **195**:788.

Himeno, M., Takahashi, J., and Komano, T., 1979, *Agric. Biol. Chem.* **43**(6):1285.

Karlson, P., 1963, *Angew. Chem. Int. Ed. Engl.* **2**:175.

Kensler, T. W., and Mueller, G. C., 1978, *Life Sci.* **22**:505.

Kramer, K. J., Sanburg, L. L., Kézdy, F. J., and Law, J. H., 1974, *Proc. Natl. Acad. Sci. USA* **71**:493.

Lezzi, M., and Wyss, C., 1976, in: *The Juvenile Hormones* (L. Gilbert, ed.), pp. 252–269, Plenum Press, New York.

Maroy, P., Dennis, R., Beckers, C., Sage, B. A., and O'Connor, J. D., 1978, *Proc. Natl. Acad. Sci. USA* **75**:6035.

Metakovskii, E. V., Kakpakov, V. T., and Gvozdev, V. A., 1975, *Dokl. Akad. Nauk SSSR* **221**:960.

Metakovskii, E. V., Cherdantseva, E. M., and Gvozdev, V. A., 1976, *Mol. Biol. (Moscow)* **1**:158.

Mitsuhashi, J., and Grace, T., 1970, *Appl. Entomol. Zool.* **5**:182.

Moir, A., and Roberts, D., 1976, *J. Insect Physiol.* **22**:299.

O'Connor, J. D., Maroy, P., Beckers, C., Dennis, R., Alvarez, C. M., and Sage, B. A., 1980, in: *Gene Regulation by Steroid Hormones* (A. K. Roy and J. H. Clark, eds.), pp. 263–277. Springer-Verlag, New York.

Reinecke, J. P., and Robbins, J., 1971, *Exp. Cell Res.* **64**:335.

Richard-Mollard, C., and Ohanessian, A., 1977, *Wilhelm Roux Arch. Entwicklungsmech. Org.* **181**:135.

Riddiford, L. M., 1976, *Nature (London)* **259**:115.

Riddiford, L. M., and Mitsui, T., 1978, in: *Comparative Endocrinology* (P. J. Gaillard and H. Boer, eds.), p. 519, Elsevier, Amsterdam.

Roberts, D., 1975, *Curr. Top. Dev. Biol.* **9**:167.

Robertson, J., and Pipa, R., 1973, *J. Insect Physiol.* **19**:673.

Rosset, R., 1978, *Exp. Cell Res.* **111**:31.

Schmialek, P., Boroski, M., Geyer, A., Miosga, V., Nündel, M., Rosenberg, E., and Zapf, B., 1973, *Z. Naturforsch. Teil C* **28**:453.

Schneider, I., 1972, *J. Embryol. Exp. Morphol.* **27**:353.

Stevens, B., Alvarez, C. M., Bohman, R., and O'Connor, J. D., 1980, *Cell* **22**:675.

Wyss, C., 1976, *Experientia* **32**:1272.

Yund, M. A., King, D. S., and Fristrom, J. W., 1978, *PNAS* **75**:6039.

PART II

VERTEBRATES

CHAPTER 9

Survey of Chordate Metamorphosis

JOHN J. JUST, JEANNE KRAUS-JUST, AND
DEBORAH ANN CHECK

1. INTRODUCTION

The survival of any species depends on its producing enough viable offspring so that on the average a single reproductive adult is replaced by
another during its lifetime. In this regard, the phylum Chordata has utilized two approaches. The first and probably more primitive approach is
the production of large numbers of unprotected eggs. This reproductive
strategy is used by some species in all chordate classes excluding birds
and mammals.

The second approach provides parental protection for a relatively
small number of eggs. This strategy has frequently arisen in the course
of chordate evolution. The modes of protection are diverse, ranging from
simply guarding fertilized eggs against environmental fluctuations or
predators (found in fish, amphibians, birds, and mammals) to retaining
fertilized eggs within the mother and nurturing them throughout development by means of a placenta. True viviparity (placenta formation) is
present among tunicates, fish, and mammals.

Other complex parental protective mechanisms have evolved among
chordates which are rather curious and diversified. Ovoviviparity, in
which development of the young takes place in the reproductive tract of

JOHN J. JUST, JEANNE KRAUS-JUST, and DEBORAH ANN CHECK · Physiology Group,
T. H. Morgan School of Biological Sciences, University of Kentucky, Lexington, Kentucky
40506.

the female without a true placenta, has been extensively covered in natural history literature, and is found among tunicates, fish, and amphibians. Originally ovoviviparity implied that no nutrients were supplied to the developing young from the mother. However, in many cases the young obtain nutrients, and a near-continuum exists between true ovoviviparity and viviparity among fish and amphibians.

The curious use of other body parts of either male or female fish or amphibians for protection of their young has been reported. Parts of the digestive system, such as the buccal cavity and even the stomach, provide protection and may also supply nutrients to the developing young. In others, skin modifications are used to protect the eggs or young, the most common being the marsupial pouch; this form of protection occurs among fish, amphibians, and mammals. In amphibians, outgrowths of skin, particularly from the dorsal surface, permit the attachment of eggs. The embryo develops in these pouches until hatching.

Paleological evidence covering a period of about 500 million years indicates that whether or not chordate animals protected their young was not the sole reason for their survival or extinction. Today there are nearly 45,000 species distributed among the various chordate groups, with about half of these among fish, while the smallest number is present in the cephalochordates (Fig. 1).

No claim has been made that metamorphosis occurs in amniotes, but it is known that metamorphosis occurs in other major chordate groups, particularly in those species that offer little or no parental protection to the young, and where reproduction occurs in water. Unfortunately, the term "metamorphosis" is used to mean various things, depending on the author. This necessarily makes any survey of chordate metamorphosis a personally biased study. No agreement has been reached as to what constitutes metamorphosis in tunicates, cephalochordates, fish, and amphibians. Adoption of a common terminology would both eliminate the confusion and further the cause of investigation. The following three criteria are offered as a standard approach to metamorphosis.

1. There must be some change in form of nonreproductive structures between the time the embryo hatches from the egg and before it reaches sexual maturity. With this criterion, all structural changes having to do with embryonic development, sexual maturation, and aging are eliminated.
2. The form of the larva enables it to occupy an ecological niche different from that of the adult, clearly setting it apart from the embryo or adult. This criterion ensures that the change in form does not represent late embryonic changes or developmental changes in the young organism, but aids species survival by permitting the exploitation of new ecological niches.
3. The morphological changes that take place at the end of larval life (climax) depend on some environmental cue, either external

(e.g., light, salinity, temperature, food supply) or internal (e.g., hormonal changes, changes in yolk reserves).

That the change in form has received the most attention in the literature is not remarkable, considering that it is the most obvious aspect to study in terms of gross morphology, histology, and molecular changes. The other two criteria remain to be investigated. It is recognized that metamorphosis in any one taxonomic group may not meet all criteria; however, experimental procedures should be initiated within each group to ensure that more than one of the criteria are met before naming the process "metamorphosis."

Once a common definition of metamorphosis is accepted, the next urgent need is the use of standardized stages to describe the process.

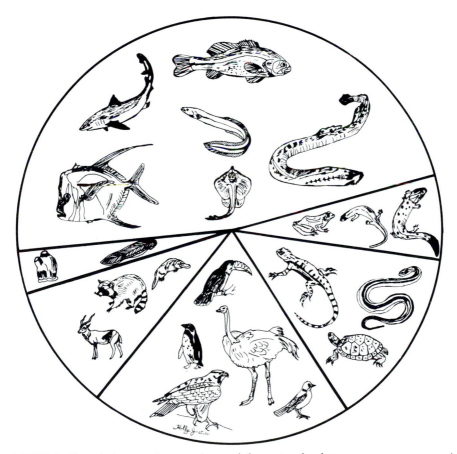

FIGURE 1. The relative species abundance of the major chordate groups among approximately 45,000 living species, starting clockwise with fishes (47.4%), amphibians (6.4%), reptiles (13.5%), birds (19.9%), mammals (9.4%), and urochordates and cephalochordates (3.4%).

Comparisons between published results become difficult within a single species and are nearly impossible when crossing species lines if authors ignore published gross morphological terminology or refer to it improperly. While such publications may be of some use to a specific discipline, such as biochemistry, cell biology, or ecology, they contribute little to the overall understanding of metamorphosis.

It is imperative that all published literature refer to standardized morphological stages and that if such stages are not available effort should be expended to establish them. While standardized morphological stages are available for amphibian and agnathan metamorphosis, there is still a great need for description of metamorphic stages among tunicates, cephalochordates, and fish. All morphological changes occurring during larval life should be referred to as "metamorphosis," and published stages should not restrict the use of this term to the last dramatic changes that occur during the final transition from larvae to young adults. These last dramatic changes in form should be referred to as "transformation" or "climax" stages. The latter term will be used in this review with the intention that it will be accepted as the common terminology not only in anuran metamorphosis but also in all chordate metamorphosis.

Many terms have been used to denote organisms that can reproduce while still retaining most of their larval characteristics (paedogenesis, paedomorphosis, progenesis, neoteny). Historically, these terms were coined to distinguish at least two ways in which larvae with reproductive capabilities could have arisen (Pierce and Smith, 1979; Gould, 1977). First, the development of the nonreproductive larval structures is delayed relative to that of the reproductive structures; in the second, development of the reproductive structures is accelerated relative to that of the nonreproductive structures. In practice one cannot tell how this reproductive capacity arose in most chordates that retain their larval characteristics. The term "neoteny" will be used in this review to refer to all chordate larvae that have or appear to have reproductive capacity while still retaining larval characteristics. This term will be used because it has been used in all chordate groups to describe such organisms, was used in the very early literature, and is the most common term.

The following survey will describe the major metamorphic events occurring in chordates. The subphyla Urochordata, Cephalochordata, and Vertebrata are included. Among the seven living classes of vertebrates, metamorphosis has been ascribed to three—Agnatha, Osteichthyes, and Amphibia. Metamorphosis is not thought to occur in the other four classes—Chondrichthyes, Reptilia, Avia, and Mammalia. Included in the survey will be a discussion of the basic biology of each larval group and the type of ecological niche they occupy. The triggering mechanism initiating metamorphic climax will be discussed in reference to each of the chordate groups. Finally, the major unresolved questions will be discussed.

2. UROCHORDATES

The subphylum Urochordata (tunicates) is composed of three living classes (Appendicularia, Thaliacea, and Ascidia), all of which are marine. The vast majority (over 90%) of the 2000 living species of tunicates comprise Ascidia. While claims have been made that they are the only class to undergo metamorphosis (Young, 1962), evidence would suggest that all three classes metamorphose. Adult ascidians are benthic, sedentary organisms, either solitary or colonial. Adult Appendicularia (Larvacea) are pelagic, solitary organisms that swim via currents generated by their tails. Adult Thaliacea are pelagic, solitary or colonial animals that float and swim via water currents generated by muscle bands around their bodies.

2.1. Appendicularia

Appendicularia have received much attention recently due to the major role of the adults in ocean food chains. The adult form includes a trunk, tail, and house (Fig. 2D). The adult trunk epidermis secretes an external gelatinous house containing a complicated filter-feeding apparatus. These gelatinous houses are abandoned four to eight times daily, presumably because the filtering apparatus becomes clogged with plankton. Discarded and occupied houses are an important food source for fish and crustacean larvae (Alldredge, 1977).

All but one species of Appendicularia are hermaphroditic. Fertilization occurs externally and the adult dies after spawning. Spawning can begin as early as 3 days after metamorphic climax, and few adults live more than 20 days. The tadpole hatches about 3 hr after fertilization (Fig. 2C). Metamorphic changes occur over the next 5 hr of its larval life span and include the development of a mouth and cerebral ganglia in the trunk and a 200% increase in tail length. In the third hour after hatching the caudal appendage and cuticle are lost from the tail and swimming commences. For the last 2 hr of larval life the animal swims actively, tail fins appear, and siphons open in the trunk. At metamorphic climax swimming activity intensifies and the tail shifts from the posterior end of the animal to the side, this shift being completed in a few seconds. The second event of climax is the production by the trunk epithelium of a gelatinous house with an elaborate filtering apparatus. This house building begins shortly after the tail shifts, and takes less than 25 sec. When complete the house has a volume 300 times larger than the trunk (Fenaux, 1977).

After metamorphic climax all cell division stops and growth depends on an increase in the size of individual cells. The DNA content increases by endomitosis. Depending on their function, different cells have different amounts of DNA, ranging from 50 to 500 times the amount of DNA

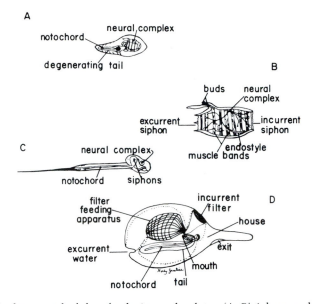

FIGURE 2. The larvae and adults of pelagic urochordates. (A, B) A larva and adult, respectively, of the class Thaliacea (order Doliolidae) (after Kükenthal and Krumbach, 1956). The larva is shown at the time of tail degeneration; the adult is the asexual reproducing form showing the large pharyngeal area. (C, D) A larva and adult, respectively, of the class Appendicularia. The tail and neural complex are easily recognized in the larva. The adult retains the tail and is surrounded by a relatively large, complex gelatinous house containing, among other things, an incurrent filter.

in diploid cells. Similar cases of increased ploidy have been observed during development of other members of the animal kingdom: scale formation in the wings of the Ephestia moth; giant chromosomes of various insect tissues; and nerve cells in fish and amphibians. No explanation is available for either the cause of the polyploidy or its function during appendicularian growth.

2.2. Thaliacea

The three orders of Thaliacea undergo both sexual and asexual reproduction (Kükenthal and Krumbach, 1956). Asexual reproduction in all three occurs by budding from the mature organism, and the offspring either stay attached to the parent or are released to start an independent existence. In sexual reproduction of Salpida and Pyrosomida, fertilization is internal. In Salpida the developing embryo forms a placenta and its development is viviparous. In Pyrosomida development is ovoviviparous with the zygote developing in the adult cloaca. The embryo uses egg yolk for sustenance before it is released by the adult as a newly metamorphosed juvenile.

In the third order, Doliolidae, an adult form called the gonozooid sheds sperm and ova into the water, where fertilization takes place. The fertilized egg develops into a free-swimming tadpole (Fig. 2A). During larval life it is a nonfeeding zooplankton in the open ocean, but at metamorphic climax a feeding apparatus develops while the tail is lost. The resulting adult is called the oozoid (Fig. 2B). Throughout its life span the oozoid gives rise to a series of buds that eventually detach. During the free-swimming stage some of these buds develop into gonozooids, thereby repeating the life cycle.

2.3. Ascidians

Adult ascidians are sedentary organisms covered by a noncellular tunic through which two external siphons project, the branchial and atrial siphons (Fig. 3). Filter-feeding occurs by passage of water through the branchial siphon to the branchial sac, where food is trapped by mucous secretions from the endostyle. Trapped food passes from the pharynx down the esophagus through the stomach to a short intestine that opens into the atrial cavity. Water in the branchial sac also enters the atrial cavity and is expelled with feces through the atrial siphon.

All but one species of ascidians are hermaphroditic. In solitary ascidians fertilization is external, the gametes being released through the atrial siphon. Development from zygote to tadpole takes less than 3 days, and can be as rapid as several hours. Except for one species, fertilization is internal in colonial ascidians. The zygote develops into a tadpole in various organs of the adult, depending on the species. These organs include special brood pouches, oviducts, the atrial cavity, and even the noncellular tunic. Zygotes are released through the atrial siphon except in those species in which development occurs in the tunic. Tadpoles are expelled at various stages of development, and in some cases completely

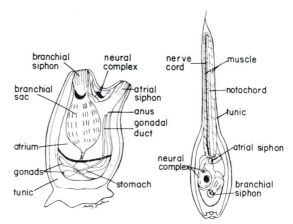

FIGURE 3. Typical adult (left) and larval (right) ascidians. The adult is a solitary form and contains a large branchial filter-feeding apparatus. The tadpole has a rudimentary, nonfunctioning filter-feeding apparatus in its trunk and a large tail.

metamorphosed animals are released. Release of gametes from solitary adults, and of tadpoles from colonial adults, occurs after exposure to light following a period of darkness (West and Lambert, 1976).

While there are external and internal structural differences between species, all larvae possess a tail and trunk (Fig. 3). The tail is composed of a notochord constructed of vacuolated cells, and a dorsal hollow neural tube. Two bands of striated muscle lie on either side of the notochord and neural tube. Each band is composed of a number of muscle rows, the number depending on species and location in the tail. At the ultrastructural level, striated muscles of the tadpole tail appear similar to vertebrate striated muscle, except that the sarcoplasmic reticulum does not always have a transverse tubular system associated with it (Burighel *et al.*, 1977). An epidermal cell layer and external tunic cover the entire tail and trunk.

The trunk of the tadpole contains many organs, the most conspicuous being a sensory vesicle situated on the right side. This vesicle contains two sense organs, a statocyte or presumptive balancing organ used to detect gravity, and an ocellus or photoreceptor. The ocellus resembles a vertebrate eye in that it contains a retina, lens, and pigment cup (Barnes, 1974). A hollow neurohypophysis, a larval visceral ganglion, and a solid definitive adult ganglion are located on the left side of the tadpole. These neural structures are situated between the atrial and the branchial siphons. The siphons are filled with tunic, making it impossible for the tadpole to feed. The branchial siphon connects to a branchial basket with its associated gill slits and endostyle. The atrial siphon has a developing atrium associated with it. A rudimentary circulatory system and adhesive glands are also located at the anterior end of the trunk (Cloney, 1979).

There is great variety in tadpole structure from species to species (Millar, 1971). The most striking difference is in length, which ranges from 0.5 to 10.0 mm. The largest tadpoles are produced by colonial forms, and since all dimensions increase in these tadpoles, we calculate that total individual tadpole volume may vary by as much as 10,000. Not only are tadpoles of the solitary species smaller, but they also exhibit less maturity in their organ systems during the early part of their free-swimming stage, examples being the heart, pharyngeal gill slits, atrium, ocellus, and statocyte. Some colonial larvae may be so mature that they start asexual reproductive budding before metamorphic climax. Irrespective of their structural maturity, tadpoles from both solitary and colonial adults have a free-swimming period.

The free-swimming period does not permit the tadpole itself to exploit new ecological niches, but does aid in the geographic dispersal of the species. This dispersal is dependent upon the life span and swimming ability of the tadpole. Depending on the species, the larval life span may range from a few minutes to as long as several days. Even within a single species there is considerable variability. Free-swimming tadpoles do not feed, surviving on stored nutrients. While one might therefore expect species with larger eggs to have a longer larval period, there is no correlation between egg size and larval life span. The swimming rate of the larvae increases linearly with an increase in the length of the tail, 0.6 cm/sec/1.00-mm tail length (Millar, 1971). In most species the tadpole tail is two-thirds of the total body length. Therefore species with larger tadpoles have a greater potential for dispersal than do species with smaller tadpoles.

During the free-swimming period, larvae undergo periods of activity and inactivity. When active, the tadpole swims toward the ocean surface, exhibiting positive phototropism and negative geotropism. During inactive periods tadpoles sink toward the ocean floor. Just prior to metamorphic climax, tadpoles become negatively phototropic and positively geotropic. This enables larvae to find suitable dark spots for attachment. It is not known what initiates the change in behavior at climax (Crisp and Ghobashy, 1971). Tadpoles lacking an ocellus are known to respond to light, and thus the change in behavior cannot be attributed solely to a change in the ocellus. Swimming activity is regulated by two systems in the tadpole tail: the muscles and nerves stimulating the rhythmic swimming activity, and electrical impulses arising from the epidermis of the tail inhibiting swimming activity (Mackie and Bone, 1976). It is not known if the relative strength of the inhibitory or stimulatory components changes just before climax to help account for a change in tadpole activity.

Many authors have published descriptions of metamorphic climax in single tunicate species, and Cloney (1978) reviewed the events of metamorphic climax in ascidians. The following description is a compilation of the 10 criteria of Cloney and a description of metamorphosis by other authors. The first event in climax is the expulsion of a sticky cementing substance from the adhesive papillae. This substance serves to attach the tadpole to the substrate. After expelling this material, the papillae evert and are retracted back into the trunk of the tadpole. Tail resorption is the next major event, and various mechanisms have been identified in various species. The first mechanism involves the contractile properties of the epithelial layer, causing the first phases of tail shortening. In certain species the notochord cells develop contractile properties that effect tail shortening. Finally, in other species, tail resorption is caused by striated tail muscle contraction (Cloney, 1978; Anderson et al., 1976). Total tail resorption can take place in as little as 90 min, but may require 24 hr in some species.

The tadpole is attached by its anterior end, and during climax the tadpole rotates its body axis so that the two siphons become situated on the new anterior end (Fig. 4). Depending on the species, this shift in siphons and internal reorganization takes from 2 hr to 2 days. Only two hypotheses exist to account for the shift in internal organs. The first is that activity of the smooth trunk muscles causes the organism to shift inside the tunic. The second and more likely explanation is that differential rates of growth and cell death cause a reorientation of internal

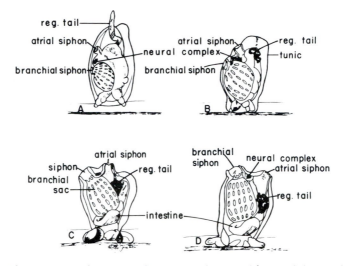

FIGURE 4. The sequence of metamorphic events after attachment of the ascidian tadpole (*Distaplia occidentalis*) to the substrate (after Cloney, 1978). The metamorphosing animal in A is shown at 5 min after attachment to the substrate; B, C, and D show the same animal at 3.5, 18, and 44 hr after attachment, respectively. The illustrations show the completion of tail regression and the rotation of both siphons. Initially the branchial siphon is parallel to the substrate but by the completion of metamorphosis (D) it has rotated approximately 90° so that it is perpendicular to the substrate.

organs. After adult organ systems attain their proper orientation, the young continue development. Sometime after attachment, but before feeding commences, a number of organ systems are phagocytized by blood cells. These include all of the cell types associated with the tail, and most of the larval nervous system, including sensory vesicles containing the ocellus and statocyte and visceral ganglion. Siphons are cleared and the young ascidian begins to feed within 2 weeks of attachment, with some species beginning to feed as early as 1 day after attachment. It must be emphasized that most authors have reported that under laboratory conditions tadpoles undergo some of the climax changes but not all. It has been repeatedly observed that attachment on a substrate and tail resorption are not required to produce a young feeding ascidian.

2.4. Triggers of Metamorphic Climax

While factors that initiate ascidian metamorphic climax have been studied since the 1920s (Lynch, 1961), there is no information on such factors in Appendicularia and Thaliacea. The most widely used agents to initiate ascidian climax include vital dyes and heavy metals. When tested in nonlethal doses, vital dyes (Janus green B, neutral red, Nile blue sulfate, methylene blue) and heavy metal ions (Cu, Fe, Ni) decrease the larval span drastically and lead to the formation of normal-appearing young.

Copper ions are the most thoroughly studied of these compounds. All authors have been able to decrease the larval span in a linear fashion by adding increasing amounts of copper salt to seawater. The involvement of copper ions in initiating metamorphic climax has been challenged by using phenylthiourea, a chelating agent (Whittaker, 1964). Since phenylthiourea-treated animals undergo metamorphic climax at the normal time, it was concluded that Cu ions are not involved in metamorphosis. However, one does not know what other ions aside from copper are chelated by phenylthiourea in seawater, nor is the ratio between chelated and free copper known. The site of copper action and its binding capacity for copper remain to be established. Therefore the conclusion that copper is not involved in initiating ascidian climax may be unwarranted.

Other factors accelerating metamorphosis have been found, including other ions (Na, Ca, I), brief exposure to hypotonic or hypertonic seawater, extracts of some tissues, and certain amino acids (Lynch, 1961). Cloney (1978) recently reported that dimethylsulfoxide and acetylcholine can initiate metamorphic climax. Inhibitors of climax have also been found, including certain ions (Mg), cyanide, urethane, low pH, and thiourea.

Since larvae normally change their behavior just before climax and so many different factors affect metamorphic rate, one might expect the nervous system to play a central role in regulating the rate of metamorphic climax. The potential involvement of the nervous system was tested by neural ablative experiments on a large ascidian tadpole, *Ecteinascidia turbinata* (Just, unpublished observations). The first experiment involved the removal of two sensory organs, the statocyte and the ocellus. The second involved the removal of the visceral ganglion, definitive ganglion, and neurohypophysis. A sham operation was performed by removing parts of the trunk between the base of the tail and the atrial siphon. Table 1 reveals that the removal of neural tissues did not prevent metamorphic climax as judged by both tail resorption and a shift in the relative position of siphons. Although these results suggest that the neural tissue in the trunk is not involved in metamorphic climax, they do not prove it, for these experiments were performed after tadpoles became free-swimming, and tadpoles could have released neurohumoral factors before the operation or the operation itself could have caused a release of neural factors. We can only concur with recent observations of Cloney (1978) that more detailed studies of the nervous system and sensory organs are needed in tadpoles before and during metamorphic climax.

It is widely known that administration of thyroid gland extract or iodine accelerates metamorphic climax, while thiourea inhibits it. The implication is that thyroid hormones (TH) are involved in climax; however, immersion of tadpoles in a wide range of thyroxine (T_4) concentra-

TABLE 1. THE EFFECTS OF NEURAL TISSUE ABLATION ON METAMORPHIC CLIMAX OF THE ASCIDIAN TADPOLE, *ECTEINASCIDIA TURBINATA*[a]

Experimental manipulation	Number of animals	Number of metamorphosed animals[b]	Number of deaths[c]
Statocyte and ocellus removed	43	40	7
Ganglion and neurohypophysis removed	35	24	3
Sham-operated	45	40	5
Normal	60	55	5

[a] Unpublished observations (Just).
[b] Total number of animals that completed metamorphosis 3 days after experimental manipulation.
[c] Total number of animals that died during a 3-day period, both metamorphosed and nonmetamorphosed animals.

tions caused no acceleration of the process. Injections and immersions of amphibian tadpoles with different TH analogs have shown that the analogs are not equally effective in inducing climax. The amphibian results suggest that new experiments should be performed in ascidians to assess the possible role of TH in metamorphic climax. Larger ascidian tadpoles could be used in injection experiments, while a tadpole of any size may be used in immersion studies. Whether immersion or injection studies are undertaken, it is crucial that various TH be tested and that various tadpole groups be subjected to the experimental manipulation. These attempts seem particularly desirable in light of the fact that the endostyle of the adult ascidian produces TH (Thorp and Thorndyke, 1975).

2.5. Unresolved Questions in Urochordate Metamorphosis

In ascidian metamorphosis a universal staging method should be established. The use of tail resorption as the sole criterion for defining metamorphic climax must be discontinued for two reasons. The first is that in "abnormal" metamorphosis other climax changes occur in ascidians without a concomitant tail loss. Also, since it seems desirable to unify the concept of climax among tunicate classes, tail resorption should not be used as the sole criterion for climax in one class (Ascidia) while in another class (Appendicularia) climax changes occur universally without it. Modern descriptive and experimental work is needed in the class Thaliacea. Attempts should be made to define ascidian climax at the molecular level. The application of numerous histochemical and microbiochemical methods will yield much fruitful information. The neural bases of behavioral alterations that occur before climax and of the choice of settling site during climax need to be investigated.

A hypothesis is needed that will explain the initiation of climax in tunicates. Since numerous factors influence the rate of ascidian climax, it should be possible to find the trigger for climax. By analogy with both insect and amphibian metamorphosis, one would expect that diversified factors affecting rates of climax should involve a neuroendocrine basis. TH are produced by the adult ascidian endostyle, and neurosecretory cells have been identified ultrastructurally and functionally in adult neural tissue. It would seem that a functional investigation of the endostyle and neurosecretory activities of nervous tissue in ascidian tadpoles could be very useful in developing hypotheses to explain the factors that initiate climax.

3. CEPHALOCHORDATES

The subphylum Cephalochordata is composed of two genera, *Branchiostoma* and *Asymmetron*. Adult *Branchiostoma* have two rows of gonads and two symmetrical metapleural folds terminating in front of the anus. On the right side of the adult *Asymmetron* is found a single row of gonads, but the metapleural fold extends past the anus, joining the ventral tail fin. There are less than 30 recognized species of cephalochordates, and the larval stages of half have been studied. When one considers the historical importance of cephalochordates in the theory of chordate origins, this large a proportion is not surprising. This review and the majority of publications on Cephalochordata larvae concern tadpoles of the genus *Branchiostoma*.

All adult cephalochordate species are sexually dimorphic. Gametes are released from the adult organism through a ruptured atrial wall. Fertilization occurs externally and embryonic development of the zygote is rapid, requiring only 1–2 days. The mouth develops on the left side of the larvae soon after hatching, while the first gill slit develops in the midventral position and extends to the right side of the animal. As the animal grows in length the number of gill slits increases. Growth rate (as measured by animal length, number of gill slits, and wet and dry weight) increases continually with age (Flood *et al.*, 1978). As tadpoles grow, the number of myotomes increases until just before metamorphic climax when the species-specific number is attained. During the final days of larval life drastic morphological changes take place that convert the pelagic cephalochordate larva into a benthic adult. The most thorough study of these changes was published nearly a century ago, giving the details of the movement of the mouth from a lateral position on the left side to a median anterior position in the body, and movement of the primary gill slits from the right to the left side (Willey, 1891). Simultaneously, secondary gill slits appear on the right side. A brief description of each of these eight climax stages follows (see Fig. 5).

Stage 1. Open primary gill slits and thickenings for secondary gill slits are located on the right side of the larva. The endostyle and club-shaped gland are present. The mouth is present on the left side, and the

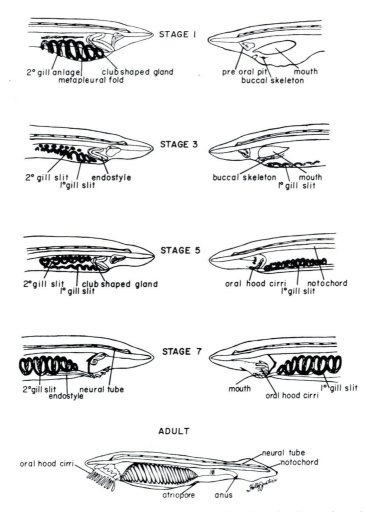

FIGURE 5. Selected larval climax stages and the adult cephalochordate. Bilateral symmetry is exhibited by the adult, while the larva is asymmetrical. In Stage 1 larva the primary (1°) and rudimentary secondary (2°) gill slits are located on the right sige of the animal, while the mouth is on the left side. Stages 3, 5, and 7 show the displacement of the mouth from the left side to the anterior midline. At the same time the 1° gill slits move to the left side and the 2° gill slits develop on the right side. After Willey, 1891.

buccal skeleton is beginning to form around the ventral surface of the mouth.

   Stage 2. Some primary gill slits are beginning to close and secondary slits begin to open. Metapleural folds start to cover the gill slits on the right side. The mouth decreases in size, and three to five buccal skeleton elements are associated with it.

   State 3. The metapleural folds have nearly covered the gill slits to form the atrium. Some primary gill slits continue to close while the openings in the secondary slits continue to enlarge. The primary gill slits start to shift to the left side and are located on the ventral surface of the pharynx. The mouth continues to decrease in size and is bent further to the anterior end of the larva. There are now five to six buccal skeleton elements associated with the mouth and they begin to form cirri.

Stage 4. The atrium is completely formed by the metapleural folds closing anteriorly leaving a posterior opening called the atriopore. The endostyle starts to grow and its location shifts posteriorly and ventrally. The mouth continues its anterior movement and the buccal skeleton elements contain cirri on their ventral surface.

Stage 5. The largest primary and secondary gill slits start forming tongue bars that ultimately divide a single gill slit into two. The endostyle extends past the club-shaped gland, which atrophies at this stage. The primary gills are predominantly on the left side. The mouth is located more anteriorly and has decreased to one-fourth the length at Stage 1. The number of cirri increases.

Stage 6. The secondary gill slits on the right side of the animal are as long as they are wide. Some of the primary gill slits are divided in two by the tongue bars. New buccal skeleton elements form at the dorsal surface of the mouth. The club-shaped gland has almost completely degenerated while the endostyle is now located more ventrally and posteriorly.

Stage 7. The gill slits on both sides are elongated in the dorsal–ventral axis and most have been cut in two by the tongue bars. The mouth is located at the anterior end of the larva and is 1/10th its original size. Tentacles start to appear around the mouth in addition to cirri. The endostyle can still be recognized as being made of two parts and is located at the ventral surface of the pharynx.

Stage 8. The mouth is at the anterior end surrounded by cirri and tentacles. The endostyle, its length nearly doubled since stage 1, is located on the ventral surface.

Willey ends his staging at this point but he describes further development. The endostyle fuses and continues its posterior growth until it lies on the entire ventral surface of the pharynx. The outpocketing of the intestine, the cecum, which first appeared at stage 6, continues to grow. Although the number of gill slits increases with the size of the larvae, different species undergo these climax stages of metamorphosis at different lengths (Fig. 6) (Wickstead, 1975). Thus, some species enter climax when they have as few as 12–14 gill slits, while other species have 20–30 gill slits. During the climax stages some of the primary gill slits are lost, and after stage 8 tertiary gill slits begin to form on both sides of the animal and continue to be added in the adult (Fig. 6).

Reported larval life spans are relatively long: 3 months for *B. lanceolatum* (Willey, 1891), $2\frac{1}{2}$–$4\frac{1}{2}$ months for *B. nigeriense* (Webb, 1958), and 3–7 months for *B. senegalense* (Gosselck and Kuehner, 1973). The discovery of large larvae with gonads ("amphioxides") in the open seas

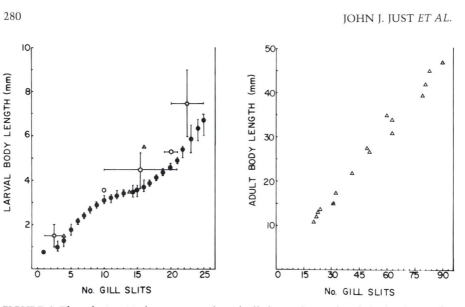

FIGURE 6. The relationship between number of gill slits and size of cephalochordates. The vertical lines through the symbols represent the range of body lengths, and the horizontal lines indicate the range of gill slit numbers. Symbols without lines indicate data from a single animal, or a range that is so small that it falls within the symbol. Data for larval species: △, *B. lanceolatum*, Lankester and Willey (1890), Webb (1969); ○, *B. belcheri*, Bone (1959), Wickstead and Bone (1959); □, *B. senegalense*, Gosselck and Kuehner (1973); ●, *B. nigeriense*, Webb (1958). Data for adult: △, *B. lanceolatum*, Courtney (1975).

suggests that some cephalochordate larvae have extended larval life spans and may even become neotenic (Wickstead, 1975).

3.1. Cephalochordate Feeding Behavior

Adult amphioxi are found filter-feeding on sandy ocean floors that are undisturbed by surface waves. If the diameter of the sand grains is too small or the silt content too high, water flow through the sediment will be too slow to permit survival. Sandy deposits with rough edges are avoided because of excessive tactile stimulation (Webb, 1975). On the Chinese coast adult populations have been known to occupy certain regions for nearly 1000 years, and nearly 35 tons of amphioxi are harvested annually as a food source. This is remarkable when one considers that a single adult weighs less than 1.0 g. The stability of long-term populations coupled with the relatively short life span of individual adults (1–8 years) implies a constant turnover within the population (Courtney, 1975).

The number of larvae found in the open sea is usually low; the largest numbers are found along coastlines where sea depths are less than 120 m. The total number of larvae found near a breeding area can be very large: in the coastal waters off northwest Africa, as many as 40,000 larvae per square meter of ocean surface were observed (Flood *et al.*, 1976). This represents an estimated production of 1.1 million tons of cephalochordate larvae from a single breeding site. This estimate represents one of the highest concentrations of any chordate larvae in any habitat. The survival rate of the larval populations is unknown. Even though adult populations in breeding grounds can be very high (9000 adults/m^2 ocean floor), it is clear that most of these larvae succumb to predators in the open ocean.

Cephalochordate larvae spend their lives swimming toward the ocean surface, then passively sinking. Some species seem to show a diurnal tendency to swim, and at night are frequently found near the ocean's surface. The majority of species are active swimmers at all times and show no diurnal variation. It is now generally agreed that larvae swim toward the ocean surface with mouth closed, and then passively sink with mouth open. The upward swimming rate is between 40 and 300 cm/min, depending on species and size, while the rate of sinking ranges from 3 to 15 cm/min, depending on size, with smaller animals sinking more slowly. All evidence is consistent with the notion that larvae feed only while sinking (Webb, 1969; Gosselck and Kuehner, 1973). Since larvae sink slower than they swim, they spend a greater proportion of their larval life in feeding than in nonfeeding activity.

Although larvae can open and close their mouths and are perhaps capable of eating animals larger than the mouth's normal size, the bulk of their food intake depends on filter-feeding. Captured larvae contain food composed mainly of phytoplankton. Some large zooplankton have been found in the larval intestine, but these may represent artifacts due to the collection of larvae in plankton nets. Food passes through the intestine by peristaltic action of smooth muscles in the gut wall (Webb, 1969).

As cephalochordate larvae are relatively small and are not powerful swimmers, ocean currents carry them away from their breeding grounds. Strong ocean currents and a long larval life span enable amphioxi to be dispersed as far as 8000 km from their original hatching site (Webb, 1975). Perhaps only a small percentage of larvae are carried such great distances, although it appears that all are transported away from the areas where they hatched. When larvae arrive at breeding sites to undergo metamorphic climax, it is not known whether they are the same organisms that were carried away by currents after hatching or whether they represent larvae from different breeding sites. A method should be devised to mark larvae after hatching in order to follow them throughout their life span. If sufficiently large numbers were marked, one could determine whether larvae return to their hatching site to mature or to that of other adults of the same species. Evidence is available that larvae are attracted to sites occupied by adults or to desirable sand deposits (Webb, 1975). From the excellent neurological studies of Bone, it is clear that larvae possess sets of motor and sensory nerves (reviewed by Barrington, 1965). The nervous and muscle systems permit larvae to actively seek out certain environments in which to metamorphose. Neither the presence of adults nor complementary sand deposits are the deciding factor that triggers metamorphic climax, as laboratory transformation occurs without benefit of either adults or sandy deposits.

3.2. Triggers for Metamorphosis

Little evidence is available to explain what triggers cephalochordate climax. By analogy with amphibian metamorphosis an early hypothesis was that TH were involved. It has been demonstrated by chemical and biological assays that adult cephalochordate endostyles produce TH (Barrington, 1965). The classic work of Willey (1891) showed that the endostyle increases in size as the animal undergoes metamorphic climax. If TH synthesis starts before climax and the rate of synthesis were the same throughout larval life, one would expect the amounts of TH to increase in cephalochordate larvae during the final days of metamorphosis. Preliminary data on a few later stages of B. lanceolatum larvae suggest that immersion of larvae into either thiouracil (an inhibitor of iodine trapping by the endostyle) or TH has no effect on the rate of metamorphosis. These results do not rule out the possibility of TH involvement in cephalochordate metamorphosis. The low sample size, use of a single stage, and the fact that none of the larvae (including controls) were taken through metamorphosis to a healthy fully formed juvenile are all factors that make interpretation of the results questionable (Wickstead, 1967).

Another potential trigger for climax might be the eyespots of the amphioxus, which are located in the spinal cord and are responsible for the initiation of swimming or burrowing responses in adults adapted to dark. The number of eyespots triples during climax stages and individual eyespots double in size (Webb, 1973). No causal relationship has been established between this dramatic increase in total photoreceptive surface and the onset of metamorphic climax. However, the eyes could act as a neuroendocrine trigger to initiate climax.

3.3. Unresolved Questions in Cephalochordate Metamorphosis

The increase in gill slit number during larval life and the movement of the mouth and primary gill slits during metamorphic climax should be reinvestigated using modern histochemical and biochemical techniques. It would be interesting to know the amount of programmed cell death and cell division involved in these changes. Aside from the morphological changes in the feeding apparatus, other organs such as the club-shaped gland and cecum are known to change during metamorphosis, but modern techniques have not been used to investigate these changes. The physiological and ecological triggers of metamorphic climax need study. Experimental approaches should be developed to test the hypothesis of Wickstead that larvae must make contact with a substratum before climax can be initiated. Such studies could include measuring the levels of TH in larvae and the effects of total darkness on metamorphic rates.

4. FISH

The living species of fish are divided among three classes: the Agnatha or cyclostomes (lampreys and hagfish); the Chondrichthyes or cartilaginous fish (rays, sharks, and chimaeras); and the Osteichthyes (bony fish). The class Osteichthyes is composed of about 20,000 species, a number that is not exceeded by any other group of vertebrates. The smallest of the three, Agnatha comprises only about 50 species. Excellent accounts of the reproduction and early development of many fish exist, but little information is available about the metamorphosis of these species (Breder and Rosen, 1966).

4.1. Agnatha

Agnatha is distinguished from all other vertebrate classes by lacking jaws and pelvic fins. The class is composed of two orders: Petromyzoniformes (lampreys) and Myxiniformes (hagfish). Most authors agree that hagfish do not undergo metamorphosis. On the other hand, much attention has been given to the metamorphosis of lamprey. Larval lamprey are blind, toothless burrowing filter feeders that metamorphose into adults with teeth, conspicuous lateral eyes, and a large sucking disc for a mouth (Manion and Stauffer, 1970; Bird and Potter, 1979).

Larval lamprey, the "ammocoete," lives buried under sand, silt, and organic debris in quiet freshwater streams. Depending on the age of the ammocoete, it may be found under different types of substrate, silt and organic debris for the youngest ammocoetes, coarse sand for the eldest. The climax stages of metamorphosis generally begin in the late summer or early fall and last 4–5 months. Upon completion of metamorphosis the young adults swim away from the burrows either to remain in freshwater streams or to migrate to the ocean. If they migrate to the ocean, they return to fresh water for breeding. About one-third of the 31 recognized species spend part of their adult lives in the sea; these species are referred to as "anadromous." It is commonly assumed that all adult lampreys are parasitic when in fact only about one-half of all recognized species are parasitic.

While attention has been focused on metamorphic climax in the lamprey, more subtle changes occur during larval life necessitating improved methods of staging earlier metamorphic events. Length–frequency histograms have been routinely used to determine the length of larval life up to climax. These are made by forming size classes from larvae caught in streams, each size presumed to represent 1 year of growth. This method is not accurate unless one can isolate and measure large numbers of larvae beginning with hatching and continuing through metamorphosis. Because of fluctuations in temperature and food availability, growth rates in organisms with long larval life spans are not the same from year to year. Differences in growth rates tend to make the length–frequency histograms difficult to interpret. Estimations of lamprey larval life spans range from 3 to 6 years (Table 2). It should be noted that some lamprey species have neotenic larvae (Zanandrea, 1957). Most authors feel that the prolonged larval life period and short adult life of nonparasitic species indicate that they have evolved from parasitic species, with both groups having approximately equal total life spans.

A detailed description of external morphology was made to identify the climax stages in *Lampetra fluviatilis* and *L. planeri* (Bird and Potter, 1979). Particular attention was paid to some internal changes that begin prior to external signs of transformation. These authors identified nine stages from ammocoete through adult. The primary method of identifying various stages is based on the appearance of the larval eye (Fig. 7). The eyes first appear as small dark oval patches at stage 1 and erupt at stage 2. The iris becomes partially visible at stage 3 and is completely visible in stage 4. On the ventral surface the most obvious morphological changes are associated with mouth development. At stage 1 there is a transverse lip across the posterior margin of the mouth; beyond the lip is a depression. The lips of the oral region increase in size at stage 2. In the next stage the lateral and ventral lips start to fuse, and the ventral depression completely disappears. At stage 4 the lips have completely fused to form a small oral disc. Presumptive teeth structures become evident around the oral aperture at stage 5. An individual in stage 6 is essentially an immature adult with large eyes, the so-called macrophthalmia. The final three stages involve maturation of the metamorphosed animals and do not pertain to larval life. An earlier description of four climax stages for the landlocked sea lamprey, *Petromyzon marinus*, was based only on

TABLE 2. THE DURATION OF LAMPREY LARVAL LIFE

Parasitic lampreys	Years (mean)	Nonparasitic lampreys	Years (mean)
Caspiomyzon wagneri	2.25	*Eudontomyzon mariae*	5
Eudontomyzon danfordi	2.5	*Ichthyomyzon fossor*	5.2
Lampetra fluviatilis	4.1	*Ichthyomyzon gagei*	3
Lampetra japonica	3.5	*Ichthyomyzon hubbsi*	4.5
Lampetra tridentata	3	*Lampetra aepyptera*	3
Mordacia mordax	3	*Lampetra lamottenii*	4
Petromyzon marinus		*Lampetra planeri*	3.1
anadromous	5	*Lampetra reissneri*	4
landlocked	5.5	*Lampetra zanandreai*	4
Tetrapleudon spp.	3	*Okkelbergia aepyptera*	3

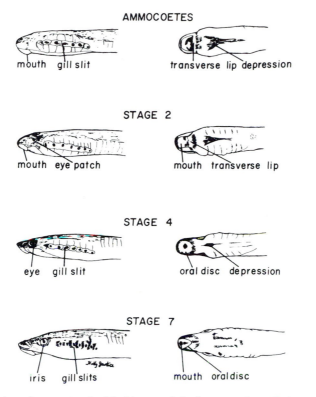

FIGURE 7. Selected stages in the life history of the lamprey. Lateral views are shown on the left and ventral views on the right. The ammocoetes represent larvae before metamorphic climax. During the larval transition stages (2, 4) the eyes make their appearance and the mouth structure changes from an oral hood into an oral disc. Stage 7 shows an immature adult lamprey. After Bird and Potter, 1979.

changes in mouth morphology (Manion and Stauffer, 1970). There also are considerable internal changes in the blood, digestive, excretory, and nervous systems. Most of these changes occur irrespective of the ecological niche occupied by the adults. Interestingly, some of the changes mimic those reported in amphibians.

4.1.1. Changes in Organ Systems during Transformation

Lamprey larvae are filter feeders existing on a varied diet of microorganisms and detritus obtained while they lie buried in freshwater streams. The caloric content derived from each is disputed. Recent evidence suggests that detritis is the more important food source (Moore and Potter, 1976). The food is trapped in mucous strands produced by the endostyle and is passed by ciliary action of the pharynx into the larval intestine, which has a single large longitudinal fold (typhlosole) extending into the lumen. This method of feeding continues until the final transformation into the adult. At this stage the esophagus develops from a solid epithelial cord in the roof of the larval pharynx. The remainder of the larval intestine decreases its lumen size markedly at transformation, feeding ceases, and the larvae exist on stored food reserves. In parasitic larvae the intestine expands near the end of metamorphic climax. The potential absorptive surface increases by the development of numerous longitudinal folds in the intestinal lumen (Youson and Connelly, 1978), the fold developing by cell division. Remodeling of the larval amphibian intestine at climax also occurs by cell division and cell death (see Fox, this volume). In nonparasitic adults the intestinal lumen remains collapsed and the adult cannot feed.

Prior to metamorphic climax larvae must store enough food reserves to permit survival for several months. In the nonparasitic species the food must last for the remainder of their lives, while in the parasitic forms stored food must be sufficient to sustain the young adult until it can find a host, from 3 to 10 months. In both parasitic and nonparasitic forms the major energy reserve seems to be lipids. Lipid accumulation is accomplished by switching from protein anabolism to lipid synthesis toward the end of larval life (O'Boyle and Beamish, 1977). Lipid content rises to about 15% of the larval wet weight immediately prior to transformation (Moore and Potter, 1976). There are many deposit sites for lipids in the larval body, including adipose tissues that lie above the nerve cord, the so-called fat column. Eight lipid deposit sites have been investigated histologically, and all show a sharp increase in content at the initiation of transformation. During the process of transformation, all sites decrease in lipid content, until lipids are detectable only in the fat column of young adults (Youson *et al.*, 1979).

L. planeri have two distinct larval and two adult hemoglobin bands as disclosed by paper and starch-gel electrophoresis. Examination of lampreys undergoing transformation revealed all four bands to be present simultaneously in the same individual (Adinolfi *et al.*, 1959). More recently, polyacrylamide gel electrophoresis of larval and adult *L. planeri* (nonparasitic) and *L. fluviatilis* (parasitic) revealed five larval hemoglobin bands. All but one fade during climax. The adult has only three bands, one of which is identical to a larval band (Potter and Brown, 1975). Identical electrophoretic patterns of hemoglobins from these species illustrate their close relationship, the main difference being that in the parasitic lamprey the hemoglobin transition is achieved faster than in the nonparasitic lamprey. The probable reason for the difference is attributed to their different habitats in the final stages of metamorphosis, for *L. planeri* remains buried in silt longer, therefore needing larval hemoglobins with greater oxygen affinity during the slower transition.

In vertebrates the switch in hemoglobin types during development is generally accompanied by a change in hemopoietic sites. In the lamprey this switch in sites was monitored histologically. Initially, hemopoiesis occurs in the blood islands located in the yolk sac of the embryo. When larvae reach a length of 6 mm the process is taken over by the typhlosole, a large dorsal infolding of the intestinal wall. When they attain a length of 20 mm an accumulation of hemopoietic cells is found in the nephric fold. Hemopoiesis continues as new tubules develop posteriorly and anterior tubules degenerate. Simultaneously, lipoblasts begin to collect to form the fat column, and when larvae are 25 mm long some blood cells are found there. With the onset of metamorphic climax the fat column is invaded by immature blood cells that begin to divide and mature. In the final stages of metamorphosis the fat column becomes the principal site of hemopoietic activity in the adult (Percy and Potter, 1976).

Strong support for this sequence in shifts of the hemopoietic sites was offered by the examination of lamprey species *L. fluviatilis* (parasitic) and *L. planeri* (nonparasitic) (Percy and Potter, 1977). In *L. planeri* the definitive adult hemopoietic site, the fat column, became active in the production of red blood cells 2 months later than it did in *L. fluviatilis*. This shift in sites mimics the timing of hemoglobin transition in these species, and therefore supports the thesis that there are separate erythrocytes for larval and adult hemoglobins. This information suggests that the shift in hemopoietic sites leads to the production of separate populations of larval and adult red blood cells. Similar shifts in the sites of production and types of erythrocytes have been observed in amphibian metamorphosis (Forman and Just, 1981; see Broyles, this volume).

The circulatory system shows remarkable adaptations to the new adult life style of the lamprey. New and larger blood vessels to the kidney and intestine appear at transformation (Percy and Potter, 1979; Youson and McMillan, 1971). Since the heart enlarges during metamorphic climax, the assumption has been that the circulatory system becomes more efficient, permitting the more active life of the adult (Claridge and Potter, 1974). Changes in the heart and the establishment of a more efficient vascular delivery system to the kidneys and intestine led to the suggestion that these structures do not have hemopoietic function in the adult. The increased blood flow through them would seem to preclude hemopoietic involvement. While this appears to be an attractive hypothesis to account for shifts in erythropoietic sites during development, unfortunately no supporting experimental data are available.

A further change seen at metamorphic climax is a variation in plasma proteins. By means of paper electrophoresis, three globulins (α, β, and γ) were isolated in the plasma of the larval lamprey (*P. marinus*). In the adult, four components have been identified, two α, one β, and one γ. In three species (*P. marinus*, *I. unicuspis*, and *L. lamottei*) polyacrylamide disc electrophoresis revealed a shift in the types of plasma proteins during metamorphosis (Uthe and Tsuyuki, 1967).

Two of the major osmoregulatory organs, the kidney and the gills, show drastic morphological changes during metamorphic climax. The kidney develops within a nephric fold. In early larval stages the kidney undergoes a continuous anterior to posterior degeneration until the process reaches about the midbody region. At the start of metamorphic climax or just prior to it, the nephric fold is composed of fat cells (95%) and a small amount of kidney tubules (5%). As climax proceeds there is a drastic increase in the amount of kidney tissue brought about by cell division. At the end of climax almost the entire nephric fold (75%) is composed of kidney tubules. In addition to the new adult structure making its appearance, the larval kidney is destroyed during climax by autolysis followed by phagocytosis (Ooi and Youson, 1979). In anadromous lampreys, superficial cells in the interlamellar area of the gills degenerate during metamorphic climax and a new type of cell appears. These new cells arise by cell division and resemble the ion-excretory (chloride) cells of marine teleosts (Peek and Youson, 1979).

The changes in kidney and gill have been fully documented in anadromous species undergoing metamorphic climax, and they are probably responsible for the onset of osmoregulatory capacity (Beamish *et al.*, 1978). Changes in the osmoregulatory capacity and an increase in plasma osmotic pressure have been demonstrated in freshwater lampreys at metamorphic climax. Increases in the plasma osmotic pressure and the appearance of an ion transport system in the skin are well documented in amphibian metamorphosis (Just *et al.*, 1977a). The osmoregulatory organs of freshwater lampreys undergoing metamorphic climax need further structural and functional investigation in light of the changes in the animals' osmoregulatory abilities.

One of the most striking changes associated with lamprey metamorphosis is the change in gross anatomy of the mouth. The muscles in the adult mouth differ from larvae irrespective of the parasitic or nonparasitic life style of the adult. The adult mouth contains a sucker and rasper, whereas the larval mouth has simple muscles and an oral hood. Both larval and adult mouth muscles are innervated by the same parts of the trigeminal motor nucleus (Homma, 1978). It is not known whether the same motoneurons make connections to muscles before and after metamorphosis or whether new neurons make connections to the adult muscles.

4.1.2. Triggers for Metamorphosis

A century ago it was discovered that the endostyle gives rise to the thyroid. The endostyle starts as a longitudinal groove in the pharyngeal floor between the first and the fifth gill arches. This groove rolls into a tube, an opening remaining in the tube at the region of the third gill arch. Inside the tube, the endostyle is composed of five cell types, at least two of which can selectively trap iodine to produce TH. At metamorphic climax the endostyle forms the thyroid by cell division of some cell types and cell death of others (Fig. 8). There is no agreement as to which of the five cell types give rise to the thyroid follicles.

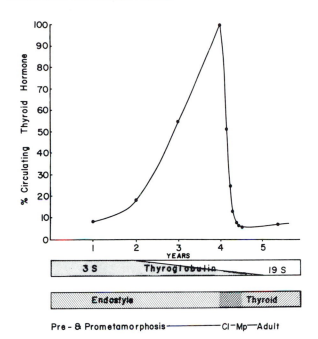

FIGURE 8. The endocrinology of TH during the lamprey life span, the ordinate giving the percent circulating TH and the abscissa the total larval life span in years. The upper rectangle illustrates the replacement of thyroglobulin of younger animals (sedimentation coefficient 3 S) with the thyroglobulin of older animals (19 S). The lower rectangle indicates the presence of the endostyle or thyroid. The period of endostyle transformation into the thyroid is marked by cross-hatching. Metamorphic progress is indicated at the bottom of the figure. Climax (Cl) requires less than 6 months, and ends with the macrophthalmia (Mp) stage. Data from authors cited in text.

The iodine trapped by the endostyle and by thyroid follicle cells is used in the synthesis of TH (Fig. 8). In these glands the TH are components of thyroglobulin. The endostyle of young ammocoetes contains a thyroglobulin with a sedimentation coefficient of 3–5 S. During larval development a 19 S thyroglobulin appears; it makes up 50% of the thyroglobulin just before climax, and 100% of that in adults (Suzuki and Kondo, 1973). During larval development the endostyle becomes more efficient in trapping iodine and producing TH (Suzuki and Kondo, 1973; Wright and Youson, 1977). After or just prior to initiation of metamorphic climax, the circulating levels of TH fall precipitously and remain low throughout adult life (Wright and Youson, 1977).

The pattern of circulating levels of TH is nearly identical to that shown in amphibian larvae (Just, 1972; see White and Nicoll, this volume), yet no clear-cut involvement of TH in lamprey metamorphosis has been demonstrated. In the last 70 years many investigators have attempted, without success, to increase the rate of metamorphosis climax by TH treatment. Treatment of larvae with goitrogens does not inhibit metamorphic climax (Barrington and Sage, 1972). Attempts to induce climax by injecting mammalian pituitaries or TSH have likewise met with failure. Administration of either goitrogen or mammalian TSH has shown that the endostyle responds to such treatment with a change in TH production. Even though biological assays for TSH in ammocoete larvae have been negative, treatment of larvae with goitrogens suggests that a TSH-like substance is present. A recent ultrastructural study of the pituitary throughout development revealed that the anterior pituitary splits into three parts at metamorphic climax, and one of these parts contains cellular granules resembling TSH granules (Percy *et al.*, 1975).

Several earlier studies published data indicating that hypophysectomy had no effect on endostyle histology or metamorphic climax. In a recent study of 10 larvae, pituitary removal prevented them from entering metamorphic climax (Larsen and Rothwell, 1972). It must be remembered that pituitary removal involves numerous hormonal changes, not only those of the pituitary–thyroid axis.

We suggest that some of the earlier experiments to manipulate TH be repeated at metamorphic climax or just prior to it. Many earlier studies may have attempted to manipulate TH concentration before the larval tissues became sensitive to it. The negative results could be due to a lack of tissue receptors for TH before metamorphic climax. In support of this hypothesis, it should be noted that the metabolic rate of larvae does not increase until late in metamorphic climax, following a decrease in circulating levels of TH (Lewis and Potter, 1976). Another reason the experiments should be repeated is that the manipulations may have been performed before the blood hormone concentrations were high enough. In amphibians it has been hypothesized that some cells degenerate because they become dependent on the high concentrations of TH present before climax. These dependent tissues die after climax because of decreasing TH concentration. To discover if such dependent tissues exist in lampreys, experiments should be performed just before climax.

Other triggers that may initiate metamorphosis have received far less attention. The clearest example of something affecting metamorphic climax in ammocoetes is the fact that pinealectomized larvae do not metamorphose (Eddy, 1969). This interesting result needs confirmation, and hypotheses should be formed and tested to account for the role of the pineal in ammocoete metamorphosis. This is particularly urgent since reports have been published stating that the pineal is not essential to anuran metamorphosis.

4.1.3. Unresolved Questions in Agnatha Metamorphosis

Mark and recapture methods should be applied to ammocoetes to determine their life span. These data are needed to more accurately determine when larvae metamorphose and the possible range of life spans for individuals produced from a breeding season. Morphological and biochemical studies should be carried out during the entire life span of larvae to determine the metamorphic changes occuring before climax. The triggers that initiate metamorphic climax should be sought aggressively. The involvement of TH should be reinvestigated with an eye to the TH receptors in various tissues. The observations of Eddy on the effects of hypophysectomy and pinealectomy need confirmation and explanation on a hormonal or neuronal basis.

4.2. Chondrichthyes

In all chondrichthian species (skates, sharks, rays, and chimaeras) fertilization is internal. The fertilized eggs are either retained in the female or enclosed in an egg case and deposited externally. Depending on water temperature and species, the incubation period ranges from 2 to 12 months. At hatching, the fish resemble the adults. When the fertilized egg is retained in the female, a full range of developmental relationships exist between embryo and mother, ranging from ovoviviparity to true viviparity. Irrespective of the relationship between mother and fetus, the young can be retained in the mother for up to 2 years. In some ovoviviparous species, the fetus loses organic material during this time and consequently loses weight. However, in other condrichthians the fetus actually gains weight while inside the mother. Three methods of nourishment are evident. In some species the mother continues to produce eggs during the gestation period, which are eaten by the fetus; in others, secretions from the uterine epithelium are ingested by the fetus. The final method of nourishing the young is through a true placenta. At birth, irrespective of mode of nourishment, the young fish resembles the adult.

Since all free-living young cartilaginous fish resemble the adult, most authors who discuss chondrichthian development do not refer to metamorphosis in this class of chordates (Breder and Rosen, 1966; Wourms, 1977). However, a brief description of their development is included for several reasons. This class closely parallels the great diversity of developmental relationships observed in amphibians, particularly in caecilians. It is possible that some species (i.e., those that feed in the oviducts) may meet some of the criteria for metamorphosis as outlined in the Introduction of this chapter. This would be true if the fetuses actually compete with one another for the available food supply in the female reproductive tract. More information is needed about the interrelationships of fetuses in the reproductive tract, the factors that control the morphological changes in the uterine tract, and the mechanism whereby intrauterine life is terminated, before one can conclude that none of these animals undergo metamorphosis.

4.3. Osteichthyes

Osteichthyes, the largest group of chordates, exhibits a great diversity of reproductive adaptations from oviparity to true viviparity. The vast majority are oviparous with external fertilization producing either buoyant or nonbuoyant eggs. Most marine fish produce buoyant eggs, while freshwater fish tend to produce nonbuoyant eggs that either are deposited on or buried in the substrate or are attached to supporting objects in the water. Very few freshwater fish produce truly pelagic eggs; however, some eggs are only slightly denser than fresh water and can be easily moved by water currents. In species with pelagic eggs, the eggs are dispersed in the plankton. In species with nonbuoyant eggs, only larvae or adults can explore new ecological niches.

In all oviparous fish, embryonic development ends when the larvae hatch out of the chorion-enclosed eggs. This can occur as early as 2 or 3 days postfertilization or as late as 5 years postfertilization (Braum, 1978; Simpson, 1979). Freshly hatched larvae usually have a notochord and myotomes, but fins and pigmentation are either not well developed or absent. Nutrients are still available to the larva through a large yolk sac. Sometime during the final stages of yolk absorption, the larvae start feeding. Larvae have been described in only one-fifth of all osteichthian species, and in many cases only a few examples are available. A complete description of all larval stages in any species is available in only 5% of the known species.

Universal staging methods are not available to describe the metamorphic progress from yolk sac larva to juvenile fish, and the great diversity of adult forms make it unlikely that a universal staging procedure can ever become acceptable (Fig. 9). Frequently, larvae bear little or no resemblance to the adults. In fact, historically some larvae with relatively long life spans were thought to represent distinct species, as was the case with eels. There is a great diversity of criteria used for demarcation of the extent of larval life. A recent series of volumes dealing with fish development suggests that the term "larvae" be restricted to that period of time immediately after hatching until the attainment of the adult fins containing the species-specific number of skeletal elements (fin rays) (Jones *et al.*, 1978; Hardy, 1978a,b; Johnson, 1978; Fritzsche, 1978; Martin and Drewry, 1978). We have adapted that terminology in our discussion of osteichthian metamorphosis.

4.3.1. Morphological Changes during Larval Life

The length of larval life varies tremendously, from as little as 2–3 weeks to as long as 3 years. Some larval fish appear to develop functional gonads, and true neoteny has been reported by a few authors (Castle, 1978). While in some species few external changes take place, in others the changes are very dramatic. In addition to changes in structure and

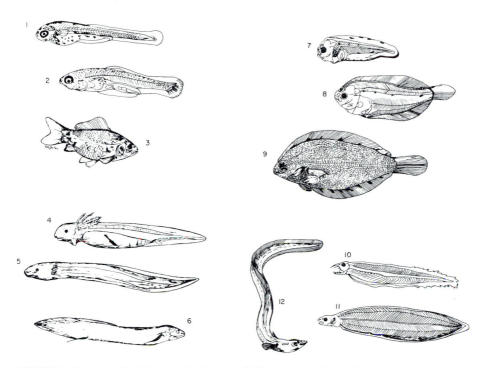

FIGURE 9. Larval and adult representatives of class Osteichthyes. Illustrations 1, 2, and 3 represent a yolk sac larva, a free-swimming larva, and an adult goldfish. In the bottom left group are lungfish as a yolk sac larva (4), an advanced larva (5), and an adult (6). Young (7) and transforming (8) larvae of the adult flounder (9) are illustrated at the upper right. A very young eel larva (10), a more advanced larva (11), and an adult eel (12) are shown at the bottom right.

number of fins, all larvae increase their pigmentation and develop external nares during metamorphosis. Most species show a gradual increase in size, the addition of fins and fin ray elements, and color changes (i.e., goldfish, Fig. 9, Nos. 1–3). Other species (i.e., the eel) undergo rather drastic changes until metamorphosis is complete, including shifts in location of fins, reduction of some fins, and changes in mouth structure. During the course of eel larval development, the external opening of the digestive and urinary systems (cloaca) changes position, shifting from a posterior to an anterior location (Fig. 9, Nos. 10–12; Castle, 1979). In lungfish the external changes are also rather dramatic, and include the disappearance of the external gills and an increase in fin size (Fig. 9, Nos. 4–6). Without a doubt the most dramatic external changes occur in the flatfish, from bilaterally symmetrical, yolk sac larva into asymmetrical adult (Fig. 9, Nos. 7–9).

In many bony fish, the respiratory surface area increases from the time of yolk sac absorption until the completion of metamorphosis (de Silva, 1974). In the initial phases of larval life, the general body surfaces, especially the fins, act as the respiratory organ. Upon completion of yolk sac absorption, the gills increase rapidly in total surface areas and become the major respiratory organ as larvae complete metamorphosis. With the exception of the lungfish, gills remain the primary respiratory organ and retain that role throughout the life span of the adults.

Along with the obvious external morphological changes, numerous internal morphological alterations occur during the metamorphosis of osteichthians. Some of the most dramatic internal changes occur in the digestive system and its derivatives. In most species the mouth opens after hatching, and the intestine increases in length throughout larval life (Bryan and Madraisau, 1977). While increasing in length, the intestine also differentiates histologically, increasing its absorptive surface by developing folds in its lumen. At metamorphic climax, it develops a pyloric valve to separate the stomach from the intestine, just as is the case in anurans (Evseenko, 1978). Two derivatives of the digestive system, the gas bladder and lungs, also undergo changes during larval development. In lungfish, lung development starts in the early yolk sac stages, and the lungs become functional according to morphological criteria during late larval life, in a manner similar to anuran tadpoles (Kerr, 1909; Atkinson and Just, 1975). The gas bladder starts as an outpocketing of the anterior portion of the intestine early in embryonic development, which enlarges and matures during larval life (Jones and Marshall, 1953; Grizzle and Curd, 1978). Near metamorphic climax, the bladder fills with gases. The cause of the initial filling seems to be species specific. Some species fill the gas bladder by swallowing air, while in others a restricted region of the blood supply to the bladder epithelium secretes gas. In species in which the bladder loses its connection to the digestive system, closing occurs around the time of metamorphic climax, while in other species the entire gas bladder is resorbed at climax. The gas bladder is retained in the remaining osteichthians, joined to the digestive system.

The skeletal system also shows dramatic alteration during metamorphosis. The most obvious of these changes, the addition of skeletal elements to fins, is the criterion adopted to define completion of metamorphosis. This change is a universal phenomenon; however, other skeletal changes also occur. Drastic skull alterations take place during larval life. In most species, bones begin to develop around protruding larval eyes, until in the adult there are well-developed bony sockets. In addition to the streamlining of the skull, overall changes occur in the branchial elements. Ossification of the branchial elements begins after yolk sac absorption, becoming almost complete at the end of larval life (Verraes, 1977; Mook, 1977). The most dramatic alterations in skull bones occur during the rotation of the eye in the flatfish. The eye is displaced either to the right or to the left side of the skull, depending on species. The brain case is continually remodeled, the eye socket forming and re-forming as the eyes migrate in front of the dorsal fin base and come to lie on the dorsal surface of the animal (Richardson and Jospeh, 1973; Evseenko,

1978). All species show another universal morphological alteration during larval life, namely the widening of the mouth. In many cases larval teeth are replaced by adult teeth at metamorphic climax.

4.3.2. Biochemical Changes during Larval Life

Aside from the internal and external morphological changes during the course of metamorphosis, there are also biochemical alterations. All stages of osteichthians (embryonic, larval, and adult) eliminate ammonia by diffusion across skin and gills. In embryonic tissues and yolk sac larvae, three nitrogenous wastes are found, ammonia, urea, and uric acid. The larvae of young fish do not have a complete urea cycle, and the urea found in the larvae probably comes from the breakdown of arginine, which arises from yolk proteins (Rice and Stokes, 1974). Mysteriously, urea disappears from the yolk sac larvae as they age. No explanation can be given for this disappearance, because no urea seems to be excreted into the environment, nor is there a known metabolic pathway that can utilize urea in larval bony fishes.

Throughout embryonic development and yolk-sac-larval stages, the young fish live on stored food. During this time their weight remains constant, yolk lipids and proteins being replaced by water. A gradual increase in the percentage of water occurs (Ishibashi, 1974; Love, 1970). After the initiation of feeding, the percent water of some larvae decreases throughout the entire larval life span, as in anuran metamorphosis. The relative decrease in water is associated with elevated inorganic and organic body constituents. The increase in inorganic constituents probably reflects an increase in the ability of larvae to regulate the internal ionic environment. In some fish, chloride cells (salt-transporting cells) develop in the skin at the time of yolk absorption. These cells remain in the skin until completion of metamorphosis (Lasker and Theilacker, 1968). During the course of larval life the gills become the major respiratory organs, and chloride cells also make their appearance in the gills to help regulate the internal ionic environment. The relative increase in organic constituents can best be shown by a percent increase in total carbon, relative to wet weight. Individual carbon compounds (carbohydrates, nitrogen-containing compounds, and triglycerides) do not show consistent changes during larval life (Ehrlich, 1974). In postlarval stages the increase in biomass is not only associated with an increase in cell number, but with an actual increase in cell size (Weatherley and Rogers, 1978). Although one would assume that the increased biomass of organic and inorganic body constituents reflects an increase in cell number during larval growth, no data are available on what causes the increased mass during metamorphosis.

As in numerous other vertebrate groups, the most investigated biochemical changes occurring in osteichthian larval development relate to the red blood cells and their major constituent, hemoglobin. Yolk sac larvae have few or no red blood cells. The number of red blood cells and the hemoglobin concentration increase steadily during larval life (de Silva, 1974). Hemoglobins isolated from larvae and adults always show a great deal of polymorphism. In all osteichthians examined, at least some larval and adult hemoglobins differ. In most species the adult fish have more types of hemoglobin molecules than do their larvae. In two species the larval hemoglobins have a greater degree of polymorphism than the adult hemoglobins, as in amphibians (Perez and Maclean, 1974; Just *et al.*, 1977b). The observed polymorphism of osteichthian hemoglobins is caused by the presence of multiple globins in individual animals. Although hemoglobins make their appearance relatively late in larval life, they are probably used to help transport oxygen to tissues. While it is true that some developing fish transport oxygen to tissues without hemoglobin, it would seem unreasonable to suggest that none of these hemoglobins are used during larval development. Oxygen is required for survival and the oxygen requirements change during development (Spoor, 1977). In addition, the multiple hemoglobins are functionally different in terms of their oxygen-carrying ability at different temperatures and osmotic pressures (Perez and Maclean, 1976). The functional oxygen-carrying capacity of the hemoglobins of the adult or larva favors the animal's use of a particular ecological niche.

4.3.3. Behavioral Changes during Larval Life

All these morphological and biochemical differences permit larvae to occupy ecological niches different from those occupied by the adults. Larvae explore these niches by modifying their behavior during larval development. The yolk sac larvae have three behavioral modes. Some hatch so early that they do not swim but remain on the substrate until most of the yolk is resorbed, as is the case in many freshwater fish. The second type of behavior includes a brief period of swimming immediately after hatching until larvae make contact with the substratum or objects, usually plant life, in the water. After making contact, dermal glands located in the head region secrete a sticky substance, thereby joining the larva to the substratum or object. These attached larvae are moved passively by water currents until yolk resorption is nearly complete, when active swimming and feeding commence. The third behavior found in yolk sac larvae is a rather active pelagic swimming. In most cases these larvae are positively phototropic and negatively geotropic, so that they are found in the upper regions of the water.

After completion of yolk sac resorption, all larvae start feeding. During the course of larval life the feeding movements become more efficient. Very young larvae succeed in about 1% of their attempts to capture prey, older larvae being successful in about 80% of their attempts (Braum, 1978). The increased feeding efficiency occurs because of changes in the mouth, eyes, and swimming rate. The mouth enlarges with the size of the animal, and since the vast majority are carnivorous, one might expect feeding efficiency to increase (Shirota, 1970). Even though some larvae can use olfaction and the lateral line system to capture prey, most larvae feed during daylight hours, locating the prey visually. As larvae age, pigmentation in the eyes, the number of optic nerve fibers, and the amount of myelin per optic nerve fiber increase. A number of larvae have a pure cone retina throughout larval life, while others exhibit an increasing number of retinal rods. All of these morphological changes result in an increased feeding efficiency in those larvae that depend on illumination for feeding (Blaxter, 1969). As larvae age they increase in size and with ever-increasing size the swimming rate increases, which makes them better predators. Not only do the older larvae swim faster, but they also increase the time spent swimming. Thus, large larvae can search larger areas for food (Braum, 1978; Blaxter and Staines, 1971).

As larvae increase in hunting efficiency, they not only capture ever-increasing amounts of prey but also change the type of prey. Older fish larvae capture larger prey organisms. It seems curious, however, that the youngest larvae actually capture a greater mass of prey compared to their weight than do older larvae (Laurence, 1977). Younger larvae probably require more food per gram tissue because of the inefficient absorption of food in their digestive systems. In laboratory experiments larvae supplied with prey inadequate to meet their caloric demands become too weak to seek food. When larvae start feeding in the laboratory, they must obtain food or large numbers will die. This is the so-called critical period. It has been argued that in the field, larvae never experience such low levels of available food because of their ability to behaviorally regulate their position sites of high prey density. Arguments have been put forth that the major factor affecting larval survival is predation and not starvation (May, 1974). A similar conclusion has recently been reached concerning the cause of death in amphibian larvae (Cecil and Just, 1979).

This review of osteichthian metamorphosis excludes a discussion of fish that undergo great physiological adaptations as young fish. Such fish include those species that live in the ocean and spawn in fresh water (anadromous) and those that live in fresh water and spawn in the ocean (catadromous). The anadromous fish include salmon, sturgeon, stickleback, and temperate bass, while the catadromous species include eels and mullets. Other authors have reviewed the final behavioral and physiological changes that the young of these species must undergo to return to the adult ecological niche as part of metamorphosis (Barrington, 1968; see Wald, this volume). A discussion of these changes as part of metamorphosis is excluded because in many instances the morphological changes occur before migration into the new environment, or the changes are relatively minor (i.e., color changes). While these migrations require many physiological adaptations, they do not necessarily imply changes associated with metamorphosis as is indicated by the fact that adults of the same species make such migrations without any gross morphological changes (Fontaine, 1975). The changes in some of these species are reversible. Salmon smolts retained in fresh water lose their new color and osmoregulatory capacity and live the remaining year as a parr, only to transform again the next year. No other animal, whether invertebrate or chordate, can reverse to the larval form after completing metamorphosis.

4.3.4. Triggers for Metamorphosis

The triggers for osteichthian metamorphosis are poorly understood and have received little experimental attention. By analogy with amphibian metamorphosis, the endocrine system, in particular the thyroid, has received the most attention as the probable causative agent for metamorphosis. Many early descriptive works of larval species included histological studies documenting changes in thyroid development. Species receiving the most attention were larvae of the catadromous and anadromous species (for reviews see Szarski, 1957, and Barrington, 1968). These histological changes were interpreted to indicate that the circulating levels of TH change during metamorphosis; however, only in the parr–smolt transformation has such an increase been demonstrated (Dickhoff *et al.*, 1978; see White and Nicoll, this volume). TH are not the only ones known to change during parr–smolt transformation. The levels of other hormones, including adrenal steroid hormones, thyroid-stimulating hormones, growth hormones, and adrenal corticotropin hormone, also change (Fontaine, 1975; Olivereau, 1975; Komourdjian *et al.*, 1976).

With this in mind, and with the probability that such changes occur in osteichthian metamorphosis, it is rather surprising how few experiments have been performed to alter the hormonal status of larvae. The circulating levels of TH during osteichthian metamorphosis and parr–smolt transformation have been experimentally altered, but the rates of metamorphosis or parr–smolt transformation have not always changed (Szarski, 1957; Hoar, 1976). The total literature does not yet support the control of metamorphosis by the endocrine glands, and Hoar concludes that evidence for hormonal regulation of parr–smolt transition is "equivocal or negative" (Hoar, 1976).

External environmental factors are known to speed up metamorphosis. In one of the earliest reports, young surgeonfish caught in open sea undergo metamorphosis in 48 hr irrespective of size (age?) after they come into contact with the substratum (Breder, 1949). Numerous other authors have directly or indirectly appealed to substrate contact by larvae of bottom-dwelling adults as the trigger for metamorphosis. Aside from contact with the substratum, temperature and light have also been shown to affect rates of metamorphosis. There is a definite correlation between parr–smolt transformation and light and temperature (Hoar, 1976; Thorpe and Morgan, 1978).

4.3.5. Unresolved Questions in Osteichthian Metamorphosis

The most urgent need in understanding osteichthian metamorphosis is for the development of a uniform nomenclature for describing the various periods of the life history of fishes. Every published work should use a standard terminology to describe development, but unfortunately many authors give only length measurements of their larvae. Since larval length is influenced by temperature, food availability, and developmental age, its use as the sole criterion for determining metamorphic progress should be discontinued. Attempts have been made by others to standardize the terminology (Balon, 1975). At present the terminology describing periods of development is varied, making comparisons between species difficult or impossible, and even the data of a single species from various laboratories under different culture conditions are difficult to correlate. The most desirable situation would be to have published, in all appropriate sources, the sequences or morphological changes (stages) for commonly used species, and agreement should be reached about the terminology used.

Considering the vast amount of literature published annually on larval life, it is surprising how little effort is directed toward the elucidation of metamorphic triggers. Perhaps the early negative results on some species have decreased the interest in this area. Vigorous attempts should be made to determine the cause of metamorphosis in all species, but particularly in the eels, flounders, bonefish, and lungfish, in which there are drastic morphological alterations. Fish culture techniques have improved greatly over the years so that even the complete metamorphosis of flounder has been achieved under laboratory conditions (Smigielski, 1979). The investigation of the triggering mechanism in selected species is justified at this time, and need not be limited to thyroid hormone investigations.

5. AMPHIBIANS

The class Amphibia is divided into three subclasses, Labyrinthodontia, Lepospondyli, and Lissamphibia. The earliest amphibians, the Labyrinthodontia and Lepospondyli, appeared about 350 million years ago, and 200 million years later became extinct. The current subclass, Lissamphibia, first appeared about 150 million years ago during the Jurassic period. Its numbers have rapidly expanded, until today there are about 3000 living species and 150 extinct species. These species are divided into three orders: Anura (Salientia), Urodela (Caudata), and Caecilia (Apoda and Gymnophiona). Of the total living species, 85% are anurans, 10% are urodeles, and only 5% are caecilians (Carroll, 1977).

Modern amphibians can be found in nearly all of the major land habitats, including deserts. The only regions not yet invaded by amphibians are those where the temperatures are extremely cold, namely the polar regions, and those at extremely high altitudes. The anurans have the broadest geographical distribution among the amphibians and are found on all continents except Antarctica. Caecilians are essentially tropical, while urodeles are restricted to the northern hemisphere with the exception of a radiation into Central and South America (Savage, 1973). This radiation accounts for a significant percentage (40%) of the existing urodele species (Wake, 1966). Even though amphibians have to remain moist as adults and most must return to water to reproduce, neither the larvae nor the adults have succeeded in establishing themselves as permanent residents in oceans or large freshwater lakes. Fertilization is external in all but two groups of anurans, the African live-bearing toads (*Nectophrynoides*) and the American tailed frog (*Ascaphus truei*). In the vast majority of urodeles, fertilization is internal, the female picking up a spermatophore (a small packet of sperm) from the substrate with her cloaca. Less than 5% of the urodeles utilize external fertilization, and these belong to two primitive families (Cryptobranchidae and Hynobiidae). All caecilians practice internal fertilization, with the male using an intromittent organ (a posterior extendable part of the male cloaca) to introduce sperm into the coaca of the female (Murphy *et al.*, 1977; Wake, 1977).

5.1. Parental Care

Regardless of the method of fertilization, the vast majority of amphibians abandon the zygote. However, examples of the zygote being retained in the reproductive tract are found in all three orders (Salthe and Mecham, 1974). Four anuran species from two genera retain the zygote internally until it hatches in the oviduct (Wake, 1978). A few urodele species also hatch their young in the oviduct, the most common being *Salamandra atra*. Probably more than 50% of the known caecilian species retain the hatched larvae in their oviducts (Wake, 1977). In all three orders the larvae that hatch in the oviduct ingest a milky secretion produced by the oviducts. Larvae in the oviducts undergo some of the metamorphic changes typical for the order, similar to that of larvae not retained in the oviducts.

Other less obvious forms of parental protection are offered to the young zygotes or larvae of all living amphibian orders. Caecilians, which do not retain the zygote, lay the fertilized eggs on moist land, and a number of reports claim that the females remaining with the zygotes offer protection to them. Nearly 60% of the urodele species belong to the family Plethodontidae, and most of these species lay terrestrial eggs. Parental care (brooding) is presumed to occur in most urodele species because one adult, usually a female, is associated with the zygotes (Salthe and Mecham, 1974). Associated adults occur in nearly all terrestrial breeding species and in some aquatic breeding species. These adults may offer protection to zygotes in any of at least four ways: (1) warding off predators; (2) preventing microbial infections of zygotes in egg cases; (3) supplying or helping to maintain moisture during periods of desiccation; and (4) agitating eggs, particularly in the aquatic environment, to prevent anoxia. There is convincing evidence that brooding adults in several urodele species play a protective role, but experimental evidence is not available on the protective role played by brooding caecilians (Forester, 1979).

The most varied parental protection is seen in anurans, with care being given by either sex. Many anuran species, particularly Leptodactylidae, lay terrestrial eggs. In numerous species with terrestrial eggs, adults have been found associated with the developing zygotes, and most authors have assumed that such brooding offers protection to the young (Salthe and Mecham, 1974). However, unlike the situation in the urodeles, little experimental evidence is available to indicate that this is of a protective nature (Woodruff, 1977). The only exception is the midwife toad (*Alytes*), in which zygotes are wrapped in egg cases around the hindlimbs of the males. The male transports these egg cases to ponds where the zygotes hatch. Many other anurans do offer protection to their young after hatching. Protection can be as simple as digging a tunnel for the tadpole to reach water (*Hemisus*), or allowing the hatched tadpoles to crawl onto their backs (dendrobatids) and then transporting them to the water (Wells, 1978).

Protective associations that ensure greater survival of the young are found in some anurans. The dorsal epidermis of some female pipids thickens during the breeding season. The male presses the fertilized eggs into her back and the epidermis then grows to nearly cover each egg within 2 or 3 days. The zygotes develop in these epidermal pockets until they hatch as advanced larvae or completely metamorphosed individuals. In some species (*Pipa*) there are no specialized embryonic structures to help supply nutrients or oxygen, while in others (*Hemiphractus*) the larvae are attached by specialized gill structures until metamorphosis is complete (Duellman, 1970; Weygoldt, 1976). Some leptodactylid adults swallow the zygotes or perhaps even the hatched tadpoles, and these young animals are retained in either the vocal sacs or the stomach of the adult until they complete metamorphosis (Corben *et al.*, 1974). Other leptodactylid tadpoles crawl into specialized pouches on the side of the adult animals and remain there until metamorphosis is complete (Ingram *et al.*, 1975). The hylids include species in which the female has a pouch on her back, into which the male inserts the fertilized eggs after mating. Therein, the embryo develops specialized gill structures, which presumably help to supply oxygen and perhaps even nutrients. The young are retained in the pouches until the mother opens them with her hindlimbs. The young leave the pouch either as totally metamorphosed froglets or as tadpoles that complete metamorphosis in ponds a few weeks later (Barrio, 1977).

5.2. Larval Morphology and Staging

Larvae hatch 2 days to 1 year after fertilization regardless of whether the zygote is protected or not. Nearly 100 years ago it was recognized that an orderly array of morphological changes could be used to measure developmental progress by arbitrarily numbering the process (staging). No single group of chordate larvae has received more attention than the anuran larvae in terms of variation in external morphology among species and different staging methods (Noble, 1931; Wright and Wright, 1949). The internal morphological changes during metamorphosis of various species will not be discussed in the present chapter, and the reader is referred to the following chapters for a discussion of the morphological and biochemical changes.

5.2.1. Anuran Larval Morphology and Staging

The greatest variability in anuran larval morphology occurs in the mouth parts, gills and associated structures, and tail. Anuran tadpoles living in ponds or quiet flowing water have broad tail fins, extending at times into the back, while tadpoles living in rapidly moving water have very narrow tail fins. The external gills of larvae anurans arise from branchial arches. All anurans form a branchial chamber containing two external gills covered by an integumentary fold (operculum) that grows from the hyoid arch. In aquatic larvae, water enters either the mouth or the nostrils, crosses the gills, and exits through an opening in the branchial chamber called the spiracle. There are four morphological variations of the branchial chambers and their openings (Starrett, 1973; Sokol, 1975). The first type is characterized by two separate branchial chambers and two external openings (xenoanura). This type of branchial morphology is seen in two anuran families (pipids and rhinophrynids). In other tadpoles two separate branchial chambers lead into a single ventral opercular tube that descends to the base of the tail, opening as a single spiracle (scoptanura). This modification is seen in the microhylids. In the third type, two branchial chambers join into a single spiracle opening anteriorly at the ventral midline (lemmanura), as seen in ascaphids and discoglossids. The final type of modification of gills occurs when branchial chambers are joined and open into a single spiracle on the left side of the tadpole (acosmanura). This type of branchial modification is most common among anurans, including bufonids, hylids, and ranids. In many anurans with terrestrial development or specialized parental protection, the tail fins or the two external gills are greatly extended, covering the developing zygote, and presumably serve as respiratory organs. No experimental evidence is available to document the degree to which such modifications actually serve as respiratory organs.

The greatest variability among anurans occurs in the mouth structure. In fact, such variations are so numerous and predictable that they are a major aid in species identification (Wright and Wright, 1949; Altig, 1970; Duellman, 1978). The width of the mouth varies from about 1/10th the body width to nearly the same width as the body. The mouth opening of most tadpoles is formed by a lower and an upper horny beak. On the exterior surface, above and below the horny beaks, are rows of labial teeth. The number and length of the rows are age and species specific. The width of the rows of labial teeth can be smaller or greater than the width of the horny beaks. As many as 10 rows can be found in certain species. The structure of individual teeth is species specific (Altig and Pace, 1974). Surrounding the teeth and beak are labial papillae.

In several families with free-living tadpoles (pipids, rhinophrynids, and microhylids) and in species without free-living tadpoles, the tadpoles sometimes have no horny beaks, labial teeth, or labial papillae around the mouth. In other species with free-living tadpoles, the mouth is extensively modified into a funnel-shaped structure for filter-feeding from the water's surface. Species living in fast-moving streams have developed an oral disc from these three structures or parts of them. The discs are used to attach tadpoles to the substrate.

The sequence of external morphological changes from anuran tadpole (Fig. 10, No. 4) to adult (Fig. 10, No. 5) has been described for several

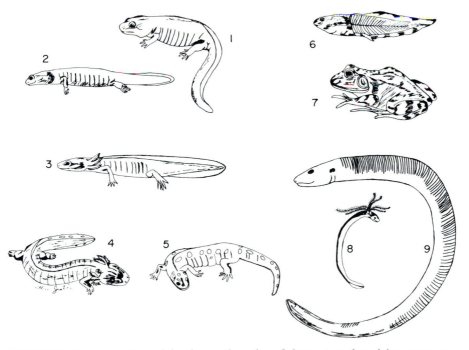

FIGURE 10. Representatives of the three orders of amphibians. Larval urodeles are seen in 2 (stream type) and 3 (pond type). Neoteny in the urodele is seen in 4, and the adults in 1 and 5. The larval anuran (6) and adult anuran (7) are shown in the upper right, and below are a very young caecilian (8) and an older free-living caecilian tadpole (9).

species. These morphological changes include early events such as differentiation and growth of hind- and forelimbs. The forelimbs may or may not be enclosed in the branchial chambers. During the final changes, the forelimbs erupt from the branchial chambers. Tail and gills are resorbed, and the alteration of the mouth structures is accomplished. The sequence of changes has usually been split into arbitrary periods referred to as stages. Table 3 lists the staging methods of various anuran species. The first three methods refer to staging methods in which the authors hoped to establish staging procedures applicable to all anurans, while the remaining 15 entries refer to staging of specific genera and species. Table 3 is included to facilitate comparisons of the literature on different species. The various staging numbers were equated primarily on the basis of hindlimb development, degree of tail resorption, and development of the adult mouth. It is difficult to precisely equate the various stages among so many staging methods, and Table 3 contains a few subjective decisions. Even though the species listed include a live-bearing animal (*Nectophrynoides*) and the midwife toad (*Alytes*), their stages can also be compared to the "normal" free-living anuran tadpole. It must be emphasized that in every case, tadpoles hatch with hindlimb buds showing, although in many publications on the biochemistry of anuran metamorphosis, one sees tadpoles incorrectly referred to as "limbless."

An abbreviated staging system is commonly used in anuran metamorphosis; that is, premetamorphosis, prometamorphosis, and climax. Unfortunately, the terms have been defined differently in the literature (Taylor and Kollros, 1946; Witschi, 1956; Etkin, 1968; Dodd and Dodd, 1976), and many authors use these terms without indicating which definitions are being used. Any of the more precise staging methods used to define metamorphic progress are preferred, such as in Table 3. The above terms should be redefined to conform to certain endocrine aspects of a given species. The term "premetamorphosis" should refer to any stages or events in larval life that are independent of thyroid function. The term "prometamorphosis" should be limited to the stages and events that are dependent on TH, while the term "climax" should be limited to those events and stages that require a tremendous surge in circulatory TH levels. It should be remembered that the degree of hindlimb differentiation varies from species to species following thyroidectomy. Premetamorphosis could then be applied only to species in which the results of thyroidectomy are known, and only to those stages that are independent of thyroid function. Similarly, climax stages and events for any species would be valid only if the surge in hormone levels had been determined. Currently the term "climax" could be used for, at most, three species: *R. pipiens, R. catesbeiana,* and *X. laevis.* Even with these three species, the term may refer to different developmental stages, because the surge in hormones may occur at different times (see White and Nicoll, this volume).

5.2.2. Urodele Larval Morphology and Staging

Urodele larvae hatched from eggs laid in water show some morphological variability, but less than that seen among anuran larvae. Their major morphological variability occurs in the tail, gills, and time of acquisition and rate of development of the limb structures (Noble, 1931; Valentine and Dennis, 1964). Urodele larvae have three sets of external gills, which remain visible throughout larval life because the operculum fails to enclose them (Fig. 10, Nos. 2–4). Larvae living in rapidly running water usually have short external gills and low fins confined exclusively to the tail (Fig. 10, No. 2). Larvae living in quiet waters have long external gills and large tail fins that in some species extend onto the backs (Fig. 10, No. 3). It has been known for nearly 60 years that the external gill size can increase in water with low oxygen tension; however, this is not the sole cause of the differences in the gill structures among the two basic larval types.

The development of limb structure varies tremendously among these larvae. Species that deposit small eggs in quiet aquatic habitats give rise to tadpoles having anterior limb buds and balancers at hatching. The balancers are bilateral skin projections from the mandibular arch and secrete mucus. This secretion enables the larva to maintain equilibrium by attaching it to the substrate. After hatching, digits form on the front legs and the balancers may elongate slightly. The hindlimb buds usually make their appearance after feeding starts, and then they differentiate. In urodeles that lay eggs in running water, the fore- and hindlimbs are formed before hatching. In these species the balancers either never form or they may be resorbed by the time of hatching. The anterior and posterior limbs of tadpoles hatched in running water begin differentiation almost simultaneously (Salthe and Mecham, 1974).

In urodeles producing terrestrial eggs, mainly the members of the lungless family (Plethodontidae), there are drastic alterations in normal larval morphology. The limbs are nearly fully differentiated in all of the animals hatched from terrestrial eggs, and the tail fins are greatly reduced or absent. The three external gills are usually modified; they may appear as highly filamentous structures, or they may fuse at the base forming a single trunk having the overall appearance of stag horns. In species that are purely terrestrial, the larvae never develop a lateral line system, while all aquatic larvae possess this sense organ. Zygotes hatch from terrestrial eggs either as immature adults or as advanced larvae, which migrate to water to complete metamorphosis.

In free-living urodele larvae, the sequence of external morphological changes associated with the transition from the larval (Fig. 10, Nos. 2, 3) to the adult form (Fig. 10, Nos. 1, 5) has been described for various species, and is arranged into arbitrary developmental stages (Table 4). All of the species listed in Table 4 belong to the pond type, facilitating comparisons even for a neotenic genera (*Necturus*). The earliest stages are

TABLE 3. COMPARISON OF LARVAL

Method	Stages											
	Limb bud growth						Toe differentiation					
Gosner[a]	26	27	28	29	30	31	32	33	34	35	36	37
Terentlev[b]	25	26	26	26	26	26	26	26	26	26	26	26
Witschi[c]	25	26	26	26	26	26	27	27	27	27	27	28
Alytes obstetricans[d]	IV_1	IV_2	IV_3	IV_4	$IV_{5,6}$	IV_7	$IV_{8,9}$	$IV_{10,11}$	$IV_{12,13}$	IV_{14}	IV_{14}	IV_{14}
Bufo[e]	—	—	—	—	—	—	—	—	—	—	—	1
Bufo bufo[f]	23	24	25	I	II	III	IV	V	VI	VII	VIII	VIII
Bufo bufo[g]	III_{10}	IV_1	IV_2	$IV_{3,4}$	IV_5	IV_6	IV_7	IV_8	IV_9	IV_9	IV_{10}	IV_{10}
Bufo melanostictus[h]	25	26	27	28,29	30	30	31	32	33	34	35	35
Bufo regularis[i]	46	47	48,49	50	51	52	53	54	54	55	55	56
Bufo valliceps[j]	26	27	28	29	30	31	32	33	34	35	36	37
Bufo vulgaris[k]	11	11	12	12	13	13	13	13	13	13	13	13
Nectophrynoides occidentalis[l]	I_b	I_b	I_b	I_b	I_b	I_b	I_b	I_b	I_b	II_a	II_a	II_a
Rana chalcoriata[m]	VIII	VIII	VIII	VIII	IX	IX	IX	IX	IX	IX	IX	IX
Rana dalmatina[n]	39	40	41	42	42	43	44	45	45	46	46	47
Rana fusca[o]	18	18	19	20	21	22	22	23	23	24	24	24
Rana pipiens[p]	I	II	III	IV	V	VI	VII	VIII	IX	X	XI	XII
Rana temporaria[q]	8	9,10	11,12	13	13	—	—	—	—	—	—	—
Xenopus laevis[r]	46	47,48	49,50	51	52	53	53	53	54	55	55	55

[a] Gosner (1960); [b] Terentlev (1950); [c] Witschi (1956); [d] Cambar and Martin (1959); [e] Schreiber (1937); [j] Limbaugh and Volpe (1957); [k] Adler (1901); [l] Lamotte and Xavier (1972); [m] Hing (1959); [n] Cambar Faber (1967).

TABLE 4. COMPARISON OF LARVAL

Method	Stages									
	Forelimb and hindlimb									
Ambystoma maculatum[a]	38	38	39	40	41	42,43	44	45	46	—
Ambystoma mexicanum[b]	—	—	—	—	—	—	—	—	—	—
Hynobius nigrescens[c]	40	41	42	43	44,45	46,47	47,48	49	50	51
Megalobatrachus japonicus[d]	26	26	27	27	27	27	27	28	28	28
Necturus maculosus[e]	31	31	32	32	33	33	33	33	34	34
Pleurodeles waltlii[f]	34	35	36	37	38	39	40	41	42	43
Triton alpestris[g]	28	28	29	29	30	—	—	—	—	—
Triton taeniatus[h]	34	35	36	23	24	39	40	41	42	43
Triturus helveticus[i]	33	34	35,36	37	38	39,40	41	42	43	44
Triturus pyrrhogaster[j]	23	23	23	37	38	24	24	24	25	—

[a] Rugh (1962) (stages of Harrison or Leavitt, improperly referred to as *A. punctatum*); [b] Marx (1935); [c] Usui and Hamasaki (1931); [j] Gallien and Bidaud (1959); [i] Anderson (1943).

ANURAN STAGING METHODS

Stages												
Rapid hindlimb growth								Tail resorption				
38	39	40	40	40	41	41	41	42	43	44	45	46
27	27	27	27	28	28	28	29	30	30	30	30	31
28	28	28	28	28	29	30	31	31	32	32	33	33
IV_{14}	IV_{15}	IV_{15}	IV_{15}	IV_{16}	IV_{16}	IV_{17}	IV_{17}	IV_{18}	IV_{19}	IV_{19}	IV_{19}	IV_{20}
1	1	1	1	1	2	2	3	4	5	6	7	8
VIII	IX	IX	X	X	XI	XI	XII	XIII	XIV	XIV	XIV	XV
IV_{11}	IV_{11}	IV_{12}	IV_{12}	IV_{13}	IV_{13}	IV_{14}	IV_{14}	IV_{15}	IV_{15}	IV_{16}	IV_{16}	IV_{17}
35	36	36	37	37	38	38	39	40	41	42	42	43
56	56	57	57	57	57	58	58	59,60	61,62	63,64	65	66
37	38	38	39	40	40	41	42	43	44	45	45	46
13	13	13	13	14	14	14	14	14	15	15	15	15
II_a	II_a	II_b	II_b	II_b	II_b	III_a	III_a	III_b	III_b	III_b	III_b	IV
IX	IX	IX	IX	IX	IX	X	X	X	X	X	X	X
48	48	49	49	50	50	50	51	51	52	53	53	53
25	25	26	26	26	26	26	27	27	28	29	29	30
XIII	XIV	XV	XVI	XVII	XVIII	XIX	XX	XXI	XXII	XXIII	XXIV	XXV
—	—	—	—	—	—	—	—	—	—	—	—	—
56	57	57	58	59	60	60	61	62	63	64	65	66

Rossi (1959); [g] Cambar and Gipouloux (1956); [h] Khan (1965); [i] Sedra and Michael (1961); and Marrot (1954); [o] Kopsch (1952) [p] Taylor and Kollros (1946); [q] Moser (1950); [r] Nieuwkoop and

URODELE STAGING METHODS

Stages																	
growth and differentiation											Gill and tail fin resorption						
—	—	—	—	—	—	—	—	—	—	—	—	—	—	—	—	—	—
—	—	—	—	—	—	—	—	—	—	—	II	III	IV	V	VI	VII	VIII
52	53	54	55	56	57	58	59	60	61	62	63	64	65	66			
28	28	28	28	29	29	30	30	31	3i	31	31	31	32	32	32	32	33
34	34	35	35	36	36	36	37	38–41	42–48	49							
44	45	46	47	48	49	50	51	52	53	54	55_a	55_a	55_b	55_b	55_c	55_c	56
—	—	—	—	—	—	—	—		—	—	—	—	—	—	—	—	—
44	45	—	—	—	—	—	—	—	—	—	55_a	55_a	55_b	55_b	55_c	55_c	56
45	46	46	47	48	49	50	51	52	53	54	—	—	—	—	—	—	—
—	—	46	47	48	49	50	51	52	53	54	55_a	55_a	55_b	55_b	55_c	55_c	56

(1939); [d] Kudo (1938); [e] Eyclesh‹ er and Wilson (1910); [f] Gallien and Durôcher (1957); [g] Knight (1938); [h] Glücksohn

characterized by balancer growth and forelimb growth and differentiation. In the midlarval stages, the hindlimbs grow and differentiate while the balancers degenerate. The final stages in urodele metamorphosis are marked by gill and tail fin regression. Although the mouth does not change as drastically as it does in anurans during the late stages, there is an eyelid development in most species during the late larval stages that is similar to that in anurans.

Universal staging procedures are not available for urodele larvae, and, unfortunately, most authors never use any staging procedures to describe metamorphic progress, but instead rely on length measurements alone. This method of course makes any comparison between species, or even within the same species in different habitats, very difficult. Some authors have given both length measurements and limb development in several species, thus demonstrating that metamorphic progress could presently be monitored by more than one method. Some authors refer to premetamorphosis, prometamorphosis, and climax without defining their terminology. In most of these cases it is impossible to determine whether they are using the definitions related to urodeles (Wilder, 1924; Marx, 1935) or those for anurans cited above. The use of these terms should be decreased in the description of urodele metamorphosis, but if used, the usage should correspond to certain aspects of the endocrinology of the species as defined earlier for anurans.

5.2.3. Caecilian Larval Morphology and Staging

For nearly 50 years caecilian development received little attention, until a recent resurgence of interest (Taylor, 1968; Wake, 1977). However, between 1830 and 1940 there were nearly 100 publications in German dealing with the caecilian development. The most noteworthy of these publications are the exquisite descriptive studies by Sarasin and Sarasin (1887–1890) and the numerous descriptive works on larval anatomy by Marcus and co-workers (Marcus, 1939).

The fertilized egg is either retained in the oviduct or deposited on land (for none are deposited in water). The developing animal is limbless and always has three pairs of external gills, which are never covered by an operculum (Fig. 10, Nos. 8, 9). Nevertheless, external morphological variations exist among these larvae in gills and teeth, as well as in the presence or absence of a lateral line system. Larvae developing inside terrestrial eggs have delicate narrow filamentous gills, restricted to the most anterior portion of the animal (Fig. 10, No. 8). In many species one of the three external gills is very rudimentary. The larvae (fetuses) developing in the oviducts have two types of gills. The fetus in terrestrial adults contains the same type of highly filamentous gills as seen in larvae developing outside the mother, while fetuses developing in the oviducts of aquatic adults have large hollow sacs on each side that presumably act as gills. In older fetuses these sacs grow to a length equal to 50–70%

of the larval length. Both filamentous and saccular gills receive blood from three aortic arches. All larvae have teeth that differ from those of the adult. Fetal teeth are present on both the upper and the lower jaws, and more than a single row of teeth are found on each jaw. The larval teeth are small and have species-specific morphology (Taylor, 1968; Wake, 1977).

There are various morphological changes in the transition from larva (fetus) to adult, particularly in the eyes, gills, teeth. While the sequence of morphological changes has never been assigned arbitrary numbers (stages) in any caecilian, very complete illustrations and descriptions are available for all phases of external morphological changes in two of the caecilian species producing terrestrial eggs (Sarasin and Sarasin, 1887–1890; Brauer, 1899). The illustration numbers in these papers could and should be used as staging numbers for any publication on the physiology or biochemistry of development in terrestrial larvae. Unfortunately, complete descriptions of larval development in the oviducts are not available, but it does appear that all developing caecilians undergo the same series of metamorphic changes.

All caecilians start with normal embryonic eye development. Sometime during late embryonic or fetal life, the eyes are covered to various degrees with skin or bone (Taylor, 1968). Certain muscles and nerves leading to the eyes degenerate just as they do in blind cave salamanders. During late fetal or larval life small tentacles develop in front of the eyes. It is presumed that these tentacles function as sensory organs. There are no major changes in the mouth structure during late fetal or larval life. Larval teeth are lost and adult teeth develop soon after the fetuses or larvae become free living. The larvae have no tail or tail fin degeneration associated with metamorphosis, but if these structures appear, they are lost very early in embryonic development.

The most noteworthy external morphological change associated with metamorphosis is gill degeneration during late embryonic, larval, or fetal life. All fetuses are born after gill degeneration, while hatching of terrestrial organisms may or may not occur before the gills have been resorbed (Wake, 1977). At hatching several species possess gills that are soon lost, leaving only a gill slit. It is not clear whether the gills are resorbed as in the other amphibian orders, or whether they are broken off during the migration to water, as claimed by Sarasin and Sarasin (1887–1890).

In the final stages of caecilian metamorphosis, other changes occur. The skin thickens and skin glands form as in other amphibian larvae. In this order small scales also form in the skin (Sarasin and Sarasin, 1887–1890). The lungs are probably not functional until after birth or hatching, and therefore metamorphosis must also entail lung maturation. It has recently been demonstrated that a hemoglobin solution made from fetal blood has a greater oxygen affinity than that of the maternal hemoglobin (Garlick et al., 1979). The difference in oxygen affinities is probably not caused by differences in the protein structure of the globin, but by the higher ATP content of adult red blood cells.

5.2.4. Morphology of Fossil Larvae

Examples of fossil larvae have been found for both the modern subclass (Lissamphibia) and an extinct subclass (Labyrinthodontia). No fossil larvae of the subclass Lepospondyls have been found. Although fossil larvae of the subclass Lepospondyli have been found. Although fossil for caecilians. The first adult caecilian fossil fragment was found less than 10 years ago.

The most complete series of a fossil larval amphibian is that of an anuran tadpole belonging to the pipid family (Estes *et al.*, 1978). Even though these tadpoles are probably 100 million years old, they show the same external morphological features as do *X. laevis*. These fossils were staged according to the system for *Xenopus* (see Table 3), and the extinct population ranged from stage 50 to stage 61. The size of these fossil larvae and the sequence of bone ossification are similar to those of *Xenopus*. These animals were preserved well enough so that numerous organs could be observed, including tail fins, eyes, optic nerves, tentacles, lungs, and thymus.

Since urodeles are relatively rare compared to anurans, and usually produce fewer offspring than anurans, it is not surprising that few urodele fossil larvae have been found. Fossil larvae belonging to the genera *Ambystoma* and *Ramonellus* have been found that are about 1 million years old (Tihen, 1942; Nevo and Estes, 1969). Their bone structures are similar to those of present-day species. A number of jaws with attached teeth indicate the same developmental sequence as that seen in living *Ambystoma* species. The jaws and other bones are so large as to suggest that this extinct population may have been neotenic.

It is not surprising that modern amphibians (Lissamphibia) metamorphosed when they first evolved. However, it is surprising that the fossil evidence indicates that an extinct subclass, Labyrinthodontia, also metamorphosed. Examination of 200-million-year-old fossils showed that Labyrinthodontia underwent metamorphosis (Boy, 1974). The smallest fossil larvae were about 3 cm long and had large eyes. These animals had three external gills without an operculum. The Labyrinthodontia larvae resembled the pond-type urodele larvae in that they developed forelimbs first and hindlimbs later. These animals had long tails that probably contained large tail fins. As these larvae underwent metamorphic changes, the sequence of bone ossification in their skulls was similar to that observed in urodele larvae, whereas the ossification sequence of the vertebral column was similar to that seen in anuran larvae. The final stages of metamorphosis were marked by gill resorption and the appearance of bony scales on the ventral surface.

5.3. Larval Life Span and Survival Rates

Larval life spans are affected by a variety of external factors, including temperature, availability of food, and amounts of trace elements, especially iodine. Life spans for various amphibian larvae are presented in Figs. 11 and 12. Many species have several different life spans published in the literature, and in such cases the average life span of all publications on that species is presented in these figures. Urodele species that show neoteny in only certain populations are recorded twice in Fig. 12, once for the average life span calculated for nonneotenic populations, and a second time for neotenic populations.

Figure 11 shows the larval life spans for 153 anuran species, or about 6% of the total living anuran species. At one extreme, 24 anuran species lay terrestrial eggs that hatch directly into froglets. Species with extremely short larval life spans hatch from eggs laid in temporary pools, which are formed after rain. The adults of these species survive under extremely arid conditions, such as the desert. The median life span of all species is 7 weeks. Less than 20% of all species have a larval life span

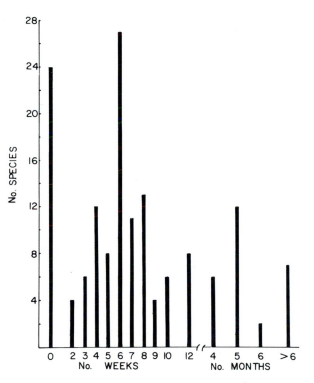

FIGURE 11. The frequency distribution of tadpole life spans for 153 anuran species. The larval age is given in weeks and months on the abscissa. Direct development of anurans within eggs is indicated at time 0. The original data are from numerous sources; see text for method of data presentation.

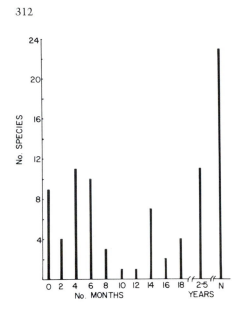

FIGURE 12. The frequency distribution of tadpole life spans for 68 urodele species. The larval age is given in months and years on the abscissa. Direct development of urodeles within eggs is indicated at time 0, while N indicates neoteny. The original data are from numerous sources; see text for method of data presentation.

greater than 12 weeks, and only seven species are known to have a larval life span greater than 6 months. In these seven species there is no indication that they ever become neotenic. Longer life spans are observed in species either at the northernmost part of their range or at high altitudes (Cecil and Just, 1979; Collins, 1979). However, low temperatures do not necessarily imply an extended larval life span, because larval anurans of northern Russia do not have unusually longer life spans (Shvarts *et al.*, 1973).

Many reports claiming a life span of 1 or more years for a species are predicated solely on the presence of bimodal size classes in tadpole populations. If such populations were not sampled frequently enough to identify either the growth and differentiation rates of the tadpoles or the time of egg deposition, it is improper to conclude that there is a relatively long larval life span. This is particularly true since several adult anurans lay eggs twice a year in the same habitat (Emlen, 1977).

Even though the total anuran larval life span varies markedly (Fig. 11), the final stages of gill and tail degeneration show far less variation. In animals with long larval life spans, the relative time spent in the final transformation is far less than that of species that have short larval life spans.

The larval life span of 68 species of urodeles is shown in Fig. 12. This represents just over 20% of all known living urodeles. Among urodeles there are two tendencies, the elimination of the larval period entirely and, at the other extreme, the elimination of the adult phase (neoteny). There is a tendency to underestimate the number of species (a total of nine) that have no free-living larval life span, because embryos have not been found for many plethodontid species, which lay fertilized eggs deep in the soil. There is a tendency to overestimate the relative percent of neoteny in urodeles (recorded for 23 species) because the larvae are long-lived and large, making them easier to find in nature. The median life span of urodele larvae is 6 months. Extended larval life spans are found in those urodeles living in cold water or at high altitudes.

There is little known about the length of fetal and larval life spans in the caecilians. Those that lay terrestrial eggs give rise to both larvae and fully formed young. The range and size of various free-swimming larvae suggest that the length of time spent as a larvae should be a month or two. While the length of fetal life span in the oviduct is not known, the large size of the fetus at birth would suggest a relatively long fetal life span of several months.

It is obvious that the average anuran and caecilian larval life span is shorter than that of most urodeles with free-living larvae. In fact, 69% of all anuran larvae and probably many caecilian larvae have a life span of 2 months or less, while only 10% of the urodeles have such a short life span. Only 5% of the anurans, and probably none of the caecilians, have a larval life span greater than 6 months.

Since most tadpoles, particularly the anurans, are more or less defenseless, one might expect them to have a relatively low survival rate. In anuran larval populations, the percent survival from hatching to the completion of metamorphosis ranges from 1 to 18%, with a single species known to have a survival rate as high as 55% (Kadel, 1975; Cecil and Just, 1979); in urodele larvae, it ranges from 9 to 20% (Anderson *et al.*, 1971; Bruce, 1972; Bell and Lawton, 1975). Zero survival rate occurs when temporary ponds dry up before a population is able to complete its metamorphosis.

One might expect that tadpoles with long larval life spans should have a lower survival rate than tadpoles with short larval life spans because of the hazards facing the tadpoles in the water. However, there is no relationship between survival rates and larval life spans. The reason that long-lived larvae do not have a lower percent survival is because amphibian larvae are most susceptible to predation immediately after hatching, when they are small, and their ability to avoid predation increases with age (Cecil and Just, 1979). Much literature suggests that tadpoles suffer high mortality during the final stages of transformation, although little experimental evidence is available to substantiate this occurrence in nature (Arnold and Wassersug, 1978).

5.4. Amphibian Larval Behavior

Of the possible behaviors that larvae exhibit, none have received more attention or are more important for the species than that of food acquisition. The feeding behavior of all living amphibian orders has been investigated.

5.4.1. Anuran Feeding Behavior

Anuran tadpoles exhibit three basic feeding methods: (1) filter-feeding on suspended particles; (2) grazing on algae or aquatic vascular plants; (3) consumption of dead or living animal matter.

As filter feeders, anuran larvae ingest anything in suspension (i.e., algae, protozoa, bacteria, fine particles of vascular plants). The size of the trapped particles ranges from as little as $0.1\mu m$ to as large as the mouth opening. The internal anatomy of the filter apparatus affects its efficiency and varies among different species (Wassersug and Rosenberg, 1979). Water entering the mouth is moved through the system by the pumping action of muscles and the cartilaginous skeleton in the floor of the mouth and finally exits through the spiracle. The velum, a movable nonmuscular flap, divides the oral cavity of all tadpoles (except the pipids) into an anterior buccal and a posterior pharyngeal cavity. Larger food particles are trapped by the anterior buccal cavity. Smaller particles are trapped in the mucus secreted by the branchial surface of the velum. The trapped food is carried by the cilia into the esophagus.

The manner, rate of obtaining, and rate of assimilating food are not constant in any one species experiencing differing amounts of food resources in the environment. As the concentration of food particles increases, the amount ingested increases until a constant level is attained. The amount of ingested food remains constant during periods of high available food levels because filter feeders decrease the buccal pumping rate and volume of every stroke (Seale and Wassersug, 1979). At very low food densities the pump muscles finally fatigue, resulting in a sharp decrease of ingested material, probably because the larvae pump only enough water to meet their oxygen demands. Some species are able to become grazers, carnivores, or even cannibals when the available food levels decrease, while others have no ability to change their method of feeding and thus starve. Other species are predominantly grazers or carnivores irrespective of the availability of food. All grazers and carnivores use horny beaks and labial teeth to break food down mechanically, and then use the filtering apparatus to trap the food for digestion.

Irrespective of how they obtain food, all anuran tadpoles have one final way to lessen the chance of starvation. The rate of food movement through the intestine decreases drastically during periods of starvation to permit higher rates of assimilation (Wassersug, 1975). Most tadpoles have several digestive enzymes, including pepsin, amylase, and lipase (Altig *et al.*, 1975). Since they have no cellulase, much plant material passes through the alimentary tract intact.

5.4.2. The Feeding Behavior of Urodele, Caecilian, and Fossil Larvae

All urodele larvae are normally carnivorous. Some start feeding during yolk resorption, while others do not feed until the yolk is completely resorbed. Feeding initiation is similar to that observed in Osteichthyes. Depending on the species, there may or may not be a difference in the type of prey taken as the larvae mature (Dodson and Dodson, 1971; Licht, 1975). The size of prey captured increases as urodele larvae age (Avery, 1968; Dodson and Dodson, 1971; Himstedt *et al.*, 1976). Prey include larval and adult arthropods, molluscs, annelids, and in some cases vertebrates.

Urodele larvae seem to be predominantly visual predators, although some species are able to use smell and/or the lateral line system to locate prey. Species that feed at night probably cannot be solely visual predators (Hassinger *et al.*, 1970). They are not completely random feeders since the food found in their intestines does not totally reflect organisms found in the environment, as is the case for anurans. Some urodele larvae can selectively catch certain species and avoid others. Younger larvae feed more in the open; as they age they change their feeding behavior, becoming more secretive.

Free-living caecilian larvae may be filter feeders (Sarasin and Sarasin, 1887–1890), but the evidence is very circumstantial, and they may also be carnivorous. A second method of nutrition is found in caecilian larvae (fetuses). The fetus uses its teeth to scrape secretions of the oviduct epithelium and sometimes the epithelium itself (Wake, 1977). The materials thus obtained are ingested. Both the fetal feeding behavior and the maternal secretions combine to provide substantial nutrition to the developing young. Depending on species, the fetus increases two- to sixfold in size after hatching and at birth can be one-third to two-thirds the length of the mother. One to twelve fetuses are born from a single female, with most females giving birth to more than a single young.

The structure of fossil anuran and urodele tadpoles suggests that feeding behaviors were identical to those of present-day species. Ossification of a fossil larval labyrinthodontian jaw suggests that the youngest animals were filter feeders (Boy, 1974). As the larvae advanced in age, jaw ossification suggests that they became carnivores before the final transformation.

5.4.3. Competition for Food by Amphibian Larvae

Because species share habitats to a degree, and food resources can be limited in any one ecological niche, it is not surprising that there is intra- and interspecific competition for food. Adults of different species tend to prevent interspecific competition by breeding at different times or at different places. They are not completely successful, because tadpoles of different species do sometimes find themselves in the same ecological habitat.

By living in different microhabitats, feeding at different times of the day, or feeding on different prey organisms, competition among urodele larvae is decreased when they inhabit the same area. Although several caecilian species with free-living tadpoles do occupy the same terrestrial habitat, no information is available on the possibility of interspecific competition among their larvae. Since most anurans are nonselective filter feeders, it is not surprising that there is evidence for interspecific food competition. Even if different species of anuran larvae end up in the same large aquatic habitat, they tend to avoid competition by occupying different areas in the same pond (Heyer, 1973). Depending on the species, tadpoles have different feeding capacities and efficiencies (Seale and Beckvar, 1980). At times of low food availability caused by high tadpole densities in the environment or by low productivity of the environment, species with better feeding capacity do have higher survival and growth rates. Tadpoles of two different species that normally occupy the same habitat have been found to compete for food in artificial enclosures in a pond (Wiltshire and Bull, 1977). The more successful species increases its survival and growth rates and metamorphic progress at the expense of the other.

Intraspecific competition occurs for a species in a given environment with insufficient food. The effects of crowding on the rates of growth and length of larval life are well documented in both field and laboratory studies. In many anuran species, crowding prevents metamorphosis universally; in some species, however, a few tadpoles will grow and undergo metamorphosis while the majority remain stunted (Salthe and Mecham, 1974). The large tadpoles can grow, differentiate, and complete metamorphosis even with low food availability because (1) they produce a substance that prevents growth in small tadpoles; (2) they are more efficient filter feeders and thus outcompete smaller larvae for food; and (3) they prevent smaller animals from eating the fecal material of the population. Even though fecal material has a low caloric value, it is an important energy source during priods of low food availability (Steinwascher, 1978).

Urodele larvae also show the effects of crowding, namely low growth rates and delayed metamorphosis (Wilbur and Collins, 1973). In some urodelé species crowding results in large larvae becoming dominant over smaller animals. It is not known how this dominance is established or maintained. From the published literature it is not possible to determine whether caecilian tadpoles (fetuses) show any intraspecific food competition. Fetuses from the same female can be of different lengths, suggesting that some form of competition may occur in the oviduct.

The most drastic inter- and intraspecific competition for food occurs when tadpoles cannibalize each other. Many urodele larvae species eat anuran larvae and may be a major predator of these larvae when they are located in the same pond (Walters, 1975). Some urodele larvae are cannibalistic under normal conditions, and all are cannibalistic under conditions of low food availability. Very few anuran larvae are normally carnivorous, but some become cannibalistic during periods of starvation. Such anuran larvae will cannibalize both anuran and urodele young

(Martin, 1967; Heyer *et al.*, 1975). It is not known whether caecilians exhibit cannibalistic behavior, but the number of fetuses born (1–12) to any given female does not preclude the occurrence of cannibalism in the oviduct. Until a thorough study is made on how many eggs hatch in a female and the subsequent number of fetuses born from her, intraspecific cannibalism cannot be excluded.

5.4.4. Microhabitat Selection by Tadpoles

No information is available on caecilian behavioral regulation, although it appears that they may regulate their position in the oviduct since they seem to be evenly spaced there (Wake, 1977). Urodele and anuran larvae are not randomly distributed in the aquatic environment, suggesting that they behaviorally regulate their positions. Larvae use various environment cues to determine their position including temperature, light, odor, and even the type of substrate.

The absence or presence of other organisms in a given ecological niche may have a profound influence on the behavior of tadpoles. Predators cause dispersal of tadpoles into hiding places, and the tadpoles presumably use both their sight and their lateral line system to detect them. Many anuran larvae and a few urodele larvae form aggregations (schools) in nature (Beiswenger, 1975; Salthe and Mecham, 1974). It is not known what triggers the initial aggregation, whether it be light intensity, temperature, or food availability. One or two species may aggregate purely on sight. Irrespective of what causes the original aggregation, the animals tend to stay in the school by the use of an optomotor response (Wassersug, 1973) In an optomotor response, an individual moves with objects in its environment, so that the individual maintains a relatively consistent position in the environment, and thus the school is maintained even during periods of movement.

Microhabitat selection based on temperature has been observed for amphibian larvae for over 70 years. Both anurans and urodele larvae aggregate at specific temperatures, although the preferred temperature of different species varies. There is a wide range of preferred temperatures among anuran tadpoles. *Bufo* tadpoles seem to prefer temperatures around 30°C, while young *A. truei* tadpoles choose temperatures below 10°C (Beiswenger, 1978). Differences in preferred environmental temperatures do exist among urodele larvae species, but they are not as great as are those for anurans (Heath, 1975; Keen and Schroeder, 1975). These preferred temperatures increase as tadpoles grow and differentiate.

The majority of amphibian larvae are attracted to light (photopositive) as young tadpoles, although a few photonegative young larvae are also known (Marangio, 1975; Jaeger and Hailman, 1976). A shift in the spectrum of preference gradually occurs, tadpoles showing a preference for the green, adults a preference for the blue (Jaeger and Hailman, 1976). By the time tadpoles undergo tail resorption, blue light is preferred. It is of obvious ecological advantage for tadpoles to be attracted to green light. In bright areas the temperatures should be elevated, which would facilitate rates of metamorphosis, while the green spectrum would identify potential food sources.

It has been known for a number of years that tadpoles choose their position not only on the basis of temperature or illumination, but also on the texture of the substrate (Altig and Brodie, 1972; Mushinsky, 1976). Tadpoles have been taught to recognize the substrate on which they live, the location of food, and the levels of illumination (Wiens, 1972; Hershkowitz and Samuel, 1973; Mushinsky, 1976; Miller and Berk, 1977). It is remarkable that these learned behavioral responses, all based on the visual system as the intial sensory input, are remembered by the adult. There are numerous changes in the visual system at transformation, including an increase in the number of rods relative to cones, an increase in the size of the outer-rod segment, an increase in the diameter of the optic nerve, and a decrease in the optic nerve length (Witkovsky *et al.*, 1976; Cullen and Webster, 1979). There are also numerous changes in the central nervous system and motor system throughout larval life (see Kollros, this volume). Since memory retention and appropriate motor output occur in adults, many of the sensory, central, and motor neuronal changes observed during metamorphosis may not be as significant as is usually assumed.

5.5. Triggers for Amphibian Metamorphosis

A new hypothesis concerning the trigger for amphibian metamorphosis has recently been developed based on the absolute size of tadpoles and their rate of growth (Wilbur and Collins, 1973). This hypothesis has become very popular among investigators studying the ecological aspects of urodele metamorphosis. One tenet of this hypothesis maintains that tadpole transformation can occur only if a certain minimum size is first attained, and cannot occur after a certain maximum size is reached. It seems improbable that minimum or maximum tadpole sizes exist for tranformation because most external morphological changes associated with transformation can be induced by TH irrespective of tadpole size. This does not preclude the possibility that tadpoles must reach a certain size before normal neural–endocrine mechanisms come into being. However, there is no evidence for an upper size limit for normal endocrine mechanisms to function.

The second important principle in the Wilbur and Collins hypothesis is that the relative growth rate of the entire body is an important determinant of transformation time. However, it is unlikely that this parameter can determine the transformation time. Over 50 years ago experimental embryologists recognized that growth and growth rate were not as important as the state of differentiation in establishing transformation.

In fact, all staging procedures (Tables 3, 4) predict transformation based on differentiation rates and not on growth rates. Animals will undergo metamorphosis after TH administration irrespective of previous growth rates.

The one indisputable trigger for transformation of free-living anuran and urodele larvae is TH (Dodd and Dodd, 1976; see White and Nicoll, this volume). The extent of the influence of TH on the metamorphosis of anurans and urodeles in terrestrial eggs, oviducts, or other pouches or cavities of the adults is not known. In the few species where attempts have been made to influence morphological changes in these animals, TH showed some accelerating effects. No experimental manipulation of metamorphosis in caecilians has been attempted, although thyroid histology indicates that there is an increased production of TH at transformation (Klumpp and Eggert, 1935).

In all neotenic urodele species there is always a malfunction of normal TH physiology. In species that are obligatory neotenic forms, the larval tissues never develop an ability to respond to TH (*Necturus*, *Proteus*). In facultative neotenic urodeles, the nonmetamorphosing population always has associated with it a curtailment of TH production. The reasons for lack of hormone production could be as simple as a lack of iodine in the environment, or as complex as a lack of pituitary stimulation by the hypothalamus (see White and Nicoll, this volume.)

5.6. Problem Areas in Amphibian Metamorphosis

Universal staging procedures are needed for caecilian and urodele metamorphosis; until they are established, morphological information in addition to length should be given to describe metamorphic progress. The available anuran staging methods must be referred to by all authors concerned with anuran metamorphosis.

Before universal principles of the effects of TH in amphibian metamorphosis can be established, the internal morphological and biochemical responses to TH must be examined in more diversified species. The vast majority of biochemical investigations (80–90%) have been carried out on only three anuran species, namely *R. catesbeiana*, *X. laevis*, and *R. pipiens*.

Many biochemical changes are dependent on the action of TH in the three species named. However, it has yet to be proven that the same organ system of other anurans, most urodeles, and all caecilians responds in a similar manner and at the same developmental stage.

If all three orders respond to thyroid hormones, it may be proven that there is an increase in the release of TH at the time of transformation. The circulating levels of TSH need to be established as well as the nature of the amphibian hypothalamic hormones. Preliminary information suggests that amphibian and mammalian hypothalamic thyrotropin-releasing hormones are different. The environmental and/or neural control of the hypothalamus must be ascertained to understand the initial trigger for amphibian metamorphosis.

6. FUTURE WORK IN CHORDATE METAMORPHOSIS

Several specific areas that need attention in individual chordate groups have been discussed. Besides these, attention needs to be drawn to the environmental and physiological factors that initiate and terminate larval life. While much has been written about the transition from larval to adult life, remarkably little information is availble about the modulation of the transition from embryonic to larval life.

Most authors agree that the time of hatching varies both within and between species, but the cause of the variability is unknown. In most chordate groups ectodermal glands release hatching enzymes that digest the egg case and permit the larvae to hatch. Until recently it was not known what triggers the release of these enzymes. New evidence suggests that the hatching time can be controlled by the oxygen tension in the embryo. Hatching of the marbled salamander (*Ambystoma opacum*) never occurs in water saturated with oxygen. When the oxygen tension is decreased, either by artificially lowering the oxygen level with nitrogen or by permitting the embryos to utilize oxygen for respiration, they hatch. Circumstantial evidence in several other chordate groups suggests that low oxygen tension may be a fairly universal trigger initiating individual larval life among chordates (Petranka, Just, and Crawford, in preparation). The decreased oxygen tension appears to cause the release of hatching enzymes from the ectodermal glands. It is not known whether this release is caused by a direct effect of oxygen tension on these gland cells or an effect of oxygen tension on the activity of the nerves that innervate the glands.

For every chordate group environmental factors are known that accelerate metamorphosis. In the different groups these factors include the ionic composition of the environment, temperature, light, nature of substrate, substrate contact, availability of food, etc. Even in a single chordate group the environmental factors that speed up metamorphosis are very diversified. Only two physiological systems (nervous and endocrine) have the capacity for translating such diversified information into usable information for the organism. Almost nothing is known of the sensory system responsible for detecting all of the environmental inputs. Little is known about the integrative center of these sensory signals, which ultimately must send the signal to initiate metamorphosis. In anurans and urodeles part of the integrating center clearly is the hypothalamus and its associated nerves. In amphibians there is clearly an intermediate organ between the integrating center and the responding systems, namely the thyroid. Most authors have assumed such an intermediate system must also exist in other chordate groups. The intermediate system has not been conclusively identified in any other group of chordates although most authors clearly expect it to be part of the endocrine system.

ACKNOWLEDGMENTS. We would like to thank numerous colleagues for providing reprints, preprints, and manuscripts that, we trust, helped to make this review more current and broadly based. Several people helped with the literature search: Sharon, Diane, and Emma Just, and the excellent library staff of the University of Kentucky, headed by Elizabeth Howard. The original artwork was expertly drawn by Holly Justice. We wish to thank several colleagues for reading parts of this manuscript: Eugene Crawford, Leo Demski, Susan Grady, Jim Petranka, and especially Susan Braen, who read the entire manuscript. All made valuable suggestions. Finally, we wish to thank the editors of this volume for their patience and encouragement. The literature search for this review ended in December 1979.

REFERENCES

Adinolfi, M., Chieffi, G., and Siniscalco, M., 1959, *Nature (London)* **184**:1325.
Adler, W., 1901, *Int. Monatsschr. Anat. Physiol.* **18**:19.
Alldredge, A. L., 1977, *J. Zool. (London)* **181**:175.
Altig, R., 1970. *Herpetologica* **26**(2):180.
Altig, R., and Brodie, E. D., Jr., 1972. *J. Herpetol.* **6**(1):21.
Altig, R., and Pace, W. L., 1974, *J. Herpetol.* **8**(3):247.
Altig, R., Kelly, J. P., Wells, M., and Phillips, J., 1975, *Herpetologica* **31**:104.
Anderson, D. T., White, B. M., and Egan, E. A., 1976, *Proc. Linn. Soc. N.S.W.* **100**(4):205.
Anderson, J. D., Hassinger, D. D., and Dalrymple, G. H., 1971, *Ecology* **52**(6):1107.
Anderson, P. L., 1943, *Anat. Rec.* **86**:58.
Arnold, S. J., and Wassersug, R. J., 1978, *Ecology* **59**(5):1014.
Atkinson, B. G., and Just, J. J., 1975, *Dev. Biol.* **45**:151.
Avery, R. A., 1968, *Oikos* **19**:408.
Balon, E. K., 1975, *J. Fish. Res. Board Can.* **32**:1663.
Barnes, S. N., 1974, *Cell Tissue Res.* **155**:27.
Barrington, E. J. W., 1965, *The Biology of Hemichordata and Protochordata*, Oliver & Boyd, London.
Barrington, E. J. W., 1968, in: *Metamorphosis: A Problem in Developmental Biology* (W. Etkin and L. I. Gilbert, eds.), pp. 223–270, Appleton–Century–Crofts, New York.
Barrington, E. J. W., and Sage, M., 1972, in: *The Biology of Lampreys* (M. W. Hardisty and I. C. Potter, eds.), pp. 105–134, Academic Press, New York.
Barrio, A., 1977, *Physis Secc. C Cont. Los Org. Terr.* **36**(92):337.
Beamish, F. W. H., Strachan, P. D., and Thomas, E., 1978, *Comp. Biochem. Physiol.* **60A**:435.
Beiswenger, R. E., 1975, *Herpetologica* **31**:222.
Beiswenger, R. E., 1978, *J. Herpetol.* **12**(4):499.
Bell, G., and Lawton, J. H., 1975, *J. Anim. Ecol.* **44**(2):393.
Bird, D. J., and Potter, I. C., 1979, *Zool. J. Linn. Soc.* **65**:127.
Blaxter, J. H. S., 1969, in: *Fish Physiology*, Volume 3 (W. S. Hoar and D. J. Randall, eds.), pp. 177–252, Academic Press, New York.

Blaxter, J. H. S., and Staines, M. E., 1971, in: *Fourth European Marine Biology Symposium* (D. J. Crisp, ed.), pp. 467–485, Cambridge University Press, London.

Bone, Q., 1959, *Q. J. Microsc. Sci.* **100**:509.

Boy, J. A., 1974, *Palaeontol. Z.* **48**:236.

Brauer, A., 1899, *Zool. Jahrb. Abt. Anat. Ontog. Tiere* **12**:477.

Braum, E., 1978, in: *Ecology of Freshwater Fish Production* (S. D. Gerking, ed.), pp. 102–136, Blackwell, Oxford.

Breder, C. M., Jr., 1949, *Copeia* **1949**(4):296.

Breder, C. M., Jr., and Rosen, D. E., 1966, *Modes of Reproduction in Fishes*, Crown, New York.

Bruce, R. C., 1972, *J. Herpetol.* **6**(1):43.

Bryan, P. G., and Madraisau, B. B., 1977, *Aquaculture* **10**(3):243.

Burighel, P., Nunzi, M. G., and Schiaffino, S., 1977, *J. Morphol.* **153**(2):205.

Cambar, R., and Gipouloux, J. D., 1956, *Bull. Biol. Fr. Belg.* **90**:97.

Cambar, R., and Marrot, B., 1954, *Bull. Biol. Fr. Belg.* **88**:168.

Cambar, R., and Martin, S., 1959, *Actes. Soc. Linn. Bordeaux* **98**:3.

Carroll, R. L., 1977, in: *Patterns of Evolution, as Illustrated by the Fossil Record*, Volume 5 (A. Hallam, ed.), pp. 405–437, Elsevier, Amsterdam.

Castle, P. H. J., 1978, *Copeia* **1978**:29.

Castle, P. H. J., 1979, *Bull. Mar. Sci.* **29**:1.

Cecil, S. G., and Just, J. J., 1979, *Copeia* **1979**:447.

Claridge, P. N., and Potter, I. C., 1974, *Acta Zool. (Stockholm)* **55**:61.

Cloney, R. A., 1978, in: *Settlement and Metamorphosis of Marine Invertebrate Larvae* (F.-S. Chia and M. E. Rice, eds.), pp. 255–282, Elsevier/North-Holland, New York.

Cloney, R. A., 1979, *Cell Tissue Res.* **200**:453.

Collins, J. P., 1979, *Ecology* **60**(4):738.

Corben, C. J., Ingram, G. J., and Tyler, M. J., 1974, *Science* **186**:946.

Courtney, W. A. M., 1975, in: *Symposia of the Zoological Society of London*, No. 36 (E. J. W. Barrington and R. P. S. Jefferies, eds.), Academic Press, New York, pp. 213–233.

Crisp, D. J., and Ghobashy, A. F. A. A., 1971, in: *Fourth European Marine Biology Symposium*, (D. J. Crisp, ed.), pp. 443–465, Cambridge University Press, London.

Cullen, M. J., and Webster, H. D., 1979, *J. Comp. Neurol.* **184**(2):353.

de Silva, C., 1974, in: *The Early Life History of Fish* (J. H. S. Blaxter, ed.), pp. 465–485, Springer-Verlag, New York.

Dickhoff, W. W., Folmar, L. C., and Gorbman, A., 1978, *Gen. Comp. Endocrinol.* **36**:229.

Dodd, M. H. I., and Dodd, J. M., 1976, in: *Physiology of the Amphibia*, Volume III (B. Lofts, ed.), pp. 467–599, Academic Press, New York.

Dodson, S. I., and Dodson, V. E., 1971, *Copeia* **1971**:614.

Duellman, W. E., 1970, *Univ. Kans. Mus. Nat. Hist. Monogr.* No. 1.

Duellman, W. E., 1978, *Univ. Kans. Mus. Nat. Hist. Misc. Publ.* No. 65.

Eddy, J. M. P., 1969, *J. Endocrinol.* **44**:451.

Ehrlich, K. F., 1974, in: *The Early Life History of Fish* (J. H. S. Blaxter, ed.), pp. 301–323, Springer-Verlag, New York.

Emlen, S. T., 1977, *Copeia* **1977**:749.

Estes, R., Spinar, Z. V., and Nevo, E., 1978, *Herpetologica* **34**(4):374.

Etkin, W., 1968, in: *Metamorphosis: A Problem in Developmental Biology* (W. Etkin and L. I. Gilbert, eds.), pp. 313–348, Appleton–Century–Crofts, New York.

Evseenko, S. A., 1978, *Zool. Zh.* **57**(7):1040.

Eycleshymer, A. C., and Wilson, J. M., 1910, in: *Normentafeln zur Entwicklungsgeschichte der Wirbeltiere*, Volume 11 (F. Keibel, ed.), Gustav Fisher, Jena.

Fenaux, R., 1977, *Proc. Symp. Warm Water Zoopl. Spec. Publ. UNESCO/N10* **1977**:497.

Flood, P. R., Braun, J. G., and De Leon, A. R., 1976, *Sarsia* **61**:63.

Flood, P. R., Gosselck, F., and Braun, J. G., 1978, in: *International Council for Exploration of the Sea, Las Palmas, April 1978 Symposium on the Canary Current; Upwelling and Living Resources*, No. 48, pp. 1–11.

Fontaine, M., 1975, *Adv. Mar. Biol.* **13**:241.

Forester, D. C., 1979, *Copeia* **1979**:332.

Forman, L. J., and Just, J. J., 1981, *Gen. Comp. Endocrinol.* **44**:1.

Fritzsche, R. A., 1978, in: *Development of Fishes of the Mid-Atlantic Bight, An Atlas of Egg, Larval and Juvenile Stages*, Volume V, Fish and Wildlife Service, U.S. Department of the Interior, FWS/OBS-78/12.

Gallien, L., and Bidaud, O., 1959, *Bull. Soc. Zool. Fr.* **84**(1):22.

Gallien, L., and Duröcher, M., 1957, *Bull. Biol. Fr. Belg.* **91**:97.

Garlick, R. L., Davis, B. J., Farmer, M., Fyhn, H. J., Fyhn, U. E. H., Noble, R. W., Powers, D. A., Riggs, A., and Weber, R. E., 1979, *Comp. Biochem. Physiol.* **62A**:239.

Glücksohn, S., 1931, *Wilhelm Roux Arch. Entwicklungsmech. Org.* **125**:341.

Gosner, K. L., 1960, *Herpetologica* **16**:183.

Gosselck, F., and Kuehner, E., 1973, *Mar. Biol.* **22**:67.

Gould, S. J., 1977, in: *Patterns of Evolution* (A. Hallam, ed.), pp. 1–26, Elsevier, New York.

Grizzle, J. M., and Curd, M. R., 1978, *Copeia* **1978**:448.

Hardy, J. D., Jr., 1978a, in: *Development of Fishes of the Mid-Atlantic Bight, An Atlas of Egg, Larval and Juvenile Stages*, Volume II, Fish and Wildlife Service, U.S. Department of the Interior, FWS/OBS-78/12.

Hardy, J. D., Jr., 1978b, in: *Development of Fishes of the Mid-Atlantic Bight, An Atlas of Egg, Larval and Juvenile Stages*, Volume III, Fish and Wildlife Service, U.S. Department of the Interior, FWS/OBS-78/12.

Hassinger, D. D., Anderson, J. D., and Dalrymple, G. H., 1970, *Am. Midl. Nat.* **84**:474.

Heath, A. G., 1975, *Herpetologica* **31**:84.

Hershkowitz, M., and Samuel, D., 1973, *Anim. Behav.* **21**:83.

Heyer, W. R., 1973, *J. Herpetol.* **7**(4):337.

Heyer, W. R., McDiarmid, R. W., and Weigmann, D., 1975, *Biotropica* **7**(2):100.

Himstedt, W., Freidanh, U., and Singer, E., 1976, *Z. Tierpsychol.* **41**(3):235.

Hing, L. K., 1959, *Treubia* **25**(1):89.

Hoar, W. S., 1976, *J. Fish. Res. Board Can.* **33**:1234.

Homma, S., 1978, *Brain Res.* **40**:33.

Ingram, G. J., Anstis, M., and Corben, C. J., 1975, *Herpetologica* **31**:425.

Ishibashi, N., 1974, in: *The Early Life History of Fish* (J. H. S. Blaxter, ed.), pp. 339–344, Springer-Verlag, New York.

Jaeger, R. G., and Hailman, J. P., 1976, *J. Comp. Physiol. Psychol.* **90**(10):930.

Johnson, G. D., 1978, in: *Development of Fishes of the Mid-Atlantic Bight, An Atlas of Egg, Larval and Juvenile Stages*, Volume IV, Fish and Wildlife Service, U.S. Department of the Interior, FWS/OBS-78/12.

Jones, F. R. H., and Marshall, N. B., 1953, *Biol. Rev. Cambridge Philos. Soc.* **28**:16.

Jones, P. W., Martin, F. D., and Hardy, J. D., Jr., 1978, in: *Development of Fishes of the Mid-Atlantic Bight, An Atlas of Egg, Larval and Juvenile Stages*, Volume I, Fish and Wildlife Service, U.S. Department of the Interior, FWS/OBS-78/12.

Just, J. J., 1972, *Physiol. Zool.* **45**:143.

Just, J. J., Sperka, R., and Strange, S., 1977a, *Experientia* **33**:1503.

Just, J. J., Schwager, J., and Weber, R., 1977b, *Wilhelm Roux Arch. Entwicklungsmech. Org.* **183**:307.

Kadel, K., 1975, *Rev. Suisse Zool.* **82**(2):237.

Keen, W. H., and Schroeder, E. C., 1975, *Copeia* **1975**:523.

Kerr, J. G., 1909, in: *Normentafeln zur Entwicklungsgeschichte der Wirbeltiere*, Volume 10 (F. Keibel, ed.), Gustav Fischer, Jena.

Khan, M. S., 1965, *Biologia (Lahore)* **11**(1):1.

Klumpp, W., and Eggert, B., 1935, *Z. Wiss. Zool.* **146**:329.

Knight, F. C. E., 1938, *Wilhelm Roux Arch. Entwicklungsmech. Org.* **137**:461.

Komourdjian, M. P., Saunders, R. L., and Fenwick, J. C., 1976, *Can. J. Zool.* **54**:544.

Kopsch, F., 1952, Die Entwicklung des braunen Grasfrosches *Rana fusca* Roesel (dargestellt in der Art der Normentafeln zur Entwicklungsgeschichte der Wirbeltiere), Stuttgart, 1952.

Kudo, T., 1938, in: *Normentafeln zur Entwicklungsgeschichte der Wirbeltiere*, Volume 16 (F. Keibel, ed.), Gustav Fischer, Jena.

Kükenthal, W., and Krumbach, T., 1956, *Handbuch der Zoologie*, Volume 5, 2nd Half, *Tunicata*, Gruyter, Berlin.

Lamotte, M., and Xavier, F., 1972, *Ann. Embryol. Morphog.* **5**:315.

Lankester, E. R., and Willey, A., 1890, *Q. J. Microsc. Sci.* **31**:445.

Larsen, L. O., and Rothwell, B., 1972, in: *The Biology of Lampreys* (M. W. Hardisty and I. C. Potter, eds.), pp. 1–67, Academic Press, New York.

Lasker, R., and Theilacker, G. H., 1968, *Exp. Cell Res.* **52**:582.

Laurence, G. C., 1977, *Fish. Bull. U.S.* **75**:529.

Lewis, S. V., and Potter, F. C., 1976, *Acta Zool. (Stockholm)* **57**(2):103.

Licht, L. E., 1975, *Can. J. Zool.* **53**(11):1716.

Limbaugh, B. A., and Volpe, E. P., 1957, *Am. Mus. Novit.* No. 1842, pp. 1–32.

Love, R. M., 1970, *The Chemical Biology of Fishes*, Academic Press, New York.

Lynch, W. F., 1961, *Am. Zool.* **1**:59.

Mackie, G. O., and Bone, Q., 1976, *J. Mar. Biol. Assoc. U.K.* **56**:751.

Manion, P. J., and Stauffer, T. M., 1970, *J. Fish. Res. Board Can.* **27**:1735.

Marangio, M. S., 1975, *J. Herpetol.* **9**(3):293.

Marcus, H., 1939. *Bio-Morphosis* **1**:355.

Martin, A. A., 1967, *Aust. Nat. Hist.* **15**:326.

Martin, F. D., and Drewry, G. E., 1978, in: *Development of Fishes of the Mid-Atlantic Bight, An Atlas of Egg, Larval and Juvenile Stages*, Volume VI, Fish and Wildlife Service, U.S. Department of the Interior, FWS/OBS-78/12.

Marx, L., 1935, *Ergeb. Biol.* **11**:244.

May, R. C., 1974, in: *The Early Life History of Fish* (J. H. S. Blaxter, ed.), pp. 3–19, Springer-Verlag, New York.

Millar, R. H., 1971, in: *Advances in Marine Biology* (F. S. Russell and M. Yonge, eds.), pp. 1–100, Academic Press, New York.

Miller, R. B., and Berk, A. M., 1977, *J. Exp. Psychol. Anim. Behav. Processes* **3**(4):343.

Mook, D., 1977, *Copeia* **1977**:126.

Moore, J. W., and Potter, I. C., 1976, *J. Anim. Ecol.* **45**:699.

Moser, H., 1950, *Rev. Suisse Zool.* **57**(2):1.

Murphy, J. B., Quinn, H., and Campbell, J. A., 1977, *Copeia* **1977**:66.

Mushinsky, H. R., 1976, *Copeia* **1976**:755.

Nevo, E., and Estes, K., 1969, *Copeia* **1969**:540.

Nieuwkoop, P. D., and Faber, J., 1967, *Normal Table of Xenopus laevis*, North-Holland, Amsterdam.

Noble, G. K., 1931, *The Biology of the Amphibia*, McGraw–Hill, New York.

O'Boyle, R. N., and Beamish, F. W. H., 1977, *Environ. Biol. Fishes* **2**:103.

Olivereau, M., 1975, *Gen. Comp. Endocrinol.* **27**:9.

Ooi, E. C., and Youson, J. H., 1979, *Am. J. Anat.* **154**(1):57.

Peek, W. D., and Youson, J. H., 1979, *Can. J. Zool.* **57**:1318.

Percy, R., and Potter, I. C., 1976, *J. Zool.* **178**:319.

Percy, R., and Potter, I. C., 1977, *J. Zool.* **183**:111.

Percy, R., and Potter, I. C., 1979, *J. Zool.* **187**:415.

Percy, R., Leatherland, J. F., and Beamish, F. W. H., 1975, *Cell Tissue Res.* **157**:141.

Perez, J. E., and Maclean, N., 1974, *J. Fish. Biol.* **6**:479.

Perez, J. E., and Maclean, N., 1976, *J. Fish. Biol.* **9**:447.

Pierce, B. A., and Smith, H. M., 1979, *J. Herpetol.* **13**(1):119.

Potter, I. C., and Brown, I. D., 1975, *Comp. Biochem. Physiol.* **51B**:517.

Rice, S. D., and Stokes, R. M., 1974, in: *The Early Life History of Fish* (J. H. S. Blaxter, ed.), pp. 325–337, Springer-Verlag, New York.

Richardson, S. L., and Joseph, E. B., 1973, *Fish Bull. U.S.* **71**:735.

Rossi, A., 1959, *Mont. Zool. Ital.* **66**:133.

Rugh, R., 1962, *Experimental Embryology*, Burgess, Minneapolis, Minn.

Salthe, S. N., and Mecham, J. S., 1974, in: *Physiology of the Amphibia*, Volume II (B. Lofts, ed.), pp. 309–521, Academic Press, New York.

Sarasin, P., and Sarasin, F., 1887–1890, in: *Zur Entwicklungsgeschichte und Anatomie der Ceylonesischen Blindwühle Ichthyophis glutinosus 1887*, Volume 2, C. W. Kreidel's Verlag, Wiesbaden, pp. 1–263.

Savage, J. M., 1973, in: *Evolutionary Biology of the Anurans, Contemporary Research on Major Problems* (J. L. Vial, ed.), pp. 351–445, University of Missouri Press, Columbia.

Schreiber, C., 1937, *Atti Accad. Naz. Lincei Rend. Cl. Sci. Fis. Mat. Nat. Rend.* **25**:342.

Seale, D. B., and Beckvar, N., 1980, *Copeia* **1980**:495.

Seale, D. B., and Wassersug, R. J., 1979, *Oecologia (Berlin)* **39**:259.

Sedra, S. N., and Michael, M. I., 1961, *CFSK Morfol.* **IX**(4):333.

Shirota, A., 1970, *Bull. Jpn. Soc. Sci. Fish.* **36**(4):353.

Shvarts, S. S., Pyastolova, O. A., and Dobrinskii, L. N., 1973, *Soc. J. Ecol.* **4**:287.

Simpson, B. R. C., 1979, *Symp. Zool. Soc. London* **44**:243.

Smigielski, A. S., 1979, *Fish. Bull. U.S.* **76**(4):931.

Sokol, O. M., 1975, *Copeia* **1975**:1.

Spoor, W. A., 1977, *J. Fish Biol.* **11**:77.

Starrett, P. H., 1973, in: *Evolutionary Biology of the Anurans, Contemporary Research on Major Problems* (J. L. Vial ed.), pp. 251–271, University of Missouri Press, Columbia.

Steinwascher, K., 1978, *Copeia* **1978**:130.

Suzuki, S., and Kondo, Y., 1973, *Gen. Comp. Endocrinol.* **21**:451.

Szarski, H., 1957, *Am. Nat.* **XCI**(860):283.

Taylor, A. C., and Kollros, J. J., 1946, *Anat. Rec.* **94**:7.

Taylor, E. H., 1968, *The Caecilians of the World: A Taxonomic Review*, University of Kansas Press, Lawrence.

Terentlev, P. V., 1950, *The Frog*, Sovetskaya Nauka, Moscow.

Thorp, A., and Thorndyke, M. C., 1975, in: *Symposia of the Zoological Society of London*, No. 36 (E. J. W. Barrington and R. P. S. Jefferies, eds.), pp. 159–177, Academic Press, New York.

Thorpe, J. E., and Morgan, R. I. G., 1978, *J. Fish Biol.* **12**:541.

Tihen, J. A., 1942, *Univ. Kans. Sci. Bull.* **28**:189.

Usui, M., and Hamasaki, M., 1939, *Zool. Mag. (Tokyo)* **51**:195.

Uthe, J. F., and Tsuyuki, H., 1967, *J. Fish. Res. Board Can.* **24**(6):1269.

Valentine, B. D., and Dennis, D. M., 1964, *Copeia* **1964**:196.

Verraes, W., 1977, *J. Morphol.* **151**(1):111.

Wake, D. B., 1966, *Mem. South. Calif. Acad. Sci.* **4**:1.

Wake, M. H., 1977, in: *The Reproductive Biology of Amphibians* (D. H. Taylor and S. I. Guttman, eds.), pp. 73–101, Plenum Press, New York.

Wake, M. H., 1978, *J. Herpetol.* **12**(2):121.

Walters, B., 1975, *J. Herpetol.* **9**(3):267.

Wassersug, R. J., 1973, in: *Evolutionary Biology of the Anurans, Contemporary Research on Major Problems* (J. L. Vial, ed.), pp. 273–297, University of Missouri Press, Columbia.

Wassersug, R. J., 1975, *Am. Zool.* **15**:405.

Wassersug, R. J., and Rosenberg, K., 1979, *J. Morphol.* **159**(3):393.

Weatherley, A. H., and Rogers, S. C., 1978, in: *Ecology of Freshwater Fish Production* (S. D. Gerking, ed.), Blackwell, Oxford, pp. 52–74.

Webb, J. E., 1958, *Philos. Trans. R. Soc. London Ser. B* **241**:335.

Webb, J. E., 1969, *Mar. Biol.* **3**:58.

Webb, J. E., 1973, *J. Zool.* **170**:325.

Webb, J. E., 1975, in: *Symposia of the Zoological Society of London*, No. 36 (E. J. W. Barrington and R. P. S. Jefferies, eds.), pp. 179–212, Academic Press, New York.

Wells, K. D., 1978, *Herpetologica* **34**(2):148.

West, A. B., and Lambert, C. C., 1976, *J. Exp. Zool.* **195**:263.

Weygoldt, P., 1976, *Z. Tierpsychol.* **40**:80.
Whittaker, J. R., 1964, *Nature (London)* **202**:1024.
Wickstead, J. H., 1967, *J. Mar. Biol. Assoc. U.K.* **47**:49.
Wickstead, J. H., 1975, in: *Reproduction of Marine Invertebrates* (A. C. Giese and J. S. Pearse, eds.), pp. 283–319, Academic Press, New York.
Wickstead, J. H., and Bone, Q., 1959, *Nature (London)* **184**:1849.
Wiens, J. A., 1972, *Anim. Behav.* **20**:218.
Wilbur, H. M., and Collins, J. P., 1973, *Science* **182**:1305.
Wilder, I. W., 1924, *J. Exp. Zool.* **40**:1.
Willey, A., 1891, *Q. J. Microsc. Sci.* **32**:183.
Wiltshire, D. J., and Bull, C. M., 1977, *Aust. J. Zool.* **25**(3):449.
Witkovsky, P., Gallin, E., Hollyfield, J. G., Ripps, H., and Bridges, C. D. B., 1976, *J. Neurophysiol.* **39**:1272.
Witschi, E., 1956, *Development of Vertebrates*, Saunders, Philadelphia.
Woodruff, D. S., 1977, *Herpetologica* **33**:296.
Wourms, J. P., 1977, *Am. Zool.* **17**:379.
Wright, A. H., and Wright, A. A., 1949, *Handbook of Frogs and Toads of the United States and Canada*, Comstock, Ithaca, N.Y.
Wright, G. M., and Youson, J. H., 1977, *J. Exp. Zool.* **202**:27.
Young, J. Z., 1962, *The Life of Vertebrates*, Oxford University Press, New York.
Youson, J. H., and Connelly, K. L., 1978, *Can. J. Zool.* **56**:2364.
Youson, J. H., and McMillan, D. B., 1971, *Anat. Rec.* **170**:401.
Youson, J. H., Lee, J., and Potter, I. C., 1979, *Can. J. Zool.* **57**:237.
Zanandrea, G., 1957, *Nature (London)* **179**:925.

CHAPTER 10

Cytological and Morphological Changes during Amphibian Metamorphosis

HAROLD FOX

1. INTRODUCTION

To prepare for terrestrial life amphibian larvae, especially anurans, undergo a complex metamorphosis, which embraces elaborate morphological, biochemical, physiological, and behavioral changes. Since it was first discovered that thyroid feeding of anuran larvae hastens metamorphic changes, a vast corpus of knowledge on the subject has accumulated. During premetamorphosis there is no effective level of hormonal secretion of the thyroid and its activity is minimal. Near the end of premetamorphosis (about TK stages X–XI, which refer to the stages of Taylor and Kollros, 1946), through prometamorphosis and climax (Etkin, 1970), larval development is under endocrinological control of high complexity (Dodd and Dodd, 1976). Such hormones operate in the larva in a milieu influenced by factors such as light, temperature, diet, iodine, or larval crowding in the ambient water.

A progressive increase in the rate of synthesis and release of thyroid hormones raises their circulating level in the larva and elicits metamor-

HAROLD FOX · Department of Zoology, University College, London WC1E 6BT, United Kingdom.

phic changes. It is likely that the measured levels are lower (Frieden and Just, 1970) than heretofore believed (Etkin, 1970). The relatively lower increased levels registered during later stages of metamorphosis are explained by the increasing propensity of larval tissues avidly to utilize the ever-increasing quantities of hormones, produced by the thyroid. Contrasting theories of endocrinological functions and their relationship to metamorphosis are discussed by Dodd and Dodd (1976).

The overt expression of anuran larval hormonal activity is morphological change. Probably most larval organs (at least after embryogenesis) are influenced to a greater or lesser extent, some to disappear completely, others to be modified or wholly remodeled. There is a wealth of information including that on the ultrastructure of larval tissues during metamorphosis. This account describes in some detail a selected group of organs and tissues, rather than sketchily and thus inadequately dealing with structural changes of a larger number of components.

A description of the key endocrine organ, the thyroid, precedes a consideration of the central nervous system, particularly the spinal cord where there is a high degree of cellular necrosis and further differentiation of surviving neurones; for completeness the associated notochord is considered. The alimentary canal and associated pancreas are almost entirely remodeled structurally and functionally by new populations of cells; the parenchyma cells of the liver are modified functionally simultaneously with changes in their ultrastructural profile; the pronephros completely disappears; and the skin shows successive additions, deletions, and modifications of a variety of different cell types, throughout larval ontogeny.

The scheme is arbitrary, in part determined by the amount and reliability of previous information—it is regretted that only absolutely essential references are included—but also a result of the author's personal involvement with specific larval structures. Related components are considered, albeit briefly, if they are relevant to the subject under consideration, and organoendocrinological relationships and biochemistry are reduced to essentials and then only if they are necessary to illuminate changes in specific organic ultrastructure.

2. THE THYROID

The organomorphological changes that occur during larval metamorphosis are closely related to the fundamental changes in cellular functional activity, required for terrestrial existence. Endocrine organs likewise grow and differentiate during metamorphosis, though some specific development could well be short term; for example, the thyroid shows features of reduction in the size of some components at the end of climax, when the circulating thyroid hormonal level is reduced (Regard, 1978).

The paired thyroid glands of larval amphibians are generally similar in appearance to those of other vertebrates. They comprise colloid-filled follicles supported by connective tissue. In *Xenopus* definitive paired anlagen are recognized at NF stage 41 [which refers to the stages of Nieuwkoop and Faber, 1956; for information on the level of anatomical development of *Xenopus* (and *Rana*) larvae, in terms of stage numbers, the reader is referred to Fig. 1 of Chapter 14], and at NF stages 46–47 the thyroid lobes are only a small mass of cells about 20 μm long. Incipient follicles form at about NF stage 51. As the anuran larva develops the thyroid steadily increases in overall size (Fox, 1966). Thus, in *Rana temporaria* during metamorphosis from CM stages 41 to 53 [which refer to the stages of Camber and Marrot, 1954], the measurements of the thyroid nuclear population and of its tissue volume increase by about 50 times, the overall volume of the thyroid by 60 times, and the volume of the colloid substance by over 200 times. The individual cell volume (volume of thyroid tissue per nucleus) is about 700–800 μm^3. At CM stage 54, the end of climax, when the thyroid is reduced in size, cell volume is also reduced to about 400 μm^3. Peripheral chromophobe droplets are largest and most plentiful at the height of climax (Fox, 1966).

Follicular cell height of the larval *Xenopus* increases during metamorphosis. The follicular cell volume of *Rana* is fairly constant, and cell height increases at climax, though it is still not reduced by the end of metamorphosis (see Figs. C and J, Fox and Turner, 1967). Follicular cells of anuran larvae thus presumably increase their secretory activity during metamorphosis by increase in number and change in shape.

The fine structure of the thyroid of premetamorphic *Xenopus* larvae (NF stages 40–52) was described by Jayatilaka (1978). Two incipient longitudinal strands of the thyroid anlagen are recognized at NF stage 40, follicles with colloid vesicles in the cells and in small lumina at NF stages 46–47, and there are better developed follicles at NF stages 48–50. The cuboidal cells contain a RER, mitochondria, and a Golgi apparatus. Colloid fills the follicular lumen and light-staining cells are present between the basement membrane and the follicular cells.

During metamorphosis anuran thyroid cells increase their amount of granular endoplasmic reticulum and Golgi complexes, changes presumably concerned with the synthesis and secretion of thyroid hormones. Intracellular droplets increase in number and dense lysosomallike vacuoles appear. At this time the serum thyroxine (T$_4$) level of *R. pipiens* increases 10-fold over that of younger larvae, which suggests that at climax there is hydrolysis of colloid droplets and T$_4$ release into the circulation. Parafollicular (C) cells, seen in the mammalian thyroid, are rarely found in the thyroid glands of metamorphosing or young toads of *X. laevis*. An occasional parafollicular cell is located basally in the epithelium, but it never has a free luminal surface. Their role in *Xenopus* is unknown.

The enzymology of the larval thyroid, at the level of the electron microscope, and the cytophysiology through metamorphosis have been reviewed by Regard (1978).

3. THE CENTRAL NERVOUS SYSTEM AND ASSOCIATED NOTOCHORD

3.1. General Development of the CNS

The anuran larval brain, typified by *Xenopus*, is generally well developed at NF stage 55 during prometamorphosis. Later development is mainly by enlargement and further differentiation, and topographical changes occur at climax. The spinal cord differentiates proximodistally, and nerves enter the tail by NF stages 33–34. At this period anterior spinal nerves originate from the ventral surface of the spinal cord, to be followed later on by those behind. Incipient dorsal root ganglia are recognized at NF stage 39; at NF stage 43, in the hinder trunk, they are 60 μm long. Large segmentally arranged Rohon-Beard cells, believed to originate from neural crest, differentiate in the dorsolateral regions of the spinal cord at NF stage 39; they degenerate at NF stage 50 and are superseded by the spinal ganglia; only a few pycnotic cells remain at NF stage 55 (Kollros, cited in Nieuwkoop and Faber, 1956). Rohon-Beard cells are retained for longer periods in hypophysectomized larvae (Hughes, 1968). T_4 treatment of *Triturus cristatus* larvae elicits their premature involution.

3.2. The Mauthner Cells

Situated on each side of the medulla, the paired Mauthner cells (which have large axons descending through the trunk cord to the tail) are visible by electron microscopy in *Xenopus* at NF stages 41–42, or in terms of location and nuclear size at NF stages 31–32 (Billings, 1972). They are seen at CM stage 41 of *R. temporaria* (larva with incipient hindlimb buds). Mauthner cells, which are considered to be special adaptations during aquatic larval life, have been reported in adult *R. temporaria*; others believe that they disappear in anurans soon after climax, possibly due to the reduction in the level of circulating thyroid hormones. Mauthner cells are still present in *R. temporaria* at CM stage 54 (the end of climax) and, though shrunken, in *Xenopus* 2 months after metamorphosis. A three-year-old female had neither Mauthner cells nor axons (Moulton *et al.*, 1968). Perhaps their retention is species specific in anurans. Mauthner cells are wholly lacking in *Bufo* larvae (Moulton *et al.*, 1968).

The perinuclear fibrillar matrix of the larval Mauthner cell at late prometamorphosis is well endowed with ribosomes and polyribosomes,

a RER, Golgi complexes, smooth vesicles, and lysosomal bodies. Mitochondria are abundant in the perikaryon and dendrites but less so in the
axoplasm. Mauthner cells probably ultimately degenerate by autolysis
utilizing lysosomal enzymes, and possible indications of postmetamorphic involution are recognizable in *Xenopus* at climax, for mitochondria are swollen and cristae poorly developed (Moulton *et al.*, 1968).
Billings (1972) showed Mauthner cells of *Xenopus* to have a high content
of RER and ribosomes throughout metamorphosis. There are numerous
microfilaments and microtubules, and gradually the number of vesicular
bodies, alveolar vesicles, and small secondary lysosomes increases. The
cells disappear at climax and have not been found in the adult.

3.3. Lateral Motor Column Cells

Lumbar ventral horn cells (lateral motor column or LMC cells) first
appear in *Xenopus* at NF stage 50, when the hindlimb rudiment is a
roundish bud. The peak number occurs at NF stages 52–53, when hindlimbs are paddle shaped, and maximum LMC cell loss is at NF stage 54
(Prestige, 1973). Brachial ventral horn neurones of *Xenopus* differentiate
between NF stages 52–53 and 57, peaking in number on each side at NF
stage 55. Degeneration reduces their number by more than two-thirds by
the end of metamorphosis. LMC cells labeled with tritiated thymidine
can be followed from their origin in the mantle layer laterally and thence
more medially in the ventral horn; the posterior ones are younger than
those situated more anteriorly (Prestige, 1973). During prometamorphosis, as cells of the LMC grow and differentiate to innervate the fore-
and hindlimbs, up to three-quarters of the LMC neurones of the lum-
bosacral region of *Xenopus* and *Eleutherodactylus* and two-thirds of the
neurones of the spinal ganglia of *Xenopus* degenerate. For every differentiated neurone of the LMC, eight to nine neuroblasts degenerate and
the chromatopycnotic cells are phagocytosed by microglial phagocytes
(see Hughes, 1968). Up to the early hindlimb bud stage, ventral horn
neurones develop independently. Afterwards they require sensory stimuli
via pathways from the limb, through the dorsal root ganglia, for their
maintenance. Cells of the ventral horn and of the associated spinal ganglia
degenerate when the developing hindlimbs are amputated (Hughes, 1968).
Similar phenomena occur with brachial ventral horn cells of *Xenopus*
after forelimb extirpation. Nevertheless, the events that occur are complicated and depend to a great extent on when the developing hindlimbs
are amputated. In general, in the ventral horn, following amputation
without regeneration, there may be an initial cell loss of brief duration;
later, or even much later, a secondary and extensive cell loss reduces cell
number far below that of a control (Kollros, personal communication).
Furthermore, if in *R. pipiens* larvae the periphery of the hindlimb available to the ninth spinal sensory ganglion is increased, by extirpation of
ganglia 8 and 10, mitotic activities and neuronal numbers increase in

ganglion 9. The total number of ventral root nerve fibers shows a similar initial rise, peak, and decline as in the case of the cells of the ventral horn, and dying fibers do not myelinate. Fiber loss is accentuated when the hindlimb bud of *Xenopus* is removed.

Thyroid hormones are essential for the growth and differentiation of LMC neurones, though cells of the CNS may also be influenced by prolactin and somatotropin. T_4 can also directly elicit substantial necrosis of neurones that fail to establish viable peripheral connections. (For further information on thyroid hormones and the nervous system during anuran metamorphosis, see Chapter 13.)

3.4. Fine Structure of LMC Cells and Their Degeneration

The fine structure of LMC neurones of *R. pipiens* larvae has been described (see Decker, 1976). Young neurones at TK stages VI–VII (Taylor and Kollros, 1946), equivalent to NF stages 52–53 of *Xenopus* before the beginning of prometamorphosis, resemble ultrastructurally and cytochemically those of older larvae previously hypophysectomized. The columns of neurones include cells separated from each other by about 15–20 μm, and they possess a high ratio of free to membrane-bound ribosomes, a Golgi complex, mitochondria, microtubules, and occasionally a dense body or autophagic vacuole. Nucleoside diphosphatase, acid phosphatase (the most common enzyme), aryl sulfatase, and cathepsinlike esterase are deposited within GERL (Golgi–endoplasmic reticulum–lysosomal complex). Occasionally acetylcholinesterase occurs within the inner Golgi saccules. Administration of T_4 results in about one-third of the neurones becoming bipolar with well-developed RER arranged in rows, as Nissl substance positive for acetylcholinesterase. Lipid accumulates, which is characteristic of aging cells. A majority of LMC cells degenerate, their nuclei become chromatopycnotic, Golgi cisternae dilate and fragment into small vesicles, and mitochondria dilate and become opaque. Autophagic vacuoles, or secondary lysosomes, positive for acid phosphatase, appear in larvae treated with low doses of T_4 (≤ 10 μg/liter) (Fig. 1). Glial phagocytic cells, which show acid phosphatase in GERL, Golgi saccules, dense bodies, and heterophagous vacuoles (these also include aryl sulfatase), phagocytose neuronal debris (Figs. 2, 3). A higher dosage of T_4 (50 μg/liter) influences the degree of hypertrophy of glial cells and their lysosomal enzyme content. In contrast, the degree of neuronal lysosomal activity seems to be directly related to the state of its differentiation, and large lysosomes develop only when low doses of T_4 are administered (Decker, 1976). T_4 administered to normal or hypophysectomized larvae of *R. pipiens* resulted in a four- to eightfold increase in lysosomal acid hydrolases within 5–6 days, and changes in their activity elicited neuronal necrosis. The lysosomal membrane becomes more labile to any treatment that disrupts membranes. It seems that T_4 influ-

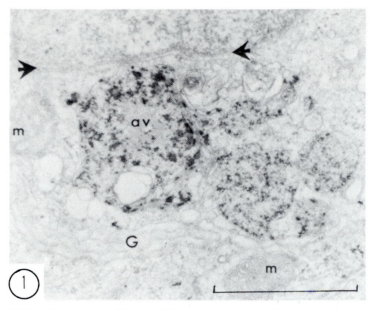

FIGURE 1. Hypophysectomized larvae of *R. pipiens* injected with T_4 (10 μg/liter) just before the onset of premetamorphosis. After 3–4 days of treatment large autophagic vacuoles (av) show cathepsinlike esterase activity in some degenerating lateral motor column neurones. The dilated Golgi vesicles (G) show no reaction product within them. The mitochondria (m) are degenerate, and a narrow chromatin band extends along the nuclear envelope (arrowed). Bar = 1 μm.

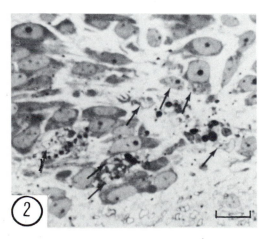

FIGURE 2. Phagocytosis of neuronal debris by phagocytes (arrows) 5–6 days after T_4 treatment of larvae of Fig. 1. Bar = 20 μm.

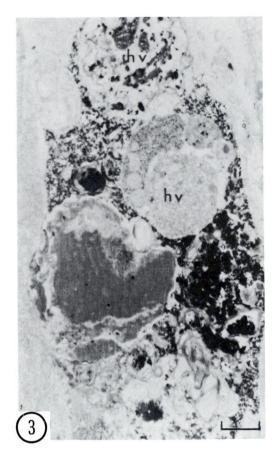

FIGURE 3. Heterophagic vacuoles (hv) positive for lysosomal acid phosphatase in a phagocytic microglial cell, after 5–6 days of T$_4$ treatment of larvae of Fig. 1. Bar = 1 μm. Figures 1–3 by kind permission of Dr. R. Decker, 1976, *Dev. Biol.* **49**.

ences the synthesis and packaging of enzymes via *de novo* production of RNA and protein (Decker, 1977).

3.5. Tail Nerve Cord

In contrast to the selective degeneration of some cellular components of the spinal cord and ganglia of the larval body, all those of the tail nerve cord degenerate during climax (Fox, 1973a). A small number of cilia and numerous microvilli line the lumen of the prometamorphic, nondegenerate tail neural tube of *R. temporaria*. Climactic tail degeneration begins at the tip, and a small circumscribed region of the distal degenerating nerve cord persists as the tail shortens in the distoproximal direction, until its final disappearance. The reduced lumen fills with collapsed neural tissue, and lipid and pigment accumulate in the cells. Autophagic vacuoles and larger cytolysomes, positive for acid phosphatase, are recognized, and a large number of membrane-bound bodies of variable diameter appear in the nerve cord cells at this time. Mesenchymal macrophages and granular cells phagocytose the degraded collagen surrounding the tail neural tissue and infiltrate the degenerating nerve cord. Presumably they ingest neuronal necrotic debris and digest it in their heterophagic vacuoles.

3.6. General Structure and Development of the Notochord

The notochord spatially is intimately associated with the larval spinal cord and provides its main skeletal support. In higher vertebrates a notochord is present only in early embryonic (or larval) stages. In anuran larvae by the end of climax the notochord has completely disappeared together with the tail and it has practically disappeared in the body. In *Xenopus* incipient notochordal vacuolation has occurred at NF stage 26 and full vacuolation by NF stages 37–38. The elastica interna and externa have appeared at NF stage 32. Degeneration of the notochord begins anteriorly in the body at NF stage 46, and by NF stage 55, during prometamorphosis, there is substantial reduction. Intervertebral compression gradually occurs, and by the end of climax there is little notochordal tissue remaining (see Nieuwkoop and Faber, 1956).

The preclimactic anuran tail notochord, which tapers distally, has intracellularly vacuolated cells enveloped at the periphery by a basement membrane and a thin elastica interna, and more proximally also by the elastica externa. Between the two elastic layers are circumferentially arranged collagen fibrils, which disappear near the tip but more proximally in the tail they bound the elastica externa. The whole is enveloped by connective tissue of fibroblasts and collagen in a structureless matrix, possibly composed either of muco- or glycoprotein or of neutral polysaccharides. Nuclei and most of the cytoplasm of the cells are at the periphery of the notochord, though some nuclei occur more centrally within fine, closely apposed cytoplasmic extensions joined by desmosomes. There are hemidesmosomes at the external surface of outermost cells situated below the basement membrane.

3.7. Notochordal Fine Structure and Its Degeneration

The perikaryon, especially of more peripheral cells, has a well-developed RER, numerous free ribosomes and polyribosomes, some SER, and a few mitochondria with well-developed cristae. Golgi complexes are frequently recognized. There is incipient fibrosity in the cytoplasm. Numerous plasmalemmal vesicles occur near or open into the intercellular spaces; less commonly they open into the space below the basement membrane. These vesicles may well be pinocytotic and participate in the formation of the intracellular vacuoles.

The fine structural features of degeneration of the notochord of the anuran (or urodele) larva during metamorphosis may well be similar to those that occur in the anuran tail at climax (Fox, 1973b). By the beginning of climax, tail notochordal cells of *R. temporaria* are extremely fibrous and large denser fibrous bodies have formed. Throughout climax tail notochordal cells, degenerating in the distoproximal direction, develop small secondary lysosomes, autophagic vacuoles and large membrane-bound cytolysomes, granular bodies, myelin figures and lamellated structures, lipid, and pigment. Mitochondria disorganize and lose their cristae to become empty vesicles or first lose their outer membranes with subsequent dispersion of their degenerating cristae amid the surrounding cytoplasm. Nuclei become chromatopycnotic, but only late in the degeneration sequence. Necrotic areas become more widespread within the cells and soon the collagenous sheaths and elastic membranes of the notochord are invaded by mesenchymal macrophages, which further disorganize and thence ingest, within heterophagic phagosomes, partially or highly degraded collagen and the necrotic debris of the degenerate notochordal cells (see Fox, 1973b). Widespread deposition of acid phosphatase occurs within the degenerating notochordal cells, and at the height of climax the hind region of the tail stub has few signs of recognizable notochordal cellular profiles. In species of *Rana* and *Rhacophorus* during tail regression, there is increasing deposition of esterase, acid phosphatase, and leucine aminopeptidase around the notochord at the tail tip.

4. THE ALIMENTARY CANAL

During anuran ontogeny, microphagous (suspension) feeding of the larva is superseded by macrophagous (solid) feeding in the postmetamorphic froglet and adult. These functional changes are reflected in significant histomorphological alterations and remodeling of the intestinal tract, including the pharynx. During anuran metamorphosis the gut shortens considerably, and there is striking cellular degeneration and regrowth in different regions (Bonneville and Weinstock, 1970; Hourdry and Dauca, 1977). On the basis of earlier investigations by light microscopy, the alimentary canal of anuran larvae is arbitrarily separated into (a) a foregut, which includes the esophagus, stomach region, and a short ciliated region, (b) a more extensive midgut, and (c) a hindgut. It is still not clear whether the larval stomach, "manicotto glandulare," is a digestive organ or merely stores food before its digestion and absorption further along the intestine. Ueck (1967) believed that the larval foregut of *X. laevis* does not store food, but that the foregut of *Hymenochirus boettingeri*, which has a shorter intestine, is a storage organ. Dodd (see Dodd and Dodd, 1976), in contrast, showed the larval foregut of *Xenopus* to store food-laden mucous strings. Both authors deny any digestive function. Examination by electron microscopy of various regions of the larval gut of *X. laevis* suggested use of the terms "gastrointestinal region" and "duodenum" for those parts of the foregut intestine following the esophagus, by virtue of their structure and relationship with the liver and pancreas (Fox *et al.*, 1970, 1972). Perhaps the gastrointestinal region functions partly as a temporary storage organ and also serves for modest luminal digestion of ingested food, and the cellular apical granules (vide infra), seen also in climactic larvae and the postclimactic froglet, may be involved. In the larva, cilia drive food products toward the duodenum, where numerous luminal microvilli provide a large surface for further digestion and absorption.

The pharynx of *Xenopus* at NF stage 43 shows little cellular differentiation, and lipid and yolk still occur in the cells at NF stage 44 (Fox and Hamilton, 1971; Fox *et al.*, 1972), whose luminal cells now have apical granules, probably containing mucus to be released into the lumen; ciliary-microvillous cells line the pharyngeal dorsolateral grooves. By NF stage 47, however, the cells contain well-developed mitochondria, a smooth and granular endoplasmic reticulum, free ribosomes and polyribosomes, microfibrils, and Golgi bodies, though lipid may still be present (Fox and Hamilton, 1971). Ciliary-microvillous cells, numerous in the dorsolateral grooves, still remain in *Xenopus* and *Rana* just before climax, though by the end of metamorphosis cilia have disappeared in these regions (Fox *et al.*, 1972).

The esophageal cells of *Xenopus* at NF stage 44 have luminal microvilli and occasional cilia. Most microvilli have disappeared at NF stage 47, when ciliated and vesicular cells alternate to line the lumen (Fig. 4). Cilia are still numerous at NF stages 57–59 (the end of prometamorphosis); at climactic NF stage 61 there are cells with small microvilli, ciliated cells are fewer, and there are forerunners of goblet cells. Near the end of climax the esophageal lumen is lined entirely by microvilli (Fig. 5), the so-called striated border of light microscopy (Fox *et al.*, 1970, 1972).

Descriptions of the ultrastructure of the gastrointestinal region (Fox *et al.*, 1970, 1972) and of the comparable "manicotto glandulare" (Ueck, 1967) of *Xenopus*, during prometamorphosis, are broadly in agreement. The wall of this region is several cells thick. Before climax some luminal cells have cilia (Fig. 6), which disappear by the end of metamorphosis. Gastrointestinal cells have numerous mitochondria, and the large number of apical dense granules, most plentiful at the end of climax (Fig. 7), are similar in appearance to those in the intestinal cells of larval *R. catesbeiana* (Bonneville and Weinstock, 1970). By the end of climax the gastrointestinal lumen is bounded by stublike microvilli with fuzzy coats (Fox *et al.*, 1972).

The luminal surface of the duodenum of *Xenopus* at NF stage 47 includes ciliary, ciliary-microvillous, and large numbers of columnar microvillous cells. The wall of the duodenum is usually two to three cells thick. Like cells of the gastrointestinal region, those of the duodenum have numerous mitochondria but few apical dense granules (Fig. 8), which suggest that digestive substances are not manufactured and secreted into the lumen at this stage. The single hepatopancreatic duct

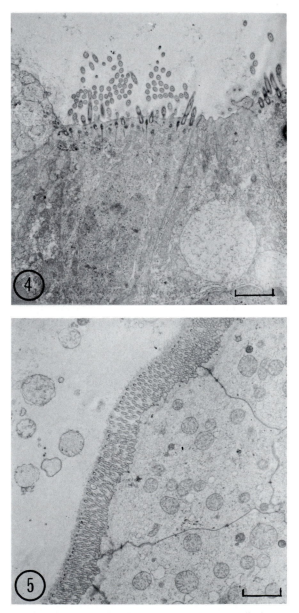

FIGURES 4, 5. Esophageal luminal cells of *X. laevis* at NF stage 47 (before premetamorphosis) and stage 65 (practically the end of climax). There are ciliated and vesicular cells at stage 47 but only surface microvilli (striated border) by the end of climax. Bar = 2 μm.

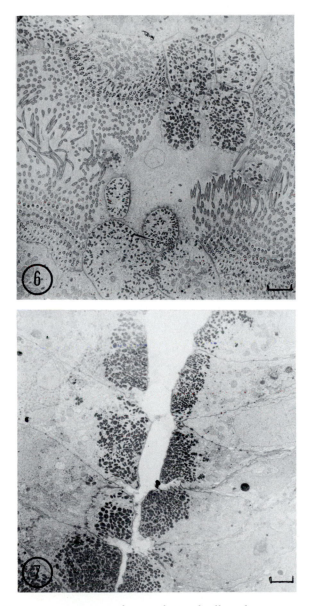

FIGURES 6, 7. *Xenopus* gastrointestinal-region luminal cells at the respective stages of Figs. 4 and 5. Cilia are present and there are less numerous dense apical granules at stage 47; cilia are absent and the cells have a heavily dense granular apical region at stage 65. Bar = 2 μm.

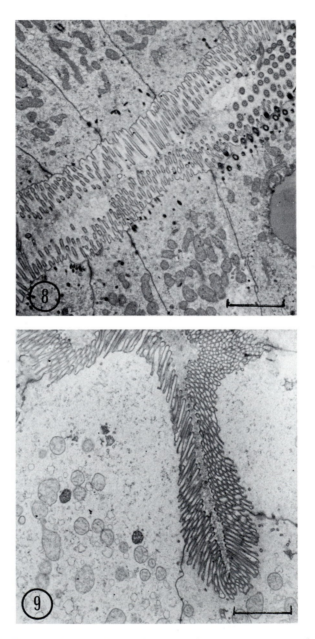

FIGURES 8, 9. *Xenopus* duodenal luminal cells at the respective stages of Figs. 4 and 5. Cilia and microvilli line the lumen at stage 47 and only profuse microvilli by the end of climax. Bar = 2 μm.

opens into the duodenum to aid digestion of food material. Cilia have disappeared from the duodenal luminal surface by the end of climax, when microvilli alone line the lumen (Fig. 9) (see also Bonneville and Weinstock, 1970).

Cells of the rectum of larval *Xenopus* at NF stages 44–47 still contain lipid and yolk. In the hinder region and the cloaca, cilia and microvilli line the lumen; only microvilli occur more anteriorly in the hindgut. Cell profiles of the hinder gut and rectum are generally similar (Fox *et al.*, 1972). By the end of climax the rectal lumen is lined only by cells with profuse short microvilli. The cells contain many small dense bodies, which are fewer in NF stage 47.

Postmetamorphic intestinal cells originate from small nests of undifferentiated larval basal cells, which proliferate and differentiate during climax as the primary surface cells degenerate. In *Xenopus*, changes in the intestinal epithelium first commence at NF stage 58, the beginning of climax. By NF stage 61 degenerating brush border cells contain numerous secondary lysosomes, and newly differentiating cells proliferate, form microvilli, and line the lumen. The entire process takes 2–3 days (Bonneville and Weinstock, 1970). Intestinal growth and degenerative changes in larvae of *Discoglossus pictus* and *X. laevis* during metamorphosis, with special reference to the role of lysosomal acid hydrolases, have been described (see Hourdry and Dauca, 1977). Acid phosphatase increases in amount in the degenerating intestinal cells. In organ culture at a low level of T_4 (10^{-8} g/ml), the granular endoplasmic reticulum of the explanted intestinal epitheliocytes of larval *D. pictus* increases in amount. At higher levels of T_4 (10^{-7} g/ml), acid phosphatase activity is recognized in lysosomes, the Golgi apparatus, and occasionally in other regions of the cell. Necrotic cells are extruded into the lumen of the explant, in a manner similar to that occurring *in vivo* (Pouyet and Hourdry, 1977).

Shortening, the loss of luminal cilia, and their replacement by microvilli are the most obvious histomorphological changes recognizable in the alimentary canal during anuran metamorphic climax.

These changes reflect the different modes of feeding by larvae and the postmetamorphic froglets and adults, when new neuromuscular mechanisms are required to deal with macrophagous feeding. Microvilli, in enormous numbers, provide the large surface necessary for the absorption of digested food.

5. THE PANCREAS

The structure, degeneration, and subsequent regeneration at climax of various functional components of the anuran exocrine pancreas and changes in shape and topography during metamorphosis have been reported (see Race *et al.*, 1966). Degeneration occurs by autolysis and remnants slough off into sinusoidal capillaries, though probably some phagocytosis ensues.

Race *et al.* (1966) attributed the profound decrease in size of the pancreas of larval *R. pipiens*, of nearly 70% in weight relative to body weight, to dehydration, for they found no evidence of widespread cellular necrosis. However, the extensive larval pancreatic cellular necrosis and regeneration recorded by earlier workers have been confirmed (see Leone *et al.*, 1976).

In the preclimactic larva of *Alytes obstetricans* before forelimb emergence, the pancreas is well developed. At climax, when forelimbs emerge, pancreatic volume decreases, mainly due to reduction of intrapancreatic tissue, partial disappearance of exocrine tissue, and slight reduction of endocrine tissue. Absolute volume reduction is about 80%. Autolysed cells are phagocytosed *in situ* or necrotic tissue is lost via the pancreatic ducts and blood capillaries. After metamorphosis the pancreas increases in size, with a relatively larger increase in the endocrine tissue due to proliferating islets of specialized type cells A($\times$ 40) and B($\times$ 12) (see Beaumont, 1977). As the pancreas regenerates, new RNA is synthesized, and it has been suggested that certain predetermined cells need to degenerate, in order to permit the remaining rudiment to develop into the adult pancreas.

At NF stage 47 the pancreas of the larval *Xenopus* already has two pancreatic ducts, joining the hepatic duct posteriorly, and the single hepatopancreatic duct opens into the upper margin of the duodenum (Fox *et al.*, 1970). Pyramid-shaped cells of the exocrine acini, separated from one another by straight intercellular junctions, surround a central lumen. Their cytoplasm includes a well-developed RER, ribosomes, mitochondria, and prominent Golgi complexes. Remnants of yolk and lipid may still remain. Most acinar cells have numerous membrane-bound, secretory zymogen granules, up to 1.4 μm in diameter, localized apically near the lumen (see Figs. 19–21) (Fox *et al.*, 1970). The walls of the pancreatic ducts are composed of a single layer of columnar cells with rounded nuclei and there are fairly straight intercellular junctions. The duct margin is bounded by short microvilli, and small zymogen granules, up to 0.8 μm in diameter, lie beneath the cell apices. There are few mitochondria and a prominent Golgi complex is situated alongside the nucleus. The outer surface of the pancreatic duct is enveloped by collagen and connective tissue cells (see Fig. 22, Fox *et al.*, 1970).

Pancreatic endocrine or insular tissue is first recognizable in *Xenopus* at NF stage 42 (Leone *et al.*, 1976), and insulin, which is synthesized throughout larval life, increases at climax. In young anuran larvae endocrine pancreatic tissue consists of a small number of islets grouped as cells A or B. At climax, when exocrine acinar tissue partially regresses, the endocrine tissue eventually proliferates and the islets become mixed. In *Discoglossus* pancreatic islet cells B are first recognized when the hindlimb buds appear, and cells A some days later, when the limbs commence growth. Lipid is reduced in the islet cells, but pigment and other granules are present. In contrast, acinar cells show only modest differentiation at this time, with only the incipient formation of zymogen granules. Acinar cells of *A. obstetricans*, cultured at stages equivalent to NF stages 50–60 of *Xenopus*, contain typical zymogen granules. Islet cells A and B have granules, probably originating from Golgi saccules. So-called D cells identified in the larval pancreas are not seen in the cultured tissue (Pouyet and Beaumont, 1975).

Using a different cellular terminology, Leone *et al.* (1976) described acinar A cells of the pancreas of *Xenopus*, at NF stage 42, to have a well-developed RER, Golgi complexes, and numerous large secretory granules; B cells were without an RER and granules, but had well-developed mitochondria. B cells may be undifferentiated A cells, or, because of their location, may possibly be similar to the centroacinar cells of the adult pancreas. Endocrine C cells in islets have small membrane-coated granules. Exocrine acini increase in number at NF stages 54–56, and necrosis occurs at NF stage 61 during climax, when acinar cells are difficult to distinguish. C cells are practically unaffected. The adult pancreas has numerous well-developed acini of A and B cells surrounding lumina, and C cells cluster in islets. Lipase activity reaches a peak in *Xenopus* at NF stages 54–56, that of amylase at NF stage 51. Both enzymes then decrease their activity to a minimum by the end of climax, corresponding to the period when little feeding occurs. After metamorphosis enzyme activity gradually increases to reach the adult level.

6. THE LIVER

During natural or T_4-induced metamorphosis of anuran larvae, the liver shows a striking increase in the production of urea-cycle enzymes, albumin, hydrolases, and various other enzymes and ribonucleic acids. The change from the excretion of ammonia to urea is well known (Frieden and Just, 1970), though the adult aquatic *Xenopus* retains the larval method of excretion (see Cohen *et al.*, 1978, for a summary of liver biochemistry and references). The various biosynthetic changes of the larval liver during metamorphosis would appear to be expressed in changes seen at an ultrastructural level (vide infra).

Aspects of the fine structure of hepatocytes, gallbladder, and the bile duct of *X. laevis* at NF stage 47 (approximately equivalent to TK stage II of *R. pipiens*) have been described (Fox *et al.*, 1970). Liver parenchyma cells form small lobules, and microvilli line the small lumina of the bile canaliculi. There is a well-developed RER and numerous ribosomes; lipid is still present.

The gallbladder has a wall of columnar cells and is about 7 μm thick. Numerous short microvilli with fuzzy coats line the cavity. Ribosomes and smooth-walled vesicles are plentiful, and many mitochondria aggregate near the apical region of the cells. Large lipid droplets are still present (see Figs. 13–15, Fox *et al.*, 1970).

Ciliated and microvillous cells, often adjacent to one another, line the lumen of the bile duct, whose wall, composed of a single layer of cells, is 6–10 μm thick. Some ciliary-microvillous cells also occur. Dense rounded bodies are frequently found at the apex of some microvillous cells, mitochondria are numerous, and ribosomes abound. Cilia are present along the entire luminal surface of the bile duct, extending from the short nonciliated neck adjacent to the gallbladder to the region near where the hepatopancreatic duct opens into the duodenum (see Figs. 16–18, Fox *et al.*, 1970).

It seems unlikely that hepatocytes significantly increase in number during anuran metamorphosis, coincident with the new biosynthesis that occurs at this time. Liver cellular proliferation has been reported to occur during early larval development, which indeed must happen, but though the hepatocyte-DNA labeling index increases during T_4-induced metamorphosis of bullfrog larvae, the total amount of NDA does not appear to change significantly; nor does the liver weight change. Nevertheless, Atkinson *et al.* (1972) reported increased biosynthesis of hepatocyte nuclear and mitochondrial DNA and the absence of polyploidy after T_3-induced metamorphosis, concluding that such changes were not consistent with a "fixed population" of cells (see Cohen *et al.*, 1978). Others have found the hepatocytes of *Rana* and *Xenopus* larvae to hypertrophy, as did Kistler and Weber (1975) in *Xenopus*, though with some (though not significant) hyperplasia. In general agreement with their results, Oates (1977) described cellular hyperplasia in the liver of *Xenopus* until NF stage 51, subsequent hypertrophy until NF stage 57, and finally during climax a decrease in cell volume, when the hepatocyte mitotic index was very low or nil.

Thus, probably fresh biosynthesis proceeds within the same, essentially nondividing hepatocytes during prometamorphosis and climax. Subsequently as the liver of the froglet enlarges, presumably hyperplasia of the original clonal population of hepatocytes resumes, such cells continuing to function biochemically in the adult manner.

Changes and massive increase in the liver RER of anuran larvae during T_4- or T_3-induced metamorphosis were first shown by Tata (1967), who suggested a relationship with the biochemical changes occurring at this time (Figs. 10, 11). Other ultrastructural changes in hepatocytes of larvae of *R. pipiens* during metamorphosis have also been reported. They include pleomorphism of the mitochondria, the presence of glycogen in nuclei of late larval stages and its disappearance in the cytoplasm, probably to form glucose. Pinocytotic vesicles occur in adjacent hepatocytes, but only in the larva. The Golgi complexes appear to decrease in size by the end of metamorphosis. RER occurs in rows of cisternae throughout metamorphosis, though the number of cisternae decreases near the end of climax, which coincides with an increase in the amount of smooth

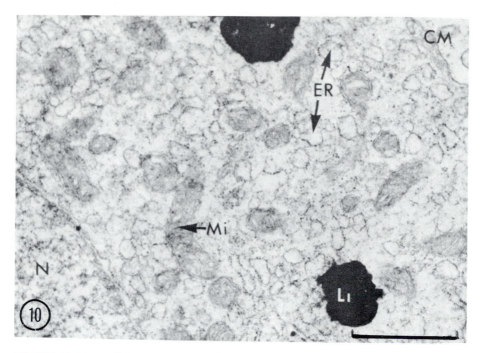

FIGURE 10. Liver cell from premetamorphic larva of *R. catesbeiana*, showing only a modest development of the granular endoplasmic reticulum (ER); mitochondria (Mi), the nucleus (N), a lipid droplet (Li), and the cell membrane (CM) are indicated. Bar = 1 μm.

endoplasmic reticulum. The adult hepatocyte resembles that of earlier larval stages. Likewise other workers found changes in nuclei, from euchromatic to heterochromatic, and an increase in size and in the appearance of mitochondria and the Golgi complexes of hepatocytes, in larvae of *R. catesbeiana* during natural and T_4-induced metamorphosis. The RER proliferated and the cisternae dilated. Such fine structural changes were presumed to be related to the increased biosynthetic activity and synthesis by the larval liver during the metamorphic cycle (see references in Dodd and Dodd, 1976).

The fine structure of the developing liver of normal *X. laevis* larvae through metamorphosis has recently been described by Spornitz (1978). In addition to changes in structure and amount of various organelles such as the RER, related to biosynthetic activity, hepatocytes also synthesize, store, and metabolize glycogen. Embryonic glycogen has disappeared by NF stage 42, but at NF stage 46, after larval feeding has begun, smaller α and β glycogen particles are formed, independent of the SER. From prometamorphosis to the end of climax (NF stage 66), the glycogen content of the liver increases from 0.2% to 10% by weight. In fact, glycogen of the adult liver of *Xenopus* may comprise 20% by weight. These results contrast with reports of glycogen depletion of liver during T_4-induced metamorphosis of *Xenopus* (Kistler and Weber, 1975). Perhaps the ex-

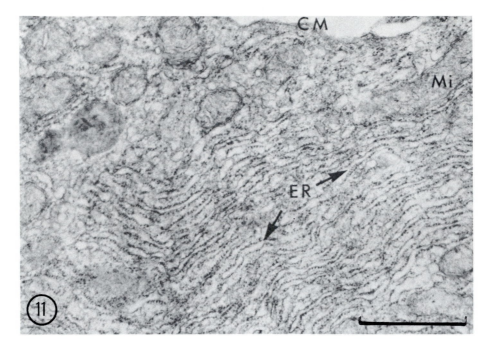

FIGURE 11. Larva of *R. catesbeiana* induced into metamorphosis after treatment with T₃ for 6 days. The hepatocyte has differentiated a highly complex lamellated granular endoplasmic reticulum. Bar = 1 μm. Figures 10 and 11 by kind permission of Dr. J. R. Tata, 1967, *Biochem. J.* **104**.

planation is to be found in the different conditions of feeding (or fasting) of the various larvae used by the different investigators.

7. THE PRONEPHROS

7.1. Structure and Development

The paired pronephroi initially comprise the sole excretory (and probably osmoregulating) system. During prometamorphosis, however, the pronephroi and mesonephroi function simultaneously, until the pronephroi disappear during climax, leaving the ureotelic mesonephroi as the adult kidneys. The pronephros functions very early in larval development and bipronephrectomized young larvae soon become edematous. Early in larval life water and other substances, which diffuse from surrounding tissues into the coelom, are the main source of fluid traversing the pronephros to the exterior. Later, filtration through the paired glomi projecting into the coelom is probably more important. The pronephric (and mesonephric) tubules of *Xenopus*, at NF stage 55, appear capable of phagocytic activity, for they ingest carbon particles and mammalian erythrocytes drawn from the body cavity.

Present available evidence strongly supports the view that pronephric growth, differentiation, and ultimate degeneration are mainly controlled by circulatory thyroid hormones. Antithyroid goitrogens and surgical hypophysectomy or thyroidectomy inhibit anuran larval development, including that of the pronephros and mesonephros, and T_4 stimulates their development, or at later larval stages degeneration in the case of the pronephros (see Fox, 1963; Fox and Turner, 1967; Oates, 1977). T_4 circulating concentration, temperature, and the larval stage when thyroid hormones are active, are of supreme importance for metamorphic change including that of the pronephros. It is not surprising, therefore, that histological changes in the thyroid seem to be related to the degree of pronephric degeneration. However, whether thyroid hormones influence pronephric cells exclusively at the level of the genome, or indirectly, is not clear. Hyperfunction of the pronephros can increase the size of various pronephric components, and hypofunction, or reduction of intratubular fluid tension, leads to a reduction of luminal volume of the pronephric tubules and duct (see Fox, 1963). Inhibitory growth factors, or chalones, have been suggested to regulate growth of the larval pronephros (see Oates, 1977).

Examination by light microscopy of the amphibian larval functional pronephros (Fig. 12) reveals a complex tightly arranged mass of tubules, somewhat oval-shaped in form, with three (anurans) or usually two (urodeles) ciliated nephrostomial tubules opening by nephrostomes into the coelom. The nephrostomial tubules join a common proximal convoluted tubule, which leads into a short ciliated intermediate segment followed by a nonciliated distal tubule, and then the pronephric duct. The pronephros is situated within the postcardinal sinus and receives oxygenated blood from the dorsal aorta; individual spinal nerves lead over the nephrostomial tubules, reflecting an ancestral segmental relationship. Alongside the inner margin of the pronephros a glomus comprises a small capillary mass originating from the dorsal aorta. The basic arrangement of the mesonephric nephron is similar to that of the pronephros (Fox, 1963).

RNA is detectable in the pronephric proximal tubule of R. pipiens at Shumway stage 23 (equivalent to NF stage 42 of Xenopus), as tubules differentiate. RNA levels stay high until pronephric differentiation is completed. PAS staining of brush borders and protein droplet resorption are similar to the reactions found in mesonephric and metanephric kidneys. Alkaline phosphatase activity increases in the brush border of pronephric tubules of Bombina orientalis during larval development, but decreases to eventually disappear simultaneously with pronephric degeneration at climax. Alkaline phosphatase activity of the mesonephric tubules, however, remains high throughout metamorphosis.

Oates (1977) found the mitotic index of pronephric tubule cells of *Xenopus* to decrease at NF stage 51, though luminal volume and tubule length increase between NF stages 45 and 55. The pronephros has decreased in size by NF stage 61. In *R. temporaria* from CM stages 29 to 47 the nuclear population of the pronephros and its length, tissue volume, luminal volume, and thus overall volume and the tubular internal surface area increase. Measurements of these components at CM stage 49 (about the beginning of climax) and thereafter showed them to be substantially reduced, until the final disappearance of the pronephros by the end of climax at CM stage 54. The pronephric tubular individual cell volume ranged between 4000 and 7000 μm^3, but as tubules regress cell volume reduces to between 2000 and 5000 μm^3. The maximum number of cells forming a pronephros (judged from its nuclear population) is about 10,000 at CM stage 47. During climactic CM stages 51–52, the number has reduced to about 2000 and at CM stage 53 to about 400. The glomus reaches a maximum length of 0.3 mm at CM stage 47 and is reduced to a vestige by the end of climax. The mesonephros lengthens threefold from CM stages 29 to 50. It appears to shorten slightly at climax, perhaps due to the overall body shortening at this time. Afterwards the mesonephros increases in size to attain adult form (see Fox, 1963).

7.2. Fine Structure

Examination of the fine structure of functional, preclimactic, proximal pronephric tubules of *R. temporaria* (Fox, 1970) showed them to be bounded externally by a basement membrane and some collagen fibrils. There is substantial infolding of the plasma membrane, and lateral interdigitations occur between tubule cells. Presumably this arrangement provides a large surface area to facilitate transport of substances between cells and into (or from?) the posterior cardinal sinus. Cells have a large round or oval-shaped nucleus, often with a recognizable nucleolus, and there are numerous mitochondria situated basally (Fig. 15). A Golgi complex near the nucleus is not commonly seen in *Rana* but is more prominent in *Xenopus* (Fox and Hamilton, 1971) as it is in *Ambystoma*. The proximal tubule lumen is lined by microvilli 2 μm long, and there are about 6000 of them per millimeter length of luminal surface. Pinocytotic vesicles occur between the microvillous bases and within the cells. The rest of the cell includes a RER and SER, ribosomes and polyribosomes, and small fat droplets. The distal tubule is similar to the proximal tubule in appearance, except for the absence of microvilli and intercellular junctions are less infolded. The short intermediate tubule is ciliated at the luminal surface. The cells have a high ribosomal content, and lipid and pigment occur.

In *Xenopus* at NF stage 47 (Fox and Hamilton, 1971) the cells of the nephrostomial tubules have straight intercellular junctions, mitochondria are irregular in shape and variable in number. The tubule wall is one

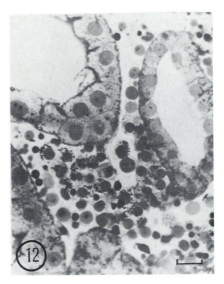

FIGURE 12. Fully developed nondegenerate functional pronephric tubules of *R. temporaria* at CM stages 45–46 (about the beginning of prometamorphosis). There are numerous erythrocytes, granulocytes, and lymphocytes between the tubules. Bar = 20 μm.

cell thick and cilia line the lumen. The pronephric duct has a particularly high degree of infolding of the plasma membrane, similar to that found in *Triturus*. Its luminal surface is fairly smooth, with only a few short projections. Generally its cellular ultrastructure is similar to that of the distal tubules.

7.3. Degeneration

In *Xenopus* pronephric degeneration is stated to begin at NF stage 53, with the first signs of atrophy at NF stage 54 (Nieuwkoop and Faber,

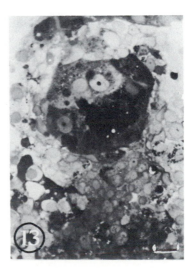

FIGURE 13. Degenerate nonfunctional pronephric strand of *R. temporaria* from CM stages 51 to 53 (height of climax). The autolysing degenerating tissue is enveloped by blood cells and other phagocytic cells. Bar = 20 μm.

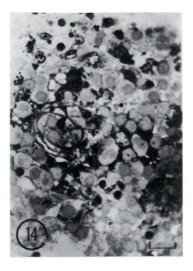

FIGURE 14. Remnants of degenerate pronephric tissue of *R. temporaria*, from the same CM stages as in Fig. 13, now reduced to a mass of tissue unrecognizable as tubular or strandlike, and enveloped amid erythrocytes, granulocytes, lymphocytes, and other phagocytic cells. Bar = 20 μm. (Figures 12–14: 1-μm-thick Araldite sections of tissue fixed in osmic acid and stained with toluidine blue.)

1956). Indeed, acid phosphatase activity may occur in pronephric tubules of *Xenopus* at NF stage 51. However, Oates (1977) described well-formed tubule lumina at NF stage 55, though incipient connective tissue bordered the coelom near the nephrostomes. The luminal border of the tubule cells was ragged at NF stage 57. In *R. sylvatica* pronephric degeneration has been reported to begin simultaneously with the rapid growth

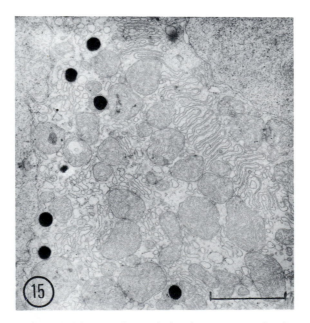

FIGURE 15. Area of proximal pronephric tubule of prometamorphic larva of *R. temporaria*. Note the numerous mitochondria and highly convoluted intercellular junction. Bar = 1 μm.

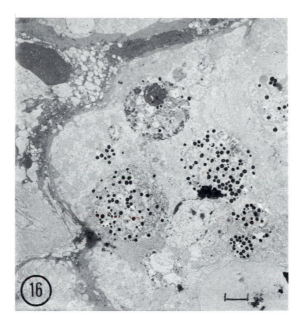

FIGURE 16. Outer region of a degenerating pronephric tubular strand of *R. temporaria* at climax. A number of large pigmented cytolysomes represent areas of intense autolytic necrosis, found to be positive for acid phosphatase. Enveloping the pronephric strand is a large "dark" cell, probably phagocytic in nature. Bar = 2 µm. Figures 12–16 courtesy of the Company of Biologists; see Fox, 1970.

of the hindlimbs, but in *R. temporaria* maximum pronephric size occurred at CM stage 47, when the hindlimbs were well developed, and degeneration became apparent between CM stages 47 and 49 (equivalent to NF stages 55–56 of *Xenopus* and TK stages XII–XV of *R. pipiens*). In general, incipient pronephric involution of anuran larvae seems to commence around the middle of prometamorphosis, with massive breakdown at climax.

Pronephric degeneration of *R. temporaria* is uneven and variable in its progress (Fox, 1970). Adjacent cells may differ in their rate of necrosis; even within a cell different (or similar) organelles vary in their state of degeneracy, some being almost wholly, others hardly, degenerate. Nevertheless, through climax pronephric involution is continuous, and the organ gradually disappears amid surrounding extrarenal cells of the postcardinal sinus (Figs. 12–14). Large autophagic vacuoles, secondary lysosomes, or cytolysomes, termed degeneration bodies (Fox, 1970), often up to 8 µm in diameter, are recognized in degenerating pronephric cells (Fig. 16). They are centers of intense necrosis and include mitochondria at varied degrees of degeneration, vesicles, myelin figures, dense osmiophilic bodies, lipid, and pigment. The background matrix is of a disorganized granular composition. These areas are delimited usually by one or several delicate membranes and acid phosphatase activity is registered within them. Pronephric nuclei become chromatopycnotic, and DNA synthesis decreases in a pronephros about to degenerate. Eventually the tubular basement membrane and overlying collagen disappear, plasma infoldings are lost together with luminal microvilli and cilia, and tubules regress to solid necrotic strands whose debris ultimately becomes unrecognizable amid surrounding postcardinal cells (Fig. 14).

During prometamorphosis pronephric tubules in the postcardinal sinus are surrounded by blood cells, which increase in number. There is a high preponderance of erythrocytes to leukocytes. It has been claimed that erythrocyte stem cells in the intertubular areas of the pronephros and mesonephros, of *R. pipiens* larvae, develop *in situ*. Lymphoid histogenesis occurs in the thymus and pronephros of *R. pipiens* at TK stage I (equivalent to NF stage 46 of *Xenopus*), and the pronephros and mesonephros are important centers of hemopoiesis during larval life. Whether renal anlagen include lymphocytic stem cells, however, is not certain. It is possible that erythrocyte and lymphocyte stem cells originate elsewhere and later on concentrate between and around pronephric and mesonephric tubules. Oates (1977) reported the first appearance of interpronephric tubular lymphocytes of *Xenopus* at NF stage 53, and erythrocytes were numerous at NF stage 55. Agranular lymphocytes, granulocytes, and erythrocytes, recognized surrounding pronephric tubules of prometamorphic *R. temporaria* at CM stages 45–46 (equivalent to NF stage 54 of *Xenopus*), increase in number around climactic degenerating tubules (Fox, 1970). In *R. temporaria* characteristic "dark" cellular extensions, often nucleated, appear to invade degenerating tubule cells; they are probably phagocytes (Fig. 16). The role of the granulocytes and lymphocytes intimately associated with the involuting pronephros is not clear. Probably pronephric necrotic debris, derived by autolysis, is ingested by the "dark" phagocytes and granulocytes.

8. THE SKIN

The larval amphibian epidermis and dermis comprise a complex of many different cellular components, each of which differ in structure, function, and topographical relationship, and frequently in their time of origin and longevity. The skin progressively becomes more elaborate and cosmopolitan in its cellular composition as development proceeds. However, some types of epidermal cells originate early in larval life and disappear before the onset of prometamorphosis. They subserve specific functions during early larval ontogeny. Other cell types are recognized throughout larval and adult life. Presumably their function is the same in an aquatic larva or a terrestrial frog. Nevertheless, though the ultimate fate of some specialized cells is not clearly known, it is likely that apart from the germinative cell layer, most if not all epidermal cells (and much of the dermis too) ultimately degenerate and disappear, some to be regularly replaced throughout life, though obviously tail skin is not renewed after tail involution at climax. Amphibian skin is thus a dynamic cellular system adapted to a complex and changing life cycle.

8.1. Early Epidermal Cellular Differentiation

Ciliary and nonciliated mucus-containing cells differentiate very early in amphibian larvae. About one-third of the surface cells of the embryonic *Ambystoma mexicanum* are ciliated. They are first recognized at the tail bud surface of *R. temporaria* at CM stages 22–23 (equivalent to NF stages 29–33 of *Xenopus*). A truly demarcated bilaminar epidermis is formed here slightly later (vide infra). Ciliary cells eventually disappear, though vestiges remain even at NF stage 43 (Fox and Hamilton, 1971). They are recognizable in the body epidermis of *R. temporaria* at CM stage 34 (equivalent to NF stage 44 in *Xenopus*; Fox and Whitear, unpublished).

Elongate, bottle-shaped cells of the hatching glands (HGC) differentiate in the epidermis of the frontal region, and to a lesser extent along the dorsal midline to the level of the ear vesicles, in prehatching NF stage 22 of *Xenopus* and Shumway stage 17 of *R. chensinensis* (equivalent to about NF Stage 24 of *Xenopus*). During hatching the two outer jelly layers surrounding the larval *Xenopus* rupture, mainly due to imbibition of water. The HGC secrete a hatching enzyme(s), probably a protease, which partially degrades the fertilization envelope. Subsequent movements of the larva rupture the weakened membrane to permit hatching.

HGC, about 45 μm high in *R. chensinensis*, have prominent, apical microvilli, and cytoplasmic RER and mitochondria. Apical granules formed in the Golgi complex at NF stage 24 of *Xenopus* subsequently increase in size and number and are secreted during NF stages 24–38. In prehatching stages apical granules of Golgi bodies contain polysaccharides other than glycogen. In hatching stages (hatching of *Xenopus* begins at about NF stages 35–36), acid phosphatase was found in these components. In posthatching stages acid-phosphatase-rich granules fuse with membrane-bound bodies within the cell to form an elaborate phagolysome, an expression of cellular degeneration. Probably a functional change occurs in HGC soon after hatching, from the secretion of carbohydrate-rich granules to those containing acid phosphatase or lysosomes. HGC slowly diminish in size and in *Xenopus* degenerate by autolysis at NF stage 39 after hatching. DNA-dependent RNA synthesis probably influences granular development in early larvae (see Yoshizaki, 1976).

The anlage of the future paired pear-shaped cement glands, or oral suckers of *Xenopus*, first appears in the ventroanterior region of the early neurula at about NF stage 15. Maturity and activity of the cells (CGC) are at their peak at NF stages 35–36 (Picard, 1976); they degenerate after NF stage 40 and have disappeared before NF stage 50 (Nieuwkoop and

Faber, 1956). In *Hyla regilla* the suckers are first adhesive soon after the late tail bud stage, when secretory granules are recognizable. The cuboidal CGC may possibly be derived directly from epithelial cells. Fine un-myelinated nerve endings, 1 μm or less in diameter, lead from the tri-geminal ganglion to the cement gland. The CGC subsequently develop an extensive membrane system producing a mucouslike secretion, which is elaborated in the Golgi apparatus and packaged in secretory granules. The secretory adhesive mucin is probably a glycoprotein. The cytoplasm also includes microtubules and microfilaments. In the elongated CGC of *Xenopus* Picard (1976) described an apical zone (5% of cell zonation) with α and β vesicles; a transit zone (30%) of many microtubules, with secretory vesicles migrating to the cell apex; a zone of biosynthesis (40%) with concentric regions of RER and an extensive Golgi apparatus; a zone containing an elongated nucleus (15%) and little lipid and yolk; and a storage zone (10%) with lipid and yolk, which may occupy up to 50% of the cell volume. Type B cells (20%) stain weakly with methylene blue; type A cells (80%) stain more strongly, have fewer organelles and more clear vesicles, and at NF stage 35 their greater preponderance suggests that cellular involution has commenced. Degenerating CGC develop large autophagic vacuoles and are ultimately ingested by phagocytes. Cement glands of *Hyla* are poorly differentiated after neurulae are treated with actinomycin D, probably due to interference with DNA-dependent RNA synthesis.

Among anuran tadpoles some epidermal cells early on modify to fashion a horny beak, which functions throughout larval life and disap-pears at climax. The horny jaws of *R. pipiens* larvae are clearly apparent early before premetamorphosis or from Shumway stage 24 to TK stage III, and though smaller with fewer keratinized cells they are similar in TK stages V–XV. Below each serrated edge of the beak of *R. pipiens* there is a column of cells, which are flatter near the base. Apically, cone-shaped cells nestle into each other; at the top of the column the sharply angled cone cells are keratinized. Together the cells of the columns make up a palisade, and adjacent cells interdigitate and join by desmosomes. Apical cell loss, which occurs throughout the existence of the beak, is accel-erated by thyroid hormones.

8.2. General Structure of Larval Skin

The epidermis proper is formed when the outer epithelial layers of the skin are delimited by a basement membrane. This membrane, and the first wisps of collagen of the future basement lamella, first line the inner margin of the second epithelial layer of the early developing tail of larval *R. temporaria*, at CM stage 25 (equivalent to about NF stage 39 of *Xenopus*) (Fox and Whitear, unpublished). Body and tail epidermis of larval anurans comprises two to three layers of cells, increasing to five to six layers in the body at climax, when outermost cells may slough

(Fox, 1977). In general, epithelial cells of tail and body epidermis are similar in fine structure. Among other organelles present the lysosomes become larger and more numerous in outer epithelial cells of the epidermis at late prometamorphosis and climax. Tonofilaments are more profuse in the epidermal cells of older larvae, and the inner margin of the basal epidermal cells is lined by hemidesmosomes, with which are associated the masses of tonofilaments that form the Figures of Eberth. Underlying the epidermis are the adepidermal space with its lamellated bodies and adepidermal membrane, and then the orthogonally arranged collagen fibrils of the basement lamella within a viscouslike ground substance. In the tail anchoring fibrils and larger anchoring fibers lead from the adepidermal membrane and traverse the basement lamella, the anchoring fibers ultimately joining subdermal collagen surrounding muscle tissue. In addition to epithelial cells, larval tail and body epidermis may include surface mucous goblet cells and intraepithelial Leydig cells, which are more common in urodeles. In *Taricha torosa* the large Leydig cells, numerous in midlarval stages, disappear at metamorphosis as in the axolotl. They probably secrete mucus into subsurface extracellular compartments of the epidermis to prevent desiccation (see Fox, 1977).

Merkel cells are present in the epidermis of larval and adult amphibians (Fox and Whitear, 1978). In *R. temporaria* they are first definitely recognizable in the epidermis of the tail at CM stages 33–34 and of the body at CM stage 35 (equivalent to NF stages 44–45 of *Xenopus*). In *Xenopus* they were first found at NF stages 49–50 (tail), stage 50 (hindlimb digits), and stage 52 (body) (Fox and Whitear, unpublished; see also Tachibana, 1979, for *R. japonica*). The characteristic Merkel cells are believed by some workers to be derived from the neural crest and to migrate to the skin, but as Merkel cells still occur in the epidermis of larval *A. maculatum* following removal of the neural crest and all other neural tissues, this is unlikely, at least in amphibians (Tweedle, 1978). Merkel cells appear to originate and subsequently differentiate from rounded, undifferentiated, precursor or interstitial cells, located between and distinct from the epidermal epithelial cells (Fox and Whitear, unpublished; Tachibana, 1979). The fate of Merkel cells is unknown. As they occur in amphibian larvae and adults in apparently similar rarity (about 0.3% of the epidermal cell population), then either the same cells survive throughout life, which seems unlikely, or old Merkel cells involute and are replaced by new ones from preexisting nonsenile Merkel cells or from epidermal cells of the basal germinative layer. Merkel cells have reciprocal synapses with nerve terminals, and function possibly as sensory touch receptors, or mechanoreceptors of some kind or mediate a trophic influence (see brief discussion in Fox and Whitear, 1978).

Stiftchenzellen occur in the epidermis in several species of *Rana* but only in their larvae. The youngest larva of *R. temporaria* found to possess a clearly recognizable Stiftchenzelle (in the dorsal tail fin epidermis) was at CM stages 28–29 (equivalent to about NF stage 41 of *Xenopus*) (see

description of the fine structure by Whitear, 1976). The fact that Stiftchenzellen are not found in adult frogs led Whitear to suggest that they
degenerate near (or at) climax, which would account for the high proportion of degenerate-looking examples found at this time. These cells
may function as chemoreceptors. Mitochondria-rich cells are also occasionally found in the surface layer of the epidermis of the larval tail and
body; they were recognized in body epidermis of *R. temporaria* larvae at
CM stages 33–34 (equivalent to about NF stage 44 of *Xenopus*) (Fox and
Whitear, unpublished). In the adult the flask cells of the epidermis are
also mitochondria rich (Masoni and Garcia-Romeu, 1979); they react
strongly for carbonic anhydrase and possibly are involved in osmoregulation or ion transport. Mitochondria-rich cells also occur in the epithelium of the bladder and of the palate. There are also immigrant melanophores, polymorphonuclear neutrophils, granulocytes, and mesenchymal
macrophages, all without desmosomes, and distributed nerve fibers. Flask
cells (vide supra) probably originate *in situ* and are first recognized in
body epidermis of *R. temporaria* at climax; they are prominent and comprise about 10% of adult epidermal cells.

All amphibian larvae and adult perennibranchiate urodeles possess
epidermal neuromasts, which disappear at climax in anurans but are
retained in the aquatic *Xenopus*. The origin of amphibian lateral line
organs from pre- and postauditory placodes and their subsequent migration and differentiation (and association with nerves) have been described
in *Triturus pyrrhogaster* (Sato and Kawakami, 1976). It is unlikely that
they are of neural crest origin. Receptor cells each with a surface kino-
cilium and stereocilia and supporting mucus-containing type I and cu-
pola-forming type II cells develop. Lateral line organs are fully developed
at the time of hatching.

Below the basement lamellar collagen the dermis of an anuran climactic larval body skin includes melanophores, xanthophores, and iridophores, granulocytes and other leukocytes usually in capillaries, mesenchymal fibroblasts and macrophages, and multicellular mucous and
granular (serous) glands, which open by ducts at the skin surface at climax. There are also muscle and nerve components including Schwann
cells. Most of these components with the exception of the glands are
present in the dermis of the larval tail. Iridophores and xanthophores are
rarely if ever present in tails of *R. temporaria* and *X. laevis*, though
iridophores do occur in the tail of larval *R. catesbeiana*. Nevertheless,
what appeared to be an incipiently developed pre-iridophore was recognized in a tail of *R. temporaria* at CM stages 33–34, and preiridophores
and prexanthophores were found in the body dermis of larval *R. temporaria* at CM stage 33 (Fox and Whitear, unpublished).

8.3. Degeneration of Larval Tail and Body Skin

In general, skin of the larval tail and body degenerates in a similar
fashion, except that outer layers of the epidermis of the body, shed at

climax, are replaced by new ones derived from the basal germinative layer, and in the dermis various components, such as collagen, substantially increase in amount to reach adult proportions. At climax outermost epidermal cells of the distal tail region, as it progressively shortens to a stump, and likewise those of the body, include fibrous or other dense, rounded bodies, which are probably lysosomes. Such cells become electron dense, flattened, dehydrated, and cornified as keratin is deposited within them. Cell components autolyse and are eventually unrecognizable (Fox, 1977). Ultimately only tonofilaments, fused together in a homogeneous mass, remain within the cell. An outer thickened dense membrane under the plasma membrane (slightly thicker in adults) envelopes larval keratinized cells, which are finally shed as desmosomes are degraded, probably by lysosomal enzymes. Inner epidermal cells likewise show features of autolysis and may develop large cytolysomes before they cornify at the skin surface and are shed. It is unlikely that there is any significant macrophagic activity during tail degeneration. Keratin synthesis in larval skin of *Xenopus* normally occurs at about NF stage 57, near the beginning of climax when 40 days old. It has been induced precociously *in vivo* or *in vitro* by T_3 at a developmental age corresponding to NF stages 50–52 (NF stage 50 is about 15 days old). The fine structure of the nondegenerate prometamorphic skin of the operculum, which encloses the chamber within which the forelimbs develop, is similar to that of the rest of the larva. During metamorphic climax it degenerates and disappears in a manner similar to that of tail skin (Fox, 1977), except that surprisingly opercular epidermal cells apparently do not keratinize.

During later stages of prometamorphosis larval epidermal cells show acid phosphatase in small lysosomes, autophagic vacuoles, and cytolysomes. The primary lysosomes probably originate from Golgi cisternae or GERL. At the height of climax there is heavier deposition of acid phosphatase in larger cytolysomes, although the outermost highly cornified cells show little of the enzyme for most of the organelles, apart from tonofilaments, have disappeared by this time. Epidermal cellular degeneration of the tail, body, and external gills of larval *R. temporaria* and *Xenopus*, and of their adults, is similar except that involuting larval external gill filaments do not keratinize. Probably all types of amphibian postclimactic epidermal cells, apart from those of the basal germinative layer, whatever their derivation, eventually degenerate and are shed, to be replaced. Details of these processes in the case of some specialized cells are not known.

At metamorphic climax the dermal basement lamellar collagen of the anuran larval tail is invaded by mesenchymal macrophages, which engulf the collagen within heterophagic vacuoles utilizing lysosomal enzymes. The fate of other dermal components, such as melanophores, at this time is not clear. They may well survive in the newly metamorphosed young froglet. In *R. pipiens* larvae treated with T_4, macrophages similarly invade the basement lamella of back skin and presumably phagocytose collagen fibrils, and the same process seems to occur in *Xenopus*. A new stratum spongiosum develops from mesenchymal cells, and the stratum compactum, or adult lamella, forms from collagen fibrils and mesenchymal cells. Polymerization of new dermal collagen continues throughout the frog's life, and the stratum compactum may well be continuously remodeled, for phagocytosis of its collagen by macrophages occurs in the back skin of the adult *R. temporaria* (see Fox, 1977).

In general, therefore, larval epidermal cellular degeneration occurs by autolysis, demonstrated by autophagy and the presence of cytolysomes. Some necrotic cells, such as those of hatching and cement glands, are probably phagocytosed by neighboring macrophages. Likewise, probably most of the remnants of autolysed cells of the external gill filaments suffer phagocytosis. Epidermal cells autolyse, keratinize, and are shed from the body surface. Perhaps a small number of such cells in the tail and body are phagocytosed, but this feature would appear to be of minor significance. Flask cells of the postclimactic froglet and adult do not keratinize and are shed with the slough.

Acid phosphatase and other lysosomal enzymes are recognized in degenerating larval epidermal cells near and at climax (also in degenerating hatching gland cells, vide supra), and in surface presloughing and sloughing epidermal cells of adults. These enzymes occur in a variety of degenerating larval tissues (see references in Fox, 1977). However, the exact role of such enzymes during cellular degeneration is still not clear. They may indeed have a variable function in different tissues, as for example tail muscle. The levels of lysosomal enzymes increase in the larval tail toward climax, and protein synthesis seems to be necessary for tail regression. Deposition of acid phosphatase is heavier in necrotic tail tissues and in macrophages near and at climax (Weber, 1969), and collagenase continues to be synthesized in the tail during its regression. Furthermore, T_4 induces increased hyaluronidase activity in back sin of *R. catesbeiana* larvae during metamorphosis. However, Smith and Tata (1976) conclude that T_3-induced tail regression could well result from the activation of "proteolytic cascades," for they found no significant synthesis of new protein during the first 3–4 days of treatment of tails in culture.

It would seem reasonable to suppose, therefore, that autophagy of degenerating larval tissues and the heterophagy by macrophages, function by the utilization of lysosomal enzymes, though other nonlysosomal-proteolytic enzymes could well be involved.

8.4. Thyroid Hormones and Larval Skin

The influence of thyroid hormones on the origin, development, and subsequent degeneration of specific larval skin cells is variable, though probably most skin components of the tail degenerate at climax when the circulating thyroid hormonal level is at a maximum. Whether the

same endocrine influences operate in larvae and adults, to activate epidermal cell death, is not clearly understood. Immersion of young larvae
of *R. pipiens* in T_4 (up to 300 µg/liter) at TK stages III and IX–XI (equivalent to later premetamorphic and early prometamorphic NF stages 49–50
and 54–55, respectively, of *Xenopus*) can elicit changes in the mitotic
index and growth fraction of different layers of epidermal cells of the
hindlimb bud and limb. Implantation of T_4-impregnated pellets into prometamorphic larvae of *R. pipiens* causes localized molting of body skin.
Implantation into the operculum elicits degeneration of opercular skin
and the internal gills and the transformation of adjacent body skin into
the postclimactic adult type. These effects are thus presumably a direct
result of thyroid hormones. Nevertheless, thyroidectomy does not inhibit
molting in adult anurans, in contrast to urodeles where there is piling
up of unshed keratinized layers of cells. Molting is inhibited in both
anurans and urodeles in adults after hypophysectomy (see Larsen, 1976).
Epidermal cellular proliferation in the adult newt *Notophthalmus viridescens* seems mainly to be influenced by prolactin.

Anuran larval premetamorphosis is a period of low thyroid activity
and rapid body growth; indeed, the thyroid shows only incipient differentiation at this time (Etkin, 1970), and the circulating thyroid hormonal
level is extremely low. Probably premetamorphosis is a period almost
independent of thyroid function, for *Xenopus* larvae hypophysectomized
at about NF stages 33–36 (tail bud stages) and at NF stages 49–50 (first
trace of hindlimb bud) can develop as far as NF stage 52 and 54, respectively. They barely, if at all, commence prometamorphosis.

Among the different larval skin cells, epidermal ciliary, hatching
gland, and cement gland cells of *Xenopus* have differentiated and subsequently disappeared by NF stage 50. Differentiated Stiftchenzellen of
R. temporaria and differentiated Merkel cells of *Xenopus* and *Rana* are
first found in larvae before premetamorphosis (vide supra). It is of interest
that they are well developed with a substantial content of microfilaments
in the tail of a giant *Xenopus* larva (Fox and Whitear, unpublished), which
had been maintained at NF stage 54 for 18 months immersed in the
goitrogen propylthiouracil. Thus, presumably the presence of thyroid
hormones in the larva is not essential for the origin, maintenance, and
perhaps longevity of Merkel cells. Mesenchymal fibroblasts, which probably synthesize procollagenous fibrils of the dermal basement lamella,
are recognized in the tail of *R. temporaria* at CM stage 25 (equivalent to
NF stage 39 of *Xenopus*). In *Xenopus* melanophores first appear dorsally
in the head of the larva at NF stages 33–34 and xanthophores in the
peritoneum and outer surface of the eye cup at NF stage 46. Lateral line
organs have differentiated into individual sensory organs at NF stage 41,
and are conical at the skin surface at NF stage 51 (Nieuwkoop and Faber,
1956). They have been reported to be highly differentiated at NF stage
54, but this structural differentiation is recognizable earlier (Fox and
Whitear, unpublished). Epithelial cells of the external gill filaments of

TABLE 1. VARIOUS CELLULAR COMPONENTS OF AMPHIBIAN SKIN

	Larva		Adult	Cell location	
	Tail	Body	Body	Epidermis	Dermis
Epithelial cells	+	+	+	+	−
Surface keratinocytes	+	+	+	+	−
Keratinized beak cells	−	+	−	+	−
Hatching gland cells	−	+	−	+	−
Cement gland (oral sucker) cells	−	+	−	+	−
Ciliary cells	+	+	−	+	−
Mucous surface cells	+	+	+	+	−
Merkel cells	+	+	+	+	−
Stiftchenzellen (anurans)	+	+	−	+	−
Mitochondria-rich cells	+	+	+ [bladder (*Bufo*) and palate (*Rana*)]	+	−
Flask cells	−	+	+	+	−
Goblet cells (found in *Xenopus*)	+	+	−	+	−
Leydig cells (mainly in larval urodeles)	+	+	−	+	−
Melanophores	+	+	+	+	+
Xanthophores	−	+	+	−	+
Iridophores	+ (*R. catesbeiana*)	+	+	−	+
Mesenchymal macrophages	+	+	+	+	+
Mesenchymal fibroblasts	+	+	+	−	+
Granulocytes	+	+	+	+	+
Polymorphonuclear leukocytes	+	+	+	+	+
Nerve fibers	+	+	+	+	+
Schwann cells	+	+	+	−	+
Neuromast organs [sensory cells (with stereocilia and kinocilia) and supporting cells]	+	+	+ (*Xenopus*: perennibranch urodeles)	+	−
Muscle tissue (striated)	+	+	+	−	+
Muscle tissue (smooth)	+	+	+	−	+
Mucous and granular glands	−	+	+	+ (neck)	+ (neck)

Xenopus and *Rana* also differentiate early in larval development and are merely remnants before the end of premetamorphosis.

Hypophysectomized larvae of *R. pipiens*, at the tail bud stage, do not subsequently develop skin glands, and goitrogens such as potassium perchlorate inhibit glandular development. McGarry and Vanable (1969) concluded that T_4 treatment of cultured forelimb skin of *Xenopus* stimulates glandular cell division, and in some way assists their development. At later larval stages T_4 only seems to influence the rate of maturation of the epidermal glandular rudiment. These authors found older larval skin glands to have an increased sensitivity to thyroid hormones, a conclusion confirming that of previous workers.

In summary, therefore, it seems that the life cycles of ciliary, hatching gland, cement gland, and epithelial external gill filament cells are probably independent of thyroid hormones, and so at least are the origin and differentiation of Stiftchenzellen, Merkel cells, melanophores, preiridophores, prexanthophores, mitochondria-rich cells (vide supra), cells of the neuromasts, and mesenchymal fibroblasts. Whether the degeneration of Stiftchenzellen is influenced by the high level of circulating thyroid hormones at climax is not known. The origin and differentiation of the epidermal flask cells and of the skin glands and in addition the ultimate loss of the beak cells appear to depend on a threshold level of thyroid hormone.

ACKNOWLEDGMENTS. I am grateful to D. Franklin, Margaret Keenan, R. Mahoney, and E. Perry for their assistance in the preparation of this work. Dr. Mary Whitear kindly read the manuscript and suggested a number of improvements.

REFERENCES

Atkinson, B. G., Atkinson, K. H., Just, J. J., and Frieden, E., 1972, *Dev. Biol.* **29**:162.

Beaumont, A., 1977, in: *Mecanismes de la rudimentation des organes chez les embryons de vertébrés*, CNRS No. 266 (A. Raynard, ed.), CNRS, Paris, pp. 113–124.

Billings, S. M., 1972, *Z. Anat. Entwicklungsgesch.* **136**:168.

Bonneville, M. A., and Weinstock, M., 1970, *J. Cell Biol.* **44**:151.

Cambar, R., and Marrot, B., 1954, *Bull. Biol. Fr. Belg.* **88**:168.

Cohen, P. P., Brucker, R. F., and Morris, S. M., 1978, in: *Hormonal Proteins and Peptides,* Volume 6 (C. H. Li, ed.), pp. 273–381, Academic Press, London.

Decker, R. S., 1976, *Dev. Biol.* **49**:101.

Decker, R. S., 1977, *Brain Res.* **132**:407.

Dodd, M. H. I., and Dodd, J. M., 1976, in: *Physiology of the Amphibia*, Volume 3 (B. Lofts, ed.), pp. 467–599, Academic Press, London.

Etkin, W., 1970, *Mem. Soc. Endocrinol.* No. 18, 137.

Fox, H., 1963, *Q. Rev. Biol.* **38**:1.

Fox, H., 1966, *J. Embryol. Exp. Morphol.* **16**:487.

Fox, H., 1970, *J. Embryol. Exp. Morphol.* **24**:139.

Fox, H., 1973a, *J. Embryol. Exp. Morphol.* **30**:377.

Fox, H., 1973b, *Z. Zellforsch. Mikrosk. Anat.* **138**:371.

Fox, H., 1977, in: *Comparative Anatomy of the Skin* (R. I. C. Spearman, ed.), *Symp. Zool. Soc. London* No. 39, pp. 269–289, Academic Press, London.

Fox, H., and Hamilton, L., 1971, *J. Embryol. Exp. Morphol.* **26:**81.

Fox, H., and Turner, S. C., 1967, *Arch. Biol.* **78:**61.

Fox, H., and Whitear, M., 1978, *Biol. Cell.* **32:**223.

Fox, H., Mahoney, R., and Bailey, E., 1970, *Arch. Biol.* **81:**21.

Fox, H., Bailey, E., and Mahoney, R., 1972, *J. Morphol.* **138:**387.

Frieden, E., and Just, J. J., 1970, in: *Biochemical Actions of Hormones*, Volume 1 (G. Litwak, ed.), pp. 1–52, Academic Press, London.

Hourdry, J., and Dauca, M., 1977, *Int. Rev. Cytol. Suppl.* 5, 337.

Hughes, A. E. W., 1968, *Aspects of Neural Ontogeny*, Logos Press, London.

Jayatilaka, A. D. P., 1978, *J. Anat.* **125:**579.

Kistler, A., and Weber, R., 1975, *Mol. Cell. Endocrinol.* **2:**261.

Larsen, L. O., 1976, in: *Physiology of the Amphibia*, Volume 3 (B. Lofts, ed.), pp. 53–100, Academic Press, London.

Leone, F., Lambert-Gardini, S., Sartori, C., and Scapin, S., 1976, *J. Embryol. Exp. Morphol.* **36:**711.

McGarry, M. P., and Vanable, J. W., 1969, *Dev. Biol.* **20:**426.

Masoni, A., and Garcia-Romeu, F., 1979, *Cell Tissue Res.* **197:**23.

Moulton, J. M., Jurand, A., and Fox, H., 1968, *J. Embryol. Exp. Morphol.* **19:**415.

Nieuwkoop, P. D., and Faber, J., 1956, *Normal Table of Xenopus laevis (Daudin)*, North-Holland, Amsterdam.

Oates, C. L., 1977, Control of growth and degeneration in the amphibian kidney with special reference to larval development, Ph.D. thesis, University, Newcastle-upon-Tyne, England.

Picard, J. J., 1976, *J. Morphol.* **148:**193.

Pouyet, J. C., and Beaumont, A., 1975, *C.R. Seances Soc. Biol. Paris* **169:**846.

Pouyet, J. C., and Hourdry, J., 1977. *Biol. Cell.* **29:**123.

Prestige, M. C., 1973, *Brain Res.* **59:**400.

Race, J., Robinson, C., and Terry, R. J., 1966, *J. Exp. Zool.* **162:**181.

Regard, E., 1978, *Int. Rev. Cytol.* **52:**81.

Sato, A., and Kawakami, I., 1976, *Annot. Zool. Jpn.* **49:**131.

Smith, K. B., and Tata, J. R., 1976, *Exp. Cell Res.* **100:**129.

Spornitz, U. M., 1978, *Anat. Embryol.* **154:**1.

Tachibana, T., 1979, *Arch. Histol. Jpn.* **42:**129.

Tata, J. R., 1967, *Biochem. J.* **104:**1.

Taylor, A. C., and Kollros, J. J., 1946, *Anat. Rec.* **94:**7.

Tweedle, C. D., 1978, *Neuroscience* **3:**481.

Ueck, M., 1967, *Z. Wiss. Zool. Abt. A*, **176:**173.

Weber, R., 1969, in: *Lysomes in Biology and Pathology*, Volume 2 (J. T. Dingle and H. B. Fell, eds.), pp. 437–461, North-Holland, Amsterdam.

Whitear, M., 1976, *Cell Tissue Res.* **175:**391.

Yoshizaki, N., 1976, *Dev. Growth Differ.* **18:**133.

CHAPTER 11

Hormonal Control of Amphibian Metamorphosis

BRUCE A. WHITE AND CHARLES S. NICOLL

1. INTRODUCTION

The phenomena of development and metamorphosis have fascinated biologists and naturalists alike for centuries, and comparative physiologists, biochemists, and endocrinologists have systematically studed these processes for almost a century now. Accordingly, a large body of information has accumulated on various aspects of vertebrate biology relating to growth and transformation. Space limitations require that we restrict our coverage in this chapter primarily to recent work on the endocrine control of development and metamorphosis in amphibians. Hence, we will not be able to consider environmental factors that affect these processes, or the recent findings on the control of transformations in other vertebrates, such as smoltification in salmonids. In addition, we will not be able to cite many pertinent references on the control of growth and metamorphosis even in the amphibians. Readers are referred to the excellent and comprehensive reviews of Frieden and Just (1970) and Dodd and Dodd (1976), and to other chapters in this volume for references on some of the earlier literature not cited by us.

In the present chapter we have reviewed the more recent work on the control of amphibian development and metamorphosis by the thyroid, pituitary, and adrenal glands and by the hypothalamus. We have at-

BRUCE A. WHITE and CHARLES S. NICOLL · Department of Physiology–Anatomy, University of California, Berkeley, California 94720. Dr. White's present address is: Laboratory of Cellular Gene Expression and Regulation, Memorial Sloan-Kettering Cancer Center, New York, New York 10021.

tempted to synthesize the current information and to consider it in relation to available theories of control of metamorphosis. The recent evidence indicating that metamorphosis and osmoregulation are intimately related is also discussed.

2. THE THYROID

Etkin (1935) made one of the first attempts to describe, both qualitatively and quantitatively, the development of thyroid activity and its relationship to larval anuran development. He characterized the normal sequence of external metamorphic changes in *Rana pipiens* and then attempted to reproduce normal metamorphosis in thyroidectomized and intact tadpoles by immersing them in dilute solutions of thyroxine (T_4). The natural progression of metamorphosis could be best approximated by exposing tadpoles to increasingly concentrated solutions of the hormone. From these studies, Etkin (1935) proposed that thyroid hormone (TH) levels need to be low for an extended period. During this time, called premetamorphosis, the tadpole grows considerably, but shows little other morphological change. The subsequent phase, called prometamorphosis, is characterized by accelerating hindlimb growth. Since growth of the hindlimb is thyroid dependent, prometamorphosis should be accompanied by gradually increasing levels of TH. Etkin (1935) concluded that the final phase of rapid transformation, called metamorphic climax, would require rapidly increasing levels of thyroid activity and secretion.

2.1. Thyroid Function during Premetamorphosis

A summary of some of the actions of TH in larval and neotenic amphibians is presented in Table 1. Although the actions of these hormones are not discussed fully within this chapter (see Frieden and Just, 1970; Dodd and Dodd, 1976, for more detailed discussion), it is apparent from Table 1 that TH have broad developmental influences that drive differentiation toward the adult form. In this sense, TH are classically characterized as having no positive role in regulating growth and development of larval structures. Nevertheless, there is convincing evidence that they are synthesized very early in development. Kaye (1961) reported that thyroidal organification of iodine occurs in *R. pipiens* at the beginning of the larval period. Hanaoka *et al.* (1973) demonstrated the ability of the thyroid to synthesize triiodothyronine (T_3) and T_4 at a developmental stage prior to the appearance of thyroid follicles.

Steinmetz (1954) inhibited growth in premetamorphic tadpoles by T_4 treatment and increased growth by administration of the goitrogen propylthiourea (PTU), indicating that biologically active levels of circulating TH existed during this period. These findings suggest that TH have an inhibitory action on larval growth. In contrast, Fox and Turner (1967)

TABLE 1. SOME MORPHOLOGICAL AND FUNCTIONAL CHANGES INDUCED BY THYROID HORMONES DURING AMPHIBIAN DEVELOPMENT AND METAMORPHOSIS

Skin
  Formation of dermal glands
  Degeneration of skin over tail
  Proliferation of skin over hindlimb
  Sodium transport
  Differentiation of Leydig cells
  Formation of forelimb skin "window"
  Degeneration of operculum
  Formation of nictitating membrane

Nervous system
  Growth of Mauthner cells
  Growth of mesencephalic V nucleus
  Growth of cerebellum
  Growth of lateral motor column cells
  Growth of hypothalamic nucleus preopticus
  Increase in retinal rhodopsin

Muscle
  Degeneration of tail muscle
  Growth of limb muscle
  Growth of extrinsic eye muscles

Connective and supportive tissues
  Degeneration of tail
  Degeneration of gill arches
  Restructuring of mouth and head
  Calcification of axial and appendicular skeleton
  Development of hypophysial portal system and median eminence

Kidney
  Pronephric resorption
  Induction of prolactin receptors

GI tract and associated structures
  Regression of gills
  Intestinal regression and reorganization
  Induction of urea-cycle and other enzymes in liver
  Pancreatic reduction and restructuring

reported that goitrogens inhibited growth in *R. temporaria* and *Xenopus laevis* tadpoles. Another study with findings conflicting with those of Steinmetz (1954) is that of Just and Kollros (see Frye *et al.*, 1972) in which immersion of hypophysectomized tadpoles in dilute T_4 solution (80 ng/ml) stimulated growth. Other evidence indicating that TH may have anabolic effects on larval structures comes from studies on Mauthner neurons. These cells reside in the myelencephalon and, since they innervate the tail, represent a larval feature. Accordingly, one might expect them to regress in response to TH administration. Indeed, early studies showed that T_4 implantation accelerated Mauthner cell demise. Later experiments, however, indicated that Mauthner cell size was reduced by thyroidectomy and increased by T_4 exposure (Pesetsky, 1962). Thus, it is worth repeating the cautionary note made by Frye *et al.* (1972) that our understanding of the role of TH in larval anuran growth is incomplete. An accurate assessment of this role must include consideration of the possibility that TH exist in the blood at very low but active levels during premetamorphosis and that the overall effect of these hormones is dose dependent as has been shown in other developmental systems (see below).

2.2. Thyroid Function during Prometamorphosis and Climax

The early observations of Etkin (1935) suggest that TH secretion increases during prometamorphosis and a "surge" of secretion occurs just before metamorphic climax. Other studies such as threshold experiments performed by Kollros (1961) have also led to the expectation that TH secretion must increase for completion of metamorphosis. This expectation was partially confirmed by Kaye (1961), who observed a marked rise in ^{131}I uptake by *R. pipiens* thyroids at the onset of prometamorphosis. This increase coincided well with the TH-dependent initial acceleration of hindlimb growth in these animals.

Just (1972) used plasma protein-bound iodine (PBI) as an index of circulating TH in the same species. His results are presented in Fig. 1 along with data from Etkin (1968) on changes in hindlimb length relative to body length. Plasma PBI levels begin rising in early prometamorphosis and then increase dramatically to reach a peak in late prometamorphosis. It is evident from Fig. 1 that the rise in PBI values also coincides well with the TH-dependent accelerated growth of the hindlegs. Thus, the

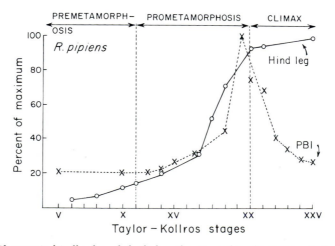

FIGURE 1. Changes in hindleg length/body length ratio and in plasma protein-bound iodine (PBI) during development and metamorphosis of *R. pipiens* tadpoles. The respective data are from Etkin (1968) and Just (1972) and are plotted as a percent of the maximum recorded value.

increase in thyroid activity during prometamorphosis, as measured either by ^{131}I uptake (Kaye, 1961) or by PBI (Just, 1972), and the surge in late climax, conform to the predictions of Etkin (1935) and Kollros (1961). However, the peak in PBI occurs about one stage prior to the onset of climax [i.e., Taylor and Kollros (TK) stage XX] and about two stages before significant tail regression can be measured (Figs. 1 and 3). It could be argued that the peak in PBI in TK stage XIX initiates the events that occur later during climax. Indeed, even though the PBI levels are declining during climax, they are still higher than the levels during prometamorphosis.

Recently, radioimmunoassay (RIA) procedures have been used to measure T_3 and T_4 levels in the blood of bullfrog tadpoles. Because of the high sensitivity of these procedures, substantial clarification of questions on the control of metamorphosis was anticipated with their application. However, although the RIA data from R. catesbeiana are consistent with the events of metamorphic climax, they are not compatible with the findings of Kaye (1961) or Just (1972) of increased thyroidal activity during prometamorphosis in R. pipiens.

These RIA results for R. catesbeiana are presented in Fig. 2, along with the PBI data of Just (1972), which show a surge between TK stages XVIII and XIX, followed by a gradual decline through climax (Fig. 2A). Miyauchi et al. (1977; Fig. 2B) recorded a surge in T_3 between stages XIX and XX and one for T_4 one stage later. The levels of both hormones then declined gradually. On the other hand, Regard et al. (1978; Fig. 2C) found that the levels of T_3 and T_4 rose gradually after stage XVII, then both hormones increased rapidly between stages XXII and XXIII. The blood levels of both hormones then declined, but at different rates. The pattern of blood T_4 recorded by Mondou and Kaltenbach (1979; Fig. 2D) differs from those of the other two groups in some respects but has similarities in others. Plasma T_4 became detectable in stage XIX but did not surge until between stages XX and XXI. The blood T_4 concentration remained elevated until stage XXIII and then fell rapidly.

The reasons for the discrepancies between the PBI data on the one hand and the RIA data on the other, and for the lack of consistency among the three RIA studies are not apparent. They may be due to species differences and to different experimental conditions and inconsistencies in staging tadpoles, etc. Nevertheless, the patterns of TH measured by Miyauchi et al. (1977) and by Mondou and Kaltenbach (1979) are similar and more consistent with the metamorphic changes of climax than are the data of Regard et al. (1978).

It is curious that circulating TH levels are not detectable by the very sensitive RIAs in tadpoles before late prometamorphic stages. Although most of the anatomical transformation occurs during climax when TH levels are measurable by RIA, TH-induced changes start at the beginning of prometamorphosis when the rate of hindleg growth accelerates to outpace body growth. Morphometric data on gut regression (Carver and

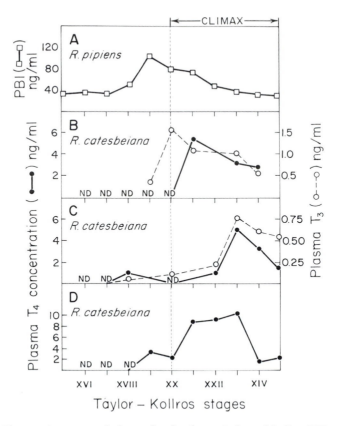

FIGURE 2. Changes in serum of plasma levels of protein-bound iodine (PBI; panel A) or thyroid hormones (T_3 and T_4; panels B–D) in grass frog or bullfrog tadpoles during late prometamorphosis and metamorphic climax. The data presented in panels A–D are from Just (1972), Miyauchi *et al.* (1977), Regard *et al.* (1978), and Mondou and Kaltenbach (1979), respectively. ND, not detectable.

Frieden, 1977), which showed loss of gut weight and length beginning about mid-prometamorphosis, also suggest that TH are present in the blood at levels sufficiently high to induce changes prior to their becoming detectable by RIA. However, it is possible that TH levels increase during prometamorphosis but remain lower than the detection limits of available RIAs. Furthermore, increasing tissue sensitivity to TH may be an important feature of the prometamorphic period (see below).

The RIA data from bullfrog tadpoles (Fig. 2B–D) have been combined and replotted as a composite in Fig. 3, along with morphometric data on *R. catesbeiana* tadpoles from Krug (1980). It should be noted that the changes in tail length/body length and in hindleg length/body length measured in bullfrog tadpoles are in excellent accord with the same measurements made by Etkin (1968; Fig. 2) in *R. pipiens* larvae. This display of the data clearly shows that the rise in blood TH to RIA-detectable levels (stage XIX) is delayed by about seven stages from the initial rise in the rate of hindleg growth (stage XII) and by two stages from the rapid increase that occurs after stage XVI. However, the rise of blood TH in late prometamorphosis, and the sustained elevation during most of climax, correlate well to tail regression. Thus, although the RIA data do not conform to predictions during prometamorphosis, they are consistent with the morphological changes occurring during climax. Conversely, the PBI data of Just (Fig. 2) are more compatible with the events of prometamorphosis than with those occurring during climax. Clearly, the nature of the PBI measured by Just (1972) needs to be elucidated and similar measurements should be conducted in *R. catesbeiana*. Could the PBI be thyroactive material other than T_3 or T_4? RIA measurements of TH in *R. pipiens* may help to clarify the question.

The RIA measurements of Miyauchi *et al.* (1977) and Mondou and Kaltenbach (1979) showed a peculiarity that was also noted by Just (1972) with his PBI measurements. Although the highest T_3 and T_4 levels were measured by RIA in transforming animals, some tadpoles undergoing climax had nondetectable T_4 levels. On the other hand, T_3 was detectable in all samples from tadpoles of TK stages XX and beyond. Similarly, Just (1972) found a wide range in PBI values in the stages with the highest mean values.

Just (1972) suggested that because of a high metabolic clearance rate, the levels of TH after a surge would be detectable for only a few hours. This may be a reasonable explanation since Miyauchi *et al.* (1977) reported almost no binding of T_4 by plasma proteins of bullfrog tadpoles.

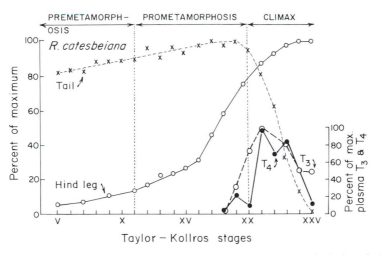

FIGURE 3. Changes in the ratios of tail length, and hindleg length to body length during development and metamorphosis (data from Krug, 1980) compared to changes in plasma T_3 and T_4 levels in tadpoles of *R. catesbeiana*. The data on plasma TH levels are a composite of the results presented in Fig. 2, panels B–D. All results are plotted as a percent of the maximum recorded value.

Perhaps most of the output of the thyroid, including T_4, is cleared rapidly enough to account for such variation in PBI and T_4 levels from climaxing tadpoles. T_3, which does not vary as much as T_4, might be retained longer due to the presence of a high-affinity plasma binding protein ($K_d = 1.4 \times 10^{-10}$ M; Miyauchi *et al.*, 1977). However, Regard *et al.* (1978) and Galton (1980) did observe binding of T_4 by plasma proteins.

The data of Regard *et al.* (1978) differ from those of Just (1972), Miyauchi *et al.* (1977), and Mondou and Kaltenbach (1979) in that less variation was found in hormone levels during climax. Regard *et al.* (1978) stated that ". . . all *R. catesbeiana* tadpoles at stages XXIII (corresponding to the middle of metamorphic climax) had higher levels of plasma T_3 and T_4 than did any animal at the onset of metamorphic climax." However, their values were mostly from pooled samples. Thus, the possibility remains that some of the pools may have included individual samples with low or nondetectable T_4 levels.

Of course, plasma levels of hormones may not reflect accurately their cellular concentrations. In this respect, T_3 is also probably retained better than T_4 by cells due to the presence of cytosolic proteins that bind T_3 250 times more avidly than T_4 (Yoshizato *et al.*, 1976; Kistler *et al.*, 1977). In fact, the role of cytosolic receptors for TH in tadpoles may be to ensure adequate retention of T_3 by the cells during metamorphosis. However, tissue concentrations of T_3 and T_4 have not been measured in tadpoles at different developmental stages.

Leloup and Buscaglia (1977) measured plasma PBI and TH (by RIA) in *X. laevis* tadpoles (Fig. 4). Plasma T_4 became detectable in late prometamorphosis, and the levels rose steadily through prometamorphosis to reach a peak in midclimax. Then the T_4 levels fell to moderate values

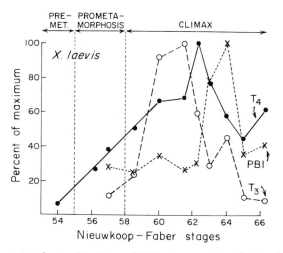

FIGURE 4. Changes in plasma T_3, T_4, and PBI concentration during development and metamorphosis of tadpoles of *X laevis*. Data from Leloup and Buscaglia, 1977.

by late climax. T_3 was not detectable until late prometamorphosis, but its plasma concentration rose rapidly during climax to reach a maximum before the T_4 peak. The T_3 then fell rapidly to low levels by the end of metamorphosis. Thus, although there are similarities between the RIA data on TH in *R. catesbeiana* and *X. laevis* during climax, there are substantial differences prior to climax. In *X. laevis*, blood T_4 levels do conform to expectations during prometamorphosis, and the T_3 and T_4 concentrations relate well to the morphological changes occurring during that period.

The pattern of changes in plasma PBI measured in *X. laevis* (Fig. 4) differs markedly from that found by Just (1972) in *R. pipiens* (Fig. 1). Unlike the latter species, *Xenopus* showed relatively constant low levels of PBI until midclimax. A surge in PBI occurred after Nieuwkoop and Faber stage 62, to reach a peak by stage 64. Then the PBI fell rapidly to reach presurge levels by late climax. It is interesting that in *Xenopus*, the surge in PBI came after the peaks in T_3 and T_4. Obviously, the nature of the PBI in this animal also warrants clarification.

These RIA data of Leloup and Buscaglia (1977) on plasma TH levels are consistent with ultrastructural and radiochemical data on thyroid development in *Xenopus*. Synthetic activity of the thyroid appears early in the larval period, and ultrastructural evidence indicates a general sequence of low activity in premetamorphosis, a steady increase during prometamorphosis up to climax, and then a rapid decrease in apparent activity after midclimax (see Dodd and Dodd, 1976). The thyroid epithelial cells at metamorphic climax are columnar, rich in rough endoplasmic reticulum and cytoplasmic granules, and have a highly vesiculated Golgi apparatus. Using an electron microprobe, Hourdry and Regard (1975) reported an increase in stable, protein-bound iodine in the colloid of *X. laevis* thyroids up to midclimax. Larvae reaching the end of climax had a relatively decreased iodine content. Similarly, Dodd and Dodd (1976) reported that thyroidal uptake of ^{131}I and the ratio of thyroidal $[^{131}I]$-T_4 to total body ^{131}I increased up to midclimax in *X. laevis*. These results on the development of thyroid function in *Xenopus* are essentially parallel to those recorded in ranids (see Dodd and Dodd, 1976).

2.3. Thyroid Function in Larval Urodeles

Thyroid activity in urodeles seems to conform generally with their developmental predisposition and the stage of the species. In obligate neotenes (e.g., *Necturus maculosus*), the uptake of ^{131}I by the thyroid was low, ranging from 1.3 to 5.6% (see Norris and Platt, 1973). In *Ambystoma tigrinum*, which is a facultative neotenic species, thyroidal ^{131}I-uptake values of approximately 4, 20, and 5% were obtained for neotenic, metamorphosing, and adult animals, respectively. These changes resemble the situation in anurans inasmuch as the thyroid is active during the time of transformation and it resumes a state of relative inactivity in the postmetamorphic animals. In *A. gracile*, both iodine uptake by the thyroid and circulating T_4 levels, as measured by RIA, were significantly greater in animals undergoing metamorphosis than in either larval or transformed animals (Eagleson and McKeown, 1978). Another similarity to anurans was reported in this study. Thyroid activity was increased in animals that were in the size range typical of transforming animals but which were not undergoing anatomical tranformation. This finding suggests the presence of a "prometamorphic period" in urodeles.

2.4. Metabolism of Thyroid Hormones

In mammals, the ratio of total plasma T_4/T_3 is about 50. This contrasts with the much lower ratio found in amphibians. For example, during early climax in *X. laevis* tadpoles, mean T_3 and T_4 levels are essentially equivalent at about 5 ng/ml (Leloup and Buscaglia, 1977; Fig. 4). Amphibians have the lesser T_4/T_3 value because their T_4 levels are much lower than those found in mammals, whereas plasma T_3 concentrations are about equivalent in the two vertebrate classes. Although the reasons for the low T_4 levels in amphibians have not been established, there are several possible causes.

It seems unlikely that the low plasma T_4 levels are due to a failure of the thyroid to synthesize and secrete T_4 in adequate amounts. Thyroidal content of T_4 exceeds that of T_3 in most vertebrates. Although the intrathyroidal concentrations of iodothyronines have not been determined in any of the anurans in which RIA measurements on blood TH levels have been made (Fig. 2), we have no reason to believe that their thyroidal T_4/T_3 ratio would be unusually low. In fact, preliminary RIA data from experiments with thyroids of bullfrog tadpoles in organ culture indicate that the glands from late-stage animals secrete large amounts of T_4 and very little T_3 (Krug and Nicoll, unpublished).

Peripheral 5'-monodeiodination of T_4 may be an important source of T_3 in anurans, as it is in mammals. In fact, Leloup and Buscaglia (1977) proposed that the temporal separation of the T_3 and T_4 peaks in *Xenopus* (Fig. 4) argues for peripheral conversion of T_4 to T_3. Miyauchi *et al.* (1977) also found a separation of plasma T_3 and T_4 peaks in bullfrog tadpoles (Fig. 2B). In *X. laevis*, a peroxidase type of deiodinating activity exists in the liver and tail during the larval period (Robinson and Galton, 1976). Total deiodinating activity of both tissues increased at metamorphic climax. Contrary to the expectations of Leloup and Buscaglia (1977), 5'-monodeiodination of T_4 to T_3 was not observed by Robinson and Galton (1976). However, T_4 was deiodinated more quickly than T_3, suggesting that low circulating T_4/T_3 ratios may result from T_4 being metabolically cleared more rapidly than T_3.

Evidence for deiodination per se in *R. catesbeiana* tadpoles is largely negative. With respect to the metabolism of $[^{125}I]$-T_4, Ashley and Frieden

(1972) state, "No deiodination products were detected in liver, bile and peripheral tissues from tadpoles . . ., or in juvenile or adult bullfrogs." However, the question of T_4 deiodination in bullfrog tadpoles remains unsettled. Yoshizato et al. (1976) observed a limited capacity of bullfrog tail fin cytosol to deiodinate $[^{125}I] \cdot T_4$. On the other hand, Galton (1980) reported a considerable amount of deiodination of $[^{125}I]$-T_4 by bullfrog tadpoles, as evidenced by more than 50% of the label being recovered as inorganic iodine in the serum 24 hr after injection. However, no $[^{125}I]$-T_3 was found after injection of $[^{125}I]$-T_4 in either serum or isolated liver nuclei. Thus, there is no compelling evidence for the peripheral conversion of T_4 to T_3 in larval anurans. Nevertheless, the RIA data showing chronologically separate T_4 and T_3 peaks indicate the need for further research on this problem.

2.5. Tissue Sensitivity to T_3 and T_4

Discussion of peripheral deiodination of T_4 to T_3 naturally leads to the consideration of which hormone is the biologically active form in larval amphibians. In mammals, T_4 can be considered as a prohormone of T_3, its peripheral deiodination to T_3 representing an activation step in the formation of the hormone. T_3 is the active form by virtue of the fact that one class of nuclear receptors exists that has an affinity for T_3 more than 10-fold greater than the affinity for T_4.

In bullfrog tadpoles, T_3 is more potent in inducing metamorphic changes such as tail regression and urea excretion (see Kistler et al., 1977). In X. laevis, the induction of regression of tail tips in vitro has a shorter lag period and more rapid response with T_3 than with T_4 (Robinson et al., 1977). Several attempts have been made to explain the difference in potencies between T_3 and T_4 in amphibians by differences in receptor binding, but conflicting observations have been made. Nuclear receptors were studied in vitro in tail fins and isolated liver cells (see Kistler et al., 1975) from bullfrog tadpoles. The results of both studies contrast with the findings on TH receptors in mammals. Competitive displacement of $[^{125}I]$-T_3 and $[^{125}I]$-T_4 by T_3 and T_4 indicate that T_3 binds to two sites, but that T_4 binds to only one. Scatchard analysis revealed more sites available for T_3 than for T_4, but each hormone bound to nuclear receptors with equal affinity. Thus, it seems that the greater biological activity of T_3 may be due, in part, to a higher number of T_3 sites rather than a higher affinity of receptors for T_3. In contrast, Galton (1980) reported that hepatic nuclei from bullfrog tadpoles showed more sites for T_4 than for T_3 in vivo, but that the affinity of these sites was greater for T_3 ($K_d = 2.4 \times 10^{-10}$ M) than for T_4 ($K_d = 4.1 \times 10^{-10}$ M). From these data, and from the inability to detect conversion of T_4 to T_3 in her study, Galton (1980) concluded that T_4 may be more important in larval amphibians than it is in mammals. Furthermore, the finding that T_3 affects adenylate cyclase in tadpole tails *in vitro* raises the possibility of extranuclearly mediated actions of TH (Stuart and Fischer, 1979). Whether such extranuclear actions discriminate between T_3 and T_4 remains to be studied.

Sensitivity to TH, like thyroid function, appears very early in development in most tissues. For example, *X. laevis* becomes responsive to either T_4 or T_3 by 40–60 hr postfertilization (see Dodd and Dodd, 1976). Tata (1970) observed the appearance of a temperature-sensitive binding of T_4 and T_3 to whole embryos concomitant with their becoming responsive to the hormones. The findings of Tata are supported by the study of Robinson *et al.* (1977) in which *X. laevis* tail tips became sensitive to thyroid hormones, as measured by regression *in vitro*, at 60 hr postfertilization. A gradual increase in sensitivity followed, as evidenced by more rapid rates of tail regression. Additionally, the total and specific activity of T_3-induced acid phosphatase at the time of 40–60% tail regression increased with developmental stages. These observations suggest that TH appear early in development and that receptor number and/or cellular competence to respond to T_3–receptor interaction increase during the larval period in *X. laevis*.

Kollros and associates have gathered data from several tissues that further support the idea that tissue sensitivity to TH increases gradually during development (e.g., Chou and Kollros, 1974). Etkin (1968) argued that often such studies were complicated by different lengths of tissue exposure to endogenous hormones. In addition, Dodd and Dodd (1976) emphasized the need for consideration of possible changes in the levels of endogenous growth hormone or prolactin (see below), which may influence apparent tissue sensitivity to TH.

A detailed study of nuclear binding sites showed that, at least in the tail of bullfrog tadpoles, the apparent increase in sensitivity is probably based on an increase in T_3 binding to nuclear receptors (Yoshizato and Frieden, 1975). Although the apparent affinity of the receptors decreased somewhat during late prometamorphosis, the number of binding sites doubled. Yoshizato and Frieden (1975) suggested that the increase in T_3 receptors may be a result of stimulation by slowly increasing levels of TH. Because changes in tissue sensitivity to TH during development do occur, and these changes may play an important role in allowing completion of metamorphosis, the regulation of nuclear T_3 and T_4 receptors by TH and other hormones deserves further investigation.

2.6. Pituitary Regulation of Thyroid Function

The pituitary control of the thyroid in larval amphibians is primarily stimulatory, and early studies indicate that the progression of development and metamorphosis is dependent on an intact pituitary. More recent studies, which have examined in detail the effects of hypophysectomy on thyroid activity and structure, have identified several aspects of thyroid development, hormone synthesis and release that are under pituitary control (see Dodd and Dodd, 1976). Replacement therapy in hypophysectomized larvae further confirmed the existence of a TSH, although when employed it consisted of either mammalian TSH or amphibian pituitary extract (Dodd and Dodd, 1976). Amphibian TSH probably resembles mammalian TSH, since injection of antiserum raised against bovine TSH into tadpoles blocked spontaneous and T_4-induced metamorphosis (Eddy and Lipner, 1976). Recently, a partially purified glycoprotein from bullfrog pituitaries was isolated and characterized as a TSH (MacKenzie et al., 1978).

Early development of the thyroid in larval amphibians appears to be partially independent of the pituitary. Hanaoka (1967a) found that after removing the preprimordium of the hypophysis, development was arrested in premetamorphosis. Although the thyroid appeared hypoplastic in these operated larvae, the thyroid epithelium had assumed the normal follicular arrangement. Subsequently, Hanaoka et al. (1973) were able to isolate T_4 and thyroglobulin from the thyroids of hypophysectomized B. bufo tadpoles. Kaye (1961) observed that thyroids from hypophysectomized tadpoles begin iodine transport at the same developmental stage (late embryonic) as controls. The operated animals fell behind the controls in relative ability to concentrate iodine. These studies indicate that: (1) the follicular arrangement, which occurs in late embryonic stages (see Dodd and Dodd, 1976), develops independently of pituitary stimulation; (2) the onset of thyroglobulin and thyroid synthesis, which occurs in late embryonic stages, is also independent of TSH; and (3) normal thyroid function becomes dependent on TSH by the first few larval stages.

The developmental pattern of thyrotropic activity during the larval period is not as well delineated as is that of the thyroid, as homologous RIAs for amphibian TSH are not yet available. It appears that the thyrotrophs are the first basophilic cells to develop in the pars distalis of X. laevis (Kerr, 1966). Morphometric analysis of thyrotropic cell granules showed that release of the presumably mature granules begins during mid-premetamorphosis and is greatest during early climax (see Dodd and Dodd, 1976).

Dodd and Dodd (1976) developed a sensitive bioassay for measuring pituitary content of TSH based on radioiodine uptake by X. laevis tadpoles that were hypophysectomized in early climax. They found a high TSH content in crude whole pituitary extracts during early climax, and the levels fell by midclimax. Pituitary TSH levels increased again and then decreased gradually as the animals completed transformation. The decline in pituitary TSH content in early climax correlates with the sharp rise in circulating T_3 observed by Leloup and Buscaglia (1977) in the same species (Fig. 4). Considering the data for the three hormones (T_4, T_3, and TSH), the initial fall in pituitary TSH content was likely due to a rapid surge in secretion of the hormone. This was then reflected by an increase in circulating TH (especially T_3) levels. While this bioassay provides the first report of TSH content in tadpole pituitaries, it may be confounded by a thyroid response to other pituitary hormones (e.g., prolactin). Still, despite limited data, the activity of the thyrotropic cells generally conforms to the changes in thyroid gland function during metamorphosis.

2.7. Control of Thyrotropic Function by Thyroid Hormones

In mammals, there is evidence for a direct negative feedback of TH on TSH synthesis and/or secretion by the pituitary. This mode of control is also apparent in larval anurans. Kaye (1961) reported a decline in iodine uptake by the thyroids of R. pipiens tadpoles after exposure to T_3 or T_4. She also noted an increased basophilia in the pars distalis in response to T_3 treatment. This feedback sensitivity of the pituitary to TH developed by mid-premetamorphosis. In X. laevis, thyroidectomy decreased pituitary TSH concentration, which, based on the morphological reactions of remnant follicles, was probably the result of increased TSH secretion (Dodd and Dodd, 1976).

Both the pituitary and the hypothalamus have been implicated as the site of negative feedback by TH. By rendering tadpoles hypothyroid with the goitrogen PTU, Hanaoka (1967b) observed that the thyroid enlarged four- to sixfold after 80 days and the pars distalis twofold. These effects of PTU on pituitary and thyroid mass were significantly diminished by hypothalamic extirpation. Accordingly, Hanaoka (1967b) suggested the hypothalamus as the main site of negative feedback by TH. A small increase in thyroid follicular cell size in hypothalectomized tadpoles treated with PTU over that in hypothalectomized-only tadpoles suggested some direct feedback on the pituitary. However, this study is complicated by the facts that removal of hypothalamic stimulation would diminish the ability of the adenohypophysis to respond to TH deficiency and that the gland probably becomes sensitized to the inhibitory effects of TH. The possibility exists, therefore, that low levels of TH in hypothalectomized and PTU-treated animals effected some direct inhibition of thyrotropic function. Similar experiments were performed in X. laevis (see Goos, 1978). Again, a direct feedback on the pituitary was indicated by the fact that PTU-treated animals displayed some thyroid stimulation and hyperactive thyrotrophs in the absence of the hypothalamus.

2.8. Control of Thyrotropic Function by the Brain

Studies such as the hypothalectomy experiments of Hanaoka (1967a,b) have established that the amphibian pituitary and, thus, amphibian development, are controlled by the central nervous system. The structures required for neural control of pituitary thyrotrophs appear during larval development. These consist of the following: (1) hypothalamic peptidergic neurosecretory neurons that reside in the paired nucleus preopticus

(NPO) and which send axons toward the median eminence, and (2) a neurohemal organ, the median eminence, from which hypothalamic releasing factors enter the hypophysial portal systems and are conveyed to the adenohypophysis. The neurosecretory neurons, which were identified in *X. laevis* tadpoles by the pseudoisocyanin staining technique (see Goos, 1978), appear during early larval stages at the same time that the thyroid epithelium assumes a follicular arrangement. The number and nuclear size of these neurons increase throughout development up to climax. In fact, the development of the magnocellular neurons of the dorsal NPO closely parallels that of pituitary thyrotrophs and thyroid epithelium (see Goos, 1978).

The median eminence in *R. pipiens* begins to develop in early prometamorphosis with a thickening of the infundibular stalk and internalization of the capillary bed located on its ventral surface. During late prometamorphosis, the stalk becomes further thickened and the insinking capillaries develop into the primary portal plexus, which becomes contiguous with the pars distalis via portal veins (see Etkin, 1968).

Evidence for the dependency of normal metamorphosis on neuroendocrine stimulation was first obtained by Etkin (see Etkin, 1968), who observed that hypothalectomized tadpoles bearing an ectopic pituitary transplant could, at best, undergo a slow, protracted development. This work was repeated (Etkin and Lehrer, 1960) after the advances in neuroendocrine physiology were made, and the lack of metamorphic transformation in tadpoles bearing an autotransplanted pituitary was explained as a dependency of the thyrotrophs on an intact median eminence–hypophysial vascular connection. Several laboratories have confirmed that the activation of thyrotropic activity requires a stimulus from the hypothalamus (see Goos, 1978).

The chemical nature of the hypothalamic thryotropin releasing hormone (TRH) in amphibians is presently unknown. In mammals, TRH is a tripeptide, (pyro) Glu–His–Pro–NH_2. Large amounts of mammalian TRH have been found in the hypothalami of *A. mexicanum*, *A. tigrinum*, and *R. pipiens* (Taurog *et al.*, 1974) and in the blood and peripheral tissues of *R. pipiens* (Jackson, 1978). Several studies indicate mammalian TRH has no TSH releasing activity in amphibians and other poikilotherms (see Taurog *et al.*, 1974; Clemons *et al.*, 1979). In addition, Sawin *et al.*, (1978) reported no changes in TRH levels in the blood and several other tissues during T_3-induced metamorphosis in *A. mexicanum*. Accordingly, amphibian TRH must be different from mammalian TRH.

In addition to the probable peptidergic stimulation from the NPO–median eminence pathway, the pars distalis may be under direct monoaminergic control. Employing the Falck–Hillarp fluorescence technique, Aronsson (1976) described the development of aminergic fibers in the pars distalis of *R. temporaria* during spontaneous metamorphosis. These fibers, which are less dense than those in the pars intermedia, arise early in premetamorphosis and disappear at climax. Subsequent work

(Aronsson, 1978) showed that these fibers always persisted until climax and then regressed even in tadpoles in which the duration of the larval period was experimentally prolonged or abbreviated. Of course, the observations of Aronsson do not distinguish between correlative and causal relationship of the presence of aminergic fibers and the onset of metamorphic climax, nor do they elucidate which pituitary cell type, if any, is under the influence of these fibers. Yet their regression at climax, which is an event primarily controlled by the thyroid axis, raises the possibility of monoamines directly controlling TSH secretion in some fashion. Monoamines have been implicated further in direct or indirect control of adenohypophysial function by the examination of median eminence ultrastructure (see Dodd and Dodd, 1976), which showed marked degranulation of both peptidergic and monoaminergic fibers at metamorphic climax in *R. temporaria*.

2.9. Effects of Thyroid Hormones on Hypothalamo–Hypophysial Development and Function

Most investigators would agree that activation of the hypothalamo–hypophysial–thyroid axis is required for metamorphic climax. Etkin (1968) proposed that TH exert a "positive feedback" on the hypothalamus and median eminence through their maturational effects. As mentioned above, the maturation of the median eminence occurs relatively late during the larval period. Etkin (1968) observed that tadpoles whose metamorphic progress was blocked by thyroidectomy showed little development of a median eminence despite their large size. Also, the appearance of a functional median eminence could not be induced by TH as precociously as could other developmental changes. Thus, the development of the median eminence is thyroid dependent, but its acquisition of sensitivity to TH is delayed. From these findings, Etkin (1968) reasoned that during premetamorphosis the pituitary receives essentially no stimulation from the hypothalamus and consequently is very sensitive to negative feedback by thyroid hormones. As the median eminence matured, more hypothalamic stimulation (i.e., thyrotropin releasing factor) would reach the pituitary and the activity of the pituitary–thyroid axis would increase. Since increasing levels of TH would induce more median eminence maturation, the system would be self-accelerating (hence, the term "positive feedback"). This procession of events would continue up to the surge of TH secretion just prior to climax. Once metamorphosis neared completion, the negative feedback of TH would predominate, driving the system back down to the low level that is characteristic of adult amphibians.

A maturational action of TH on brain structure was also reported by Goos and associates (see Goos, 1978) in *X. laevis*. In these experiments, exposure of the animals to PTU inhibited the development of pseudois-ocyanin-positive cells in the ventral NPO, which normally appear prior to climax. Furthermore, the development of the external zone of the median eminence was blocked. However, these experiments also indicated that the TH inhibit the activity of the hypothalamus once it matures. Prometamorphic tadpoles that had well-developed neurosecretory neurons in the dorsal NPO displayed extensive degranulation of these neurons in reponse to PTU (see Goos, 1978). Cessation of PTU treatment or treatment with T_4 produced reaccumulation of pseudoisocyanin-positive material in the dorsal NPO neurons. These cellular responses to manipulation of the thyroid state were similar to those observed in presumed thyrotrophs and lend strong support to the existence of a negative feedback by TH at the hypothalamic and hypophysial level. Thus, it could be argued that the morphological evidence accumulated by Goos and associates (see Goos, 1978) does not support Etkin's (1968) hypothesis of a positive feedback by TH on the hypothalamus—median eminence complex during prometamorphosis.

Other evidence indicates that TH exert a negative feedback effect on thyroid function in anurans essentially throughout development. Kaye (1961) found that T_3 or T_4 inhibited ^{131}I uptake by the thyroids of bullfrog tadpoles in pre- and prometamorphosis and in early climax. Dodd and Dodd (1976) obtained similar results with *X. laevis* in prometamorphosis and in climax. They also found that thyroidectomy of *Xenopus* tadpoles in prometamorphosis or climax caused a depletion of pituitary TSH content. However, we would caution that the existence of negative feedback does not preclude the coexistence of positive feedback, as discussed below.

Two recent reports support the suggestion that TH can exert a positive effect on TSH and thyroid secretion in amphibians. Norris and Gern (1976) found that metamorphosis of *A. tigrinum* could be induced by intracisternal injection of T_3 at a dose that was ineffective when given peripherally. These results indicate that T_3, by a direct action on the hypothalamus, can induce maturational and/or functional changes in the hypothalamus that activate the pituitary and the thyroid. Eddy and Lipner (1976) studied the effects of an antiserum to bovine TSH on T_4-stimulated hindleg growth in bullfrog tadpoles. A single injection of T_4 alone stimulated a considerable degree of hindleg growth in the animals. However, injection of the antiserum to TSH along with the T_4 reduced the growth response. Thus, much of the leg growth response to the exogenous T_4 was mediated by endogenous TH. Presumably, the T_4 stimulated TSH secretion by an action on the hypothalamus and/or the adenohypophysis.

The fact that positive and negative long-loop feedback mechanisms between the hypothalamo–hypophysial complex and the peripheral target glands are not mutually exclusive has been well documented by studies on the control of ovarian function in mammals. For example, ovariectomy in the rhesus monkey produces a 10-fold increase in circulating luteinizing hormone (LH) concentration. Replacement of estradiol to these animals quickly reduces LH levels, demonstrating negative feedback. Similarly, maintenance of supranormal circulating estrogen levels during the early follicular phase of the menstrual cycle reduces LH levels. If, however, the increase in estrogen is of sufficient duration and magnitude, the drop in LH is followed by a surge of LH secretion. Furthermore, the LH surge appears to be induced even under conditions in which stimulation of the pituitary by LH-RH is experimentally held constant (Knobil *et al.*, 1980). In addition, Turgeon (1979) has shown that the positive and negative feedback actions of estrogen on LH secretion can occur simultaneously in the rat. Hence, opposing kinds of feedback can coexist and are not mutually exclusive, but one type can predominate over the other at different times. It seems possible, therefore, that TH could be responsible for an increase in TSH secretion at climax even though these hormones can exert negative feedback effects starting very early in development, and continuing through climax.

3. PROLACTIN

Two more recently discovered participants in the endocrine regulation of amphibian metamorphosis are the pituitary hormones, prolactin (PRL) and growth hormone (GH). These hormones appear to direct growth and development both by actions on peripheral organs and by modulating the activity of the thyroid gland.

Vertebrate PRL and GH are secreted by similar (acidophilic) cell types of the pars distalis, have similar molecular structures, and have overlapping activities (see Nicoll, 1974; Clarke and Bern, 1980). Ample evidence has shown that GH and PRL are separate and distinct molecular species in the adenohypophysis of adults of several amphibians (see Nicoll, 1974). In addition, RIA and biochemical evidence also indicates the existence of PRL and GH as distinct entities in the pituitary and blood of *R. catesbeiana* tadpoles (Clemons and Nicoll, 1977b). Furthermore, Guyetant *et al.* (1977) found two acidophilic cell types in the pituitary of *Alytes obstetricans* tadpoles. In contrast to these findings, *X. laevis* tadpoles appear to have only one functionally differentiated pituitary acidophil (Kerr, 1966; Pehleman and Bade, 1976). During metamorphic climax, cells in the pituitary of *Xenopus* undergo considerable degranulation and reorganization and subsequently a second type of acidophil appears. Thus, while *X. laevis* tadpoles may elaborate both GH and PRL, these histological findings raise the possibility that the pituitary may secrete distinct larval and adult hormones during the life history of an animal.

3.1. PRL, GH, and Growth

Prolactin was first mentioned in connection with the control of larval anuran growth by Etkin and Lehrer (1960). These investigators observed that growth rate and development in R. pipiens tadpoles were slowed significantly by hypophysectomy. If the pituitary was autotransplanted to the tail musculature, normal or excessive growth rates were maintained. The growth rate of autotransplant recipients surpassed that of the controls particularly in later stages, when controls displayed decreasing growth rates presumably due to increasing thyroid activity. Growth in autotransplant-bearing tadpoles persisted to the extent that abnormally large tadpoles were produced. Etkin and Lehrer (1960) concluded that the secretion of a growth factor by the in situ pituitary must be under inhibitory hypothalamic control, especially at later stages of development. Accordingly, an excess of a growth-promoting factor must have been secreted by the ectopic pituitaries. Since PRL is under inhibitory hypothalamic control in mammals and it possesses some intrinsic somatotropic potency (see Nicoll, 1974), the hormone is a reasonable candidate for the growth-promoting factor in R. pipiens tadpoles.

It should be emphasized that interpretation of Etkin and Lehrer's results is complicated by the likely deficiency of TH in the operated animals. The complexity of this problem is underlined by the apparent biphasic growth response to thyroidectomy in their experiments. Growth rates were depressed initially by thyroidectomy, but exceeded contols at later stages. It should also be stressed that not all species show an increased growth rate after autotransplantation of the pituitary (Dodd and Dodd, 1976).

Prompted by the findings of Etkin and Lehrer (1960), several laboratories examined the effects of PRL and GH from mammalian (usually sheep or cow) species on the growth rate of larval amphibians (see Dodd and Dodd, 1976). Consistent with the effects of autotransplantation, the early studies on R. pipiens and R. catesbeiana showed that PRL was the more somatotropic hormone and supported the possibility that PRL was the larval growth hormone. For example, doses of PRL as low as 1.0 μg/animal per day for 2 weeks significantly increased both body weight and length in R. pipiens tadpoles. However, once metamorphosis is complete, mammalian PRL becomes much less potent than GH in promoting growth (see Dodd and Dodd, 1976).

The tail of anuran and urodele larvae is the most completely studied target organ of PRL. The hormone produces an increase in both length and height of the tail at low doses. Yoshizato and co-workers have emphasized the increase in the connective tissue (see Yoshizato et al., 1972) component of the tail fin as the predominant result of PRL administration. Their studies indicate that PRL increases the synthesis and decreases the catabolism of the major components of extracellular matrix, collagen and hyaluronic acid. Similarly, Yamaguchi and Yasumasu (1978) reported that PRL decreased collagen breakdown in tadpole fin, as measured by ^{14}C-labeled hydroxyproline release. More recently, Kikuyama et al. (1980) found that PRL isolated from bullfrog pituitaries had similar effects as ungulate PRL on tadpole tail collagen synthesis and that it also inhibited the effects of TH on tail shrinkage in vitro.

Eddy (1979) reported an increase in total tail water and sodium content and a decrease in total tail potassium in bullfrog tadpoles in response to PRL treatment with no change in sodium or water concentration in tail muscle or plasma. From these observations, she concluded that PRL caused an expansion of extracellular space in the tail. Similar increases in tail sodium and water content have been observed in *A. tigrinum* (see Platt *et al.*, 1978) after PRL injection. Thus, the increased synthetic activity of tail fibroblasts may occur in response to or be facilitated by a PRL-induced expansion of extracellular space. This raises the question of whether PRL stimulates tail growth by a direct action on the tail fibroblasts or by an indirect inductive effect through fluid and electrolyte shifts, or both. Although receptors for PRL have been found in the tail (White and Nicoll, 1979) and the hormone inhibits TH-induced tail resorption *in vitro* (see Dodd and Dodd, 1976), no study has demonstrated convincingly a direct *in vitro* growth response of the tail to PRL. A study by Rothstein *et al.* (1980) also raises the possibility that PRL and GH may effect growth in some tissues in the bullfrog by stimulating the production of somatomedins by the liver.

Another exclusively larval structure whose growth is stimulated by PRL is the gill. Prolactin increased gill length in larval *A. tigrinum* (Platt and Christopher, 1977). Prolactin may favor anuran gill growth as well, but no studies have examined this possibility. Branchial binding of ^{125}I-labeled PRL has been reported in tadpoles of *R. catesbeiana* (White and Nicoll, 1979).

Evidence for an influence of PRL on the growth of structures that are not exclusively larval is contradictory, but generally favors some role of the hormone. Several short-term studies showed that hindlimb growth is not affected by PRL. However, long-term treatment with the hormone decreased hindlimb growth (Wright *et al.*, 1979). Consistent with this inhibitory effect are the findings of Eddy and Lipner (1975) and Clemons and Nicoll (1977a) that injections of antiserum to ovine or bullfrog PRL caused increased leg growth in *R. catesbeiana* tadpoles.

An anabolic action of PRL on the tadpole hindlimb has also been reported by Yamaguchi and Yasumasu (1978). They claim that injections of the hormone enhanced labeled amino acid incorporation into isolated thigh bones from bullfrog tadpoles *in vitro*. Prolactin also increased incorporation of [^{14}C]proline into the thigh *in vivo* while depressing collagen catabolism, as evidenced by a reduction in [^{14}C]hydroxyproline release. The effects of PRL were observed in premetamorphic animals and were abolished by T_4, which increased the synthesis and breakdown of collagen. Thus, although PRL may stimulate fibroblast and/or osteoblast activity in the hindlimb while inhibiting osteoclast activity, the hormone does not promote real bone growth.

PRL produces none of the metabolic effects in tadpoles that are characteristic of GH effects in mammals. Thus, while PRL may be a potent anabolic and somatotropic agent in specific tissues, it may do little to drive general metabolism in a direction most suitable for overall growth. Snyder and Frye (see Frye *et al.*, 1972) suggested that a selective advantage is provided by the employment by larval anurans of a hormone that promotes growth in certain tissues, but which does not mobilize important energy stores (e.g., fat body) for the promotion of body growth, as growth must be followed by a considerable energy expenditure required for structural transformation. However, the extent of the regulation by PRL of systemic metabolism in larval amphibians deserves further investigation. PRL decreases both ammonia and urea excretion (see Dodd and Dodd, 1976), but the actual mechanism for this effect is unknown. PRL may favor growth in larval anurans by stimulation of intestinal absorption of amino acids, glucose, etc., as it does in mammals (see Clarke and Bern, 1980).

The studies on *R. catesbeiana* and *R. pipiens* have prompted some investigators to make the generalization that PRL may be the larval growth hormone in other vertebrates and to speculate on the intriguing possibility that PRL may be a growth-promoting factor during the fetal and neonatal periods in mammals. However, the available evidence does not support the role of PRL as the exclusive stimulator of growth in all larval amphibians. Enemar *et al.* (1968) found that ovine GH was consistently more effective than ovine PRL in *R. temporaria* tadpoles, and some investigators have found that both hormones are growth promoting in the larvae of other anuran species (see Dodd and Dodd, 1976). Similarly, GH, but not PRL, stimulated growth in larval *T. torosa* (Cohen *et al.*, 1972) and acted synergistically with TH to promote hindleg growth in bullfrog tadpoles (Bern *et al.*, 1967).

3.2. Antimetamorphic Actions of PRL

In addition to promoting growth in several species of adult and larval amphibians, PRL antagonizes the actions of TH in certain larval organs. This was first observed by Bern *et al.* (1967) with T_4 induced tail regression and by Etkin and Gona (1967) in spontaneously regressing tails in *R. catesbeiana* tadpoles. Since these studies, PRL has been shown to specifically antagonize TH-induced increases in the activity of tail β-glucuronidase (Jaffe and Geschwind, 1974), acid hydrolase (Blatt *et al.*, 1969), and collagenase (Yamaguchi and Yasumasu, 1978) *in vivo*. PRL similarly reduces tail regression and the elevation of hydrolytic enzymes *in vitro* (Derby, 1974), and it antagonizes thyroid promotion of hindleg growth (Wright *et al.*, 1979), gill regression (Platt, 1976), water and sodium loss in certain larval tissues (Jaffe and Geschwind, 1974; Platt and Christopher, 1977; Platt *et al.*, 1978), and nitrogen excretion (Medda and Frieden, 1972).

Since metamorphosis proceeds at the expense of growth, one might expect the antimetamorphic and growth-promoting actions of hormones to be intimately related. However, there is evidence for a separation of these actions. For example, in larval *T. torosa*, ovine GH is somatotropic but not antimetamorphic (Cohen *et al.*, 1972). On the other hand, ovine PRL clearly antagonized T_4-induced tail regression, but displayed very low somatotropic potency.

Although the evidence for a somatotropic role of PRL among different species is not consistent, there exists indirect but compelling evidence that an endogenous prolactin-like hormone is responsible for suppressing metamorphic transformation in three species of anurans and in one urodele. This evidence has been obtained by methods that reduce the levels of biologically active PRL in the circulation. Eddy and Lipner (1975) first showed that injection of an antiserum to ovine PRL accelerated T_4-induced metamorphosis in premetamorphic tadpoles. These findings were subsequently substantiated by Clemons and Nicoll (1977a), who used an antiserum raised against purified bullfrog PRL. These authors also reported that an antiserum to bullfrog GH had no effect on metamorphosis, again indicating that the antimetamorphic actions may be more specific to PRL than are growth-promoting effects. More recently, Kikuyama *et al.* (1980) reported that bullfrog PRL retarded TH-induced regression of toad tail *in vitro*.

The ergot alkaloids, such as ergotamine and bromoergocryptine (CB-154), are potent and relatively specific inhibitors of pituitary PRL secretion in mammals. Platt (1976) first observed that ergocornine treatment accelerated both spontaneous and T_4-induced transformation of neotenic *A. tigrinum*. This treatment did not affect thyroidal ^{131}I uptake. Thus, the effects of the drug were presumably mediated by suppression of PRL secretion, rather than through stimulation of thyroid activity. Seki and Kikuyama (1979) also reported that ergot alkaloids accelerated metamorphosis in *B. bufo japonicus*. Utilizing a hormologous RIA for bullfrog PRL, we have recently found that CB-154 lowered circulating PRL levels and accelerated metamorphosis in bullfrog tadpoles (White, Clemons, and Nicoll, unpublished).

It is now well established, particularly from *in vitro* studies, that PRL antagonizes TH actions at the level of the target organ. It has been suggested that the hormone also exerts a direct inhibitory action on the thyroid gland. Gona observed that PRL inhibited TSH-induced metamorphosis but not that induced by T_4 (see Gona, 1968). However, several factors are problematic in this study. High doses of both T_4 and PRL (2–4 mg) were used by Gona. Other investigators have observed an antagonism between these hormones when used at lower doses. Also, metamorphic changes, such as tail regression, were used as parameters of thyroid sensitivity to TSH, rather than measuring thyroid function directly. Subsequently, Gona (1968) reported that thyroidal ^{131}I uptake and serum $PB^{131}I$ were depressed by high doses of PRL. This effect was observed in prometamorphic tadpoles but not in climaxing ones. Similarly, Kracht and Weber (1978) reported that PRL reduced the efficacy of TSH in restoring thyroid epithelial activity in hypophysectomized adult *Noto-*

phthalmus viridescens. However, PRL administration alone prevented total epithelial collapse after hypophysectomy. Although some evidence indicated that PRL does not affect thyroid function in *A. tigrinum,* as measured by thyroidal iodine uptake (Norris and Platt, 1973), Norris (1978) subsequently reported that PRL injections reduced circulating T_4 levels in this species.

The ability of PRL to antagonize TH-induced changes is limited. The hormone does not inhibit several responses to TH, including pancreatic resorption (Blatt *et al.,* 1969), increased urea-cycle enzymes (Blatt *et al.,* 1969; Jaffe and Geschwind, 1974), hepatic dehydration (Jaffe and Geschwind, 1974; Platt and Christopher, 1977), or reduction in retinal 3,4-retinol dehydrogenase activity (Crim, 1975) in certain species. This organ specificity of PRL in its antagonism of TH may be an important factor in determining the correct sequence and rate of tissue regression or differentiation during metamorphosis.

3.3. PRL–Thyroid Interactions in "Second Metamorphosis"

Certain salamandrids such as *N. viridescens* return to water for breeding (water drive) and, in doing so, they redevelop characteristics (e.g., tail fin) that are appropriate for an aquatic existence. This transformation has been termed "second metamorphosis" and, as in larval-to-adult metamorphosis, is under the control of both TH and PRL (Dent, 1975; Dodd and Dodd, 1976). The behavioral component of second metamorphosis, called "water drive," was the first aspect to be discovered as being under the control of PRL. It is now well established that PRL and placental lactogens induce water drive behavior (see Dodd and Dodd, 1976). Initially, it appeared that the same PRL–TH antagonism operates in the control of second metamorphosis as exists in larval-to-adult metamorphosis inasmuch as treating water-adapted newts with T_4 induced them to seek dry land. However, Gona *et al.* (1973) reported that PRL failed to stimulate water drive in hypophysectomized–thyroidectomized red efts. When low doses of T_4 (10 pg per injection) were added on alternate days to PRL treatment, 100% of the operated animals exhibited water drive. In a subsequent study, Gona and Gona (1976) reported that combining PTU with PRL reduced the efficacy of PRL in promoting water drive behavior.

Integumental changes accompany the water drive of metamorphosis. The skin changes from a rough, dry texture in land-adapted efts to a smooth, slimy one in the water-adapted eft. As in the control of water drive behavior, PRL appears to be the primary endocrine signal for the integumental changes associated with second metamorphosis. These cutaneous changes are also dependent on an intact thyroid gland (Dent *et al.,* 1973).

Thus, in contrast to the clear PRL–TH antagonism that exists in larval-to-adult metamorphosis, there is consistent evidence for a

PRL–thyroid synergism in the control of second metamorphosis, at least at low levels of TH. The possibility of such a synergism between PRL and low levels of TH in anuran growth during premetamorphosis was alluded to earlier (Section 3.1). This dependency of PRL on some level of TH may be based, in part, on the fact that TH promote PRL receptor binding in bullfrog tadpole tissues (White and Nicoll, 1979; White et al., 1980). It is interesting to note that Singh and Bern (1969) found that high levels of T_4 reduced the efficacy of PRL in promoting lobulo-alveolar development of mouse mammary glands in vitro, whereas lower T_4 levels synergized with PRL. Vonderhaar and Greco (1979) similarly found that T_4 was a prerequisite for normal ductal and alveolar development of the mouse mammary gland. As in larval bullfrogs, TH promoted PRL receptor binding in mouse mammary epithelium in vitro (Bhattacharya and Vonderhaar, 1979).

3.4. Control of PRL Secretion and PRL Levels during Metamorphosis

Several studies involving ectopic transplantation of the pituitary gland have provided convincing evidence that PRL secretion in some amphibian species is primarily under inhibitory control (e.g., Etkin and Lehrer, 1960; Dent et al., 1973). A hypersecretion of PRL from pituitary autotransplants would, of course, be consistent with the fact that PRL is under inhibitory control in mammals and fish (see Clarke and Bern, 1980). However, in contrast to Etkin and Lehrer's findings in R. pipiens tadpoles, autotransplantation of pituitaries did not result in excessive growth rates in X. laevis (see Dodd and Dodd, 1976) or R. temporaria (Enemar, 1978). Enemar (1978) states that these data ". . . clearly show that the two congeneric species (R. pipiens and R. temporaria) differ considerably in the growth response to autotransplantation of the adenohypophysial primordium. . . . Obviously the properties of R. pipiens tadpoles in this respect should not yet be considered valid for anuran or amphibian tadpoles in general, as has sometimes happened in the last decade." As mentioned above, the growth-promoting actions of PRL may be less universal in larval amphibians than its antimetamorphic actions. Thus, PRL may be under inhibitory control by the hypothalamus in R. temporaria and X. laevis, but may have weak somatotropic potencies in these species.

Dopamine is the main inhibitory hypothalamic regulator of PRL secretion in several species of vertebrates. Dopamine agonists appear to lower circulating levels of PRL in larval amphibians, as indicated by an acceleration of metamorphosis (Platt, 1976; Seki and Kikuyama, 1979) and by measurement of serum PRL levels by homologous RIA in bullfrog tadpoles (White, Clemons, and Nicoll, unpublished data). Zuber-Vogeli (1978) has also obtained ultrastructural evidence for an inhibitory effect of a dopamine agonist on PRL secretion in the anuran, Nectophrynoides *occidentalis.* More recently, Kikuyama and Seki (1980) presented evidence from *in vivo* and *in vitro* studies that dopamine and ergot drugs could inhibit PRL secretion in bullfrogs. These pharmacological studies are interesting, but should be considered with reservations until more physiological measurements, such as pituitary or hypothalamic dopamine concentrations, have been made.

Combining the facts that PRL is antimetamorphic and possibly under inhibitory hypothalamic control, some investigators have suggested that the levels of the hormone would be high during early larval stages, but decrease with the development of hypothalamic control over the pituitary (Bern *et al.*, 1967; Etkin, 1968). Low levels during late prometamorphosis and climax would allow TH to exert their full influence over metamorphic changes. This model for the profile of PRL blood levels was supported by the observations of Derby (1970). In his experiments, pituitaries taken from tadpoles at different developmental stages delayed tail resorption *in vitro* with different potencies, such that glands from prometamorphic tadpoles produced the greatest inhibition. Similarly, transplantation of bullfrog tadpole pituitaries into hypophysectomized–thyroidectomized recipients showed that glands from late-premetamorphic donors produced the greatest growth in the recipients. Pituitaries from prometamorphic and climaxing tadpoles produced progressively less growth (Hsü *et al.*, 1976). Although these studies suggest that the PRL content of the pituitary is greatest during late premetamorphosis and prometamorphosis, they do not address the question of hormone secretion by an *in situ* pituitary.

The need for further examination of PRL secretion and plasma level profiles is emphasized by the preliminary report of Clemons and Nicoll (1977b), in which PRL levels were measured by homologous RIA in bullfrog tadpoles. Pituitary and serum PRL levels were high at early climax and increased rather than decreased during late climax. These findings have been confirmed in a subsequent study (Clemons, Krug, White, and Nicoll, unpublished). In light of these results, it is interesting to recall the observations of Aronsson (Section 2.8). Perhaps the aminergic fibers that were found in the pars distalis are regulating (tonically inhibiting?) PRL release up to climax. The surge of PRL secretion in late climax (Clemons and Nicoll, 1977b) may thus be due to release from inhibitory influence of these fibers, which regress at the onset of metamorphic climax.

4. ADRENOCORTICAL HORMONES

Previous studies indicated that the adrenocortical hormones may play some role in the regulation of anuran metamorphosis (see Dodd and Dodd, 1976). Injections of either mineral or glucocorticoids accelerated tail regression and caused other metamorphic changes in tadpoles. Recently, three of the principal interrenal steroids have been tentatively identified and quantified by RIA in the serum of bullfrog tadpoles (Krug et al., 1978; Fig. 5). Aldosterone (A) levels, nondetectable in early development, became measurable by TK stages VII–VIII. The serum levels of aldosterone remained low through prometamorphosis and into climax, but began to rise in late climax to reach the slightly elevated levels found in froglets and adults. Thus, the pattern of serum aldosterone concentration does not relate to any developmental or metamorphic changes in the tadpoles.

Corticosterone (B) was very low during the latter half of premetamorphosis; subsequently the serum concentration of the steroid rose rapidly during early prometamorphosis to reach relatively high levels by stage XVII. The levels of corticosterone then fell in late prometamorphosis and remained at an intermediate level through midclimax, after which the plasma concentration rose again to become appreciably elevated in froglets and adults.

Of considerable interest was the finding that cortisol (F), a steroid generally considered not to be present in amphibians in significant amounts, was found in appreciable concentrations in the bullfrog larval stages. Cortisol was detectable at the earliest stage tested (V), and its concentration rose gradually from mid-prometamorphosis. After stage XIV, serum cortisol levels increased more rapidly to reach a peak in midclimax (stage XXII), after which the blood levels fell rapidly to reach low concentrations in froglets and adults.

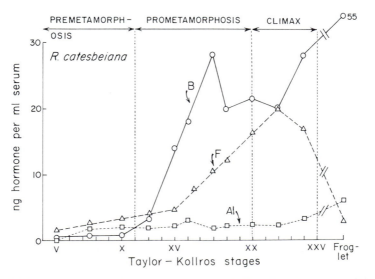

FIGURE 5. Changes in serum aldosterone (A), cortisol (F), and corticosterone (B) concentrations during development and metamorphosis of bullfrog tadpoles. The steroids were tentatively identified and quantified by radioimmunoassay only. Data from Krug et al., 1978, and Krug, 1980.

The patterns of changes in the serum glucocorticoid levels suggest that adrenal tissue is activated to function in a manner that differs from that of the thyroid. Corticosterone secretion shows a bimodal profile with a prolonged surge beginning in early prometamorphosis, and a second one occurring in late climax that extends into the adult. The pattern of corticosterone serum levels does not clearly relate to any external morphological changes that occr during either prometamorphosis or climax. However, the rise in corticosterone levels during prometamorphosis is associated with TH-induced gut regression (Carver and Frieden, 1977; Krug, 1980).

The profile of serum cortisol levels coincides very well with the pattern of hindleg growth found in bullfrog tadpoles during pre- and prometamorphosis (Fig. 3). In addition, the fact that the peak levels of cortisol are associated with the beginning of rapid tail resorption is of interest (Fig. 3). In view of the previous reports that adrenal steroids can act synergistically with TH to promote tail regression (see Dodd and Dodd, 1976), it seems possible that endogenous cortisol, and possibly corticosterone, could facilitate TH-induced metamorphosis.

5. OSMOREGULATION AND METAMORPHOSIS

Recently, an association has been made between the osmoregulatory actions of PRL and TH, and their functions in controlling development and metamorphosis in amphibians. In *A. tigrinum*, TH caused a loss of water and sodium from larval tissues, whereas PRL promoted their retention (Platt and Christopher, 1977; Platt *et al.*, 1978). Likewise, Eddy (1979) found that PRL injections increase the extracellular space in the tail fins of *R. catesbeiana* tadpoles (see Section 3.1). The fluid-retaining (i.e., edematous) effects of PRL in larval amphibians have been noted by other investigators (Hsü *et al.*, 1976).

Other results suggesting that water and electrolyte balance are related to the regulation of development and metamorphosis were obtained by White *et al.* (1978). They found that keeping bullfrog tadpoles in tapwater containing 85 mM $NaCl_2$ or 10–15 mM $CaCl_2$ enhanced T_4-induced tail regression. Furthermore, Platt and LiCause (1980) reported that high doses of oxytocin accelerated tail fin regression and body weight loss in larval *A. tigrinum*. They suggested that these effects of oxytocin may be partly attributable to the diuretic actions of oxytocin.

The available evidence indicates that fluid loss from larval tissue, which occurs in response to TH, may mediate some of the metamorphic actions of these hormones. In contrast, PRL promotes fluid retention, and this effect may be involved in the antimetamorphic (antithyroidal) actions of the hormone. Thus, the increased blood levels of PRL seen during climax in bullfrog tadpoles (Clemons and Nicoll, 1977b) may serve to restrain the dehydrating and metamorphosis-inducing actions of TH, which are also elevated during climax. Indeed, the fact that T_4 induces renal PRL receptors (White and Nicoll, 1979) suggests a mechanism whereby TH may also exert a restraining effect on their own actions. If these renal receptors for PRL are involved in water and/or electrolyte retention, their induction by TH during climax, when endogenous PRL levels are increasing, would serve to retard fluid loss from the larvae. Thus, the combined renal actions of PRL and TH during climax would ensure that metamorphosis does not progress too rapidly.

In view of the apparent connection between the control of metamorphosis and osmoregulation, and the fact that aldosterone is of importance for NaCl retention in adult amphibians, it is of interest to compare the changes in serum aldosterone levels with other changes that are related to osmoregulation. Figure 6 shows a comparison of the profile of serum aldosterone levels with the changes in plasma osmolarity, reported by Just *et al.* (1977). In addition, data on the development of renal PRL receptors, as determined by White and Nicoll (1979), are included in this figure. It is clear that the pattern of plasma aldosterone concentration does not relate to either the changes in plasma osmolarity or the rise in renal PRL binding. Possibly the increase in serum aldosterone that occurs between the midclimax tadpole and the adult relates to the onset of skin and/or bladder Na^+ transport. The rise in plasma osmolarity that occurs in late prometamorphosis and extends into midclimax does relate to the rise in renal PRL receptors during the same period. It is significant that the rise in these two physiological parameters also relates to the surge of TH, as measured by RIA (Fig. 3).

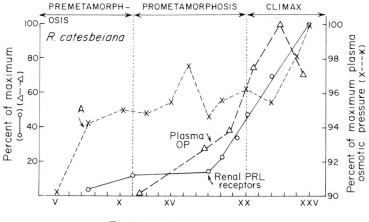

FIGURE 6. Changes in serum aldosterone (A) concentration (from Krug *et al.*, 1978, and Krug, 1980), plasma osmotic pressure (OP; from Just *et al.*, 1977), and renal prolactin (PRL) receptors (from White and Nicoll, 1979) during development and metamorphosis of bullfrog tadpoles. Note differences in scales of the right and left ordinates.

6. CONTROL OF ANURAN DEVELOPMENT AND ACTIVATION OF METAMORPHIC CLIMAX

The data that have been collected during the past few years permit a reassessment of existing theories and concepts of the control of amphibian development and metamorphosis. The most complete model of anuran metamorphosis to be proposed thus far is that of Etkin (1968). Dodd and Dodd (1976) have proposed an alternative theory, which is discussed below. The essential features of the Etkin model are as follows:

1. During premetamorphosis, the median eminence of the hypothalamus is undeveloped and the brain exerts little or no control over adenohypophysial functions. Consequently, PRL secretion is high, and TSH secretion (and thus TH levels) is low. Therefore, PRL can promote larval growth without interference from TH. Negative feedback of TH on TSH secretion is operative.

2. During prometamorphosis, TH secretion increases slowly, as reflected by growth of the hindlegs. The increased TH secretion presumably results from rising TSH levels, and this increase probably reflects gradual development of hypothalamic influence on the adenohypophysis.

3. The rising TH levels also promote development of the median eminence and the establishment of portal vascular connections between the hypothalamus and the adenohypophysis. As this process progresses, more TRH is able to reach the pituitary to stimulate increased TH secretion. The increased TH promote further development of the median eminence. Thus, a positive feedback loop is established.

4. While hypothalamic control of pituitary function is developing, PRL secretion comes under inhibitory control and the circulating levels of PRL decrease progressively. Thus, PRL antagonism of TH action on peripheral tissues would be reduced, allowing developmental changes to occur more rapidly.

5. Late in prometamorphosis, the median eminence and its vascular connections with the hypophysis have developed substantially, and a rapid increase in TSH secretion occurs. This surge causes a massive increase in TH secretion, which in turn causes the rapid and complete transformation of the animal (i.e., climax). Blood levels of PRL are greatly reduced during this period, reflecting maximum hypothalamic inhibition. Thus, inhibition of TH action by PRL is minimized.

6. During climax, Etkin (1968) further proposed that the positive feedback interactions of the hypothalamo–hypophysial–thyroid axis would be lost, leaving only the negative feedback loop. No mechanism was proposed to account for the loss of the positive feedback loop.

Some of the results discussed in previous sections are compatible with some features of the Etkin hypothesis but others are not. The plasma PBI data of Just (1972; Fig. 1) in *R. pipiens* are in excellent qualitative and temporal agreement with the Etkin model. Kaye's (1961) results on thyroidal ^{131}I uptake in the same species are also consistent with the model. Plasma PBI and ^{131}I uptake rise during prometamorphosis in a manner that relates well to the accelerating growth of the hindlegs. The PBI data showed peak plasma concentration in late prometamorphosis, as predicted by Etkin (1968), and they are consistent with his suggestion of a positive feedback loop operating during this period. As discussed previously (Section 2.9), the results of Eddy and Lipner (1976) with antiserum to TSH are also consistent with the existence of a positive feedback loop in anurans. However, neither the PBI data of Just (1972) nor the results of Eddy and Lipner (1976) with TSH antiserum prove that such a positive feedback mechanism is operative. Nevertheless, the existence of negative feedback of TH during all stages does not preclude the co-existence of a positive feedback loop (see Section 2.9).

The RIA data on plasma T_3 and T_4 levels in *R. catesbeiana* tadpoles (Fig. 2) agree with Etkin's suggestion that a surge in TH levels would be involved in metamorphic climax, but there is a temporal discrepancy. The surge in TH levels occurs almost entirely during climax, rather than in late prometamorphosis as Etkin predicted. This surge in T_3 and T_4 is concomitant with the onset and progression of the rapid transformations that occur during climax. Thus, the occurrence of the TH surge coincides with actual metamorphosis better than does the surge in PBI in *R. pipiens* (Just, 1972). However, the fact that plasma PBI levels are high during prometamorphosis in *R. pipiens* but no T_3 or T_4 can be measured by RIA during most of this period in *R. catesbeiana* is disturbing.

As discussed previously (Section 2.2), the rise in plasma T_4 levels during prometamorphosis in *Xenopus* is also consistent with Etkin's predictions of the control of development during prometamorphosis. However, the surge in T_3 and the peak in T_4 levels occur during climax in this species also.

Measurement of plasma PRL levels during climax (Clemons and Nicoll, 1977b) showed that the hormone increased, rather than decreased as predicted by Etkin, and by Bern *et al.* (1967). We have confirmed these results in subsequent studies (Clemons *et al.*, unpublished). As discussed above and elsewhere (White and Nicoll, 1979), the elevated PRL levels during climax may be important for retarding TH-induced transformations in tissues whose responsiveness to PRL is also being reduced by the TH. In addition, the high PRL levels may be of significance for the control of organs that are acquiring sensitivity to PRL, such as the kidney.

Dodd and Dodd (1976) have proposed an alternative hypothesis to account for what they feel are discrepancies between observations in the literature and the Etkin (1968) model. They are particularly disturbed by the findings that thyroidal ^{131}I uptake can reach high values earlier in prometamorphosis than the peak in plasma PBI, as measured by Just (1972). The temporal discrepancy between the peaks in ^{131}I uptake, and those of T_3 and T_4 as measured by RIA is even greater than that found with plasma PBI. Accordingly, Dodd and Dodd (1976) suggested that TH secretion is relatively high during early prometamorphosis but that this is not reflected in increased PBI (or by RIA-detectable plasma T_3 or T_4 levels) because of rapid clearance of the TH. Tissue utilization or "avidity" for the hormones is proposed to be very high. The secretion rate of TH continues to increase to such a degree that late in prometamorphosis, the capacity of the tissues to bind and utilize TH is saturated. Consequently, the continually increasing output of TH will result in a surge in plasma TH levels. Although this proposal of Dodd and Dodd (1976) has some merit, there are no data on the clearance rates of TH from the blood of tadpoles in different stages of development. In addition, their assumption that increased uptake of ^{131}I by the thyroid during early prometamorphosis is accompanied by the secretion of a proportionate amount of TH remains to be substantiated.

Thus, although some of the recently acquired data are consistent with some aspects of Etkin's model, other results indicate that the theory needs to be revised. Additional studies are needed to substantiate the existence of his proposed positive feedback loop. In addition, information or hypotheses on how the hypothalamo–hypophysial–thyroid axis can switch from a positive to a negative feedback would be useful. We could speculate that the aminergic fibers that "innervate" the anuran tadpole adenohypophysis (see Section 2.8) might be involved in both the positive and the negative feedback loops. These fibers disappear during metamorphic climax. Thus, rising TH during late prometamorphosis may act on the hypothalamus to cause these fibers to increase TSH secretion, and thus account for the positive feedback. The rising TH levels may also cause these fibers to eventually degenerate. Hence, neural stimulation of TSH secretion would be lost and the inhibitory action of TH directly on the pituitary could operate unopposed. It will be of interest to learn whether these aminergic fibers have such functions.

7. CONCLUSIONS

A substantial amount of new information has been published during the past few years on the endocrine control of amphibian development and metamorphosis. Nevertheless, addition knowledge is needed to allow for more complete understanding of these processes. For example, information on the relationship between thyroidal iodine uptake and TH secretion in different developmental stages would be useful. In addition, data on changes in the clearance rates of TH during development could be important, and the relationship between plasma PBI and RIA measurements of TH needs to be clarified. The observations that plasma PBI and TH concentration (see Section 2.2) are highly variable in tadpoles during climax suggest that TH secretion (and therefore TSH secretion) may be pulsatile during this period. Hence, measurements of hormone levels in single plasma samples are of limited value, especially if diurnal variations in hormone levels also occur.

Although we do not have a new theory to propose to explain neuroendocrine control of amphibian development and metamorphosis, we would suggest that our thinking about these processes may need some new emphasis. Previous models have focused on changing plasma levels of TH and PRL (or growth-promoting hormones) to account for the developmental and metamorphic changes. However, evidence indicates that changing tissue sensitivity to hormones is a factor in these processes, and synergistic effects of other hormones are well established. These phenomena may be more important than they are generally considered to be. For example, the increasing growth of the hindlimbs during prometamorphosis may be due to increased tissue sensitivity to TH and/or to synergism with GH (see Bern *et al.*, 1967) rather than to a change in circulating levels of TH per se. Alternatively, reduced sensitivity of tissues to PRL would increase their responsiveness to available TH and/or GH. Adrenal steroids may facilitate developmental changes during prometamorphosis by altering tissue sensitivity to TH, GH, or PRL. The pattern of blood glucocorticoid levels (especially cortisol) indicates that synergism with the TH that surges in early climax may be an important feature of the final transformation of the animals. These processes (i.e., synergism and changes in tissue sensitivity) should be evaluated more carefully in future studies on amphibian metamorphosis.

ACKNOWLEDGMENTS.   We are indebted to Dr. Sharon M. Russell for suggestions on the style and grammar of the manuscript and for proofreading. Our research relating to the subject matter of the chapter was supported by NSF Grant PCM 79-04562 to C.S.N. and by funds from the committee on research of the University of California at Berkeley.

REFERENCES

Aronsson, S., 1976, *Cell Tissue Res.* **171**:437.
Aronsson, S., 1978, *Gen. Comp. Endocrinol.* **36**:497.
Ashley, H., and Frieden, E., 1972, *Gen. Comp. Endocrinol.* **18**:22.
Bern, H. A., Nicoll, C. S., and Strohman, R. C., 1967, *Proc. Soc. Exp. Biol. Med.* **126**:518.
Bhattacharya, A., and Vonderhaar, B. K., 1979, *Biochem. Biophys. Res. Commun.* **88**:1405.
Blatt, L. M., Slickers, A., and Kim, K. H., 1969, *Endrocrinology* **85**:1213.
Carver, V. H., and Frieden, E., 1977, *Gen. Comp. Endocrinol.* **31**:202.
Chou, H. I., and Kollros, J. J., 1974, *Gen. Comp. Endocrinol.* **22**:255.
Clarke, W. C., and Bern, H. A., 1980. *Hormonal Proteins and Polypeptides* **8**:105.

Clemons, G. K., and Nicoll, C. S., 1977a, *Gen. Comp. Endocrinol.* **31**:495.

Clemons, G. K., and Nicoll, C. S., 1977b, *Gen. Comp. Endocrinol.* **32**:531.

Clemons, G. K., Russell, S. M., and Nicoll, C. S., 1979, *Gen. Comp. Endocrinol.* **38**:62.

Cohen, D. C., Greenberg, J. A., Licht, P., and Bern, H. A., 1972, *Gen. Comp. Endocrinol.* **18**:384.

Crim, J. W., 1975, *J. Exp. Zool.* **192**:355.

Dent, J. N., 1975, *Am. Zool.* **15**:923.

Dent, J. N., Eng, L. A., and Forbes, M. S., 1973, *J. Exp. Zool.* **184**:369.

Derby, A., 1970, *J. Exp. Zool.* **173**:319.

Derby, A., 1974, *J. Exp. Zool.* **193**:15.

Dodd, M. H. I., and Dodd, J. M., 1976, in: *Physiology of the Amphibia*, Volume 3 (J. A. Moore, ed.), pp. 467–599, Academic Press, New York.

Eagleson, G. W., and McKeown, B. A., 1978, *Can. J. Zool.* **56**:1377.

Eddy, L. J., 1979, *Gen. Comp. Endocrinol.* **38**:360.

Eddy, L., and Lipner, H., 1975, *Gen. Comp. Endocrinol.* **25**:462.

Eddy, L., and Lipner, H., 1976, *Gen. Comp. Endocrinol.* **29**:333.

Enemar, A., 1978, *Gen. Comp. Endocrinol.* **34**:211.

Enemar, A., Essvik, B., and Klang, R., 1968, *Gen. Comp. Endocrinol.* **11**:328.

Etkin, W. N., 1935, *J. Exp. Zool.* **71**:317.

Etkin, W., 1968, in: *Metamorphosis: A Problem in Developmental Biology* (W. Etkin and L. I. Gilbert, eds.), pp. 313–348, Appleton–Century–Crofts, New York.

Etkin, W., and Gona, A. G., 1967, *J. Exp. Zool.* **165**:249.

Etkin, W., and Lehrer, R., 1960, *Endocrinology* **67**:457.

Fox, H., and Turner, C. S., 1967, *Arch. Biol.* **78**:61.

Frieden, E., and Just, J. J., 1970, in: *Actions of Hormones on Molecular Processes*, Volume 1 (G. Litwack, ed.), pp. 1–53, Academic Press, New York.

Frye, B. E., Brown, P. S., and Snyder, B. W., 1972, *Gen. Comp. Endocrinol. Suppl.* **3**:209.

Galton, V. A., 1980, *Endocrinology* **106**:859.

Gona, A. G., 1968, *Gen. Comp. Endocrinol.* **11**:278.

Gona, A. G., and Gona, O. D., 1976, *J. Endocrinol.* **68**:349.

Gona, A. G., Pearlman, T., and Gona, O., 1973, *Gen. Comp. Endocrinol.* **20**:107.

Goos, H. J. Th., 1978, *Am. Zool.* **18**:401.

Guyetant, R., Bugnon, C., Fellman, C., and Block, B., 1977, *C. R. Acad. Sci.* **285**:559.

Hanaoka, Y., 1967a, *Gen. Comp. Endocrinol.* **8**:417.

Hanaoka, Y., 1967b, *Gen. Comp. Endocrinol.* **9**:24.

Hanaoka, Y., Miyashita, K. S., Kondo, Y., Kobayashi, Y., and Yamamoto, R., 1973, *Gen. Comp. Endocrinol.* **21**:410.

Hourdry, J., and Regard, E., 1975, *Gen. Comp. Endocrinol.* **27**:277.

Hsü, C., Yü, N., and Hsü, L., 1976, *Gen. Comp. Endocrinol.* **30**:424.

Jackson, I. M. D., 1978, *Am. Zool.* **18**:385.

Jaffe, R. C., and Geschwind, I. I., 1974, *Gen. Comp. Endocrinol.* **22**:289.

Just, J. J., 1972, *Physiol. Zool.* **45**:143.

Just, J. J., Sperka, R., and Stranges, S., 1977, *Experientia* **33**:1503.

Kaye, N. W., 1961, *Gen. Comp. Endocrinol.* **1**:1.

Kerr, T., 1966, *Gen. Comp. Endocrinol.* **6**:303.

Kikuyama, S., and Seki, T., 1980, *Gen. Comp. Endocrinol.* **41**:173.

Kikuyama, S., Yamamato, K., and Mayumi, M., 1980, *Gen. Comp. Endocrinol.* **41**:212.

Kistler, A., Yoshizato, K., and Frieden, E., 1975, *Endocrinology* **97**:1036.

Kistler, A., Yoshizato, K., and Frieden, E., 1977, *Endocrinology* **100**:134.

Knobil, E., Plant, T. M., Wildt, L., Belchetz, P. E., and Marshall, G., 1980, *Science* **207**:1371.

Kollros, J. J., 1961, *Am. Zool.* **1**:107.

Kracht, J., and Weber, E. G., 1978, *Cell Tissue Res.* **187**:305.

Krug, E. C., 1980, Ph.D. dissertation, Department of Physiology–Anatomy, University of California, Berkeley.

Krug, E. C., Honn, K. V., Batista, J., and Nicoll, C. S., 1978, *Am. Zool.* **18:**614.

Leloup, J., and Buscaglia, M., 1977, *C. R. Acad. Sci.* **284:**2261.

MacKenzie, D. S., Licht, P., and Papkoff, H., 1978, *Gen. Comp. Endocrinol.* **36:**566.

Medda, A. K., and Frieden, E., 1972, *Gen. Comp. Endocrinol.* **19:**212.

Miyauchi, H., LaRochelle, F. T., Jr., Suzuki, M., Freeman, M., and Frieden, E., 1977, *Gen. Comp. Endocrinol.* **33:**254.

Mondou, P. M., and Kaltenbach, J. C., 1979, *Gen. Comp. Endocrinol.* **39:**343.

Nicoll, C. S., 1974, in: *Handbook of Physiology, Endocrinology IV, Part 2* (E. Knobil and W. H. Sawyer, eds.) pp. 253–291, American Physiological Society, Washington, D.C.

Norris, D. O., 1978, in: *Comparative Endocrinology* (P. J. Gaillard and H. H. Boer, eds.), pp. 109–112, Elsevier/North-Holland, Amsterdam.

Norris, D. O., and Gern, W. A., 1976, *Science* **194:**525.

Norris, D. O., and Platt, J. E., 1973, *Gen. Comp. Endocrinol.* **21:**368.

Pehleman, F. W., and Bade, J., 1976, *Gen. Comp. Endocrinol.* **29:**259.

Pesetsky, I., 1962, *Gen. Comp. Endocrinol.* **2:**229.

Platt, J. E., 1976, *Gen. Comp. Endocrinol.* **28:**71.

Platt, J. E., and Christopher, M. A., 1977, *Gen. Comp. Endocrinol.* **31:**243.

Platt, J. E. and LiCause, M. J., 1980, *Gen. Comp. Endocrinol.* **41:**84.

Platt, J. E., Christopher, M. A., and Sullivan, C. A., 1978, *Gen. Comp. Endocrinol.* **35:**402.

Regard, E., Taurog, A., and Nakashima, T., 1978, *Endocrinology* **102:**674.

Robinson, H., and Galton, V. A., 1976, *Gen. Comp. Endocrinol.* **30:**83.

Robinson, H., Chaffee, S., and Galton, V. A., 1977, *Gen. Comp. Endocrinol.* **32:**179.

Rothstein, H., VanWyk, J. J., Hayden, J. H., Gordon, S. R., and Weinseider, A., 1980, *Science* **208:**410.

Sawin, C. T., Bolaffi, J. T., Callard, I. P., Bacharach, P., and Jackson, I. M. D., 1978, *Gen. Comp. Endocrinol.* **36:**427.

Seki, T., and Kikuyama, S., 1979, *Endocrinol. Jpn.* **26:**675.

Singh, D. V., and Bern, H. A., 1969, *J. Endocrinol.* **45:**579.

Steinmetz, C. H., 1954, *Physiol. Zool.* **27:**28.

Stuart, E. S., and Fischer, M. S., 1979, *Gen. Comp. Endocrinol.* **38:**314.

Tata, J. R., 1970, *Nature (London)* **227:**686.

Taurog, A., Oliver, C., Eskay, R. L., Porter, J. C., and McKenzie, J. M., 1974, *Gen. Comp. Endocrinol.* **24:**267.

Turgeon, J., 1979, *Endocrinology* **105:**731.

Vonderhaar, B. K., and Greco, A. E., 1979, *Endocrinology* **104:**409.

White, B. A., and Nicoll, C. S., 1979, *Science* **204:**851.

White, B. A., Ray, L. B., Clemons, G. K., and Nicoll, C. S., 1978, *Am. Zool.* **18:**614.

White, B. A., Lebovic, G. S., and Nicoll, C. S., 1981, *Gen. Comp. Endocrinol.,* **43:**30.

Wright, Sister, M. L., Majerowski, M. A., Lukas, S. M., and Pike, P. A., 1979, *Gen. Comp. Endocrinol.* **39:**53.

Yamaguchi, K., and Yasumasu, I., 1978, *Dev. Growth Differ.* **20:**61.

Yoshizato, K., and Frieden, E., 1975, *Nature (London)* **254:**705.

Yoshizato, K., Kikuyama, S., and Yasumasu, I., 1972, *Gen. Comp. Endocrinol.* **19:**247.

Yoshizato, K., Kistler, A., and Shioya, N., 1976, *Gen. Comp. Endocrinol.* **29:**468.

Zuber-Vogeli, M., 1978, *C. R. Acad. Sci.* **286:**1379.

CHAPTER 12

Biological Basis of Tissue Regression and Synthesis

BURR G. ATKINSON

1. INTRODUCTION

The metamorphic process takes place in anuran tadpoles in an orderly fashion and with precise timing, each new development fitting into a perfectly coordinated pattern. Organs essential for an aquatic environment (tail and gills) do not begin to degenerate until the growth and development of organs essential for a terrestrial environment (legs and lungs) are nearly complete in their growth and development (Fig. 1). During the latter stages of larval development, the blood chemistry changes also manifest the functionality of the transition from gill to lung respiration.

The dependency of some of these metamorphic events on increasing circulating levels of thyroid hormone (TH) is now well established. Growth and development of organs required for terrestrial existence are initiated by considerably lower levels of exogenous TH than are required for initiating the degeneration of organs essential for an aquatic life. While each of these TH-dependent responses may require a specific, and perhaps continuous, endogenous level of this hormone for their initiation, the timely and coordinated *in vivo* pattern of response may, in fact, depend upon other hormonal influences, tissue competence, and interactions of constituents from various tissues. The purpose of this review is to present an up-to-date and ordered analysis of anuran metamorphic

BURR G. ATKINSON · Department of Zoology, Cell Science Laboratories, University of Western Ontario, London, Ontario, Canada.

studies concerned (1) with the modulation of differentiation and development of adult terrestrial organs and (2) with the mechanisms whereby organs essential for aquatic life undergo complete degeneration during this transitional period.

2. SPONTANEOUS AND THYROID-HORMONE-INDUCED DIFFERENTIATION AND DEVELOPMENT OF ANURAN TADPOLE LIMBS AND LUNGS

2.1. Introduction

The differentiation and development of functional limbs and lungs to replace tail and gills are essential for the transition from a larval aquatic environment to an adult terrestrial one. The ontogeny of these organs is continuous during larval development and is essentially complete by metamorphic climax (stage XX).[*] Although the molecular role of TH, as well as other hormones, in the growth of these organs is poorly understood, its influence of limb growth is well documented. Whether TH effects on the developmental processes in these organs are direct or are mediated by other induced factors is the central theme linking the studies reviewed in this section.

2.2. Tadpole Limbs

2.2.1. Spontaneous Limb Differentiation and Development

Growth of anuran hindlimbs is rapid in early stages of metamorphosis and becomes considerably slower during metamorphic climax in both *Rana pipiens* (Etkin, 1968) and *R. catesbeiana* (Fig. 1). During its early ontogeny the limb bud consists of undifferentiated loose mesenchyme covered by epithelial tissue. The limb primordium grows mainly at its distal end, and future limb segments are laid down in a proximodistal sequence (Tschumi, 1957) reminiscent of chick wing and limb morphogenesis. Apical growth of the limb bud is dependent upon a thickening of the epidermis around the distal rim of the developing limb bud (the apical ectodermal ridge, or AER).

The larval limb epidermis is usually three cell rows in thickness (Wright, 1973). During metamorphosis tadpole epidermis differentiates into the multilayered, cornified epithelium characteristic of terrestrial vertebrates. Wright's (1973) studies on the histogenesis of *R. pipiens* limb epidermis indicate that the surface cells of the epidermis exhibit birefringence at stage XII (signaling keratin formation), and at stage XVI

---

[*] Tadpole stages used throughout this review refer to those established by Taylor and Kollros (1946).

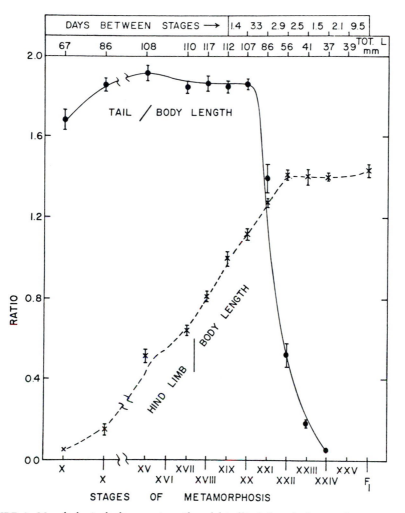

FIGURE 1. Morphological changes in tail and hindlimb length during the spontaneous metamorphosis of *R. catesbeiana* larvae. Prior to being measured, tadpoles were staged according to the Taylor and Kollros (1946) morphological criteria for *R. pipiens*. All values represent the mean of 12 or more tadpoles; vertical bars signify the standard error of the mean.

gland rudiments appear in the inner epidermal layer, descending into the stratum spongiosum of the dermis as they enlarge in subsequent stages. Concurrent with the appearance of gland rudiments, the outer and inner layers of epidermal cells exhibit divergent patterns of DNA synthesis; there is a rapid decrease in labeling index in the outer layer, which ceases entirely by stage XXII, while the inner layer peaks at stage XVI and declines slowly thereafter. As the epidermis thickens during metamorphic climax, additional rows of cells are added. By stage XXII there are four to six rows of epidermal cells.

Proximal limb bud mesenchymal cells begin to acquire specific differentiated characteristics following stage VI of tadpole development. In stage VII *R. pipiens* tadpoles, Dunlap has shown that the femur, ilium, fibula, and tibia are in early stages of chondrogenesis, and myogenesis is detectable in the thigh (Muntz, 1975). By stage IX, the cartilaginous skeleton of the hindlimb (excepting the foot) is preformed, and myogenesis is evident in the tarsals (Muntz, 1975). Ossification begins in the collagenous femur at stage X, progressing distally to the phalanges by stages XIII–XV (Kemp and Hoyt, 1969a). The proximodistal development of muscle and replacement of collagen by bone in the hindlimb are essentially complete by stage XV (Kemp and Hoyt, 1969a), and outgrowth of the tadpole hindlimbs is maximal (Fig. 1) by climax (stage XX).

2.2.2. Induced Limb Differentiation and Development

Anuran hindlimb growth and differentiation are accelerated by low concentrations of TH, indicating a low threshold of response to TH (Kaltenbach, 1953; Wright, 1973). Following thyroidectomy, hindlimb development and differentiation are arrested and ossification and growth of bones in the limbs are retarded. In hypophysectomized tadpoles minute hindlimbs appear, but do not develop further than those found in stage VII tadpoles. Administration of iodine or thyroid tissue to thyroidectomized or hypophysectomized tadpoles restores limb growth and ossification in the hindlimbs (Allen, 1925). Immersion of hypophysectomized *R. pipiens* larvae (stages IV–V) in a dilute solution of thyroxine (T_4) restores limb growth and advances the tadpoles to stage X or XI. Implementation of T_4 pellets in the shank of premetamorphic tadpoles causes growth of the hindlimbs and local ossification of the bone. Immersion of cultured tadpole limb or tadpoles in T_4 sequentially stimulates limb epidermal mitotic activity and the differentiation of glands similar to those observed in adult limb epidermis (Wright, 1973). A number of investigators have attributed these morphogenetic and cytological transitions to the direct action of T_4, and concluded that TH is probably directly responsible for similar changes that occur in the limbs during the normal sequence of metamorphic development.

Recent morphological, histological, and biochemical studies (Dhanarajan, 1979; Dhanarajan and Atkinson, 1981), employing a single injection of triiodothyronine (T_3) (3×10^{-10} mole/g body wt) to stage VI–VII *R. catesbeina* tadpoles, also strongly support the contention that this hormone plays a role in accelerating limb growth and differentiation (Figs. 2 and 3). Preceding T_3 treatment of the tadpole, paraffin sections of *R. catesbeina* hindlimbs reveal that the primordia are primarily composed of undifferentiated loose mesenchymal tissue covered by a simple, three-layered epithelium. No change in organ size is perceptible the first day after hormone administration, but gradual increase in length of the limb (doubled by 6 days after T_3 treatment) with concomitant differen-

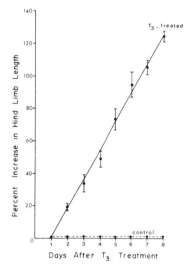

FIGURE 2. Percent increase in hindlimb length of *R. catesbeiana* tadpoles (initially stages VI–VII) following administration of T_3. Each point represents measurements on the same six tadpoles; vertical bars represent the standard error of the mean.

tiation of the mesenchymal tissue is evident each day thereafter (Figs. 2 and 3). Epidermal tissue becomes multilayered (four to six rows of cells) and by 8 days after hormone treatment exhibits well-differentiated glands (Fig. 4).

The response of some of the mesenchymal cell nuclei in the 24-hr period following T_3 administration is remarkable, and similar to that described for the early effects of T_4 on *R. pipiens* shank epidermal cell nuclei (Wright, 1973). The mesenchymal cell nuclei from hormone-treated tadpoles appear enlarged, less dense, and lightly granular when compared to the control (Fig. 5A and B). By Day 2 the majority of the nuclei in the limb mesenchyme have assumed this appearance. The major histogenetic processes induced by TH in the limb mesenchymal cells are the formation of blood vessels, cartilage, and muscle. The cytological identification of extensive blood vessel formation in areas distal to and preceding the recognizable differentiation of other cell types (Fig. 6A), as well as the red color assumed by the limb in the first 24 hr following administration of TH, implicates the early differentiation and formation of vascular tissue as a primary and sustained effect of this hormone. Differentiation of cartilage cells (chondrogenesis) is conspicuous and progresses distally from the thigh at 2 days after hormone administration to the foot region 5 days after T_3 treatment (Fig. 3). Spindle-shaped myoblasts are difficult to recognize in early stages before the production of eosinophilic differentiated proteins; the first appearance of striated muscle fibers is clearly discernible by histochemical techniques in the thigh and shank regions 4 days after treatment. Thereafter, muscle fibers appear in the more distal regions and are eventually visible (8 days after T_3 treatment) in the toe area (Fig. 6B). The general histological picture of a hindlimb from a tadpole 8 days after T_3 treatment closely resembles that

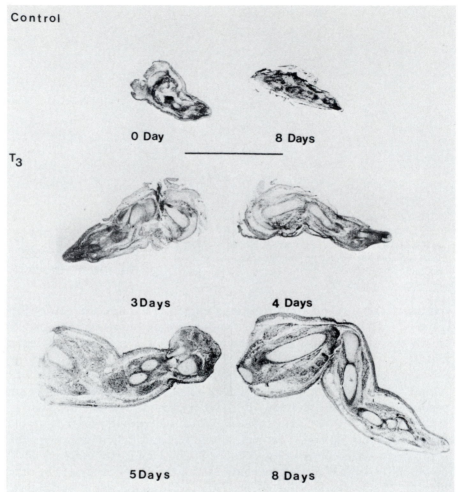

FIGURE 3. Cytological changes in *R. catesbeiana* tadpole hindlimbs (initially stages VI–VII) following T_3 administration: longitudinal sections of hindlimbs fixed in Bouin's solution and stained with alum hematoxylin and eosin. ($\times 22$.) Data from Dhanarajan and Atkinson, 1981.

of a normally developing stage XII tadpole except that the epidermis with developing skin glands is more like a stage XVI tadpole.

Immersion of stage IX *R. pipiens* tadpoles in T_4 (6.25×10^{-8} M) also results in a proximal–distal order of ossification similar to that observed in naturally metamorphosing tadpoles. This TH-induced response of collagen replacement by bone is characterized at the fine structural level by the development, after 2 days of hormone treatment, of rough endoplasmic reticulum in limb osteoblasts and the appearance of osteocytes after 9 days of hormone treatment. Kemp and Hoyt (1969b) concluded from their observations that T_4 stimulates differentiation of osteoblasts from limb perichondrial cells, and that T_4 may stimulate osteoblasts to secrete the depolymerases that prepare the osteroid matrix for mineralization along collagen fibrils.

Further studies aimed at establishing the role of hormones in tadpole limb development also clearly implicate T_4 as an accelerator of chondrogenesis and bone growth in more advanced tadpoles (stages XIV–XXII). Yamaguchi and Yasumasu's (1978) extensive *in vivo* and *in vitro* studies with *R. catesbeiana* establish that hindlimbs from T_4-treated (10^{-8} to 10^{-7} M) tadpoles of stages XIV–XXII exhibit an overall increased rate of protein synthesis, a more pronounced stimulation of [^{14}C]proline incorporation into collagen, and enhanced turnover of collagen. Prolactin treatment of similar stages did not result in specific enhancement of [^{14}C] proline incorporation but did moderately stimulate overall protein synthesis and inhibit collagen breakdown. Increasing concentrations of growth hormone administered to tadpoles earlier than at stage XVIII did not affect the incorporation of [^{14}C]proline into limb collagen, but that incorporation was markedly stimulated in tadpoles treated with growth hormone at stages XVIII or later. Yamaguchi and Yasumasu further established that the rate of collagen synthesis in thigh bones from T_4-

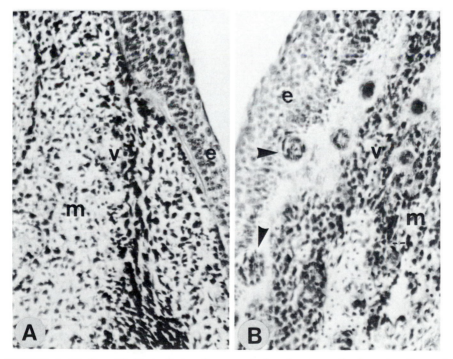

FIGURE 4. Limb epithelium during hormone-induced outgrowth, hematoxylin–eosin stained paraffin sections. (A) 5 days after T_3 treatment; (B) 8 days after T_3 treatment. The knots of cells (arrowheads) appearing at 8 days stain differentially with Giemsa, the first sign of glandular activity in the limb epithelium. e, epithelium; m, mesenchyme; v, blood vessel.

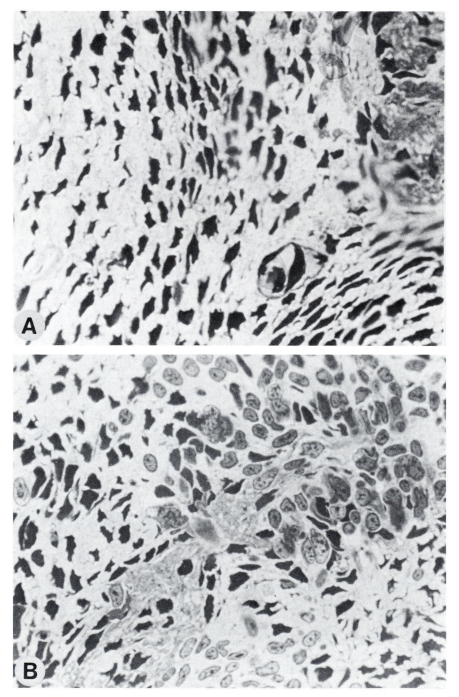

FIGURE 5. Light microscopy of Epon thick sections of tadpole hindlimb, stained with methylene blue–azure II. (A) Control, stage VII. Cells are highly condensed and stellate. (B) 24 hr after TH administration. The decondensation and rounding up of about half of the nuclei are dramatically evident.

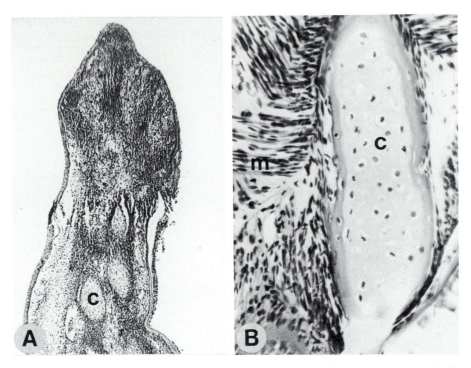

FIGURE 6. Patterns of limb morphogenesis during TH-induced metamorphosis. (A) The limb at 3 days after hormone administration. Differential staining with hematoxylin–eosin shows intense vascularization as rose-colored channels permeating the tip; oval RBCs are unmistakable. Centers of chondrogenesis (c) stain light blue. (B) Higher magnification of the toe region 8 days after hormone treatment. Striated myotubes (m) are oriented in the positions of forming muscles.

treated premetamorphic tadpoles (stages XII–XVII) is markedly enhanced by growth hormone. They conclude that the nature of the thigh bone and the proteins synthesized by it when directed by T_4 are different from those synthesized under the influence of growth hormone.

While the preceding studies of tadpole hindlimbs emphasized the role of TH in the processes of chondrogenesis and ossification, few studies have been aimed at elucidating its role in the development of limb musculature. Early research by Kaltenbach implicated TH as a stimulant for tadpole limb muscle growth and development. In vitro experiments with limb muscle minces by Strohman and co-workers indicate that the tadpole hindlimb responds directly to TH by the increased incorporation of labeled amino acid into myosin and total muscle protein (it is unfortunate that Strohman's group reported neither the developmental stage nor limb size of the tadpoles employed). More recent reports show that enhanced RNA precursor incorporation into tadpole (R. catesbeiana: Strohman, 1966; X. laevis: Ryffel and Weber, 1973) limb muscle RNA is not detectable until tadpoles have been treated for at least 86 hr with T_4. How-

TABLE 1. AMINO ACID COMPOSITION OF FROG (*RANA CATESBEIANA*)
SKELETAL MUSCLE PROTEINS

Amino acid residue[a]	Actin	Tropomyosin	M-line protein[b]	α-Actinin	Myosin heavy chain[c]
Lys	48[d]	137	64	64	40
His	17	7	17	19	3
Arg	47	35	43	51	32
Asp	89	86	93	107	114
Thr[e]	53	22	55	43	43
Ser[e]	44	27	64	42	44
Glu	112	270	123	168	198
Pro	45	0	52	27	16
Gly	78	10	59	38	50
Ala	78	109	61	78	77
CyS[f]	8	9	11	9	4
Val[g]	51	27	75	37	61
Met	31	19	8	7	8
Ile[g]	74	25	47	68	31
Leu	68	96	64	96	87
Tyr	30	11	11	6	6
Phe	28	3	35	31	25
Trp[h]	11	0	12	7	3

[a] Average of duplicate 24-hr hydrolysis on two separate preparations.
[b] Data from report by Dhanarajan and Atkinson (1979).
[c] Data from Merrifield and Atkinson (unpublished).
[d] Values are moles per 10^5 g.
[e] Extrapolated to zero time of hydrolysis.
[f] Determined as cysteic acid.
[g] Value from 72-hr hydrolysate.
[h] Determined by the method of Bencze and Schmid.

ever, Strohman noted that prior to the hormone-induced RNA synthetic response (i.e., during the 86-hr lag period), T_4 exposure resulted in the appearance of new, varied size classes of polyribosomes in the limb muscle tissue. Strohman's observations can be interpreted to mean that a primary effect of T_4 on limb muscle is occurring at the level of translation during the lag phase preceding enhanced RNA synthesis.

Our own studies support the early (within the first 24 hr after hormone administration) and sustained effect of TH on translation of muscle-specific proteins in the tadpole hindlimb. Monospecific antibodies to purified preparations of the heavy chain of myosin, tropomyosin, α-actinin, and M-line protein from frog skeletal muscle were prepared in our laboratory and used as probes for monitoring the hormone-induced appearance and the rates of synthesis of tadpole limb muscle proteins (Table 1 gives the amino acid analysis of the antigens). The presence of myofibrillar proteins in hindlimbs of stage VII tadpoles is first detectable by indirect immunofluorescence 24 hr after hormone administration (Fig. 7), and in all cases are observed initially in the proximal limb region (Fig. 8). With increasing time after hormone treatment there is a marked in-

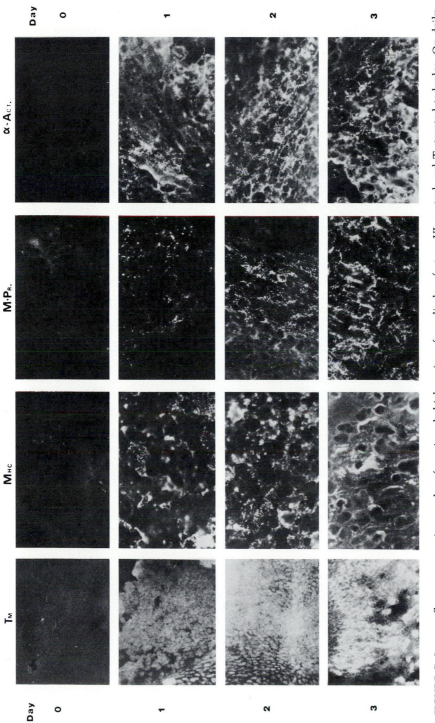

FIGURE 7. Immunofluorescence micrographs of sectioned thigh regions from limbs of stage VI control and T_3-treated tadpoles. On daily intervals thigh regions were sectioned and indirectly labeled with anti-tropomyosin (T_m), anti-myosin heavy chain (M_{HC}), anti-M-line protein (MP), and anti-α-actinin (α-Act). FITC-sheep anti-rabbit IgG was subsequently added to all fractions. (Magnification is ×200 except for T, which is ×110.) Data from Dhanarajan and Atkinson, 1981; Dhanarajan, 1979, Ph.D. thesis, University of Western Ontario, London, Canada.

Tm.   **M.Pr.**

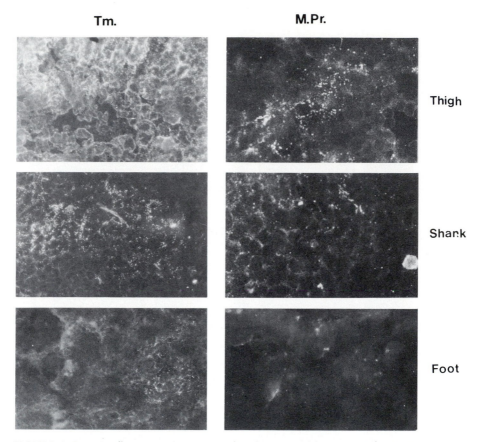

FIGURE 8. Immunofluorescence micrographs of sections from the thigh, shank, and foot regions of T_3-treated tadpoles 1 day after hormone treatment. Sections were labeled with anti-tropomyosin (T_m) and anti-M-line protein (MP) to which FITC-sheep anti-rabbit IgG was subsequently added (magnification: ×180). The proximal–distal development of immunofluorescence-visible material was also evident with anti-α-actinin and anti-myosin heavy chain. Data from Dhanarajan and Atkinson, 1981; Dhanarajan, 1979, Ph.D. thesis, University of Western Ontario, London, Canada.

crease in the amount of these proteins (Fig. 7) and their progressive appearance in the more distal regions of the limb.

Quantitative immunochemical studies (Dhanarajan and Atkinson, 1980; Dhanarajan and Atkinson, 1981) employing [^{14}C]leucine incorporation indicate a 25-fold increase in the overall synthetic rate of myofibrillar proteins within the first 24 hr after hormone treatment. The initial increase in the rate of synthesis of individual proteins (heavy chain of myosin, tropomyosin, α-actinin, and M-line protein) varied from 7- to 125-fold (Table 2). The number of molecules of each myofibrillar protein chain synthesized in control and T_3-induced tadpoles per hour per hindlimb was calculated by considering the TH-provoked changes in the leucine pool of the deskinned tadpole hindlimb (Table 3), the leucine

TABLE 2. RATES OF PROTEIN SYNTHESIS IN THE HINDLIMBS OF
CONTROL AND T_3-TREATED TADPOLES

Days after T_3 treatment	Leucine incorporation (nmoles/mg protein/hr)			
	Tropomyosin	M-line protein	α-Actinin	Myosin heavy chain
Control	0.03 ± 0.03[a]	0.04 ± 0.01	—	0.08 ± 0.03
1	0.7 ± 0.1	1.3 ± 0.2	1.3 ± 0.3	0.5 ± 0.1
2	1.5 ± 0.1	3.7 ± 0.1	3.2 ± 0.2	2.7 ± 0.3
3	3.3 ± 0.4	5.7 ± 0.1	6.0 ± 0.3	4.6 ± 0.6
4	6.4 ± 0.1	9.4 ± 0.2	9.0 ± 0.1	—
5	8.0 ± 0.1	13.4 ± 0.2	12.8 ± 0.1	5.6 ± 0.3

[a] Mean $\pm$ S.E. of three experiments.

content of each myofibrillar protein (see Table 1), and the number of polypeptide chains in one molecule of each myofibrillar protein (see Table 4A for results of calculations). Based on a previously determined amount of DNA per nucleus (11 pg per nucleus) for tadpole tissues and on DNA concentrations determined in limbs from control and hormone-induced tadpoles, the number of protein chains synthesized per minute per nucleus was calculated (Table 4B). These data clearly establish that TH administration results in a progressive enhancement in the rate of synthesis for each of these myofibrillar proteins. Determinations of the relative molar ratios for the rate of synthesis of individual proteins indicate that synthesis of these proteins is synchronously initiated and maintained at a constant molar ratio; for each heavy chain of myosin synthesized, seven to eight chains of tropomyosin, and two chains of M-line protein and α-actinin are synthesized. These quantitative data, in conjunction with the immunofluorescence and cytological results, provide compelling evidence of TH involvement in amphibian limb muscle differentiation and development.

TABLE 3. SIZE AND SPECIFIC ACTIVITY OF LEUCINE POOL IN THE
HINDLIMBS OF CONTROL AND T_3-TREATED TADPOLES[a]

Days after T_3 treatment	Total leucine content of pool (nmole)	Total [^{14}C]leucine counts in pool (dpm $\times 10^{-3}$)	Leucine content in amino acid pool (%)	Specific activity of leucine (dpm/nmole $\times 10^{-3}$)
Control	0.78 ± 0.08[b]	84.5 ± 8.0	1.5 ± 0.2	108.3 ± 4.9
1	1.25 ± 0.16	137.5 ± 15.5	1.7 ± 0.6	110.7 ± 4.2
2	3.39 ± 0.31	207.0 ± 32.6	2.7 ± 0.6	61.9 ± 11.2
3	10.37 ± 0.98	231.8 ± 16.9	3.0 ± 0.3	22.5 ± 1.1
4	23.51 ± 2.32	312.0 ± 29.1	3.6 ± 0.1	13.5 ± 1.6
5	38.50 ± 2.94	350.1 ± 17.0	4.1 ± 0.2	9.1 ± 0.3

[a] Results of pooled samples from nine tadpoles in three experiments (Dhanarajan and Atkinson, 1979).
[b] Mean $\pm$ S.E.M.

TABLE 4A. NUMBER OF POLYPEPTIDE CHAINS OF SPECIFIC
MYOFIBRILLAR PROTEINS SYNTHESIZED PER HOUR PER LIMB IN
CONTROL AND T_3-TREATED TADPOLES

Days after T_3 treatment	Tropomyosin $(\times\ 10^{-8})$	M-line protein $(\times\ 10^{-8})$	α-Actinin $(\times\ 10^{-8})$	Myosin heavy chain $(\times\ 10^{-8})$
Control	20.1	1.1	—	10.1
1	945.0	250.1	303.8	121.5
2	4,396.4	1390.2	1,466.8	630.9
3	14,498.8	3138.3	3,713.5	1713.0
4	27,729.0	4870.8	5,853.6	—
5	38,400.0	9652.0	12,182.8	4478.3

The studies to date provide support for implicating TH in initiating and enhancing differentiative and growth processes in tadpole limb bud tissue. However, the diverse differentiative processes affected in the limb bud by administration of TH to the tadpole make it highly unlikely that stimulation of each process results from the direct influence of this hormone. Following *in vivo* administration of TH to the tadpole, limb mesenchymal tissue responds rapidly by initiating myogenic and osteogenic processes, and by exhibiting enhanced chondrogenesis. The proximal–distal response patterns of these differentiative processes in the limb are preceded always by enhanced vascularization. The recent suggestion (Osdoby and Caplan, 1979) that mesenchymal cell participation in the processes of myogenesis, chondrogenesis, or osteogenesis in the developing avian limb is influenced by the positional relationship of mesenchymal cells to vascular elements raises the possibility that hormone-induced stimulation of these processes in the anuran limb may simply result from the intense vascularization occurring (Fig. 6A). The rather late differentiation of epithelial gland cells in limbs from spontaneously and TH-induced metamorphosing tadpoles suggests that the initial differentiation

TABLE 4B. NUMBER OF POLYPEPTIDE CHAINS OF SPECIFIC
MYOFIBRILLAR PROTEINS SYNTHESIZED PER MINUTE PER NUCLEUS
IN CONTROL AND T_3-TREATED TADPOLES

Days after T_3 treatment	Tropomyosin	M-line protein	α-Actinin	Myosin heavy chain
Control	26	1	—	13
1	678	180	218	88
2	3,686	1128	1190	512
3	7,660	1658	1963	965
4	13,533	2377	2855	—
5	16,626	4126	5216	1912

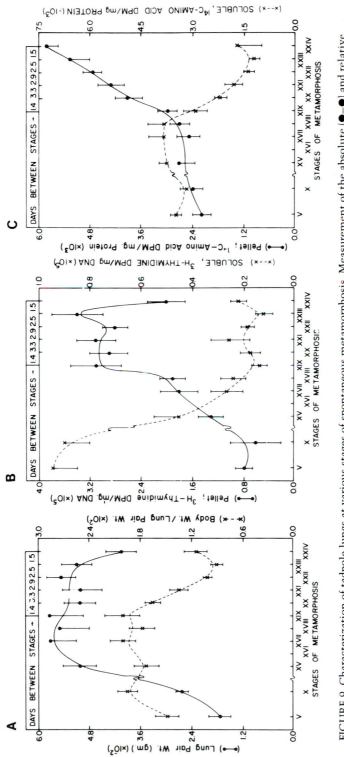

FIGURE 9. Characterization of tadpole lungs at various stages of spontaneous metamorphosis. Measurement of the absolute (●–●) and relative (x–x) lung wet weights (A), and of the *in vivo* rates of [³H]thymidine (B) and ¹⁴C-labeled amino acid hydrolysate (C) incorporation into TCA-precipitable (●–●) and TCA-soluble (x–x) fractions of tadpole lung. Data from Atkinson and Just, 1975, *Dev. Biol.* **45**:151.

of these cells may depend upon a critical number of mitoses and/or result from influences by newly differentiated adjacent tissues.

2.3. Tadpole Lungs

2.3.1. Spontaneous Lung Differentiation and Development

The anuran lung first begins to function in the late stages of spontaneous metamorphosis. The observation that the lung exhibits no weight increase after tadpole stage XV (Fig. 9A) suggests that the initiation of function in this organ is not caused by a sudden increase in lung weight (Atkinson and Just, 1975). Prior to the time that the lung normally assumes its physiological competence, the rate of thymidine incorporation into DNA is increasing (Fig. 9B) and the rate of amino acid incorporation into protein is constant (Fig. 9C). At the developmental stage in which the lung assumes its physiological function, the rate of DNA synthesis is constant and the rate of protein synthesis is markedly enhanced. The increasing rate of amino acid incorporation into protein after tadpole stage XIX, at a time when the endogenous level of TH is maximal (Just, 1972; Miyauchi et al., 1977; Regard et al., 1978), implicates a possible TH influence and suggests that the production of new proteins such as surfactant is necessary for the lung to assume a mature, physiologically functional state.

Lung morphology changes drastically during spontaneous metamorphosis. The lung of stage X animals has few internal divisions and is made up of an epithelial lining under which smooth muscle and blood vessels lie (Fig. 10A and B). During the late stages of metamorphosis the lung histology shows an increase in the potential respiratory epithelium by the formation and extension of numerous septa into the lumen of the lung (Fig. 10C and D).

While the biochemical and morphological transitions occurring in the lung cells of spontaneously metamorphosing tadpoles herald, and may be necessary for, lung physiological function, major changes in the aortic arches and in their vascularization must also occur. Indeed, it appears that lung tissue prepares biochemically and structurally to assume a respiratory role prior to vascular changes and before it actually begins its major respiratory function.

2.3.2. Induced Lung Differentiation and Development

The observation that respiratory function by the anuran lung is preceded in spontaneous metamorphosis by an increase in circulating levels of TH (Just, 1972; Miyauchi et al., 1977; Regard et al., 1978) suggests that this hormone might play a role in lung growth and development. Biochemical studies on lung from T_3-stimulated tadpoles, by Atkinson and Just (1975), were aimed at identifying lung macromolecular synthetic

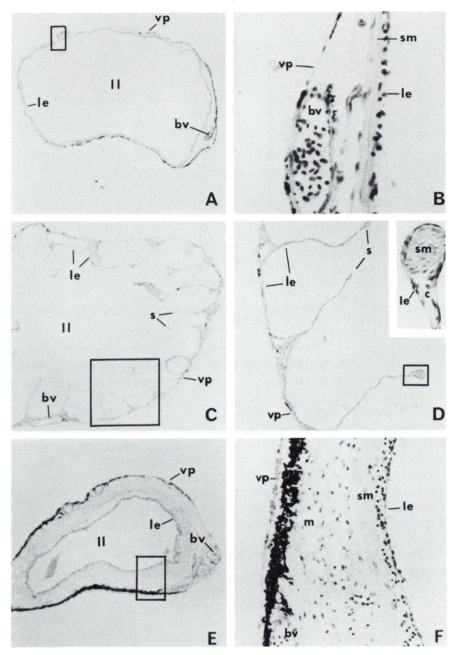

FIGURE 10. The histology of *R. catesbeiana* tadpole lungs. (A) Cross section through the midregion of right lung of stage X tadpole (×30). (B) Enlargement of rectangle in (A) (×450). (C) Cross section through the mid region of right lung of stage XXII tadpole (×15). (D) Enlargement of rectangle in (C) showing three septa (×50). Enlargement of rectangle in (D) showing the tip of the septa (×200). (E) Cross section through the midregion of right lung of stage X tadpole treated with T₃ 16 days earlier (×45). (F) Enlargement of rectangle in (E) (×170). Abbreviations: bv, blood vessel; c, capillary; ll, lung lumen; m, melanocytes; le, lung epithelium; s, septa; sm, smooth muscle; vp, visceral pleura. Data from Atkinson and Just, 1975, *Dev. Biol.* **45**:151.

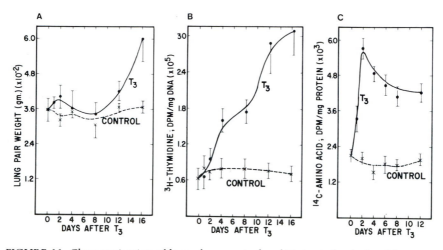

FIGURE 11. Characterization of lungs from control and T_3-treated tadpoles. Measurement of lung wet weights (A), and of the *in vivo* rates of [^{3}H]thymidine (B) and ^{14}C-labeled amino acid hydrolysate (C) incorporation into TCA-precipitable fractions of tadpole lungs. Data from Atkinson and Just, 1975, *Dev. Biol.* **45**:151.

transitions evoked by the influence of this hormone. They found that lungs in tadpoles stimulated with T_3 exhibit no significant difference in weight from control tadpoles until 12–16 days following hormone administration (Fig. 11A): the eventual weight gain is similar to the gain observed in stage XV of spontaneous metamorphosis. During the experimental period the rates of incorporation of precursors for protein and DNA remain constant in lung tissue of control animals, but increase rapidly and concurrently in lung tissue of hormone-treated tadpoles (Fig. 11B and C). The maximal rate of incorporation for each precursor is comparable to the maximal rate observed for the same precursor in lungs of spontaneously metamorphosing tadpoles (compare Fig. 11B and C with Fig. 9B and C).

   Although the aforementioned TH-induced biochemical transitions occurring in the anuran lung were similar to those observed during spontaneous metamorphosis, lung histology was dissimilar (Atkinson and Just, 1975). Lungs from TH-treated tadpoles showed few or no septa and exhibited a pronounced thickening of the lung wall (Fig. 10E and F). The thickened lung wall was composed of epithelium lining a core of abundant smooth muscle and numerous melanocytes. While these dramatic morphological changes are a result of TH administration to the tadpole and probably reflect the intense protein and DNA synthesis occurring in this tissue, they emphasize the possibility that factors (e.g., changes in aortic arch blood circulation, other hormones, tissue competence, etc.) other than TH may be also instrumental in preparing this organ for its respiratory function.

2.4. Conclusion

It is evident from the above discussion that tissues in the anuran limbs and lungs must undergo considerable cellular differentiation and development before they can assume their physiological functions. Although a direct role for TH in these developmental processes remains to be established, their influence is well documented. The diverse nature of the biochemical and morphological tissue changes occurring in these organs during this transitional period makes it highly unlikely that TH are directly responsible for initiating each change. The common thread linking the tissue transitions in both limb and lung focuses on a requirement for enhanced vascularization. A primary effect of TH on the differentiative and developmental processes in the anuran limb and lung may eventually center on modulation of the vascular elements.

3. SPONTANEOUS AND THYROID-HORMONE-INDUCED DEGENERATION OF ANURAN TADPOLE TAIL AND GILLS

3.1. Introduction

The only tadpole organs to degenerate completely during the metamorphic transition are the tail and gills. While gill degeneration has received scant attention, the dramatic degeneration occurring in the tadpole tail (Fig. 1) has led to its use as an experimental system for dissecting the processes initiating and involved in cellular death.

In classical experiments the tadpole tail was considered a homogeneous entity, but the organ consists of structural elements representing different histological types. By weight and volume the majority of the tail is skeletal muscle, consisting of metamerically arranged red and white fibers. The muscle fibers surround a core of connective tissue comprised of the notochord, neural tube, nerve fibers, and vascular elements. External to the muscle fibers and surrounding the whole tail structure is an epidermis attached to a well-developed basement lamella (Atkinson and Atkinson, 1976), with subjacent mesenchymal cells.

Atrophy of the tissues in this organ is the hallmark of tadpole metamorphic climax, which sets in concomitantly with the eruption of the forelimbs (stage XX). In all anuran species investigated so far, the retraction of the fin is the first discernible sign of tail atrophy. Shortening of the tail stem ensues, with eventual complete degradation of this organ.

Since the time of Gudernatsch's pioneering experiments, investigators have been aware that TH play a critical part in precipitating metamorphosis and degeneration of the tail and gill tissues. Studies on spontaneous and TH-induced metamorphosis have focused more on tail degeneration than on any other single metamorphic event. Until recently most investigations on anuran tail regression were aimed at identifying a common, hormone-induced mechanism for the histolytic processes occurring within this organ. Portions of a number of past reviews on anuran metamorphosis (Weber, 1969; Frieden and Just, 1970; Atkinson, 1971; Dodd and Dodd, 1976; Cohen *et al.*, 1978) highlighted general aspects of spontaneous and TH-induced tail degradation, but did not separate and comparatively evaluate *in vivo* and *in vitro* information derived from studies of whole tail and specific tail tissues. In this section a comprehensive and comparative picture of tail and gill tissue degeneration is presented and the numerous studies on tadpole tail are outlined in such a manner as to emphasize that tail degeneration is actually made up of a series of processes that are specific and perhaps different for each tail tissue.

3.2. Tadpole Tail

3.2.1. Spontaneous Degeneration of the Tadpole Tail *in Vivo*

Degeneration of the nervous system and/or depressed blood flow in tadpole tail tissue were proposed as causative agents of spontaneous atrophy of the tadpole tail and served as the basis for several early inquiries. Several early investigators suggested that the primary factor in tail tissue involution may be acidosis, caused by the accumulation of "carbon dioxide and acids of incomplete combustion," and the hypothesis was advanced that restricted blood flow might be affecting tissue pH and possibly initiating tail regression. However, results from vascular ligation experiments led Helff (1930) to suggest that the vascular rearrangement and reduced blood flow through the degenerating tail of metamorphosing tadpoles "are the result, and not the cause, of the various histolytic changes which occur during larval transformation." Studies by Brown (1946) substantiated the validity of Helff's conclusion, and also clearly established that tail muscle retraction and degeneration precede any degeneration of the neighboring nerve motor endings or capillaries. Ultrastructural studies in our own laboratory reveal that capillaries next to degenerated tail muscle fibers show no evidence of structural alteration. Brown's conclusion, that tail degeneration is not due initially to a failure of blood supply or innervation, coupled with Helff's (1930) conclusion that "the various tissues of the larval anuran's tail inherit a susceptibility to certain histolytic agents which are liberated and become functional at a certain stage of metamorphosis" led to a number of experiments aimed at identifying the histolytic agents and pinpointing their source.

Biochemical correlation between the activity of histolytic agents and protein degradation was first demonstrated by Liosner and Blacher. They found that tadpole tail tissue exhibited an 82% increase in free amino nitrogen prior to any visible morphological change. A rise in polypeptide nitrogen was reported to occur some time thereafter. That the increase in release of nitrogen-containing compounds might be due to enhanced activity of proteolytic enzymes in the tail was widely accepted with the impact of the lysosome theory (de Duve et al., 1951), which proposed a new category of cytoplasmic particles. The lysosomes, "suicide bags" endowed with acid hydrolases, were proposed to be instrumental in cell autolysis. Agrell's (1951) timely study on the enhanced activity of proteinases in the tail tissues of climactic tadpoles foreshadowed a great number of subsequent reports on the enzymatic changes accompanying regression of this organ (see reviews by Frieden and Just, 1970; Atkinson, 1971; Weber, 1969; Dodd and Dodd, 1976; Cohen et al., 1978). It soon was apparent that despite the differences in anurans and in the biochemical assays used, a common pattern of enzyme-specific changes in activity occurs during spontaneous tail regression. The involution of this organ coincides with a general decrease in the specific activity of enzymes related to energy metabolism and a general increase in the specific activity of the hydrolases, several of which are known as lysosomal enzymes, namely cathepsin,[*] β-glucuronidase, phosphatase, ribonuclease, and deoxyribonuclease. Inhibition of tail hydrolytic enzyme activity and tail resorption in spontaneously metamorphosing tadpoles by the administration of actinomycin D (Weber, 1969) suggests that the accumulation of these enzymes in the regressing tail must require new RNA synthesis, and further supports the contention that a functional correlation must exist between the enhanced activity of these lysosomal enzymes and tissue destruction.

Since changes in the rate of protein degradation (resulting presumably from the observed enhanced proteinase activity) were assumed but had not been directly measured in vivo in tails of spontaneously metamorphosing tadpoles, Little et al. (1973) devised a dual-labeling procedure (Atkinson and Little, 1972; Little et al., 1973) for measuring independent simultaneous rates of protein degeneration and synthesis in the tail in vivo. The results from these experiments establish that an increase in the rate of protein degradation is first evident in R. catesbeiana tadpole tail between stages XX and XXI, and that its detection coincides with the initial increase in cathepsin activity and loss in tail length and weight (see Figs. 1 and 12). Cell-free protein degradation studies with whole tail homogenates confirmed the in vivo findings and indicated that the in vitro rate of protein degradation is much more rapid at pH 3.3 than at 6.8, suggesting acid hydrolase involvement. The rate of protein degradation obtained for a tail of a stage XXII tadpole was calculated to be sufficient (2.2×10^{-5} mg protein/mg body wt/min) to completely degrade all of the protein in a tadpole tail in 3.7 days: the actual number of days required for a stage XXII tadpole to metamorphose to stage XXIV is 4.

---

[*] Primarily cathepsin D, since pepstatin inhibits most of the cathepsin activity in tail extracts (Sakai and Horiuchi, 1978).

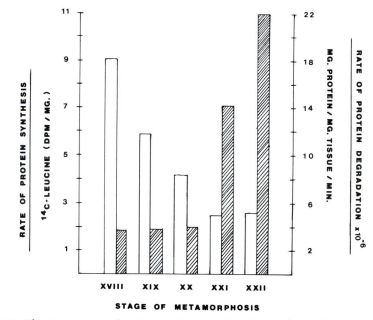

FIGURE 12. The *in vivo* rate of tadpole tail protein degradation (hatched bars) and incorporation (open bars) of [^{14}C]leucine into protein at various stages of metamorphosis. Data from Little *et al.*, 1973, *Dev. Biol.* **30**:366.

This report (Little *et al.*, 1973) also revealed that a marked decrease in the rate of protein synthesis (initially evident at stage XVII and progressively less thereafter) precedes any detectable change in the *in vivo* and *in vitro* rates of protein degradation (Fig. 12). This marked depression in the rate of protein synthesis which precedes enhanced protein degradation draws attention to the possibility that molecular events other than those concerned directly with degradation are critical to the initiation of tail resorption in spontaneous metamorphosis.

The tadpole tail structure is comprised of several different tissues and cell types. Do the changes in hydrolytic enzyme activity and protein synthesis reflect an overall reaction of the tail, or tissue- or cell-specific responses to the metamorphic hormones? Early studies (cf. Eeckhout, 1965), mainly light microscopic observations of stained tail tissue sections, supplied many details on the degenerative events in various tail tissues (leading to the hypothesis that mesenchymatous cells transform into actively phagocytosing macrophages). It is only in recent years that histochemical and fine structural studies have been directed toward pinpointing the distribution of hydrolytic enzyme activity within the tissues and cells of the tail.

Detailed studies on metamorphosing *X. laevis* and *R. Tigerina* tails indicate that acid phosphatase, cathepsinlike esterase, and β-glucuronidase activity are localized primarily in macrophages of the dermis. The enzymatic activity within these macrophages is confined to cytoplasmic inclusions containing ingested products of other tail tissues, which, by virtue of their ultrastructural characteristics, have been regarded as phagosomes (Weber, 1964). The progressive activation of macrophage activity, beginning in the connective tissue of the tail fin and spreading later to the connective tissue of the tail stem, has been the subject of several electron microscopic studies. Although it is exceedingly difficult in static two-dimensional electron micrographs to demonstrate phagocytosis by macrophages unequivocally, these studies contend that during spontaneous tail regression macrophages invade and phagocytose basement lamellar collagen (cf. Atkinson and Atkinson, 1976, for terminology), intermediate and notochordal collagen, and degenerating nerve, epidermal, and muscle cells (Weber, 1964; Fox, 1973; Lapiere and Gross, 1963; Usuku and Gross, 1965; Eisen and Gross, 1965). The highly necrotic nature of most tissues (except collagen, *incertae sedis*) prior to engulfment by the invading macrophages (Fox, 1973) suggests that macrophage activation and phagocytic activity may be stimulated by the necrotic state of the tissue (Weber, 1964). While the invading macrophages may liberate hydrolyzing enzymes to the exterior, putatively to enhance the degradation of connective tissue or aid in the further digestion of other partially degraded tissue, it is now apparent that their role and the role of their hydrolytic enzymes are to assist in the removal of tissue debris resulting from previous cell death.

Although the above studies stress the sequestration of hydrolytic enzyme activity in macrophages and their subsidiary role in tissue destruction, a series of histochemical studies from Kaltenbach and co-workers (Kaltenbach *et al.*, 1979) indicate that acid phosphatase, nonspecific esterase, aminopeptidase, alkaline phosphatase, Mg^{2+}-activated ATPase, thiamine pyrophosphatase, and 5'-nucleotidase are localized in epidermis, notochord, nerve cord, spinal ganglia, and vascular endothelium, as well as in connective tissue, and that they increase as tail resorption proceeds. Recent ultrastructural studies of specific tissues within the resorbing tadpole also disclose that the enhanced acid phosphatase activity is commonly located within primary lysosomes of degenerating tail epidermis, notochord, and nerve cord cells (Fox, 1973). Except for Mg^{2+}-activated ATPase, Kaltenback and co-workers could not detect any of the above-mentioned enzymes in regressing tail striated muscle cells, and studies by other investigators have stressed the absence of acid phosphatase reaction product within primary lysosomes of this particular tissue (Weber, 1964).

While the results from numerous biochemical studies of enzymatic activity—particularly the acid hydrolases—provided a possible lysosome-mediated, autolytic mechanism for degradation of the tadpole tail, the histochemical and electron microscopic studies of individual tail tissues point out the oversimplification of invoking a single autolytic mechanism to account for the regression of this complex organ. Prior to macrophage

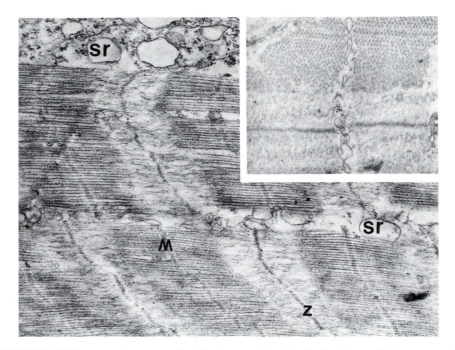

FIGURE 13. Electron microscopy of early degeneration of tadpole tail muscle (stage XXIII) of natural metamorphosis. Most of the fibrillar structure remains intact until the final stages of fiber degradation. M and Z lines are not visibly affected, although there is a frayed appearance of the thin filaments (I band) in the Z region. The sarcoplasmic reticulum (sr) is grossly vesiculated, and polyribosomes are visible in the cytoplasm. ($\times$21,500.) Inset: Early degeneration induced by TH (5 days) is comparable to the natural events, in that fraying of the thin-filament array in the I band near the Z line is one of the first signs of atrophy. The mitochondria remain intact, with a slight blurring of the membranes of the cristae. ($\times$24,000.)

engulfment or epidermal sloughing, degradation of most types of cellular tissues within the tail appears to exhibit independent, lysosome-mediated autolysis; however, the extracellular collagen framework and striated muscles—the majority of the tail tissue mass—apparently are initially degraded by other mechanisms. The orthogonally packed, collagenous layers of the tail basement lamella (Atkinson and Atkinson, 1976) are thought to be loosened initially by a hyaluronidase, secreted from mesenchymal cells, which degrades the mucopolysaccharide ground substance and releases bound water (Eisen and Gross, 1965). Further maceration of the collagen fibrils by diffusion of collagenolytic enzymes from the epidermis is presumed to permit macrophages to invade the area and engulf the partially digested collagen (Usuku and Gross, 1965; Davis *et al.*, 1975). The initial degradation of tail striated muscle cells also occurs prior to macrophage engulfment, but appears to be an autolytic process that is not dependent on intralysosomal enzymatic digestion for myofibrillar degradation. The general failure to identify primary lysosomes or autophagic vacuoles in the initial phases of myofibrillar degradation (e.g., Weber, 1964) and the failure of lysosomal proteases to partially or wholly digest myofibrillar protein (e.g., Dayton *et al.*, 1976) make the mechanism for the *initial* regression of tail muscle incompatible with the lysosomal theory of autolysis, as described by de Duve *et al.* (1951).

The autolytic mechanism involved with the initial degradation of tail muscle during normal metamorphosis remains an intriguing problem. It may well entail an initial and progressive depressed rate of protein synthesis (Little *et al.*, 1973) coupled with an activation of nonlysosomal, myofibrillar proteases such as the Ca^{2+}-activated neutral protease described in other muscle systems to preferentially hydrolyze the proteins within the Z line of the I band (e.g., Dayton *et al.*, 1976). Indeed, the loss of the I-band integrity (Fig. 13) within myofibrils has been reported as the first indication of muscle regression, occurring prior to any measurable reduction in tail length or the appearance of lysosomes (Fig. 14) (Weber, 1964). The precocious disappearance of the I bands and persistence of the A bands just prior to metamorphic climax (Weber, 1964) suggest a differential chemical erosion of the myofibrillar proteins, which may result from decreased myofibrillar protein synthesis, increased nonlysosomal protease activity (such as a Ca^{2+}-activated neutral protease), or both.

The numerous studies on spontaneous tail regression suggest a process that may be divided into three distinct phases. The initial phase, beginning at stage XVIII or XIX in *Rana*, is characterized solely by a marked and progressive decrease in the rate of general tail protein synthesis (Little *et al.*, 1973). This initial phase encompasses the stages of tadpole development at which Just (1972) was first able to detect an increase in the protein-bound iodine of tadpole plasma, and the stages of development at which both Miyauchi *et al.* (1977) and Regard *et al.* (1978) first detect slight increases in tadpole plasma T_3 levels. The second phase is marked by degenerative changes in the individual tail tissues, enhanced hydrolase activity, and tissue-specific histolysis. While the histolysis occurring in this phase appears to involve lysosome-mediated autolytic mechanisms for most tissues, the extracellular collagen framework and striated muscles are initially degraded by other mechanisms. The third phase, which temporally overlaps considerably with the second phase but is probably dependent on it to some extent, involves an activation and enhanced phagocytic activity of the macrophages. The processes in the second and third phases of tail degradation occur in stages of tadpole metamorphosis (XX–XXIII) at which maximal levels of protein-bound iodine, T_3, and T_4 occur in tadpole plasma (Just, 1972; Miyauchi *et al.*, 1977; Regard *et al.*, 1978).

3.2.2. Induced Degeneration of the Tadpole Tail *in Vivo*

The importance of the thyroid gland for the initiation of such metamorphic responses as tail degeneration was discovered over 60 years ago

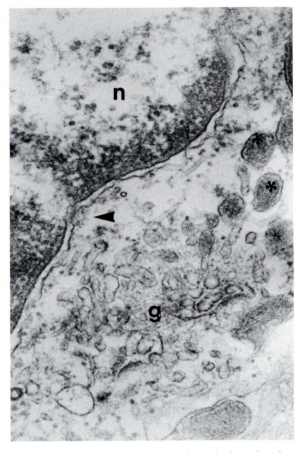

FIGURE 14. Golgi complex in a muscle fiber of the tadpole tail 5 days after hormone administration. The location in muscle is corroborated by adjacent micrographs shot in tandem. The dark bodies bounded by a single membrane are seen in other photographs apposed and sometimes conjoined to mitochondria. Their origin from the Golgi and their presence near degenerating mitochondria strongly imply that these are primary lysosomes. n, muscle nucleus; g, Golgi complex; asterisks, putative primary lysosomes; arrowhead, nuclear pore. (×70,000.)

by Gudernatsch, who found that tadpoles fed on mammalian thyroid material responded by precociously metamorphosing. Later experiments established that removal of the thyroid anlage at tail bud stages prevented tail degeneration and other metamorphic transitions, and resulted in giant tadpoles. Subsequent feeding on thyroid material caused the thyroidectomized tadpoles to undergo metamorphosis (Allen, 1925). The dramatic nature of tadpole tail resorption and indications that hormones from the thyroid gland expedite this event gave rise to a number of experimental studies aimed at elucidating the mechanism whereby these hormones initiate cell death and tissue destruction.

Studies demonstrating that both T_3 and T_4 are synthesized in tadpoles and frogs prompted investigators to use T_3 as well as T_4 to invoke tail degeneration. Exogenous supplementation with these hormones is generally conducted either by immersing the tadpoles in a hormone solution continuously or by injecting them intraperitoneally. In some cases, TH-containing cholesterol pellets have been implanted into various tissues. While a number of controversies rightfully persist regarding the amount of hormone administered and its subsequent fate, and over the manner of administering these hormones to effect *in vivo* tail degeneration in a seemingly normal manner (see reviews by Frieden and Just, 1970; Etkin, 1968), most reported investigations have attempted to avoid these controversial issues by using at least one established parameter of spontaneous tail degeneration as their index of comparison. The correlation of TH-induced tail degeneration with spontaneous metamorphosis is complicated by the fact that a number of laboratories have shown that the *in vivo* rate of hormone-induced tail degeneration is dependent on the species and age of the tadpole, environmental temperature, hormone (T_3 or T_4), hormone concentration, and manner of administering the hormone. Underlining all of these complexities is the fact that tadpole tail is made up of a variety of tissues and cell types, each of which may

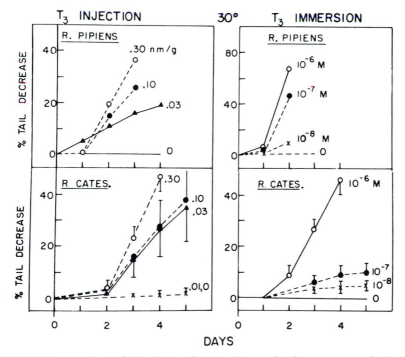

FIGURE 15. Comparison of injection and immersion on the dose-response relationship of *R. pipiens* and *R. catesbeiana* tadpoles at 30°C. Data from Ashley *et al.*, 1968.

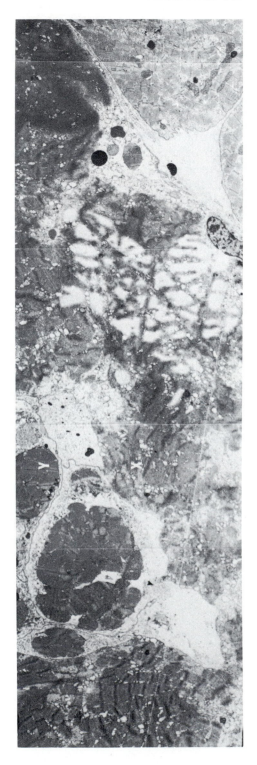

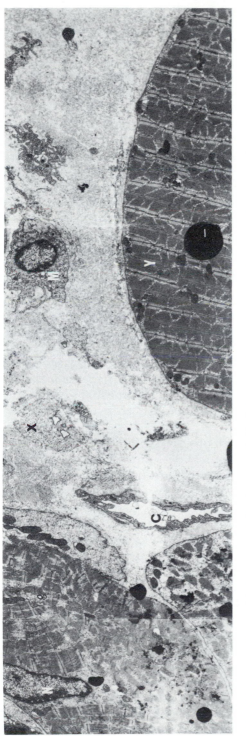

FIGURE 16. Montage of electron micrographs showing degeneration of tadpole tail muscle after T₃ administration. Study of grouped photographs clarified the relationship of highly atrophied fibers (X) amid apparently normal ones (Y). C, lumen of capillary; M, macrophage; N, myonucleus.

require a particular and different level of TH to exhibit and maintain a response. A number of recent *in vivo* and *in vitro* studies attempt to assess the hormonal requirements and analyze the responses in particular tail tissues. Until we have gained a more complete understanding of the stoichiometry of the molecular interactions of TH with individual target cells and their constituents, disagreement about hormone concentrations and delivery will persist.

Administration of exogenous T_3 or T_4, by immersion or injection, to stage VII–XII *R. pipiens* or *R. catesbeiana* tadpoles results in the detectable regression of tadpole tail tissue in a dose-dependent manner (Fig. 15). Most anuran tadpoles respond similarly to these hormones. As in spontaneous degradation, the first grossly observable sign of this is the darkening and retraction of the tail fin, followed shortly thereafter by a progressive decrease in length. Electron microscopic studies on the regressing tail fin (Usuku and Gross, 1965) demonstrate that both T_4-induced and spontaneously degenerating tail fin exhibit the same sequence of ultrastructural transitions. Our own studies (Atkinson and Atkinson, unpublished) on T_3-induced tail muscle degeneration also indicate that this muscle exhibits an ultrastructural pattern of degeneration similar to that detected in spontaneously regressing tail muscle. From our fine structural studies (Fig. 16), it is evident that tail muscle from T_3-treated tadpoles degenerates in a distal to proximal manner, that adjacent myofibers exhibit different degrees of degeneration, and that vascular elements in highly necrotic areas remain intact. As in spontaneous tail muscle degeneration (Weber, 1964), the loss of myofibrillar I-band integrity appears to be an initial and conspicuous sign of myofibrillar degeneration (see inset in Fig. 13).

The initial decrease in tail length observed in hormone-induced tadpoles is concurrent with the biochemical detection of enhanced hydrolytic enzymatic activity in this organ (see previous reviews: Frieden and Just, 1970; Weber, 1969). Although the pattern of hydrolytic enzymatic activity and its distribution among various tail tissues (Kaltenbach *et al.*, 1979) are similar to those observed in spontaneously regressing tails, the increase in biochemical activity is rarely as great. However, prior to any morphological signs of degeneration or measurable enhanced hydrolytic activity, tails from both spontaneously metamorphosing and hormone-induced tadpoles exhibit a similar *in vivo* decreased rate of protein synthesis (Tonoue and Frieden, 1970; Little *et al.*, 1973). The possibility that this early protein synthetic response to enhanced endogenous and exogenous TH levels might trigger subsequent degenerative changes has led to similar *in vivo* protein synthetic studies with particular tail tissues. Administration of T_3 or T_4 to premetamorphic tadpoles was subsequently found to cause an early and continuous depression in the rate of protein synthesis both in tail muscle (Fig. 17) and in tail fin (Saleem and Atkinson, 1978; Yamaguchi and Yasumasu, 1978; Kistler *et al.*, 1975). The rapid response of these different tail tissues to TH suggests that its initial effect

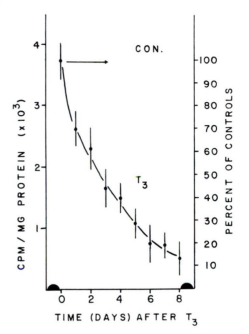

FIGURE 17. Effects of T_3 treatment on the *in vivo* rate of incorporation of [^{14}C]leucine into TCA-precipitable fractions of tadpole tail muscle homogenates.

on protein synthesis may be common to all competent tissue and required for subsequent tissue-specific histolytic processes.

A variety of transcriptional and translational mechanisms may be envisioned for TH depression of the *in vivo* rate of protein synthesis in tail muscle. We initially focused our attention on determining if this hormonal effect is due to decreased efficiency of the muscle translational apparatus. Our *in vitro* studies with discrete size classes of tail muscle polyribosomal complexes (Fig. 18) from 2- and 5-day hormone-treated animals demonstrated that the polysomes are less active *in vitro* than comparable quantities of similar-sized polyribosomal complexes from control tadpole tail muscle (Saleem and Atkinson, 1978). The decreased *in vitro* translational efficiency of the polyribosomes from tail muscle of hormone-treated tadpoles is caused by an inhibitory factor present in the pH 5 fraction of the postribosomal supernatant from tail muscle of hormone-treated tadpoles (Saleem and Atkinson, 1978, 1979). This inhibitory factor is heat labile and trypsin sensitive, and inhibits 40% of the synthetic activity of control polyribosomes when assayed in a reconstituted, cell-free translational system. This inhibitor was shown indirectly to inhibit mRNA translation by preventing formation of new initiation complexes (Saleem and Atkinson, 1980), and was proposed to selectively inhibit the *in vivo* translation of particular muscle proteins.

In a study aimed at examining the effect of this hormone on the *in vivo* synthetic rate of individual tail muscle proteins, Merrifield and Atkinson (1979) found that TH administration causes a fluctuation in

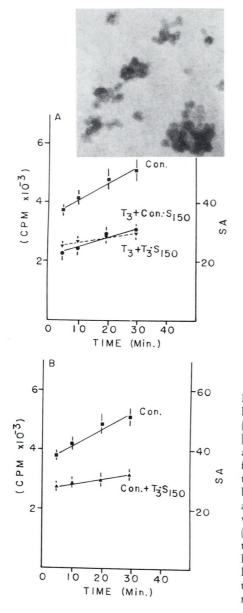

FIGURE 18. Effect of the tadpole tail muscle high-speed supernatant (S_{150}) source (Con.·S_{150} or T_3·S_{150}) on the time course of [^{14}C]leucine incorporation *in vitro* by heavy and medium-sized polyribosomal fractions from tail muscle of control (Con.) and T_3-treated (T_3) tadpoles. "Con." *not* followed by a designated source of S_{150} refers to preparations in which polyribosomes and S_{150} were both obtained from control tadpoles (see Saleem and Atkinson, 1978). Inset photograph (Atkinson and Atkinson, unpublished) is an electron micrograph of the polyribosomes used in these studies. In general, the polyribosomes appeared to contain 12–60 ribosomes.

the amino acid pool, a significant and progressive decrease in the rates of myosin, actin, and tropomyosin synthesis, and only a transitory decrease in total sarcoplasmic protein synthesis (Fig. 19). While these observations certainly agree with the earlier evidence for a selective translational inhibition of myofibrillar protein mRNAs (Saleem and Atkinson, 1978), they also support proposals suggesting that TH may act in tadpoles as selective repressors/activators of gene transcription (Cohen *et al.*, 1978).

In order to establish the relative contribution of transcriptional and translational events to the changes in myofibrillar and sarcoplasmic protein synthesis, we characterized the mRNA populations of tail muscle from control and T_3-treated tadpoles by *in vitro* translational analysis (Merrifield, 1979; Merrifield and Atkinson, 1979). Since these analyses were conducted under conditions where product formation was proportional to mRNA concentration, it was possible to determine the relative concentrations of individual mRNAs as well as the diversity of total mRNA populations from tail muscle of control and T_3-treated tadpoles. An examination of the macromolecular constituents of tail muscle demonstrates that the mRNA content (determined from poly-A^+ RNA/g muscle) of tail muscle from T_3-treated tadpoles is approximately 93% of that found in control animals (Table 5). Moreover, *in vitro* translational analysis of the poly-A^+ RNA populations from muscle of control and T_3-treated tadpoles revealed that the decrease in myofibrillar protein synthesis (specifically actin and tropomyosin) observed *in vivo* is, relative to other mRNAs (e.g., creatine phosphokinase), *not* correlated with a decrease in the amount of translatable poly-A^+ RNA for actin and tropomyosin.

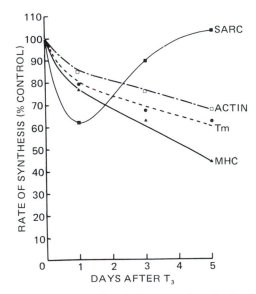

FIGURE 19. Effect of T_3 treatment on the relative rates of synthesis of myosin heavy chain (M), tropomyosin (T), actin (A), and sarcoplasmic (S) proteins in tadpole tail muscle *in vivo*. Data from Merrifield and Atkinson, 1979; Merrifield, 1979, Ph.D. thesis, University of Western Ontario, London, Canada.

TABLE 5. MACROMOLECULAR COMPONENTS OF TAIL MUSCLE FROM
NORMAL AND 5-DAY T$_3$-TREATED TADPOLES

	Normal	5-Day T$_3$-treated	Treated/normal
RNA (mg/g wet wt muscle)	1.32 ± 0.09[a]	1.35 ± 0.06	1.03
DNA (mg/g wet wt muscle)	0.528 ± 0.09	0.599 ± 0.04	1.14
Protein (mg/g wet wt muscle)	166.4 ± 4.4	170.0 ± 8.2	1.03
RNA/DNA	2.50	2.25	0.90
Protein/DNA	315.2	283.9	0.90
mg poly-A$^+$/mg RNA[b]	0.030	0.028	0.93
mg poly-A$^+$/g wet wt	0.041	0.038	0.93

[a] Values are the mean from six animals ± S.E.M.
[b] Average of the amount of total RNA bound to oligo dT from two different preparations of cytoplasmic RNA isolated from normal and 5-day T$_3$-treated animals (Merrifield, 1979; Merrifield and Atkinson, 1979).

This firmly establishes that the T$_3$-induced decrease in the synthesis of these myofibrillar proteins is partially, if not wholly, a result of inhibition at the translational level.

In order to detect differences in the *in vitro* translational products of poly-A$^+$ RNA from muscle of control and T$_3$-treated animals, the newly synthesized proteins were separated by two-dimensional gel electrophoresis and analyzed by fluorography. The major translational products synthesized by comparable amounts of poly-A$^+$ RNA from control and T$_3$-treated tadpoles are qualitatively and quantitatively similar (Fig. 20), but a number of subtle differences are evident (see circled numbers in Fig. 20). These differences result from either *de novo* (proteins 4, 6–9) or selectively increased synthesis (proteins 1–3, 5) of a few proteins in the translational mixture primed with RNA only from T$_3$-treated animals. These results imply that the poly-A$^+$ mRNA population of tail muscle from T$_3$-treated tadpoles is more diverse than that from control animals, and uphold the idea that TH may also influence the transcription of *specific* genes.

Reports dealing with *in vivo* TH-induced regression of the anuran tail substantiate that this process is TH dependent and is comparable to spontaneous tail regression. Regression involves at least three distinct phases of cellular activity for completion: initiation of the second and third phases of the degenerative process in tail muscle (as well as whole tail) is preceded by an initial phase characterized as an early, sustained depression in the rate of protein synthesis. The depressed *in vivo* rate of protein synthesis in tail muscle of hormone-treated tadpoles is specific for several myofibrillar proteins, and results from synthetic inhibition at the translational level. Sustained *in vivo* synthesis of sarcoplasmic proteins and the diversity of the *in vitro* translated products of poly-A$^+$ RNA from muscle of hormone-treated tadpoles suggest that TH may also direct the transcription of specific genes.

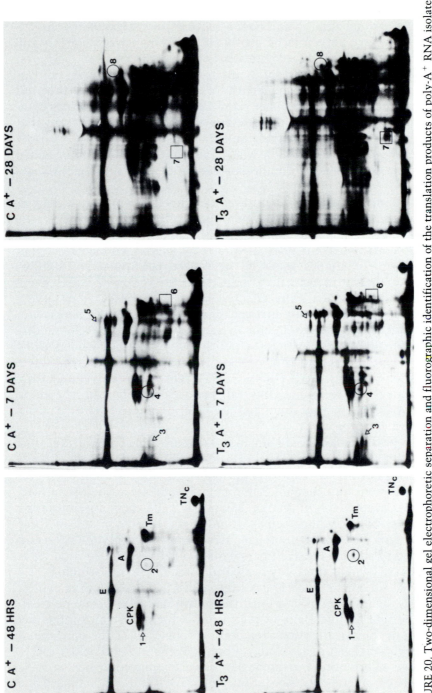

FIGURE 20. Two-dimensional gel electrophoretic separation and fluorographic identification of the translation products of poly-A$^+$ RNA isolated from tail muscle of control (CA$^+$) and T$_3$-treated (T$_3$A$^+$) tadpoles. Fluorograms were exposed at −70°C for 48 hr, 7 days, and 28 days to distinguish the major and minor translational products. Numerals designate qualitative and quantitative differences between CA$^+$ and T$_3$A$^+$ products (see text). Tentatively identified proteins include creatine phosphokinase (CPK), actin (A), tropomyosin (T), and troponin c (TN$_c$). Data from Merrifield and Atkinson, 1979; Merrifield, 1979, Ph.D. thesis, University of Western Ontario, London, Canada.

3.3.3. Induced Degeneration of the Tadpole Tail *in Vitro*

Of the various amphibian larval structures investigated, whole tadpole tail tips have proven to be highly suitable for *in vitro* analysis and elucidation of the TH-induced regression of this organ. Shaffer (1963) first demonstrated both a method for maintaining tadpole (*X. laevis*) tail tips in culture and a hormone dependence for the involution of cultured tail tips. Subsequent studies by Hauser and Lehmann described a less complex, nutrient-free medium (Holtfreter's solution) for tail culture that also permitted TH-induced regression of tail tips (Weber, 1969). A variety of nutrient-free media in which whole tadpole tail tips are maintained and respond to thyroid hormone supplements were subsequently described.

Morphometric analysis of cultured tail tips has established that incubation temperature affects the maintenance of control tail tips, the timing of the TH response, and the magnitude of the response (Lindsay *et al.*, 1967). Although the original studies by Shaffer were conducted at 26°C, subsequent studies generally have been performed at 18–20°C. While a number of investigators induce the resorption of isolated tail with T_4, T_3 has proven more effective than T_4 (Lindsay *et al.*, 1967; Robinson *et al.*, 1977) or T_4 analogs* (Lindsay *et al.*, 1967). Sensitivity of cultured tail tips to both T_3 and T_4 increases with the developmental stage of the tadpole from which the tail tips were taken, as evidenced by a shortened lag period before the onset of regression and an increased rate of regression (Shaffer, 1963; Hickey, 1971; Tata, 1966; Smith and Tata, 1976; Robinson *et al.*, 1977). Thus, tail tips are competent to respond to TH at early developmental times, but the kinetics of the response changes as development progresses.

When freshly amputated tadpole tail tips are placed in culture, they are generally allowed 2 or 3 days to heal before the medium is supplemented with hormones. Tail tips cultured wholly in the absence of hormone supplement form a small, heteropolar regenerate at the cut surface, which is composed of migrating epidermal cells, and differentiating neural tube and notochord cells. Although the muscle tissue in these isolated organs may undergo some reduction in size, the myomeres retain their integrity and the muscle fibers remain organized and striated for at least 12 days. Two- or three-day healed cultured tail tips to which T_4 (2.7 $\times$ 10^{-7} M; see Hickey, 1971) has been added generally do not exhibit any visible signs of regression until the fourth day after hormone treatment. The first visible signs of regression, which is a thickening and darkening of the tail fin edges, represent a transition of the epithelium from a single cell layer to a multilayer lining (Weber, 1969; Hickey, 1971). By 4–6 days of hormone treatment the isolate is reduced in size by 50%, the fins are completely absorbed, the connective tissue is condensed and

---

* The range of T_4 is generally 1.3×10^{-6} to 1.3×10^{-7} M. The range of T_3 is generally 1.5×10^{-6} to 4×10^{-9} M.

filled with macrophages and melanophores, and the various cellular tissues are highly necrotic.

In studies with healed *X. laevis* and *R. temporaria* tail tips cultured in media supplemented with upper-threshold amounts of T_3 or T_4, significant drops in total tail DNA, RNA, and protein occur (Weber, 1969; Tata, 1966; Eeckhout, 1965; Hickey, 1971) after the fourth day of hormone treatment. By 12 days of hormone treatment an 85% loss in total RNA or DNA and a 70% loss in total protein are evident. However, when the amounts of DNA, RNA, and protein extracted from the hormone-treated tails are reported on an organ weight basis, they appear to remain constant until the eighth day. Then a striking increase in each of these macromolecules is evident, suggesting that dehydration of these tissues is even greater than the degradation of macromolecular constituents (Tata, 1966).

TH-induced resorption of whole tail pieces in culture has also been characterized by two- to fourfold increases (beginning at about the fourth day) in the specific activity (per unit protein) of acid phosphatase (Filburn and Vanable, 1973; Robinson *et al.*, 1977; Hickey, 1971; Eeckhout, 1965; Tata, 1966), cathepsin (Tata, 1966, 1970; Hickey, 1971), DNase (Tata, 1970; Eeckhout, 1965; Hickey, 1971), RNase (Tata, 1966), β-glucuronidase (Eeckhout, 1965), and β-*N*-acetylglucosaminidase (Eeckhout, 1965). Although the total enzymatic activity (activity per tail tip) of cathepsin, DNase, β-glucuronidase, and β-*N*-acetylglucosaminidase apparently also increases in regressing tail tips, only Filburn and Vanable (1973), Robinson *et al.* (1977), and Hickey (1971) report an increase in total acid phosphatase activity. The enhanced acid phosphatase activity evident in tail tip regression is accompanied by shifts in the activity levels of its isozymes and a change in pH optimum and heat sensitivity (Robinson *et al.*, 1977; Filburn and Vanable, 1973). In general, the pattern of enhanced hydrolytic activity in hormone-induced regressing tail tip isolates resembles, less dramatically, the activity in regressing tails from TH-induced or spontaneously metamorphosing tadpoles.

Neither the cellular source nor the intracellular distribution of the enhanced hydrolytic activity in tail tip isolates has been specified. The ability of these isolates to exhibit enhanced hydrolytic activity excludes leukocytic invasion as a primary source of this activity. The molecular mechanism responsible for the enhanced hydrolytic activity in this organ is thought to involve either activation of latent enzymes, the selective protection of hydrolases against degradation, or the synthesis of new hydrolytic enzymes. Experiments designed to test both the "neosynthesis" theory and establish whether RNA and protein synthesis are required for hormone-induced tail tip regression and enhanced hydrolytic activity initially employed metabolic inhibitors of RNA and protein synthesis. In tail tip isolates of *R. temporaria* a number of these inhibitors (actinomycin D, puromycin, cycloheximide, and ethionine) were found to antagonize the effect of TH by abolishing [³H]uridine and ¹⁴C-labeled amino acid incorporation into RNA or protein, thereby inhibiting hydrolase activity and preventing involution of the tail tips (Tata, 1966, 1970). The ability of these inhibitors to block hydrolase activity and tail tip regression in other species is variable (Eeckhout, 1965). Although Eeckhout (1965) cautioned that the toxic effects of the inhibitors complicated interpretation of his results, they have been used to support the neosynthesis theory (Tata, 1966, 1970). The inconsistent effects of these inhibitors have been explained by differences in penetration into tail tissues (Weber, 1969). Recently, Smith and Tata (1976) reevaluated Tata's (1966) original inhibitor studies and noted that use of these inhibitors produces "visible tissue damage." Their subsequent proposal that "these experiments merely show that the structural integrity and organization of tail cells is necessary for the hormone to produce its effect" negates Tata's previous interpretation, and leaves the interpretation of results from similar inhibitor studies open to question.

In an attempt to directly establish whether enhanced RNA and protein synthesis are required for the TH-induced process of tail tip regression, Tata (1966) studied the incorporation of [^{3}H]uridine into RNA and ^{14}C-labeled amino acids into protein of isolated tail tips. He reported that hormone-induced regression was accompanied by accelerated synthesis of both RNA and protein prior to visible resorption. Although the results he obtained supported the conjecture that new proteins may be required for initiating regression, they have not been substantiated by results from investigations of spontaneous or TH-induced tail regression in intact tadpoles (Tonoue and Frieden, 1970; Little et al., 1973; Kistler et al., 1975; Saleem and Atkinson, 1978, 1979; Merrifield and Atkinson, 1979), by results from studies with cultured tadpole tails in other laboratories (Kawahara and Yamana, 1977), or by more recent results on isolated tail tips from his own laboratory (Smith and Tata, 1976). The tail tip incorporation study Tata (1966) drew his conclusions from is one of the few instances in the anuran tail tip literature in which freshly transected tail tips were not allowed to heal prior to being stimulated with TH. The design of these particular incorporation experiments makes it difficult to attribute the reported RNA and protein synthetic responses to the influence of TH on the intracellular mechanisms wholly involved with controlling tail regression. Since the freshly transected tail tips in Tata's experiments were recovering from the trauma of transection and were in the process of healing or regeneration, the observed synthetic effects of TH could have been attributed as easily to its effects on the process of regeneration. It is possible that incubation of freshly wounded organs in the presence of TH favors the continuous production of enzymes, such as collagenase, normally associated with wounding and regeneration and prevents the modulation of enzyme expression associated with hormone-dependent tail tip regression.

Results from more recent cultured tail tip RNA and protein incorporation studies (Kawahara and Yamana, 1977; Smith and Tata, 1976) assessed the effect of TH on healed tail tip synthetic activity, and differ from those originally obtained by Tata. Kawahara and Yamana, studying the effects of T_3 supplementation on healed tail tips in culture, established that under these conditions the rates of transcription of total RNA, ribosomal RNA, and poly-A^+ RNA do not change during hormone-induced tail regression. Some posttranscriptional changes, such as the rapid processing of ribosomal precursor RNA and degradation of ribosomal and transfer RNAs, take place at late stages of regression. In some unusually designed experiments Smith and Tata (1976) were able to demonstrate involvement of the procollagenase/collagenase activator system in the T_3-induced regression of cultured tail tips* but were unable to demonstrate any quantitative or qualitative changes in the proteins synthesized by the hormone-induced tail tips. While the analytical aspect of their work was precise, the design of their protein synthetic studies suffers from the same assumptions used in Tata's earlier work. In this study, immediately prior to a 20-hr labeled amino acid incorporation period, they sliced each tail tip into eight pieces and interpreted the resulting incorporation in terms of hormone-dependent responses. Since both the T_3-supplemented and the control tail tips were subjected to the same trauma and probably required the same intracellular repair processes to be activated for wound healing, it is not surprising that their results do not reveal any protein synthetic changes resulting from the hormone treatment. Transection probably favors the continuous production of proteins associated with wounding and healing at synthetic levels so enhanced as to prevent detection of any proteins possibly associated with hormone-dependent tail tip regression. Their conclusions may ultimately prove correct, but the design of the experiments upon which their conclusions are based makes the present results subject to several different interpretations.

3.2.4. Induced Degeneration of the Tadpole Tail Fin *in Vitro*

At about the same time that Shaffer (1963) was demonstrating that the tadpole tail can be maintained as a surviving organ culture and exhibit TH-dependent resorption, Gross and co-workers attempted similar studies with less complex pieces of the tadpole tail, the tail fin. Although Gross and co-worker (Lapiere and Gross, 1963) were unsuccessful in early attempts to demonstrate absolute TH-dependent resorption of cultured tail fins (the tissue resorbed regardless of the hormonal or developmental state of the donor tadpole), they did demonstrate that actively resorbing cultured tail fins produce a diffusible neutral collagenase and a diffusable hyaluronidase (β-*endo*-N-acetylhexosaminidase) and that cultured tail fins from TH-induced tadpoles produce more of these enzymes than

---

* See Davis *et al.* (1975) for T_4 effects on collagenase activity in cultured tail fins.

controls. Subsequent studies by Eisen and Gross (1965) established epithelial cells as the major source of the collagenolytic enzyme and mesenchymal cells as the primary source of hyaluronidase. Knowledge about
the source and diffusible nature of these enzymes led to the hypothesis
of cooperation between two cell types to account for the resorption of
fin extracellular connective tissue (Eisen and Gross, 1965). The production and diffusion of collagenase from the epidermal tissue into the underlying basement lamella resulted in partial and specific degradation of
collagen, depressing its melting temperature and rendering it susceptible
to attack by nonspecific proteases. Mesenchymal cells are proposed to
concurrently release a hyaluronidase that breaks down the collagenous
matrix and allows the mesenchymal cells to invade the basement lamella
and engulf the partially degraded collagen.

In 1968 Derby successfully demonstrated that tail fin disc explants
cultured at 22°C undergo complete reepithelization of their transected
surfaces within 24 hr, maintain normal tissue organization in subsequent
culture, and exhibit TH-dependent resorption. Tail fin from either *R.
pipiens* or *R. catesbeiana* can be maintained in culture at 22 or 37°C,
once reepithelized at 22°C, without spontaneous resorption and without
further collagenase production. Once healed, the tissue clearly is capable
of regressing and producing collagenase in response to T_4 stimulation
(Davis *et al.*, 1975). Healing also occurs when freshly cut explants are
cultured at higher temperatures (such as the temperatures utilized by
Gross and co-worker); however, tissue degradation occurs (Derby, 1968),
and catabolic enzymes, such as collagenase, appear to be expressed independently of the presence of T_4 (Davis *et al.*, 1975). In establishing
criteria for the *in vitro* use of healed tail fin discs as a means for elucidating the action of T_4, Derby noted that resorption is dependent upon
the continuous presence of TH and that the resorptive response of tail
discs varies directly with the concentration of T_4 and the age of the
tadpole donor.

Light and electron microscopic studies of healed cultured tail fin
discs disclose that the sequence of morphological events in the process
of TH-induced resorption (Gona, 1969) is closely comparable to the *in
vivo* process described by Usuku and Gross (1965). Four days after hormone treatment is initiated, the first effects observed are found in the
basement lamella, evidenced by a loosening of the collagen layers and
macrophagic invasion. At about the same time epidermal cells exhibit
large lysosomal bodies and some signs of atrophy. After 6 days of hormone
treatment the epithelial cells contain autophagic vacuoles and the epidermis is transformed, by detachment of the basal epithelial cells from
the basement lamella, from a two- to three-cell-layered epithelium to a
multilayered one. By the eighth day, epidermal cells are extensively necrotic, phagocytosis of the basement lamella is complete, and macrophages are clumped into masses (Gona, 1969).

The similarity in the ultrastructural and morphological response of tail fin tissue to T_4 in vitro to that of regressing fin in vivo is reflected by changes in the activity patterns of various acid hydrolases (Stuart and Fischer, 1979) and in the collagenase activity recovered from tissue media (Lapiere and Gross, 1963; Davis et al., 1975). While an enhancement of acid phosphatase, hyaluronidase, and β-N-acetylglucosaminidase specific activities and a constant lactic dehydrogenase activity are evident in vitro and in vivo, the matnitude of the changes is considerably less in cultured tail fins. Recent studies (Seshimo et al., 1977), employing several known proteinase inhibitors, demonstrate that pepstatin (an inhibitor of cathepsin D) is effective, in a dose–response manner, in blocking the resorption of T_3-induced cultured tail fin. Although indirect, these results strongly support the hypothesis that the quantitative changes in acid hydrolase and in collagenase activities that occur in synchrony with TH-induced fin resorption are directly associated with each other, and that these enzymes are concerned with the breakdown of the extracellular connective tissue components and possibly with the cellular histolysis occurring during tail fin resorption.

The established ultrastructural and enzymatic responses of the cultured tadpole tail fin system to TH and the cellular simplicity of the fin have prompted its use as an in vitro model for elucidating the subcellular mechanism(s) by which TH initiates the histolysis of specific fin tissues. Tail fin tissue has been used by several groups to demonstrate TH uptake and the mapping of its intracellular binding sites. Much of the evidence accumulating from mapping studies (reviewed elsewhere in this volume) supports an interaction of TH with the nuclear material, and suggests a direct hormonal effect on transcription. Other evidence aimed at elucidating the mechanism of the histolytic activity of TH suggests that some aspects of TH-stimulated fin resorption are mediated by cyclic AMP (cAMP). Enhanced cAMP levels (Stuart and Fischer, 1979) and collagenase (Davis et al., 1975) activities result from TH stimulation of tail fin cultures. Since addition of cAMP to tail fin cultures not supplemented with TH stimulates hyaluronidase activity (Harper and Toole, 1973; Stuart and Fischer, 1979) as well as activates the conversion of procollagenase to collagenase (resulting in enhanced collagenase activity: see Harper and Toole, 1973), cAMP probably plays a major role in the mechanism of this hormonal response. The results from studies (Stuart and Fischer, 1979) in which known inhibitors of the adenylate cyclase system were tested, support this contention and imply that TH initially stimulates the tail fin adenylate cyclase system, which subsequently induces the appearance of enhanced hydrolytic enzyme activity.

3.3. Tadpole Gills

3.3.1. Spontaneous Degeneration of the Gills in Vivo

In the normal course of amphibian metamorphosis the tadpole gills degenerate completely (Fig. 21A) or become vestigial. The histology of

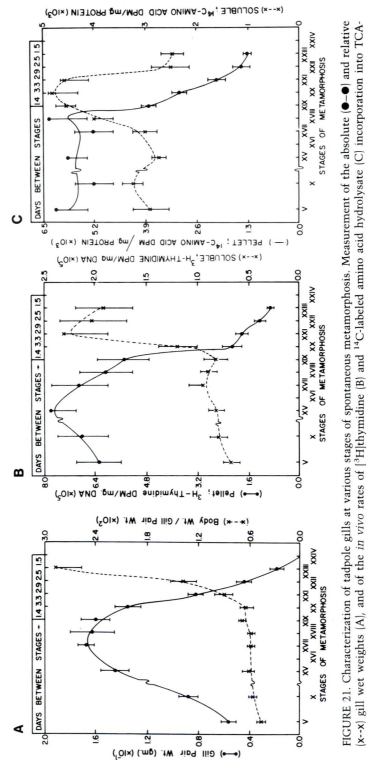

FIGURE 21. Characterization of tadpole gills at various stages of spontaneous metamorphosis. Measurement of the absolute (●—●) and relative (x—x) gill wet weights (A), and of the *in vivo* rates of [³H]thymidine (B) and ¹⁴C-labeled amino acid hydrolysate (C) incorporation into TCA-precipitable (●—●) and TCA-soluble (x—x) fractions of tadpole gill. Data from Atkinson and Just, 1975, *Dev. Biol.* **45**:151.

R. catesbeiana gills during spontaneous metamorphosis (Atkinson and Just, 1975) indicates that the gills of stage X animals are highly vascularized, and most of the gill epithelium is in close association with blood vessels (Fig. 22A and B). In young tadpoles few melanocytes are visible in the gill tissue. As gill weight decreases during metamorphic climax (stage XX–XXI), gill morphology changes markedly: gill epithelium is no longer in close proximity to the vascular network and the ratio of melanocytes to epithelium increases (Fig. 22C and D). Few necrotic cells are ever evident during spontaneous gill degeneration, but by stage XXIII the gills are black in appearance and in subsequent stages melanocytes occupy the former gill region.

Spontaneous gill degeneration in *R. catesbeiana* tadpoles is characterized biochemically by a decrease in the rate of thymidine incorporation into DNA (Fig. 21B), a decrease in the rate of amino acid incorporation into protein (Fig. 21C; Atkinson and Just, 1975), and increases in cathepsin C activity (Wang and Frieden, 1973) and in the *in vivo* rate of protein degradation (Little and Atkinson, unpublished). Atkinson and Just's (1975) observation that the DNA and protein content per gram wet weight remains constant as the total organ weight drastically decreases suggests that necrotic cells are continually being removed from the gill in a manner similar to that suggested for degenerating tail epithelial cells (Fox, 1973). This contention is supported by the observation that few or no necrotic cells are found in degenerating gills (Fig. 22C and D).

3.3.2. Induced Degeneration of the Gills *in Vivo* and *in Vitro*

Gills from *R. catesbeiana* tadpoles (stages IX–XII) induced *in vivo* with a single injection of T_3 (Atkinson and Just, 1975) undergo a significant weight loss beginning on the fourth day after hormone administration (Fig. 23A). Eight days following hormone treatment, the decrease in gill weight is maximal and thereafter remains constant. Decreased gill weight is accompanied by an equal and constant loss of DNA and protein (Atkinson and Just, 1975), suggesting that necrotic cells are continually being removed from the atrophying gill, as in spontaneous metamorphosis.

Although gills from T_3-induced tadpole do not exhibit enhanced cathepsin C activity (Wang and Frieden, 1973), the *in vivo* rate of protein degradation is enhanced (Little and Atkinson, unpublished observations), which coincides temporally with their initial loss in weight. However, prior to any loss in gill weight or enhanced degradative activity, the rate of incorporation of thymidine into DNA (Fig. 23B) and of amino acids into protein (Fig. 23C) drastically decrease (Atkinson and Just, 1975; Tonoue and Frieden, 1970). In later periods, when gill weight has stabilized, the rate of thymidine incorporation increases to levels equivalent to control values, suggesting that the gill tissue no longer responds to the injected hormone, or its response is incomplete. Histological examination

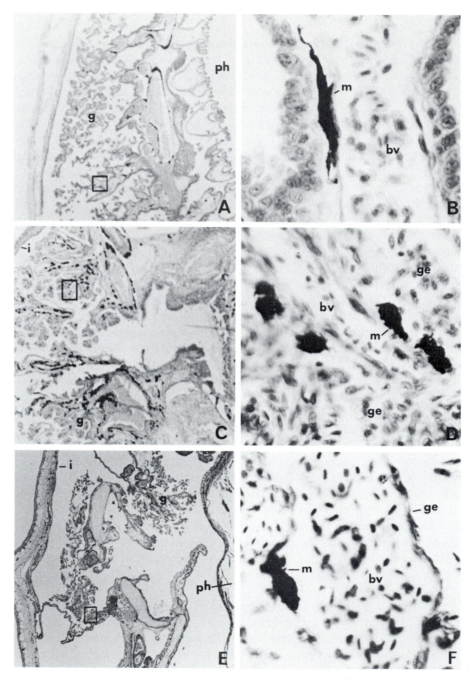

FIGURE 22. Histology of *R. catesbeiana* tadpole gills. (A) Longitudinal section through the midregion of right gill of stage X tadpole ($\times 20$). (B) Enlargement of rectangle in (A) ($\times 550$). (C) Cross section through the midregion of right gill of stage XXII tadpole ($\times 20$). (D) Enlargement of rectangle in (C) ($\times 550$). (E) Cross section through the midregion of right gill of stage X tadpole treated with T_3 8 days earlier ($\times 20$). (F) Enlargement of rectangle in (E) ($\times 550$). Abbreviations: bv, blood vessel; ge, gill epithelium; g, gill tissue; i, integument; m, melanocytes; ph, pharynx. Data from Atkinson and Just, 1975, *Dev. Biol.* **45**:151.

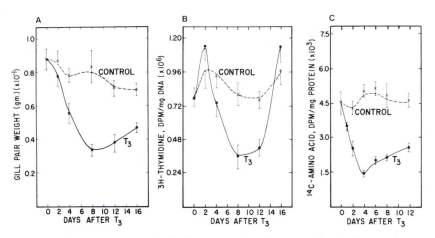

FIGURE 23. Characterization of gills from control and T_3-treated tadpoles. Measurement of gill wet weights (A), and of the *in vivo* rates of [^{3}H]thymidine (B) and ^{14}C-labeled amino acid hydrolysate (C) incorporation into TCA-precipitable fractions of tadpole gills. Data from Atkinson and Just, 1975, *Dev. Biol.* **45**:151.

of gill tissue from tadpole 8 days after hormone treatment (Fig. 22E and F) substantiates the absence of a complete or sustained hormone effect, in that the vascular tissue is intimately associated with gill epithelium and the ratio of melanocytes to epithelial cells does not change.

The transitory decreased rate of incorporation of macromolecules, the absence of marked histological changes, and the lack of complete degeneration in gills from tadpoles treated *in vivo* with a single injection of T_3 are considered evidence that some spontaneous events in normally metamorphosing tadpoles are subject to the following variables.

1. The tissues must be competent to respond fully (Kollros, 1961; Etkin, 1968).
2. Some tissues require a gradual increase (Kollros, 1961; Etkin, 1968) or the continuous presence (Derby, 1968) of TH.
3. Full response may involve the influence of other hormones in addition to TH (Atkinson and Little, 1972; Just and Atkinson, 1972; Wang and Frieden, 1973; Atkinson and Just, 1975).

Results from recent *in vitro* studies with TH-supplemented tadpole gill explants (Derby *et al.*, 1979) give new insight regarding these theories. Utilizing gill tissue from *R. catesbeiana* tadpoles of stages similar to those used by Atkinson and Just, Derby *et al.* (1979) demonstrated that gills from these animals are competent to respond to TH, that gill degeneration does not necessitate an influence by other hormones, and that complete gill degeneration is dependent upon a continuous level of TH (either T_3 or T_4). Comparison of the results from both studies implies that a single injection of TH is sufficient to initiate gill degeneration, but not sufficient to sustain the degradative processes. This conclusion un-

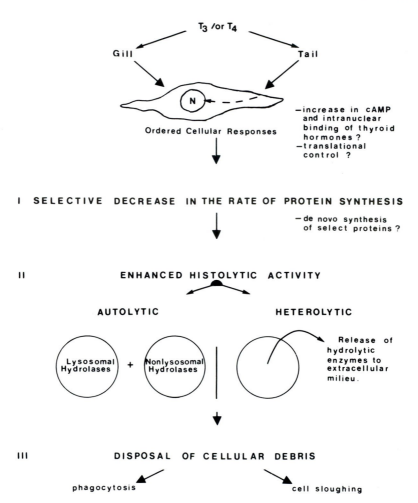

FIGURE 24.   Thyroid hormone-induced degenerative processes in the anuran tadpole.

derscores the rapid and marked depression in the rate of protein synthesis induced in gill tissue by TH administration, and suggests that these synthetic events are critical to the initiation of gill resorption.

As explanted gill tissues supplemented with TH degenerate, collagenase increases progressively in the culture medium, and a concomitant loss of tissue collagen occurs (Derby *et al.*, 1979). Although hormone-induced *in vitro* gill degeneration results in an increase in the specific activity of acid phosphatase, *total* acid phosphatase activity decreases (Derby *et al.*, 1979). As a consequence of these experiments and those by Wang and Frieden (1973), the causal relationships between lysosomal enzyme activity and gill regression remain in question.

3.4. Conclusion: Model for Hormone-Modulated Degeneration

Experimental studies of spontaneous and TH-induced anuran gill and tail atrophy suggest that degeneration in these organs encompasses at least three discrete phases of cellular activity. In the first phase, a selective decrease in the rate of protein synthesis occurs; the second phase is highlighted by enhanced histolytic activity; and in the final phase cellular debris, produced during the second phase, is removed (Fig. 24). It is also apparent that each tissue of these organs participates in each phase in a characteristic manner. Most reported studies have described the histo-lytic processes in the second and third phases of degeneration. The pri-macy of the initial phase and the way in which the selective synthesis of certain proteins is deployed in each tissue are critical to our under-standing of ontogenetic sequences where organ, tissue, and cellular death is a normal process.

ACKNOWLEDGMENT.   Editorial ultimata necessitated reducing research citations to minimum. I would like to acknowledge the many corrobor-ative studies, which are accessible through the literature cited in key references, leading to the synthesis of this review.

REFERENCES

Agrell, I., 1951, *Acta Physiol. Scand.* **23**:179.

Allen, B. M., 1925, *J. Exp. Zool.* **42**:13.

Ashley, H., Katti, P., and Frieden, E., 1968, *Dev. Biol.* **17**:293.

Atkinson, B. G., 1971, in: *Proceedings of the 7th Conference on the Thyroid and En-docrinology* (R. P. Breitenbach and A. D. Kenny, eds.) pp. 48–82, University of Missouri Press, Columbia.

Atkinson, B. G., and Just, J. J., 1975, *Dev. Biol.* **45**:151.

Atkinson, B. G., and Little, G. H., 1972, *Mech. Age Dev.* **1**:299.

Atkinson, K., and Atkinson, B. G., 1976, *Dev. Biol.* **53**:147.

Brown, M. E., 1946, *Am. J. Anat.* **78**:79.

Cohen, P. P., Brucker, R. F., and Morris, S. M., 1978, in: *Hormonal Proteins and Peptides,* Volume VI (C. H. Li, ed.), pp. 273–381, Academic Press, New York.

Davis, B. P., Jeffrey, J. J., Eisen, A. Z., and Derby, A., 1975, *Dev. Biol.* **44**:217.

Dayton, W. R., Goll, D. E., Zeece, M. G., Robson, R. M., and Reville, W. J., 1976, *Biochemistry* **15**:2150.

de Duve, C., Berthet, J., Berthet, L., and Applemans, F., 1951, *Nature (London)* **167**:389.

Derby, A., 1968, *J. Exp. Zool.* **168**:147.

Derby, A., Jeffrey, J. J., and Eisen, A. Z., 1979, *J. Exp. Zool.* **207**:391.

Dhanarajan, Z. C., 1979, Ph.D. thesis, University of Western Ontario, London, Canada.

Dhanarajan, Z. C., and Atkinson, B. G., 1980, *Can. J. Biochem.* **58**:516.

Dhanarajan, Z. C., and Atkinson, B. G., 1981, *Dev. Biol.* **82**:317.

Dodd, M. H. I., and Dodd, J. M., 1976, in: *Physiology of the Amphibian,* Volume III (B. Lofts, ed.), pp. 467–599, Academic Press, New York.

Eeckhout, Y., 1965, Contribution a l'etude de la metamorphose caudale des amphibiens anoures, these, Universite Catholique de Louvain.

Eisen, A. Z., and Gross, J., 1965. *Dev. Biol.* **12**:408.

Etkin, W., 1968, in: *Metamorphosis: A Problem in Developmental Biology* (W. Etkin and L. I. Gilbert, eds.), Appleton–Century–Crofts, New York, pp. 313–348.

Filburn, C. R., and Vanable, J. W., 1973, *Arch. Biochem. Biophys.* **159**:694.

Fox, H., 1973, *Folia Morphol.* (Warsaw English Translation) **21**:109.

Frieden, E., and Just, J., 1970, in: *Biochemical Action of Hormones*, Volume 1 (G. Litwack, ed.), Academic Press, New York, pp. 1–52.

Gona, A. G., 1969, *Z. Zellforsch. Mikrosk. Anat.* **95**:483.

Harper, E., and Toole, B. P., 1973, *J. Biol. Chem.* **248**:2625.

Helff, O. M., 1930, *Anat. Rec.* **47**:177.

Hickey, E. E., 1971, *Wilhelm Roux Arch. Entwicklungsmech. Org.* **166**:303.

Just, J. J., 1972, *Physiol. Zool.* **45**:143.

Just, J. J., and Atkinson, B. G., 1972, *J. Exp. Zool.* **182**:271.

Kaltenbach, J. C., 1953, *J. Exp. Zool.* **122**:21.

Kaltenbach, J. C., Fry, A. E., and Leius, V. K., 1979, *Gen. Comp. Endocrinol.* **38**:111.

Kawahara, A., and Yamana, K., 1977, *Cell Struct. Funct.* **2**:187.

Kemp, N. E., and Hoyt, J. A., 1969a, *J. Morphol.* **129**:415.

Kemp, N. E., and Hoyt, J. A., 1969b, *Dev. Biol.* **20**:387.

Kistler, A., Yoshizato, K., and Frieden, E., 1975, *Dev. Biol.* **46**:151.

Kollros, J. J., 1961, *Am. Zool.* **1**:107.

Lapiere, C. M., and Gross, J., 1963, in: *Mechanisms of Hard Tissue Destruction* (R. Sognnaes, ed.), pp. 663–694, Publ. No. 75, AAAS, Washington, D.C.

Lindsay, R. H., Buettner, L., Wimberly, N., and Pittman, J. A., 1967, *Gen. Comp. Endocrinol.* **9**:416.

Little, G., Atkinson, B. G., and Frieden, E., 1973, *Dev. Biol.* **30**:366.

Merrifield, P., 1979, Ph.D. thesis, University of Western Ontario, London, Canada.

Merrifield, P., and Atkinson, B. G., 1979, *Proc. Can. Fed. Biol. Soc.* **22**:8.

Miyauchi, H., LaRochelle, F. T., Suzuki, M., Freeman, M., and Frieden, E., 1977, *Gen. Comp. Endocrinol.* **33**:254.

Muntz, L., 1975, *J. Embryol. Exp. Morphol.* **33**:757.

Osdoby, P., and Caplan, A. I., 1979, *Dev. Biol.* **73**:84.

Regard, E., Taurog, A., and Nakashina, T., 1978, *Endocrinology* **102**:674.

Robinson, H., Chaffee, S., and Galton, V. A., 1977, *Gen. Comp. Endocrinol.* **32**:179.

Ryffel, G., and Weber, R., 1973, *Exp. Cell Res.* **77**:79.

Sakai, J., and Horiuchi, S., 1978, *Comp. Biochem. Physiol.* **62B**:269.

Saleem, M., and Atkinson, B. G., 1978, *J. Biol. Chem.* **253**:1378.

Saleem, M., and Atkinson, B. G., 1979, *Gen. Comp. Endocrinol.* **38**:441.

Saleem, M., and Atkinson, B. G., 1980, *Can. J. Biochem.* **58**:461.

Seshimo, H., Ryuzaki, M., and Yoshizato, K., 1977, *Dev. Biol.* **59**:96.

Shaffer, B. M., 1963, *J. Embryol. Exp. Morphol.* **11**:77.

Smith, K. B., and Tata, J. R., 1976, *Exp. Cell Res.* **100**:129.

Strohman, R. C., 1966, *Biochem. Z.* **345**:148.

Stuart, E. S., and Fischer, M. S., 1979, *Gen. Comp. Endocrinol.* **38**:314.

Tata, J. R., 1966, *Dev. Biol.* **13**:77.

Tata, J. R., 1970, in: *Ciba Foundation Symposium on Control Processes in Multicellular Organisms* (G. E. W. Wolstenholme and J. Knight, eds.), pp. 131–150, Churchill, London.

Taylor, A. C., and Kollros, J. J., 1946, *Anat. Rec.* **94**:7.

Tonoue, T., and Frieden, E., 1970, *J. Biol. Chem.* **245**:2359.

Tschumi, P. A., 1957, *J. Anat.* **91**:149.

Usuku, G., and Gross, J., 1965, *Dev. Biol.* **11**:352.

Wang, V., and Frieden, E., 1973, *Gen. Comp. Endocrinol.* **21**:381.

Weber, R., 1964, *J. Cell Biol.* **22**:481.

Weber, R., 1969, in: *Lysosomes in Biology and Pathology* (J. T. Dingle and H. B. Fell, eds.), pp. 437–461, North-Holland, Amsterdam.

Wright, Sister, M. L., 1973, *J. Exp. Zool.* **186**:237.

Yamaguchi, K., and Yasumasu, I., 1978, *Dev. Growth Differ.* **20**:61.

CHAPTER 13

Transitions in the Nervous System during Amphibian Metamorphosis

Jerry J. Kollros

1. INTRODUCTION

The tadpole as a whole can be considered to be an aggregate of mosaic territories, each responding in a distinctive way to the increased titers of thyroid hormones (TH) responsible for progressive metamorphic changes. Similarly, the separate elements of the nervous system can be considered, at least as a first approximation, to be composed of a series of mosaic territories. They differ as to the stages when they are first morphologically distinguishable, when they first acquire capacity to respond to TH, in the rates of increase in such sensitivity, i.e., in reduction of threshold concentrations required to elicit detectable change, and in the rates of response to given concentrations. They probably also differ with respect to the hormone concentrations required to elicit maximal rates of tissue response. Some elements of the nervous system grow and differentiate; others shrink, die, and disappear. The extent to which there may be elements of the nervous system that are unresponsive to TH remains largely to be established.

2. GROSS CHANGES

Gross morphological features of anuran brains change during metamorphosis. Allen (1918, 1924) and Schulze (1924) stressed the retention

---

JERRY J. KOLLROS · Department of Zoology, University of Iowa, Iowa City, Iowa 52242.

of larval characteristics of the brains of tadpoles following thyroidectomy or hypophysectomy, i.e., the larger size of ventricles and the thinner walls of the cerebellum and medulla oblongata, compared to those in metamorphosed control animals. Cooksey (1922) and Schulze (1924) both reported transformation of gross brain morphology from larval to adult types following the feeding of thyroid gland materials, e.g., the widening and shortening of the diencephalon and the narrowing and shortening of the fossa rhomboidalis of the medulla. Pesetsky (1966) confirmed that cross-sectional areas of both the gray and the white matter of the medulla oblongata of thyroidectomized tadpoles are substantially less than those recorded in smaller but metamorphosing control animals.

3. MESENCEPHALIC FIFTH NUCLEUS

Many studies of metamorphic changes in the nervous system have focused on individual cells or cell groups. One such example is that of the proprioceptors of the jaw musculature, the cells of the mesencephalic nucleus of the trigeminal nerve (see Kollros, 1977, for a review). In *Ambystoma maculatum* (Piatt, 1945) and *Rana pipiens* (Kollros and McMurray, 1955), these cells appear in the optic tectum just as feeding begins. In both species their numbers increase rapidly with time and developmental stage. In *R. pipiens* cell counts just prior to metamorphic climax are larger than at any other stage, suggesting that some cells may die during the climax phases, just as cells of this nucleus do in late embryonic stages in the chick (Rogers and Cowan, 1973) and hamster (Alley, 1974). In *Ambystoma* (Bibb, 1963) cell sizes *decrease* with time, and are unaltered by the immersion of larvae in strong solutions of thyroxine (T_4). In *R. pipiens*, in contrast, the sizes of these cells remain small for much of the larval period, begin to increase at about stage XV (of the Taylor–Kollros, 1946 series), grow even more rapidly as metamorphic climax is approached and reached, and then more slowly postmetamorphically. That this growth is directly hormone dependent, at least in part, has been demonstrated in a variety of ways. When cells first are identifiable (at stage I) their size is not alterable by immersion of tadpoles in strong T_4 solutions. At stages II–III such immersion results in slight growth response, and by stage V a more vigorous response is elicitable, though almost surely slower and less complete than at still later stages (Kollros and McMurray, 1956). The growth may represent either a direct response of the cell to T_4 or an indirect response through alterations of the jaw musculature (although most such musculature changes occur *late* in metamorphic climax, well after much of the late larval growth of these cells has been accomplished). That the response *can* be direct has been demonstrated by greater growth of mesencephalic V nucleus cells in the immediate vicinity of T_4–cholesterol pellets than at a distance from such pellets (Kollros and McMurray, 1956). While the mesencephalic V nucleus cells of advanced larvae can respond rather quickly to increases in T_4 titers, apparently the threshold concentration for such response is relatively high. This conclusion is suggested, first, by the failure of cell growth, including reduction in nucleocytoplasmic (N/C) ratios, prior to stages XV–XVI. Second, in hypophysectomized tadpoles brought slowly to stages of advanced preclimax and early climax (concentrations of T_4 ≤ 0.6 μg/liter, for periods up to 230 days), the mesencephalic V nucleus cells at stages XVII–XX+ were significantly smaller than in control animals matched as to size and stage; these smaller cells displayed N/C ratios appropriate for stages I through XIV rather than for the more advanced stage attained by external tadpole features.

The larval mesencephalic V nucleus is of special interest in metamorphic studies since cell characteristics can be shown to be dependent on TH titer, i.e., despite the large size of hypophysectomized animals, mesencephalic V nucleus cell sizes and N/C ratios are characteristic of early control stages. Treatment with TH at 20 μg/liter results in cell growth and an appropriate reduction in the value of the N/C ratio. If, once this stimulated condition has been reached, hormone treatment is stopped, and the tadpoles are maintained for several additional weeks, the mesencephalic V nucleus cells become noticeably smaller and the N/C ratios become noticeably larger, both of these values approaching those observed in unstimulated hypophysectomized siblings (Kollros and McMurray, 1956). Not surprisingly, therefore, the stoppage of metamorphic change in normal tadpoles in late preclimax stages, by treatment with thiourea, results in a shrinkage of cell size and alteration of the N/C ratio to values characteristic of younger stages. These modifications have been somewhat more fully shown in *Xenopus laevis* than in *R. pipiens*. Immediately after metamorphosis, the very young *Xenopus* juveniles demonstrate an attenuated response to prolonged treatment with thiourea, whereas those animals 3 or more months after metamorphosis fail to show any such response (Kollros, 1957). Of some interest in the several studies just reported are the increase in numbers of mesencephalic V nucleus cells in the hormone-stimulated animals, and the later decrease in such cell counts following withdrawal from the T_4 solutions. It is not known if the reduction in these cell counts is accomplished by cell death [as it appears to be in the chick, according to Rogers and Cowan (1973), and in the hamster, according to Alley (1974)] or by alteration of cell characteristics so as to make these cells indistinguishable from the nonmesencephalic V cells of the tectum.

In summary, the mesencephalic V nucleus cells, at the time of their initial appearance, are insensitive to TH, just as they appear to be, again, several months after metamorphosis. Onset of sensitivity begins in early larval life (stage III) and increases rapidly. Nonetheless, as gauged by T_4 threshold requirements for cell and nuclear growth, and for reduction of the N/C ratio, T_4 concentration requirements are much higher than for other changes, e.g., leg growth, nictitating membrane formation, and plical gland development.

4. MAUTHNER'S NEURON

The Mauthner cell (M cell) is a large neuron in the medulla oblongata at the level of entry of the eighth cranial nerve. A single M cell on each side is present in many fishes and amphibians, including larval anurans. The M cell receives diverse sensory inputs, and each cell produces a very large axon that decussates and runs posteriorly in the spinal cord into the tail. The large diameter of the axon is correlated with a high conduction velocity, and with the cell's role in the "startle response," at least in fish (see Faber and Korn, 1978, for a review). According to Baffoni and Catte (1950, 1951), Stefanelli (1951), and Weiss and Rossetti (1951), the M cell undergoes atrophic changes by the end of metamorphosis, and such changes can be initiated by treatment with T_4, i.e., implying, at least, that the increase in TH titer in the blood during metamorphic climax is responsible for the decrease in M cell size. A different conclusion was initially presented by Pesetsky and Kollros (1956), following T_4–cholesterol pellet implantations unilaterally adjacent to the hindbrain, although they, like Weiss and Rossetti, also observed growth of non-M-cell nuclei in the vicinity of the M cell. In 13 animals stimulated moderately, M cell sizes on both left and right sides were virtually identical, despite the greater concentration of T_4 on the pellet side. However, in 6 animals stimulated very strongly (enough hormone was released from the pellet to achieve skin window rupture and forelimb emergence within 8 days), there were bilateral differences, but such that the M cells on the side with the pellet were *smaller* than the contralateral ones; additionally, on both sides the M cells were, on average, much smaller than in the moderately stimulated animals. Subsequent work by Pesetsky (1962, 1966) demonstrated that thyroidectomized larvae have smaller than normal M cells, while euthyroid larvae stimulated moderately by immersion in T_4 solutions display M cell growth (see also Moulton *et al.*, 1968). These results, together with reevaluation of the earlier studies, led Pesetsky to propose that M cells normally grow in the late preclimax stages as TH titers in the tadpole increase, and that such growth and the maintenance of attained size and differentiation are dependent upon continued moderate or high TH concentration. When TH concentrations decline, as they do late in climax phases, the cells shrink, atrophy, and perhaps much later die. Thus, increased T_4 titer may result in cell addiction— dependence on the elevated titers for M cell maintenance. The several studies that deal with M cell size can generally be interpreted as conforming to this view. Further, the descriptions of cytological changes in the M cells just prior to and in early climax phases (Moulton *et al.*, 1968) are consistent with increased synthetic activity.

5. LATERAL MOTOR COLUMN

A nerve cell group thoroughly studied in its relationship to TH is that of the lateral motor column (LMC), whose cells provide motor innervation to the limbs. At lumbar levels the LMC is first evident at stage IV+ in *R. pipiens* (Reynolds, 1963), with perhaps 8000 cells on each side (Kollros, unpublished). By stage VII over 10,000 cells may be counted, the number then declining progressively to metamorphic climax, when cell numbers stabilize near a value of 2000 per side. At comparable stages in *Xenopus*, a nearly comparable set of changes occurs (Hughes, 1961; Prestige, 1967). Recruitment of new cells into the LMC continues well beyond the period of peak numbers. There is thus a long period of overlap when LMC cells are dying while others are still being recruited from the ependyma, a relationship first reported by Hughes (1961) in *Xenopus*. The last cells to be produced for the LMC by the ependyma can be labeled with an injection of tritiated thymidine at stage XI (Pollack and Kollros, 1975), but the time delay before the division product is seen in the LMC is not known exactly, but is probably not earlier than stage XIII (5–7 days), and for some cells may be considerably later. Both the role of TH in enhancing lysosomal development in LMC cells and its influence on disruptability of the lysosomal membranes have been reported by Decker (1974a,b).

The recruitment of LMC cells appears to be independent of TH, inasmuch as hypophysectomized *R. pipiens* tadpoles, which ordinarily do not develop past stage VII or VIII, develop normal cell numbers when they reach stages IV+ and V, and thereafter show an excess of cells for their stage. The long retention of a supranormal number of cells was reported by Race (1961) and Kollros (1968). Of interest is the likelihood that a very slow decline in LMC cell numbers occurs after 6–18 months (E. Pollack, personal communication); thus, TH need not be involved in some minimal, slow, cell loss, although its usual involvment in such cell loss has long been known (Kollros and Race, 1960; Race, 1961; Kollros, 1968). The immersion of tadpoles of various early stages into various concentrations of TH (from about 0.01 µg/liter upward) demonstrates:

a. A premature increase in prospective LMC cells, using late embryonic and early larval stages ["prospective" inasmuch as in the early stages of the growth response, the future LMC cells have not yet migrated laterad from the central gray matter, and larger-than-normal nuclei are seen clustered at the edges of the central gray, lateroventrad, according to Reynolds (1963)].

b. A premature initiation of cell loss. Using the same animals as in (a), cell degeneration can be observed at attained larval stages earlier than stage VI, i.e., clearly prematurely.

   c. A distinctly smaller total cell count than in comparable stages in control tadpoles, particularly among the oldest of this experimental group. This may in part be occasioned by the premature cell loss, but the magnitudes of the differences in the cell counts are suggestive of a premature cessation of new LMC cell production as well. In central levels of the LMC, e.g., at the eighth to ninth segment boundary in the spinal cord, sections of 10-μm thickness possess only 30 LMC cells each, in contrast to 45–50 such cells present in controls.

If T_4 treatment of hypophysectomized tadpoles is carried out by immersion, at hormone concentrations designed to bring about very slow leg growth, i.e., below 0.6 μg/liter at 20°C, LMC cell numbers, as gauged by counts of 10 alternate sections from the middle of the lumbar column, are virtually identical, stage for stage, with those of control animals; however, as gauged by measurements of cross-sectional areas of the cell nuclei, cell sizes are uniformly well below those of controls (Kollros, 1968). Thus, there must be some differentiation of factors responsible for cell recruitment and cell loss on the one hand, and those for individual cell growth on the other. Since limb sizes appear to have been normal, the differences seen are not attributable to variations in the volumes of the limbs to be innervated. Limb function was not tested in sophisticated ways, but appeared to be normal as to the stages of appearance of the first movements, leg kicking, leg posture, etc. (Kollros, unpublished).

The capacity of the LMC cells to respond directly to TH has also been demonstrated, by unilateral implants of T_4–cholesterol pellets adjacent to the spinal cord at lumbar levels (Beaudoin, 1956). On the pellet side LMC cell nuclei were significantly larger than those on the nonpellet side. Comparable T_4–cholesterol pellets were implanted into the upper leg on one side (Beaudoin, 1956), following the earlier work of Kaltenbach (1953a). As expected, the implanted leg developed faster than did the opposite control leg, as measured both by total length and by its external appearance (attained stage). The LMC of the pellet side almost uniformly possessed cells of a slightly "older" stage of differentiation, and of significantly larger nuclear cross-sectional area. Thus, it is clearly demonstrable that the LMC responds directly to TH changes, as well as indirectly through the growth of the limb following TH treatment.

It has been shown, additionally, that differences in temperature may have significant effects on LMC development, in the range of 10–22°C (Decker and Kollros, 1969). In general, the lower the temperature, the slower is tadpole growth, but the larger are the tadpoles at each stage. It is also seen that the size of the leg increases more than does tadpole size as a whole, and that the reduction in LMC cell loss is at least equally marked, i.e., LMC cell numbers at given stages increase stepwise for each decrease in temperature (22–18–14–10°C). Thus, a positive correlation exists at each stage between total LMC cell number and leg volume. It must be understood that lower temperatures require higher hormonal concentrations (thresholds) than otherwise to permit particular metamorphic events to occur, aside from the effect of temperature on developmental rates (Kollros, 1961). Further, at temperatures at or below 15°C, there may be reduced T_4 levels present, because of reduced transfer of thyrotropin-releasing factors to the pituitary gland, and hence less stimulation of the thyroid gland (Etkin, 1966).

While it is not of direct relevance here to record that limb amputation in *R. pipiens* tadpoles results in reduced LMC cell loss on the operated side compared to the control side, for approximately 2 weeks at 20°C, it is necessary to report that immersion in T_4 solutions results in accelerated LMC cell loss on both sides, yet the relative differences between amputated and intact sides are maintained. TH treatment does not eliminate entirely whatever stimulus from the act of amputation impinges on the LMC to reduce the rate of LMC cell death (Kollros, 1979, and unpublished).

In summary, TH can be shown to:

a. Accelerate LMC cell growth once the LMC has formed.
b. Initiate LMC cell growth in stages prior to the migration of prospective LMC cells to a position lateroventrad of the central gray.
c. Accelerate the rate of cell death, i.e., to initiate regressive changes in more cells at any given stage than would appear in control animals.
d. Accelerate LMC cell growth indirectly, through enhancing limb growth.
e. Compensate in part for the delayed cell death that follows limb amputation.
f. Play a probable role in prematurely terminating the formation of new LMC cells from the ependyma.

Additionally, in the absence of TH, as in hypophysectomy, cell growth can be halted while cell death also comes virtually to a halt, and at very low TH levels cell death can go on at rates proper to ensure LMC cell numbers appropriate for each attained stage while cell growth is virtually stopped, i.e., the threshold requirement of TH for substantial cell growth is well in excess of that to control rate of cell death.

6. DORSAL ROOT GANGLIA

The larval history of the cells of the dorsal root ganglia is much like that of the LMC in that their numbers increase during early larval stages and decrease in later ones, with significant overlap in these two aspects (for *Xenopus* see Kollros, 1956; Prestige, 1965, 1967; for *Rana* see Bibb,

1977), i.e., with cell turnover. Growth of individual cells of the lumbar ganglia follows after the early phases of LMC development. That these cells are directly responsive to TH has been shown by their responses to implants of T_4–cholesterol pellets along medial aspects of the ganglia unilaterally. The implants, which contained 10% T_4, were made at stages V–VII, and fixed 7 days later at stages VII–XII, an average advance of 2.35 stages. Controls received cholesterol pellets at the same stages, and when fixed 7 days later were at stages VII–X, for an average advance of 1.80 stages. Presumably hormone from the T_4–cholesterol pellet had entered the blood and had effects widely in the animal, in addition to more profound local effects. In the hormone-treated animals the cells on the pellet side displayed average nuclear cross-sectional areas 8% greater than on the opposite side ($P < 0.03$); these were also 21% greater than for cells on the pellet side in the control animals ($P < 0.01$). In control animals, no differences in size were seen ($P \geq 0.3$). In the experimental animals the blood levels of T_4 had been sufficient to stimulate growth of ganglion cells on both sides, but the hormone concentration and growth stimulus were clearly greater for the ganglia adjacent to the pellet than for those on the opposite side of the animal (Norman, 1966).

7. ROHON–BEARD CELLS

The fate of the cells of Rohon–Beard may be related to the development of the spinal ganglia. These special primary sensory cells, subserving both exteroceptive and proprioceptive functions before they are assumed by the spinal ganglia, are located just to either side of the dorsal midline of the spinal cord. They are first detectable in tail bud embryos (Shumway stage 19 of *R. pipiens*, stage 33/34 of *Xenopus*), being first evident at anterior trunk levels and only later in the posterior trunk and then the tail. They are present in maximum numbers just before feeding begins (Shumway stage 25 and Taylor–Kollros stage I in *Rana*), when about 260 cells may be counted in trunk levels of the cord (Suter, 1966) in animals 8 mm long. At stage I+ (12 mm) cell numbers reduce to 185, and at II+ (20 mm) to 115. Thereafter, cell loss proceeds gradually, being somewhat more complete anteriorly than posteriorly, with only 7–17 cells being identified at stages XV–XX, and only 2–8, all degenerating, at stage XXV. Degeneration of most of the Rohon–Beard cells occurs substantially earlier in *Xenopus* (Kollros, 1956) than in *Rana*. Since the major cell loss is so early in larval life, it is unlikely that the loss is related directly to TH, and more likely may be related to the early development of the spinal ganglia, which take over the sensory functions initially assumed by the Rohon–Beard cells. The delay of the final cell loss until the end of metamorphic climax, however, does suggest that there may be some involvement of TH, and the work of Stephens (1965), who obtained premature reduction in Rohon–Beard cell number and cell size after treatment of *R. pipiens* tadpoles by immersion in T_4 solutions of 1 or 10 μg/liter for up to 32 days, supports such a view. Whether hormone acts directly on the cells or indirectly by influencing either the spinal ganglia or the sensory periphery is unknown. Trevisan (1973) suggests that in the urodele it is probably indirect.

8. CEREBELLUM

The cerebellum of the tadpole remains in a relatively early developmental stage during the premetamorphic period (to stage XI), shows slow and then more rapid developmental progress in the succeeding prometamorphic period (to stages XVIII–XIX), and then very rapid maturation during metamorphic climax. These maturational changes include a substantial reduction in the number of external granule cells, the migration of most of these cells into the internal granular layer, and the conversion of small, immature Purkinje cells into large, mature ones between the two granular layers (Larsell, 1925; Gona, 1972, 1973, 1977). Gona (1977) has demonstrated a local effect of T_4 on these developmental maturational changes by implanting T_4 crystals adjacent to the optic tectum and the cerebellum.

9. SENSE ORGANS

Larsell (1925) recorded some loss of lateral line organs in climax phases of *Hyla regilla* metamorphosis after more than half of the tail had been resorbed, and the beginning of corresponding cell losses in the cerebellum, specifically of cells receiving synaptic input from the lateral lines. Still later, after 80–85% of tail loss had occurred, chromatolysis of these cells was more prominent than earlier, and the same was true of the cells of the lateral line portions of ganglia VII, IX, and X. Lateral line tracts in the brain showed losses that might be expected on the basis of the cellular changes. Because of the cellular losses, the auricular lobe of the cerebellum decreased in relative size as the final phases of metamorphosis were completed.

The eyes also permit a series of hormone-related changes to be identified. The anuran tadpole has two corneas for each eye, external and internal ones, separated by intraorbital fluid. These corneas fuse at metamorphic climax (Harms, 1923; Kollros, 1943; Kaltenbach, 1953b). Further, the extrinsic ocular muscles are relatively thin and weakly developed in the tadpole, growing enormously during late preclimax and climax stages, and simultaneously shifting their point of origin mediad on the parasphenoid bone. This growth of musculature is at least partially responsible for the bulging of the eyes of the postmetamorphic frog. Also developing late is the tendon of the nictitating membrane (Kollros, 1942).

That all of these changes can be caused directly by T_4 was demonstrated by Kaltenbach (1953b), utilizing T_4–cholesterol pellet implants into the orbit unilaterally, with the result of marked ipsilateral metamorphic changes.

Different amphibian species have either porphyropsin or rhodopsin as their visual pigment, or sometimes both, as larvae and adults. In those in which the larval pigment is porphyropsin, there is commonly conversion, entirely or in part, to rhodopsin as the adult pigment (review in Weber, 1967). According to Wald (1946), such conversion is seen in *R. catesbeiana*, the change occurring during late preclimax and climax stages (Wilt, 1959). Implants of T_4–cholesterol pellets into one eye of the tadpole resulted in more complete change of these pigments than in the contralateral eye. Later, Wilt and others (Ohtsu *et al.*, 1964) concluded that the increase in rhodopsin production is correlated with a TH-induced decrease in vitamin A_2 aldehyde production (and thus a decrease in porphyropsin).

10. MITOTIC ACTIVITY

TH influence a variety of sites in the nervous system to increase their apparent mitotic rate. Champy (1922) reported such an increase in the ora serrata of *R. temporaria* several days after the start of treatment. That this effect can be attributed to direct action of TH has been shown by Kaltenbach and Hobbs (1972), utilizing T_4–cholesterol pellet implants into the vitreous humor; they reported significantly higher mitotic rates in the experimental eye than in the control eye, not only in the ora serrata but also in the epithelia of the lens and of the external cornea. In only a small fraction of the cases was there a significant stimulation of metamorphic changes in the adnexa of the eye, and generalized metamorphic stimulation was even less obvious, implying only slow leakage of hormone from the eyeball. Beach and Jacobson (1979) have demonstrated that the retinal margin in *Xenopus* larvae also responds to T_4 by increasing mitotic activity, with the ventral margin showing distinctly greater responsivity than the dorsal margin. Postmetamorphically the response was greatly attenuated. Similarly, Weiss and Rossetti (1951) reported a large (sixfold) increase in mitotic rate in the tadpole hindbrain following implantation of TH sources into the fourth ventricle or adjacent to the choroid plexus. Spread of the effect was noticed to the anterior levels of the spinal cord. It was subsequently reported by Baffoni and Elia (1957) that just before and in metamorphic climax there is a normally occurring increase in mitotic activity in both alar and basal plates of the metencephalon of the toad, an increase that can be simulated prematurely in young tadpoles by immersion in strong T_4 solutions (Baffoni, 1957). In *Triturus*, also, at climax, the alar plate displays increased mitotic activity.

That late-stage embryos of *R. catesbeiana* are also capable of showing supranormal mitotic activity in the medulla oblongata, following immersion in T_4 solutions, was shown by Ferguson (1966). These various studies present virtually no data to suggest that the increased mitotic rates produce excess brain growth, and it is possible that the hormone treatment merely triggers a mitotic event earlier than it would otherwise have occurred, and is then followed by a period of somewhat reduced mitotic activity "due to growth and synthetic processes which must occur in daughter cells prior to the next division," as suggested by Ferguson (1966). The role of T_4 in promoting growth effects in mammal brains as well is reviewed in Grave (1977) and Jacobson (1978).

11. BEHAVIORAL CHANGES

In addition to the morphological changes associated with metamorphosis are significant behavioral ones, e.g., limb activity, air breathing, the corneal reflex, etc. Hypophysectomized or thyroidectomized *R. pipiens* develop limbs and lateral motor columns only to prefunctional stages. Both of these systems develop to functional levels only as TH levels increase. The specific sequence of changes in limb behavior has been described in *Xenopus* by Hughes and Prestige (1967), while the coincident changes in the lateral motor columns were described by Kollros (1968) and Hughes (1968). Little has been explored as to the causal relationships in change from gill to air breathing, but Hughes (1977) sketches the onset of pulmonary respiration in *Eleutherodactylus*.

The onset of the corneal reflex was described in detail in both anurans and urodeles by Kollros (1942). In *Ambystoma* and other urodeles the onset was shown, ordinarily, to be closely related to preclimax or climax events, but separable from them readily by adjustments of nutrition and growth rate. Further, complete independence from metamorphic climax was demonstrated by the development of the reflex in nonmetamorphosing, hypophysectomized larvae. In contrast, in all anurans studied, the reflex developed only in the few days before or after forelimb emergence from the opercular chamber (stage XX); it never developed in hypophysectomized tadpoles (Kollros, 1942). It was possible, however, to modify the time interval between forelimb emergence and the onset of the reflex by T_4 treatment alone, and in T_4-treated hypophysectomized tadpoles in conjunction with adjustments of temperature (Kollros, 1958). It was, in fact, possible to get reflex development to precede forelimb emergence by more than 100 days, rather than by the usual 0–10 days. Treatment of euthyroid tadpoles with 30 µg/liter DL-T_4 for 6 or 7 days, followed by removal from the T_4 bath, was at times successful in bringing about reflex onset in stages much younger than when forelimb emergence might have been expected (Kollros, 1966). Recently (Kollros, unpub-

TABLE 1. CHANGES IN THE NERVOUS SYSTEM INDUCED BY THYROID HORMONE TREATMENT OR WITHDRAWAL[a]

Responding elements	Reduced TH concentration (i.e., thyroidectomy, hypophysectomy)	Increased TH concentration (i.e., treatment of thyroidectomized or hypophysectomized tadpoles)
Diencephalon	Remains narrow	Widens and shortens
Rhomboid fossa of medulla oblongata	Remains wide	Narrows and shortens
Mesenphalic V nucleus	Cells remain small	Sensitivity to TH first evident at stages II–III[*]
		Sensitivity to TH reduced after metamorphosis
		Cells grow; nucleocytoplasmic ratio reduces; withdrawal of hormone leads to reversal of changes
Mauthner's neuron	Cell size reduced	Cell growth; habituation to TH (TH withdrawal leads to cell shrinkage)
Lateral motor column	Cell size remains small; cell number remains large, i.e., in excess of controls of same stage	Reduction in cell number, to a value appropriate for the attained stage; premature cessation of new cell formation[*]; cell sizes increase[*]
Dorsal root ganglia		Increased cell size
Rohon–Beard cells	A few persist indefinitely	Premature size reduction, and cell loss of that small Rohon–Beard cell population surviving into late larval stages[*]
Cerebellum		Growth and maturation of Purkinje cells; internal granule layer development accelerated
Eye		Fusion of external and internal corneas[*]
		Growth of extrinsic ocular muscles
		Growth of nictitating membrane
		Change of retinal pigments[*]
Mitotic activity		Accelerated in various sites, at differing times or stages
Corneal reflex	No onset in Anura	Corneal reflex develops
Median eminence	No growth or maturation	Onset of sensitivity to TH at about stage XI; growth and maturation

[a] Asterisks indicate those circumstances in which relatively high concentrations of TH are required to achieve the change, or in which only relatively high concentrations have been successful in achieving the change.

lished), long-term treatment at relatively low dosage levels (8 μg/liter or less) has resulted in the initiation of the reflex in hypophysectomized animals long before forelimb emergence; subsequent removal from the bath halted further metamorphic change while retaining the capacity to show the reflex (for up to 25 days, when the study was terminated). Again, as argued earlier (Kollros, 1958), early events in metamorphosis can be presumed to have lower threshold concentration requirements for TH than later events, and careful manipulation of concentrations and temperatures permits the early event to be elicited while being just subthreshold for later events.

12. MEDIAN EMINENCE

The median eminence of the hypothalamus is an element in the pathway initiating metamorphosis, since it transmits thyrotropin-releasing factors from the hypothalamus to the pituitary gland. Formation of the eminence appears to require the presence of the anterior pituitary (Etkin, 1966), as well as TH, though the rudiment of the median eminence is apparently unresponsive to TH until the start of the prometamorphic period (about stage XI). Thereafter, its continuing maturation ordinarily results in increased release of thyrotropin-releasing factors, hence stimulation of anterior pituitary release of TSH, and thus increased activity of TH, which are effective throughout the body, including the further speeding of the maturation of the median eminence itself.

13. CONCLUSION

The changes in neural structures and behaviors that have been presented (summary in Table 1) are but a limited subset of those that can probably be attributed to the increases in T_4 titers responsible for the events of amphibian metamorphosis. Many neural structures and systems have yet to be studied with respect to modifications that occur during development, and particularly during or near metamorphic climax. The strict dependence of such changes upon increases in TH concentrations will have to be established, and whether the hormone acts directly or through an intermediary system will likewise need to be known. The structural, physiological, and biochemical studies that need yet to be launched to explore the neural correlates of metamorphosis are still extraordinarily extensive, rewarding, and challenging.

ACKNOWLEDGMENT. The unpublished research reported herein was supported by Grant BNS 77-24888 from the National Science Foundation.

REFERENCES

Allen, B. M., 1918, *J. Exp. Zool.* **24:**499.

Allen, B. M., 1924, *Endocrinology* **8:**639.

Alley, K. E., 1974, *J. Embryol. Exp. Morphol.* **31:**99.

Baffoni, G. M., 1957, *Atti Accad. Naz. Lincei Rc. Ser. VIII* **23:**495.

Baffoni, G. M., and Catte, G., 1950, *Atti Accad. Naz. Lincei Rc. Ser. VIII* **9:**282.

Baffoni, G. M., and Catte, G., 1951, *Riv. Biol.* **43:**373.

Baffoni, G. M., and Elia, E., 1957, *Atti Accad. Naz. Lincei Rc. Ser. VIII* **22:**109.

Beach, D. H., and Jacobson, M., 1979, *J. Comp. Neurol.* **183:**615.

Beaudoin, A. R., 1956, *Anat. Rec.* **125:**247.

Bibb, H. D., 1963, The mesencephalic V nucleus in *Ambystoma jeffersonianum*, M.S. thesis, University of Iowa, Iowa City.

Bibb, H. D., 1977, *J. Exp. Zool.* **200:**265.

Champy, C., 1922, *Arch. Morphol. Gen. Exp.* **4:**1.

Cooksey, W. B., 1922, *Endocrinology* **6:**391.

Decker, R. S., 1974a, *J. Cell Biol.* **61:**599.

Decker, R. S., 1974b, *Dev. Biol.* **41:**146.

Decker, R. S., and Kollros, J. J., 1969, *J. Embryol. Exp. Morphol.* **21:**219.

Etkin, W., 1966, *Neuroendocrinology* **1:**293.

Faber, D. S., and Korn, H., 1978, *Neurobiology of the Mauthner Cell*, Raven Press, New York.

Ferguson, T., 1966, *Gen. Comp. Endocrinol.* **7:**74.

Gona, A. G., 1972, *J. Comp. Neurol.* **146:**133.

Gona, A. G., 1973, *Exp. Neurol.* **38:**494.

Gona, A. G., 1977, in: *Thyroid Hormones and Brain Development* (G. D. Grave, ed.), pp. 107–117, Raven Press, New York.

Grave, G. D., (ed.), 1977, *Thyroid Hormones and Brain Development*, Raven Press, New York.

Harms, W., 1923, *Zool. Anz.* **56:**136.

Hughes, A. F., 1961, *J. Embryol. Exp. Morphol.* **9:**269.

Hughes, A. F., 1968, Aspects of Neural Ontogeny, Logos Press Limited, London.

Hughes, A. F., 1977, in: *Frog Neurobiology* (R. Llinás and W. Precht, eds.), pp. 856–863, Springer-Verlag, Berlin.

Hughes, A. F., and Prestige, M. C., 1967, *J. Zool.* **152:**347.

Jacobson, M., 1978, *Developmental Neurobiology*, Plenum Press, New York.

Kaltenbach, J. C., 1953a, *J. Exp. Zool.* **122:**21.

Kaltenbach, J. C., 1953b, *J. Exp. Zool.* **122:**41.

Kaltenbach, J. C., and Hobbs, A. W., 1972, *J. Exp. Zool.* **179:**157.

Kollros, J. J., 1942, *J. Exp. Zool.* **89:**37.

Kollros, J. J., 1943, *J. Exp. Zool.* **92:**121.

Kollros, J. J., 1956, in: *Normal Table of Xenopus laevis (Daudin)* (P. D. Nieuwkoop and J. Faber, eds.), pp. 67–73, North-Holland, Amsterdam.

Kollros, J. J., 1957, *Proc. Soc. Exp. Biol. Med.* **95:**138.

Kollros, J. J., 1958, *Science* **128:**1505.

Kollros, J. J., 1961, *Am. Zool.* **1:**107.

Kollros, J. J., 1966, *Am. Zool.* **4:**553.

Kollros, J. J., 1968, in: *The Emergence of Order in Developing Systems* (M. Locke, ed.), pp. 272–305, 27th Symposium of the Society for Developmental Biology, Academic Press, New York.

Kollros, J. J., 1977, in: *Thyroid Hormones and Brain Development* (G. D. Grave, ed.), pp. 119–136, Raven Press, New York.

Kollros, J. J., 1979, *Am. Zool.* (abstract) **19:**979.

Kollros, J. J., and McMurray, V., 1955, *J. Comp. Neurol.* **102:**47.

Kollros, J. J., and McMurray, V., 1956, *J. Exp. Zool.* **131:**1.

Kollros, J. J., and Race, J., Jr., 1960, *Anat. Rec.* **136**:224.
Larsell, O., 1925, *J. Comp. Neurol.* **39**:249.
Moulton, J. M., Jurand, A., and Fox, H., 1968, *J. Embryol. Exp. Morphol.* **19**:415.
Norman, L. K., 1966, Effects of thyroid hormone on spinal ganglia in *Rana pipiens*, M.S. thesis, University of Iowa, Iowa City.
Ohtsu, K., Naito, K., and Wilt, F., 1964, *Dev. Biol.* **10**:216.
Pesetsky, I., 1962, *Gen. Comp. Endocrinol.* **2**:228.
Pesetsky, I., 1966, *Z. Zellforsch. Mikrosk. Anat.* **75**:138.
Pesetsky, I., and Kollros, J. J., 1956, *Exp. Cell Res.* **11**:477.
Piatt, J., 1945, *J. Comp. Neurol.* **82**:35.
Pollack, E., and Kollros, J. J., 1975, *J. Exp. Zool.* **192**:299.
Prestige, M. C., 1965, *J. Embryol. Exp. Morphol.* **13**:63.
Prestige, M. C., 1967, *J. Embryol. Exp. Morphol.* **18**:359.
Race, J., Jr., 1961, *Gen. Comp. Endocrinol.* **1**:322.
Reynolds, W. A., 1963, *J. Exp. Zool.* **153**:237.
Rogers, L. A., and Cowan, W. M., 1973, *J. Comp. Neurol.* **147**:291.
Schulze, W., 1924, *Arch. Mikrosk. Anat. Entwicklungsmech.* **101**:338.
Shumway, W., 1940, *Anat. Rec.* **78**:139.
Stefanelli, A., 1951, *Q. Rev. Biol.* **26**:17.
Stephens, L. B., Jr., 1965, *Am. Zool.* **5**:222.
Suter, J., 1966, Numbers and distribution of Rohon–Beard cells in selected larval stages of *Rana pipiens*, M.S. thesis, University of Iowa, Iowa City.
Taylor, A. C., and Kollros, J. J., 1946, *Anat. Rec.* **94**:7.
Trevisan, P., 1973, *Atti Accad. Naz. Lincei Rend. Cl. Sci. Fis. Mat. Nat. Sez. III* **53**:217.
Wald, G., 1946, *Harvey Lect.* **42**:148.
Weber, R., 1967, in: *The Biochemistry of Animal Development*, Volume 2 (R. Weber, ed.), pp. 227–301, Academic Press, New York.
Weiss, P., and Rossetti, F., 1951, *Proc. Natl. Acad. Sci. USA* **37**:540.
Wilt, F. H., 1959, *J. Embryol. Exp. Morphol.* **7**:556.

CHAPTER 14

Changes in the Blood during Amphibian Metamorphosis

Robert H. Broyles

1. INTRODUCTION

Metamorphosis is a developmental period in which marked changes occur in various tissues and organ systems. The anatomical changes are particularly dramatic in the anurans. Underlying these visible alterations are cellular and molecular changes that have attracted many investigators interested in biological regulatory mechanisms.

The molecular and cellular changes that occur in the blood during metamorphosis are dramatic. Of particular interest to developmental, cellular, and molecular biologists and to experimental hematologists is the switch in the type of hemoglobin (Hb) in the circulating red blood cells (RBCs). All vertebrates undergo Hb transitions during development. One such switch or transition that has received much attention is the switch from fetal to adult Hb in humans and other mammals. Elucidating the mechanisms by which this Hb switch is accomplished is important to the understanding of certain diseases in humans. The metamorphic Hb transition in amphibians is an important model system for the various Hb transitions occurring during the life cycles of all classes of vertebrates.

The metamorphic Hb transition of amphibians is accomplished by a switch from larval-type to adult-type Hb synthesis. As discussed in more detail below, these different Hb types have different structural and functional properties. The oxygen-binding properties of these Hbs are of

ROBERT H. BROYLES · Department of Biochemistry and Molecular Biology, University of Oklahoma at Oklahoma City, Health Sciences Center, Oklahoma City, Oklahoma 73190.

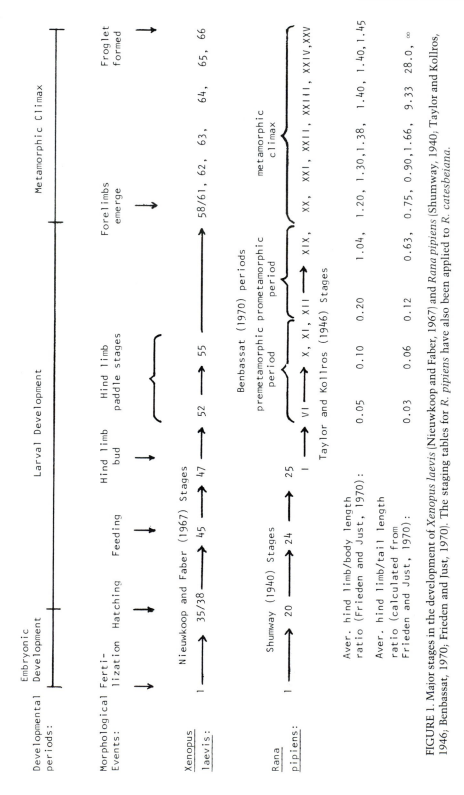

FIGURE 1. Major stages in the development of *Xenopus laevis* (Nieuwkoop and Faber, 1967) and *Rana pipiens* (Shumway, 1940; Taylor and Kollros, 1946; Benbassat, 1970; Frieden and Just, 1970). The staging tables for *R. pipiens* have also been applied to *R. catesbeiana.*

adaptive significance in the transition the animal makes from water to land (Frieden, 1968).

In anurans, the metamorphic Hb transition is rapid; the switch in Hbs in bullfrogs (*Rana catesbeiana*) is usually completed within a period of 3 weeks (Just and Atkinson, 1972). As described below, the larval and adult Hbs of bullfrogs have globin subunits with different amino acid sequences and thus are the products of different structural genes. These and other facts about the metamorphic Hb transition make it an attractive system for investigating how certain genes become activated while other genes become inactivated during cellular differentiation.

Most of my discussion will pertain to the Hb transition. Changes in other proteins, including some serum proteins, will be discussed primarily as they relate to the Hb transition. Since previous reviews (Frieden, 1968; Frieden and Just, 1970; Sullivan, 1974) have surveyed many of the earlier findings, the discussion will concentrate on the more recent work. A main aim of this chapter is to pose several interesting questions pertaining to the Hb transition and indicate how answers to them are important to our understanding of red blood cell differentiation.

The changes in serum proteins, Hbs, and RBCs are most often described in relation to other developmental events. A brief summary of several systems used to denote developmental stages in anurans is given in Fig. 1. Nieuwkoop and Faber (1967) have given a more complete comparison of the various staging tables for anurans.

2. AN OVERVIEW OF CHANGES IN SERUM PROTEINS AND IRON METABOLISM

2.1. Changes in Blood Osmolarity, Serum Protein Concentration, and Serum Albumin

During metamorphosis in *Rana* species, the concentration of serum proteins more than doubles (Feldhoff, 1971; Ledford and Frieden, 1973; Just *et al.*, 1977). At least 20% of this increase in *R. catesbeiana* is due to the increase in serum albumin, which rises more than 10-fold (Feldhoff, 1971). In some species, such as *R. heckscherii*, the increase in albumin is even more dramatic because of the extremely low concentrations in the tadpoles (Frieden, 1968). By measuring the *in vivo* incorporation of radioactive amino acids into albumin, Ledford and Frieden (1973) obtained evidence that the metamorphic increase in *R. catesbeiana* is due, at least in part, to an increased rate of synthesis beginning at Taylor and Kollros (TK) stage XVIII. The incorporation into albumin was found to peak at TK stage XXI at a value about 10-fold greater than the incorporation at stage XVII. After stage XXI, the incorporation into albumin declined; at stage XXV, the value was almost as low as that at stage XVII. Since the total albumin concentration remains high in froglets (Feldhoff,

1971), Ledford and Frieden proposed that there is a decreased relative rate of albumin degradation or a sparing of albumin in postmetamorphic animals.

Ledford and Frieden (1973) found that 3,3',5-triiodo-L-thyronine (T_3) had a rapid, marked effect on incorporation of labeled amino acids into albumin in TK stage XII or earlier tadpoles. Injection of 2×10^{-10} mole T_3/g body wt led to a threefold increase in incorporation by the first day after hormone treatment, a fivefold increase by 3 days, and a decline almost to the control value by 6 days. Constant immersion in T_3 (1×10^{-7} M) resulted in a similar pattern of increase of albumin synthesis, but there was no decline in the rate of incorporation between 3 and 6 days. Froglets were also reported to exhibit an increased relative rate of albumin synthesis upon treatment with T_3. Ledford and Frieden (1973) concluded that the increase in the relative rate of albumin synthesis during metamorphosis results from increased endogenous thyroid hormone levels, and that the subsequent decline in the measured incorporation results from decreasing thyroid hormone. They also proposed that the observed effect of T_3 on albumin synthesis may be a summation of two effects: a direct effect of T_3 (presumably on liver cells) and a stimulation by amino acids from the resorbing tail of the animal.

The role of serum albumin in the maintenance of blood volume through osmotic regulation and the adaptive significance of the marked increase of albumin and other plasma proteins during metamorphosis have been previously noted (Frieden, 1968). From a quantitative analysis of plasma osmotic pressure during metamorphosis of *R. catesbeiana*, Just *et al.* (1977) found a change of 20 mosmol/liter (from about 180 to 200 mosmol/liter) between TK stages II and XXV, most of the change occurring between stages XVII and XXIII. Of the various constituents in plasma, Na^+ accounts for about one-half of the total osmotic pressure and about one-half of the increase with development, while the total organic constituents (including protein) contribute only about 6% of the total plasma osmotic pressure and an even smaller percentage of the increase (Just *et al.*, 1977).

2.2. Changes in the Circulating Levels of Thyroid Hormones and Binding Proteins

Precise data on the amounts of bound and free T_3 and thyroxine (T_4) in tadpole blood at various stages of metamorphosis have been difficult to obtain for several reasons, the small blood volumes available in some species being one. Miyauchi *et al.* (1977) used radioimmunoassays to measure T_3 and T_4 concentrations in plasma of *R. catesbeiana* tadpoles. The radioimmunologic analyses showed that T_3 was detectable in the plasma at TK stage XIX and reached a peak by stage XX, while T_4 was detectable only in stages XXI through XXIV. Although the T_4 concentrations, when measurable, were higher than T_3, the T_3/T_4 ratio was much higher than in mammalian sera. These data, the plasma binding studies (discussed below), and previous observations that showed that exogenous T_3 is 5- to 10-fold more active than T_4 in inducing metamorphosis led Miyauchi et al. (1977) to conclude that T_3 plays a more significant role than T_4 in the metamorphosis of this species.

With regard to the binding proteins of tadpole plasma, Miyauchi et al. (1977) found virtually no specific binding of T_4 as compared to ample T_3-binding proteins. Binding of radioactive T_3 reached a maximum at TK stage XX, and then declined to 80% of the peak value by stage XXV. Displacement experiments with unlabeled T_3 revealed a difference in the T_3-binding properties of serum from stage XXV animals as compared to earlier stages, suggesting an actual change in, or at least some difference in, the T_3-binding protein(s) at the end of metamorphosis.

2.3. Changes in Iron Transport, Metabolism, and Storage

As the principal copper protein of serum, ceruloplasmin is a favored copper donor to cells for the synthesis and assembly of cytochrome oxidase and other copper proteins (Frieden, 1979). Ceruloplasmin is also a molecular link between copper and iron metabolism. As shown in Fig. 2, the ferroxidase activity of ceruloplasmin is important in the mobilization of iron from ferritin, an intracellular, high-molecular-weight iron-storage protein.

Ceruloplasmin undergoes dramatic changes during anuran metamorphosis (Frieden, 1968). In R. grylio, the most dramatic increase begins at about TK stage XVII, and the measured oxidase activity of ceruloplasmin reaches a peak at stage XX (Frieden, 1968). Thus, the increase in ceruloplasmin begins before the major increase in serum albumin synthesis.

Various other components of the scheme for mammalian iron metabolism have been investigated in R. catesbeiana undergoing metamorphosis. Osaki et al. (1974) measured changes in total liver iron, liver ferritin reducing activity, and RBC uptake of labeled iron from bullfrog transferrin, as well as the increase in gallbladder biliverdin, at different stages of natural metamorphosis. Theil and co-workers have made a thorough study of ferritin in larval and adult RBCs and liver, and changes in iron compartments during metamorphosis. The picture of iron metabolism that has emerged from these studies is summarized below. The serum components of the system are, primarily, ceruloplasmin and transferrin, while ferritin, ferritin reducing activity, and the Hbs are intracellular components (Fig. 2).

Theil (1973) found that R. catesbeiana larval RBCs contain high amounts of ferritin (1.2% of the RBC protein), while RBCs of froglets and adults contain much lower amounts (0.08% of the RBC protein). The decline in RBC ferritin content during natural metamorphosis was found to closely parallel the decline in larval Hbs described by Just and Atkinson

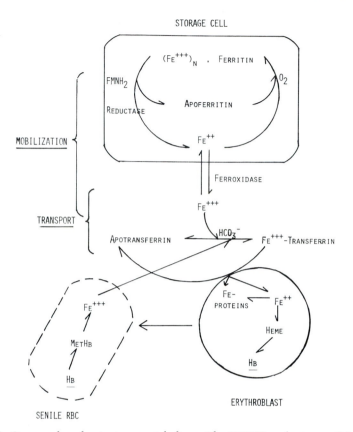

FIGURE 2. Steps and cycles in iron metabolism. The $FMNH_2$ reductase and ferroxidase (ceruloplasmin) are necessary for the retrieval of iron from storage (in ferritin) and its transfer to transferrin for transport to other cells, particularly differentiating RBCs engaged in Hb synthesis. Ferritin, $FMNH_2$ reductase, and hemoglobin metabolism are intracellular components; ceruloplasmin and transferrin are serum proteins. Met Hb, methemoglobin. Redrawn from Frieden, 1979.

(1972). By labeling prometamorphic tadpoles with [59]Fe *in vivo*, Brown and Theil (1978) found that larval RBC ferritin had more than four times the specific activity of tadpole liver ferritin. The Hbs were essentially unlabeled. By the end of metamorphosis, the situation had changed markedly. The iron label in liver ferritin had decreased by about 20%, and the Hbs contained more than three times the label of liver ferritin. (As mentioned above, RBC ferritin had dropped to very low concentrations.) The data suggest that about 60% of the iron in adult Hb during metamorphosis comes from tadpole RBC ferritin. These studies also show a shift in the primary site of iron storage from the circulating RBCs of larvae to the liver in adult frogs.

Ferritin from tadpole RBCs was found to be similar in molecular weight (478,000), subunit size (19,600), and subunit number (24) to mammalian ferritins (Theil, 1973, as cited by Brown and Theil, 1978). The ferritins of tadpole liver and adult frog liver were also found to be similar, while tadpole RBC ferritin differed from that of liver in its higher content of serine and glycine and its quantitatively different pattern of microheterogeneity when subjected to isoelectric focusing (Brown and Theil, 1978). It has been hypothesized that these differences in the structure of larval RBC ferritin may be related to its role in development, e.g., accessibility of iron for the synthesis of new Hbs during metamorphosis.

The increase in ferritin reducing activity in the liver, which becomes significant at about TK stage XXI (Osaki *et al.*, 1974), undoubtedly plays a role in mobilization of iron from that source. The marked increase in serum ceruloplasmin, which precedes all of these changes (Frieden, 1968), is probably important in the mobilization of iron from both sources (liver and tadpole RBCs) and the transfer of the iron to serum transferrin. Transferrin, in turn, is required as a donor of iron to the immature, differentiating cells that synthesize Hbs (Fig. 2).

3. CHANGES IN THE HEMOGLOBINS AND RED BLOOD CELLS

3.1. Changes in Hb Structure and Function

The most thoroughly studied amphibian Hbs are those of the common North American bullfrog, *R. catesbeiana*. Since the first study of the oxygen-binding properties of tadpole and adult bullfrog Hbs by McCutcheon in 1936, numerous investigations have been performed on the properties of these macromolecules.

The adaptive significance of the Hb transition in bullfrogs has long been recognized (Frieden, 1968). The Hbs of the tadpole have a high affinity for oxygen and are well suited for an aquatic environment where oxygen tensions are low. The adult frog Hbs have a lower affinity for O_2. Although it takes a greater O_2 tension to "load" the adult Hbs, they will release (unload) the oxygen more readily at oxygen tensions that prevail in the various tissues of the animal. Thus, the adult Hbs are well suited for an animal that is primarily air-breathing and requires a more active metabolism to support its greater muscular activity.

The differences in the behavior of these Hbs to changes in pH (the Bohr effect) and in the partial pressure of CO_2, to the presence of organic phosphates, and to chloride ion are also noteworthy. A typical Bohr effect allows hemoglobin to unload oxygen more efficiently in the peripheral tissues, where the pH is lower and the pCO_2 is higher than in the lungs, shifting the oxygen dissociation curve to the right (toward lower O_2 affinity). In mammals, organic phosphates, particularly 2,3-diphosphoglycerate (DPG), also cause a shift in the oxygen dissociation curve of Hb to the right, lowering the oxygen affinity and therefore facilitating "unloading." DPG also enhances the Bohr effect in mammalian Hbs: it takes a smaller decrease in pH to shift the oxygen dissociation curve to the right.

The oxygen affinity of both tadpole and adult bullfrog Hbs is decreased substantially in the presence of organic phosphates (Araki *et al.*, 1971). However, there is very little change in the magnitude of the Bohr effect in the presence of organic phosphates for adult frog Hbs. The tadpole Hbs exhibit an enhancement of the Bohr effect in the presence of inositol hexaphosphate (IHP) and ATP (Watt and Riggs, 1975); DPG has a much smaller effect (Araki *et al.*, 1973). The physiological significance of the effect of organic phosphates on the functioning of tadpole Hbs is open to question since Araki *et al.* (1971) reported that the concentrations of IHP and ATP in tadpole erythrocytes are too low to decrease the oxygen affinities of the Hbs *in vivo*. Thus, Araki *et al.* (1973) describe the Bohr effect for tadpole Hbs as "latent" since they could only detect it in the presence of IHP and ATP. Watt and Riggs (1975), however, have found a reversed Bohr effect for "stripped" tadpole Hbs in various non-phosphate buffers (i.e., the oxygen affinity is lower at pH 8 than at pH 6). This difference in results possibly relates to differences in the concentration of chloride ions (Cl^-) in the *in vitro* mixtures. Watt and Riggs (1975) found that the magnitude of the Bohr effect of tadpole Hbs depends on the Cl^- concentration: extrapolation of the data to zero Cl^- concentration indicated a Bohr effect close to zero.

The functional differences between tadpole and adult bullfrog Hbs arise from differences in the structures of their globin chains (Table 1), since the heme moiety of all these Hbs appears to be the same. Recent studies of these Hbs have yielded considerable, detailed information about their structures, which helps explain, in part, the functional differences noted above. For example, the first six amino acid residues of the amino terminus of the β chain of a major adult frog Hb (component C) are missing, in contrast to the β chains of tadpole and human Hb (Baldwin and Riggs, 1974), which helps explain why the allosteric effects of organic phosphates are much larger on tadpole than on adult Hbs. The amino termini of the two β chains of mammalian Hbs are the major sites for the binding of organic phosphates.

All of the α chains of both tadpole and adult bullfrog Hbs have blocked (acetylated) amino termini. The significance of this finding to the functioning of these molecules in unknown. The carboxyl-terminal residues of the α and β chains of tadpole and adult frog Hbs and of human Hb are identical (Watt and Riggs, 1975).

The origin of the heterogeneity of the larval Hbs of the bullfrog is a point that will be discussed in a subsequent section. As indicated in Table 1, there are four major larval Hbs in tadpoles, which have been numbered I to IV in thr order that they come off a column of DEAE–Sephadex A-50 containing a linear, decreasing pH gradient from pH 7.7 to 6.7 (Watt and Riggs, 1975). The order in which the Hbs come off such a column is opposite to their order of migration on electropho-

TABLE 1. *RANA CATESBEIANA* HEMOGLOBINS

Electrophoretic bands[a] Terminologies				Components isolated by column chromatography[f]	Subunit compositions (tentative designations)[g]
(1)[b]	(2)[c]	(3)[d]	(4)[e]		
Tadpole					
4[h]	ND[h]	ND[h]	ND[h]	—	—
3	Td-3	Td-3	Td-4	I	$\alpha_2^I\beta_2^I$
2a	Td-2	Td-2	Td-3	II	$\alpha_2^{II}\beta_2^{II}$
2		Td-2a	Td-2	III	$\alpha_2^{II}\beta_2^{III}$
1	Td-1	Td-1	Td-1	IV	$\alpha_2^{III}\beta_2^{III}$
Adult frog[i, j]					
				A (minor Hb)	$\alpha_2^?\beta_2^?$
				B (major Hb)	$\alpha_2^B\beta_2^C$
				C (major Hb)	$\alpha_2^C\beta_2^C$
				D (minor Hb)	$\alpha_2^?\beta_2^?$

[a] Tadpole Hb bands separated by polyacrylamide disc gel electrophoresis are numbered in the order of their migration, band number 1 being the fastest migrating. The correspondence between the electrophoretic bands of Broyles and co-workers [columns (2), (3), and (4) of the table] and components I through IV isolated by Watt and Riggs (1975) was determined by electrophoresing purified components supplied by Austen Riggs (Broyles *et al.*, 1981).

[b] From Moss and Ingram (1968a).

[c] From Broyles and Frieden (1973).

[d] From Broyles and Deutsch (1975).

[e] From Broyles *et al.* (1981).

[f] For tadpole Hbs, from Watt and Riggs (1975); for adult Hbs, from Aggarwal and Riggs (1969).

[g] The terminology used for the globin subunits is tentative. For the tadpole, the number of different globin chains is based on information derived from Watt and Riggs (1975) and from the amino acid sequences of the separated chains (Riggs, personal communication). For the adult frog, the information is taken from Aggarwal and Riggs (1969), Baldwin and Riggs (1974), and Riggs (personal communication). The globin chains of the minor adult Hbs A and D have not been thoroughly characterized (thus the question marks). However, it is known that all of the adult globin chains contain cysteine and are thus different from all of the six major tadpole globin chains, which have no cysteine (see text).

[h] A minor tadpole band. ND, no designation assigned.

[i] The adult Hbs detected by electrophoresis are mainly the major components B and C. Component C aggregates to form high-molecular-weight polymers, while component B does not (see text). These polymers of C are retarded by the sieving effect of polyacrylamide and other electrophoretic gels, allowing adult Hb C to be cleanly separated from the other bullfrog Hbs.

[j] See Broyles and Frieden (1973) for electrophoretic pattern.

resis at a basic pH. The four electrophoretic bands reported by Broyles and Deutsch (1975) and Broyles *et al.* (1981) have been found to coelectrophorese with samples of the column-purified components provided by Riggs. Thus, Hbs Td-1 to Td-4 correspond to components IV to I, respectively (Table 1).

The four major larval Hbs each have been found to be tetramers of two α-type and two β-type subunits (Watt and Riggs, 1975). There are three distinct α chains, and three major and three minor β chains. Components II and III have the same α chain, and components III and IV have the same major β chain (Table 1; Riggs, personal communication). The four larval Hbs also fall into two distinct functional groups: I and II have substantially higher oxygen affinities than III and IV. Watt and Riggs (1975) have noted that tadpoles of earlier developmental stages (TK stages III and IV; Riggs, personal communication) have as much as 90% of their total Hb as components I and II, while older, larger tadpoles (about TK stage X) have about two-thirds of their Hb as components III and IV.

Of the four adult Hbs (Table 1), B and C are major components and A and D are minor components (Aggarwal and Riggs, 1969). Component C, but not B, forms aggregates when the red cells are lysed, yielding bands of octamers (two tetramers) and higher aggregates in polyacrylamide and other electrophoresis supporting media that have a molecular sieving effect (Moss and Ingram, 1968a). It has been shown that this adult Hb polymerizes by forming disulfide bridges between tetramers (see Frieden, 1968).

With respect to the adult Hbs, Aggarwal and Riggs (1969) showed that component C has a higher oxygen affinity than component B. Saucier (1976) observed that component C is the first adult Hb to appear during metamorphosis and comprises essentially all of the Hb of young froglets that have recently completed metamorphosis. McCutcheon (1936) showed that the Hb oxygen dissociation curve of large sexually mature bullfrogs is displaced farther to the right (has a lower oxygen affinity) than that of smaller frogs.

Thus, during the life cycle of R. catesbeiana, there is a progression of changes in the expression of globin genes that results in a corresponding progression from Hbs of higher oxygen affinity to those of lower oxygen affinity.

3.2. Developmental Timing of the Hb Transition in Various Amphibians

3.2.1. Rana catesbeiana

The Hb switch in bullfrogs usually occurs rapidly (Just and Atkinson, 1972). Most of the change occurs between TK stages XXII and XXIV, a period of approximately 6–8 days (Ledford and Frieden, 1973). (See Fig. 1 for the different terminologies used for reporting developmental stages.)

Most reports indicate that adult Hb is first detectable in the early part of metamorphic climax. Hamada et al. (1966) and Moss and Ingram (1968b) report that adult Hb can be detected when the animals have a hindlimb/tail length ratio of about 0.8; others report that adult Hb can be detected at TK stage XXII (Just and Atkinson, 1972) or even stage XXI (Saucier, 1976). Polyacrylamide gel electrophoresis has been used to detect adult Hb in most of these studies. It is agreed that adult Hb first becomes detectable in the circulating RBCs shortly after the emergence of the forelimbs. In the same studies, it was generally found that the transition had reached the 50:50 point (half adult Hb, half tadpole Hb) at about TK stages XXII–XXIII, and was very nearly complete by stages XXIV–XXV (tail resorbed).

However, Banbassat (1970) reported that in R. catesbeiana, adult Hb was not detectable until well after forelimb emergence and that the transition was not completed until 4–10 weeks after metamorphosis. He also found that the Hb transition was delayed until after metamorphosis in *R. pipiens*. [As discussed later, Hollyfield (1966a) observed a transition in red cell morphology at the peak of metamorphic climax (TK stages XXII–XXIII) in *R. pipiens*.] Benbassat (1970) reported that summer tadpoles were used, but did not say whether the studies were done in early summer or late summer. Broyles and Dorn (unpublished observations) have found that as the season progresses from late spring to early fall, the Hb transition shifts to later metamorphic stages.

If it can be carefully documented that there is a seasonal shift such that the Hb transition moves out of phase with the majority of the morphological changes, such a finding will have implications for the possible mechanisms involved in mediating the Hb transition, as discussed in subsequent sections.

3.2.2. Other Anurans

Similar Hb transitions have been found to occur in other anurans, including *R. clamitans, R. esculenta, R. grylio, R. heckscherii, R. pipiens*, and *Xenopus laevis* (Frieden, 1968; Benbassat, 1970; Maclean and Jurd, 1972). Of these, the transition that occurs in *X. laevis*, the South African clawed toad, has received the most attention.

Maclean and Jurd (1971) found two major Hbs in approximately equal amounts in young *Xenopus* tadpoles [Nieuwkoop and Faber (NF) stage 45; see Fig. 1], which they labeled Hb F_1 and Hb F_2. Hb F_1 is present in about three times the amount of Hb F_2 in older tadpoles (NF stage 50). One major (Hb A_1) and one minor (Hb A_2) component were found in adults; of the total Hb, A_1 comprises about 92% and A_2 about 5%, with the remainder being divided between two very minor components that may be residual larval Hbs (Maclean and Jurd, 1971).

Adult Hb A_1 is present in tadpoles in significant amounts before the animals enter metamorphic climax. (Hb A_1 comprises about 22–23% of the total Hb at NF stage 50.) Adult Hb A_2 also becomes detectable during the late tadpole stages. The final adult pattern is established by 10 weeks postmetamorphosis.

Thus, the Hb transition begins earlier and ends later in *Xenopus* than in *R. catesbeiana*. A recent report has shown that this pattern of early onset and late completion of the Hb transition is even more exaggerated in the anuran *Bombina variegata* (Cardellini and Sala, 1979).

The developmental changes in Hbs in *Xenopus* are analogous, in part, to the transitions that occur in *R. catesbeiana*. The increase in the ratio of Hb F_1 to Hb F_2 is reminiscent of the shift in larval bullfrog Hbs reported by Watt and Riggs (1975). In both cases, there is a shift toward the faster electrophoretic forms of tadpole Hbs as larval development progresses. The transition to adult Hbs in *Xenopus* bears a resemblance to the fetal-to-adult Hb transition in humans that occurs at around the time of birth. In both cases, the adult Hb is detectable before the period of most rapid change; and the adult Hb pattern is comprised of one major and one minor component. As in *R. catesbeiana, Xenopus* larval Hbs probably have no globin subunits in common with the adult Hbs (Maclean and Jurd, 1972). In humans, the α chain is shared by fetal and by both types of adult Hbs.

3.2.3. Urodeles

Studies with the axolotl *Ambystoma mexicanum*, a neotenic animal, have shown that a larval-to-adult Hb transition can occur without anatomical metamorphosis (Maclean and Jurd, 1972; Ducibella, 1974a). A comparison of the Hb transition in this animal with that in its close relative *A. tigrinum*, which undergoes an anatomical metamorphosis, has shown that the switch from larval to adult Hb occurs between 100 and 150 days of age in both animals (Dubibella, 1974a). In *A. tigrinum*, the Hb switch is coincidental with anatomical metamorphosis. When large, neotenic adults of *A. mexicanum* are induced to undergo anatomical metamorphosis with T_4 (Ducibella, 1974a) or T_3 (Maclean and Jurd, 1972), there is no change in the Hb pattern since the larval-to-adult Hb transition has already occurred at an earlier developmental stage.

The work of Ducibella (1974b) also shows that there is a change in red blood cell types that correlates very closely with the Hb transition in *A. mexicanum*. The red cells that contain adult Hb have a different morphology than those that contain larval Hb. The red cell transition is easily shown in this species since larval RBCs crenate markedly in a medium composed of citric acid, trisodium citrate dihydrate, and dextrose, while the cells of a neotenic adult retain an almost normal morphology (Ducibella, 1974b). As described in the next section, changes in red cell type also occur in anurans.

3.3. Changes in Morphology and in Other Properties of the RBCs

Morphological differences between larval and adult RBCs have been noted in a number of anuran species, including *R. pipiens* (Hollyfield, 1966a), *R. catesbeiana* (McCutcheon, 1936), *X. laevis* (Jurd and Maclean, 1970), and *Bufo melanostictus* (Church, 1961, as cited by Hollyfield, 1966a).

Hollyfield (1966a) carefully measured RBCs from *R. pipiens* of TK stages XIX through XXV, and the average surface area of cells at each stage was calculated. The results clearly show that during metamorphosis the larger, larval RBCs are replaced by a population of smaller, adult RBCs. The cells of the new population have a lighter cytoplasm and more granular nucleus. The midpoint of the transition is reached at about TK stage XXIII. McCutcheon's measurements and observations (1936) indicate that in *R. catesbeiana*, the larval RBCs are of an oval shape and larger than the more elliptical adult RBCs.

Herner and Frieden (1961) found that in *R. heckscherii, R. catesbeiana*, and *R. grylio*, the red cell count (number of cells per cubic millimeter of blood) and whole blood hemoglobin concentration (grams per 100 ml blood) were greater for adult than for larval animals. In *R. catesbeiana* (Herner and Frieden, 1961) and *R. pipiens* (Benbassat, 1970), the amount of Hb per cell was also found to be greater for the adult than for the tadpole RBCs.

Dorn and Broyles (1978) have found that the RBCs of young froglets (TK stage XXV) and of adult frogs have a greater density than those of tadpoles in *R. catesbeiana*. The difference in density ($p = 1.085$ for tadpole RBCs and 1.105 for adult RBCs) has allowed artificial mixtures of the two cell types to be clearly separated in very shallow density gradients of colloidal silica (Fig. 3). The greater density of adult frog RBCs is consistent with their greater Hb content and smaller size.

The RBCs of tadpole and adult *R. pipiens* and *R. catesbeiana* contain adenyl cyclase activity (Rosen and Rosen, 1968). However, the adenyl cyclase of tadpole RBCs is unresponsive to stimulation by catecholamines, while that of the adult RBCs does respond with increased synthesis of cyclic AMP. The role of cAMP in these RBCs is not known.

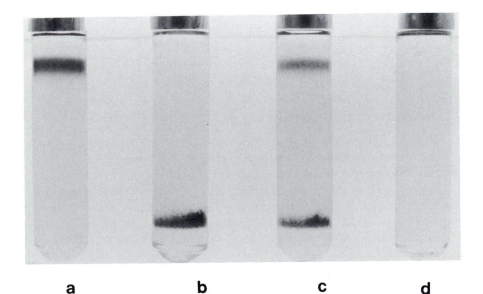

a   b   c   d

FIGURE 3. Separation of *Rana catesbeiana* RBCs on density gradients of colloidal silica. (a) Tadpole RBCs alone. (b) Adult frog RBCs alone. (c) Artificial mixture of tadpole and frog cells. (d) Tube with no cells, used to determine the shape of the gradient. Electrophoresis of Hbs from the separated bands of cells showed that there was no overlap. From Dorn and Broyles, 1978. Photo contributed by A. R. Dorn.

In amphibians, as in most other nonmammalian verbebrates, the mature, differentiated RBC retains its nucleus. This is true of the mature RBCs of both the larva and the adult. However, the nuclei of larval RBCs have a different appearance from those of the adult. Those of the *R. pipiens* tadpole RBCs are round, while the adult RBC nuclei are elliptical and have a more condensed, granular appearance (Hollyfield, 1966a; Benbassat, 1970). Benbassat (1970) observed that tadpole (*R. pipiens*) RBCs have a greater amount of reticular substance in the cytoplasm than adult RBCs.

These differences in the nuclear and cytoplasmic compartments support the view that adult RBCs are more mature cells and less active in the synthesis of nucleic acids and proteins as compared to larval RBCs. The synthetic capabilities of these cells have been studied by more direct means. Benbassat (1970) found that circulating RBCs of *R. pipiens* and *R. catesbeiana* premetamorphic tadpoles (TK stages V–XI) have a greater ability than RBCs of froglets or adults to incorporate labeled amino acids, uridine, and thymidine into macromolecules *in vitro*. Radioautography showed that in *R. pipiens*, 10–15% of the premetamorphic tadpole RBCs were labeled with [^{3}H]thymidine after a 6-hr *in vitro* incubation, while only 0.5% of the cells of TK stage XXIV animals and almost none of the RBCs from adult frogs were so labeled.

Thus, morphological and biochemical evidence shows that marked differences exist between larval and adult RBCs. The difference in the types of Hb they contain is, of course, of central importance.

4. CENTRAL QUESTIONS PERTAINING TO THE RBC–Hb TRANSITION IN BULLFROGS

4.1. Do Adult and Larval Hbs Share Any Globin Chains?

Mammalian fetal and adult Hbs have identical α chains. It was of interest to see if the larval and adult Hbs of *R. catesbeiana* also share any gene products.

Earlier and more recent studies have provided evidence predominantly in favor of the assertion that there are no globin chains in common between larval and adult Hbs. This evidence is of several types: (1) globin chains isolated from tadpole and adult bullfrog Hbs migrate differently upon electrophoresis (Moss and Ingram, 1968a); (2) there are differences in the amino acid compositions of the isolated globin chains (Aggarwal and Riggs, 1969; Watt and Riggs, 1975); (3) tadpole and frog globin chains have different NH$_2$-terminal amino acids, especially on the β-type chains (Moss and Ingram, 1968a; Watt and Riggs, 1975); (4) peptide maps from tryptic digests of the purified globin chains are different (Watt and Riggs, 1975); (5) reaction of isolated globin chains with reagents that detect sulfhydryl groups show that tadpole globin chains are devoid of cysteine while all adult globin chains contain this amino acid (Frieden, 1968); (6) antibodies prepared against tadpole and frog Hbs do not cross-react (Maniatis and Ingram, 1971b); and (7) the recently determined amino acid sequences of the globin chains of the four major tadpole Hbs and of the major adult bullfrog Hb (component C) reveal no common globin subunits between tadpole and frog Hbs (Riggs, personal communication).

Some results conflict with the above findings. Hamada and Shukuya (1966) reported that the α chains of the major tadpole and adult bullfrog Hbs have the same amino- and carboxy-terminal amino acids, and that tryptic and chymotryptic peptide maps of these tadpole and frog α chains are the same. Moss and Ingram (1968a) have pointed out and Aggarwal and Riggs (1969) have shown that bullfrog tadpoles from different sources have globin chains with different amino acid compositions, which indicates that subspecies differences exist. It is probable that this fact accounts for, at least in part, the differences between the results of Hamada and Shukuya (1966) and the bulk of the evidence cited above. Aggarwal and Riggs (1969) also found that some tadpole globin chains contain traces of cysteine (determined by conversion to cysteic acid) while other chains have none. This difference was also found to depend on the source from which the tadpoles were obtained.

4.2. Are Adult and Larval Hbs Contained in the Same RBCs during Metamorphosis?

A number of findings, including some of the differences between larval and adult RBCs reviewed above, led to the proposal that the adult Hb that appears during metamorphosis is contained within a new cell line distinct from the line that gives rise to larval RBCs. This proposal was supported by the finding of Moss and Ingram (1968b) and DeWitt (1968) that a new, microcytic cell appears in the circulating blood during natural or T_4-induced metamorphosis of *R. catesbeiana*. This new cell was classified as an immature RBC by virtue of its round shape, relatively large nucleus, lower content of Hb in the cytoplasm, and its ability to incorporate labeled precursors of globin and heme (^{14}C-labeled amino acids, δ-[4-^{14}C]aminolevulinic acid, ^{59}Fe) as revealed by radioautography (Moss and Ingram, 1968b). DeWitt (1968) concentrated the microcytic cells in density gradients of silicone fluids and found that they contained adult Hb. However, the same fraction of cells from the gradient also contained tadpole Hbs. The question of coexistence of tadpole and adult Hbs in one cell remained to be answered.

If the proposed new cell line exists, larval and adult Hbs should be found in different RBCs during the Hb transition. On the other hand, if some RBCs contain both larval and adult Hb types, at least some of the larval RBCs must be of the same lineage as the new, adult-Hb-containing RBCs.

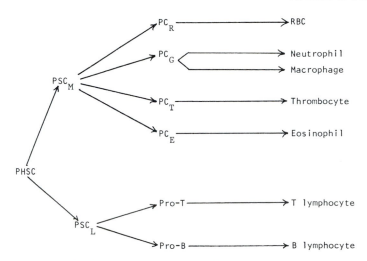

FIGURE 4. Scheme for the differentiation of all blood cell types from a common pluripotent hemopoietic stem cell (PHSC). PSC_M, pluripotent stem cell–myeloid; PSC_L, pluripotent stem cell–lymphoid; PC, determined progenitor cells committed to subsequent differentiation along the pathways shown (R, G, T, and E denote the erythroid, granuloid, thromboid, and eosinophilic progenitor cells, respectively); Pro-T, precursor to T lymphocytes; Pro-B, precursor to B lymphocytes. (See Cline and Golde, 1979.)

This is an important point because it implies something about the mechanism by which the change in gene expression occurs. If the Hb switch is accomplished by substitution of cell lines, it is implied that the choice of Hb type occurs at the level of the hemopoietic stem cell, presumably before or at about the same time the stem cell becomes committed to the erythroid pathway of differentiation (Fig. 4). But if more than one Hb type (larval and adult) is found in the same cell, it is possible that the choice of Hb type is made after the stem cell becomes committed to the erythroid pathway.

Further support for the proposed new cell line came from the electrophoresis of Hb from single cells (Rosenberg, 1970) and double immunofluorescence labeling of single RBCs with anti-tadpole and anti-adult Hb-specific antibodies (Maniatis and Ingram, 1971b,c). Both studies indicated that tadpole Hb and adult Hb are found in different RBCs at metamorphic climax in *R. catesbeiana*. However, using similar immunofluorescence techniques, Benbassat (1974) found that up to 16% of the circulating RBCs in *R. catesbeiana* contain both tadpole and adult Hbs. Jurd and Maclean (1970) found that as many as 25% of the RBCs contain both adult and tadpole Hbs at the peak of the metamorphic transition in *X. laevis*.

Thus, experiments designed to answer directly the question of whether single RBCs during metamorphic transition contain both tadpole and adult Hbs, or only one or the other, have yielded conflicting results. Species differences could account for the difference in results obtained with *X. laevis* (Jurd and Maclean, 1970) and *R. catesbeiana* (Rosenberg, 1970; Maniatis and Ingram, 1971c), were it not for the conflicting results of Benbassat (1974) with *R. catesbeiana*. Benbassat (1974) has offered two possible explanations for the difference between his results and those of Maniatis and Ingram (1971a): (1) The antisera prepared by Maniatis and Ingram may have had a greater monospecificity than those of Benbassat. A nonhemoglobin protein in tadpole RBCs could also be present in the first RBCs to appear during metamorphosis that have adult Hb, making it seem that tadpole and adult Hb appear together in some of the cells. (2) The dual-labeling method used by Maniatis and Ingram might fail to detect adult Hb in RBCs that have a lot of tadpole Hb, and vice versa, due to one fluorescent dye masking the other. The indirect method used by Benbassat is more sensitive.

The latter explanation seems less likely to be true in light of the work of Jurd and Maclean (1970); they obtained essentially the same results with both methods of labeling single cells, although they too felt that the indirect method is the preferred one. The former explanation seems more likely, although the supposed nonhemoglobin protein would have to decline to undetectable levels during maturation of adult RBCs since Benbassat (1974) checked the specificity of his antisera by immunofluorescence after cross-labeling tadpole and frog cells.

There are a number of other points about these techniques and their use that can be debated, including the methods used to ensure that the antigen is actually accessible to the labeled antibody during the reaction with cell smears or tissue imprints, and the ability of the observer to detect fluorescence of the rhodamine label (which is weaker) in the presence of fluorescein (which has a stronger fluorescence). In all three studies with tagged antibodies, a peculiarity about the fluorescence of labeled adult RBCs was noted. Whereas tadpole RBCs reacted with anti-tadpole Hb-tagged antiserum exhibited fluorescence over the cytoplasmic area of the cells, adult RBCs reacted with tagged anti-adult Hb antiserum showed a ring of fluorescence around the nucleus and little fluorescence over the cytoplasm. The explanation given for this phenomenon (Maniatis and Ingram, 1971c) was that the large amount of Hb present in the frog cell cytoplasm interferes with the fluorescence; this might be particularly true for fluorescein, which fluoresces at a wavelength near that of the absorption of heme compounds. Presumably, the cells that synthesize adult Hb during metamorphosis would be young cells with less total Hb and would not present this problem. However, Maniatis and Ingram (1971c) found that although immature cells from the liver fluoresced for adult Hb in the cytoplasm, the immature cells from the circulation exhibited the nuclear pattern noted above when adult Hb was detected. Thus, the meaning of the perinuclear fluorescence is not clear. This anomaly may contribute to the difficulties of trying to detect two types of Hb in one cell by a dual-label approach.

Although the results of Rosenberg (1970) agree with those of Maniatis and Ingram, staining Hb bands on microgels after single-cell electropho-resis is less sensitive than immunofluorescent labeling. It is possible that bands containing small amounts of Hb went undetected. Also, the di-merization of adult Hb tetramers, which is the basis of separation and detection of adult Hb in this system, does not occur at low Hb concen-trations (Moss and Ingram, 1968b).

Therefore, the question of whether larval and adult Hbs are found in the same or different RBCs during metamorphosis has not been con-clusively answered. Possibly, a conclusive answer can be obtained if monospecific antibodies to all the larval and adult Hbs are obtained and the question is reexamined using immunofluorescence techniques com-bined with a gradient centrifugation technique capable of cleanly sepa-rating RBCs of different shapes and densities (Fig. 3). Recent information on the RBC populations in these animals, which is described in the next section, emphasizes the need for further examination of this key question.

4.3. Is the RBC–Hb Transition Mediated by a Change in Erythropoietic Sites?

Related to the preceding discussion is the possibility that Hb tran-sitions are, in some cases, mediated by a change in erythropoietic sites. As reviewed by Foxon (1964), there is a considerable body of literature indicating that in anurans, larval erythropoiesis and adult erythropoiesis occur in different organs. The occurrence of erythropoiesis in the larval kidneys is well documented (see Hollyfield, 1966b). In the adult, the spleen has been reported to be the main erythropoietic site, with the bone marrow being active during the spring of the year (Jordan and Speidel, 1923a). More recent work by Maniatis and Ingram (1971a,c) has indicated that the metamorphic Hb transition is accomplished without a change in erythropoietic sites: the liver was found to be the site of RBC matur-ation in both tadpoles and young froglets. Turpen et al. (1979) have shown that erythropoiesis is the predominant hematopoietic activity in the lar-val liver of R. pipiens.

A resolution of the conflict in reports on the erythropoietic sites in R. catesbeiana tadpoles (kidneys versus liver) has been suggested by Broyles and co-workers. The work of Broyles and Frieden (1973) showed that two organs, the larval liver and the larval kidneys, participate in erythropoiesis: Short-term organ cultures of liver or kidney incorporated radioactive precursor molecules into larval Hbs separated by polyacryl-amide gel electrophoresis. Moreover, the two organs were found to syn-thesize different ratios of the various larval Hbs in short-term organ cultures (Broyles and Frieden, 1973) and to contain different larval Hbs (with no detectable overlap) after a week of organ culture (Broyles and Deutsch, 1975).It was found that the different larval Hbs disappear from the circulation at different rates during phenylhydrazine-induced anemia (Broyles and Deutsch, 1975).

This last result and the results with longer-term organ cultures suggest that the different larval Hbs elaborated by the liver and kidneys are contained within different populations of larval RBCs in the circulating blood. As reported by Benbassat (1970), tadpole RBCs are pleomorphic.

4.3.1. Two Erythropoietic Sites and Two Larval RBC Populations in Bullfrog Tadpoles

Broyles *et al.* (1981) have observed RBCs of at lease two distinct morphologies in the blood of TK stage X–XII tadpoles: One type (labeled type 1) has an oblong, oval shape and an acentric nucleus, while the other type (2) has a round-to-elliptical shape and a central nucleus. Measurements of the longest dimension (length) of peripheral blood RBCs showed that type 1 and type 2 RBCs belong to different size classes. Type 1 RBCs have a mean length of 26.85 ± 0.03 (S.E.) µm, while type 2 RBCs average 23.50 ± 0.04 µm. Histological sections of the erythropoietic organs showed that kidney is enriched in type 1 RBCs and liver is enriched in type 2 RBCs, as compared to each other and to the circulating blood. Correlations between the percentages of RBCs of the two types and the percentages of the four larval Hb types in circulating blood (correlation coefficients = 0.88) and in the erythropoietic organs of individual animals indicated that type 1 RBCs contain larval Hb Td-4 and type 2 RBCs contain larval Hbs Td-1, -2, and -3. RBCs containing Hb Td-4 were isolated from kidney tissue, and RBCs containing Hbs Td-1, -2, and -3 were isolated from liver tissue.

It was concluded that bullfrog larval RBCs with an oval shape and an acentric nucleus contain Hb Td-4 and are the product of mesonephric kidney erythropoiesis, and that elliptical RBCs with a central nucleus contain larval Hbs Td-1, -2, and -3 and are the product of liver erythropoiesis (Broyles *et al.*, 1981).

4.3.2. Erythropoietic Sites and Cell Lineages

It is now evident that at least three populations of RBCs are produced during the life cycle of *R. catesbeiana*—the two larval RBC types described above and the adult RBC population that replaces the larval RBCs during metamorphosis. The cell lineages of these RBC types are less certain. It has been suggested that all the larval RBCs are replaced by a new cell line during metamorphosis that synthesizes only adult Hb (Maniatis and Ingram, 1971c). However, as noted above, the evidence for this proposal is not conclusive.

The existence of two larval RBC populations emanating from different erythropoietic sites leads me to propose different cell lineages. One cell line would be represented by the type 1 larval RBCs emanating from the kidneys, while another would be represented by the type 2 larval RBCs emanating from the liver. I suggest that the type 2 larval RBCs are of the same lineage as the adult RBCs on the basis of the following: (1) The morphologies of type 2 larval RBCs and adult RBCs are similar. [Compare the photographs of type 2 larval RBCs by Broyles *et al.* (1981) with those published of adult RBCs—Fig. 1C, Hollyfield (1966a); Fig. 1C and D, Benbassat (1970); Figs. 3a and 4a, Maniatis and Ingram (1971a)]. (2) Type 2 larval RBCs, RBCs synthesizing adult Hb during metamorphosis, and adult RBCs of young froglets all emanate from the liver. (3) In the studies of Maniatis and Ingram (1971b,c), in which double immunofluorescence labeling of single cells was used to see if both larval and adult Hbs could be found in the same cell, it appears that the anti-tadpole Hb antibody was directed mainly toward larval Hb Td-4 and recognized the other larval Hbs to a lesser extent. The tadpole Hb fraction that they used as an antigen contained primarily tadpole component I (Td-4), with a smaller amount of components migrating between components I and IV (Td-4 and Td-1) on electrophoresis. (See Table 1 for the nomenclature of tadpole Hbs.) Component IV (Td-1) and the slowest migrating, minor Hb (which has not been characterized) were completely excluded. It is evident from Fig. 6 of Maniatis and Ingram's paper (1971b) that their anti-tadpole Hb antiserum, when reacted with the faster moving electrophoretic tadpole Hb component, fixed complement, but to an extent that was much less than the combination of the major tadpole Hb fraction (the antigen used to produce the antibody) and the antiserum. It is possible, therefore, that the anti-tadpole Hb antiserum failed to detect some tadpole Hb of a different type that could have been present in the same cells that contained adult Hb during metamorphosis.

Thus, it appears that Maniatis and Ingram were using an antibody against kidney-derived Hb and looking mainly at its adsorption to liver-derived RBCs (tissue-touch imprints). The immature cells seen in the circulation during metamorphosis also might fail to react with the anti-tadpole Hb antibody if, as the evidence of Maniatis and Ingram (1971a) indicates, those cells were released from the liver. Should this proposal be correct, it would explain how Benbassat (1974), using immunofluorescence labeling, detected both larval and adult Hb in about 16% of the RBCs at metamorphic climax. Benbassat's antiserum was directed toward a mixture of larval Hbs.

What happens within each of the erythropoietic sites when tadpoles enter metamorphic climax has not been fully elucidated. Maniatis and Ingram (1971a,b,c) have presented evidence that the switch to adult Hb synthesis is accomplished in the liver. It is likely that erythropoiesis in the kidneys declines during this period, since short-term organ cultures of young froglet kidney incorporate radioactive precursors into Hb to a lesser extent than TK stage X–XII tadpole kidney, and the only Hb that appears to be present is Td-4 in small amounts (Broyles, unpublished observations). The claim of Jordan and Speidel (1923b) that erythropoiesis switches to the spleen during metamorphosis has not been confirmed. The switch from liver erythropoiesis in froglets to spleen and/or bone marrow erythropoiesis in the mature adult is probably a postmetamorphic event (Maniatis and Ingram, 1971a,c).

4.3.3. A Role for Erythropoietic Microenvironments

Regardless of what the cell lineages are in *R. catesbeiana*, the question remains of what determines choice of Hb type in differentiating RBCs. It has been suggested that the erythropoietic microenvironments play a role in this decision (Broyles and Deutsch, 1975). In this context, an erythropoietic microenvironment is defined as any and all factors, cellular or molecular, present in the organ in the vicinity where the erythroid cells differentiate.

As noted in the preceding discussion, there are two erythropoietic organs in bullfrog tadpoles (the mesonephric kidneys and the liver) that are active throughout most of the larval period. The RBCs emanating from these two sites have different morphological appearances and express different larval globin genes. From Table 1, it can be seen that larval Hb component I (Td-4), which is contained in RBCs emanating from the kidneys, shares no globin chains with the other larval Hbs, which are contained in RBCs emanating from the liver.

In histological sections of tadpole kidney and liver, Broyles *et al.* (1981) observed that the erythropoietic foci contain nonerythroid as well as erythroid cells; the nonerythroid cells in the foci in the two organs have different appearances and the arrangements of the cells within the foci differ in the two cases.

In organ cultures, it has been found that liver co-cultured with kidney undergoes a shift in its Hb pattern; part of the shift is toward the Hb type that is the product of kidney erythropoiesis (Broyles *et al.*, 1981). The factors responsible for inducing this Hb shift have not been identified.

It is known that marked molecular and cellular changes occur in the liver during metamorphosis (Frieden and Just, 1970), but whether any of these changes relate to changes within this erythropoietic microenvironment or to the larval-to-adult Hb switch is not known.

4.4. Does the Life Span of RBCs Change during Metamorphosis?

During metamorphosis, what happens to the mature, circulating RBCs that contain tadpole Hbs? Unless the metamorphic Hb transition is delayed (Benbassat, 1970) for as yet unknown reasons, the switch is usually fully accomplished in 20 days or less in *R. catesbeiana*. Yet, Forman and Just (1976) have reported that the life span of larval *R. catesbeiana* RBCs is about 100 days. If this figure is correct, then either the conditions prevailing during metamorphosis shorten the life span of the tadpole RBCs, or larval RBCs become recognizable for premature destruction due to either changes in the RBCs themselves, changes in the reticuloendothelial system, or both.

However, it is doubtful that the figure derived by Forman and Just gives an accurate picture, for several reasons. First, the study was accomplished by determining the decrease in the number of cells with labeled nuclei, using radioautography. The label used was [^{3}H]thymidine, and no evidence has been provided that the label cannot be conserved and recycled in tadpoles. Second, the RBCs were labeled initially by injecting the [^{3}H]thymidine into tadpoles recovering from phenylhydrazine-induced anemia, since large amounts of new, synthetically active cells that will incorporate such a label are produced in such animals. Chegini *et al.* (1979) have found that once the initial, sudden release of erythroblasts into the circulation has occurred in anemic adult *Xenopus*, no further release of erythroid cells from the spleen and liver occurs until recovery from anemia is complete. Therefore, the phenylhydrazine-induced animal is likely to be one in which the normal turnover of RBCs has been altered. Third, early time-points (before 40 days after injection of [^{3}H] thymidine) are missing from the reported data, which makes the zero-time value for the percentage of labeled nuclei and the slope of the line of decrease in that percentage with time less certain. The early time-points might also have given some indication of whether or not such a line might be biphasic, which could be possible in light of the two larval RBC populations described in previous sections.

Chegini *et al.* (1979) have provided electron microscopic evidence that phagocytosis and destruction of old, damaged RBCs occur in the liver and the spleen of *Xenopus* adults treated with phenylhydrazine. Similar observations are not available for metamorphic tadpoles, but it is reasonable to expect that these organs would be involved in RBC destruction in light of their roles in the reticuloendothelial system of mammals. Saucier (1976) found that splenectomized *R. catesbeiana* tadpoles, when allowed to metamorphose, contained much more tadpole Hb than nonsplenectomized controls of a comparable stage of metamorphosis. Osaki *et al.* (1974) have shown that there is a marked increase in biliverdin in bullfrog tadpole gallbladders, beginning at TK stages XX–XXI, indicating that a marked increase in Hb catabolism is occurring during this period.

4.5. Do Thyroid Hormones Have a Direct Effect on the Switch from Larval to Adult Globin Gene Expression?

Since the early studies of the Hb transition in anurans correlated the change with metamorphosis, thyroid hormones were hypothesized to induce the switch.

Moss and Ingram (1968b) administered T_4 (5×10^{-8} M) to young *R. catesbeiana* tadpoles via the water in which the animals were kept and sacrificed the animals after various times of treatment. The animals were bled and the RBCs were incubated with radioactive precursors of Hb. Electrophoresis of whole Hbs and of globin subunits was performed to determine whether or not adult Hb synthesis had been induced. During the first week after hormone treatment, the ability of the circulating RBCs to synthesize tadpole Hb *in vitro* declined to almost undetectable levels. Adult Hb synthesis became detectable after 11–15 days, but it was necessary to add fresh, unlabeled adult Hb as a carrier for the radioactivity to correspond to the area of the gels where adult Hb aggregates normally migrate. Although the incorporation of radioactivity into adult Hb detected in this way was high compared to the amount of radioactivity in tadpole Hb, the absolute amounts of adult Hb present were apparently very small since it failed to dimerize in the absence of carrier.

Moss and Ingram (1968b) proposed that the response to *in vivo* administration of T_4 results from repression of the cell line that synthesizes tadpole Hbs and induction of the differentiation of a new cell line that synthesizes adult Hb. The lag time for detection of adult Hb synthesis would represent the time it takes the new clone of cells to undergo initial differentiation to the point of synthesizing adult Hb in detectable amounts.

Just and Atkinson (1972) injected TK stage X–XII *R. catesbeiana* tadpoles with T_3 (3×10^{-10} mole/g body wt) and studied *in vivo* Hb synthesis at various times after treatment. The main response to the hormone was an increase in tadpole Hb synthesis, which became significant by day 12. At 16 and 32 days after treatment, incorporation into adult Hb could be detected on polyacrylamide gels, but only in the presence of unlabeled, adult carrier Hb.

The results of these attempts to induce premature adult Hb synthesis with exogenous thyroid hormones have not been conclusive since the adult Hb could not be detected on polyacrylamide gels without the addition of carrier, and since the lag times for adult Hb synthesis were long (about 2 weeks). The induction of adult Hb synthesis in these experiments may result from an indirect rather than a direct effect of the hormone. The results of Benbassat (1970) and others indicating that the Hb transition does not always coincide with metamorphosis, and the results showing that the Hb transition in *A. mexicanum* is independent of metamorphosis (Ducibella, 1974a), argue against a direct role for thyroid hormones. Benbassat (1970) has stated that no effect of T_4 or T_3 was found on protein or nucleic acid syntheses by RBCs from frogs or tadpoles at any stage of natural metamorphosis when the hormone was added to the cells *in vitro*.

It can be argued that young tadpoles (TK stages X–XII) are not yet competent to respond in this way to a surge in thyroid hormones. Perhaps some other developmental events must occur (events that normally occur between TK stages XII and XIX) before adult Hb synthesis can occur in response to an increase in thyroid hormones.

The absence of data on the circulating thyroid hormone concentrations in animals in which the Hb transition is a late-metamorphic or postmetamorphic change and the absence of data that show the threshold concentrations for larval and adult erythropoietic responses, make it difficult to analyze the situation.

4.5.1. Other Humoral Factors

Experiments with anemic animals show that a humoral factor (or factors) is involved in mediating erythropoietic responses in amphibians.

When anemia is induced by phenylhydrazine treatment or by bleeding in adult *R. pipiens* (Meints and Forehand, 1977), *R. catesbeiana* (Meints and Forehand, 1977), and *X. laevis* (Maclean and Jurd, 1971), there is a partial reversion to tadpole Hb synthesis. In all three species, there is also a marked increase in the synthesis of one of the adult components (Maclean and Jurd, 1971; Meints and Forehand, 1977). It is the minor adult Hb A_2 in *Xenopus* that is induced, and it seems probable that adult component C is the one exhibiting the greatest induction in *R. catesbeiana* (Fig. 3 of Meints and Forehand, 1977), although that judgement is a tentative one. The tadpole Hb that is induced by anemia in *Xenopus* is the major larval Hb F_1 (Maclean and Jurd, 1971) and that induced in *R. catesbeiana* appears to be larval component IV (Td-1), which has been labeled "T_3" by Meints and Forehand (1977). This latter result raises the possibility that anemia rekindles liver erythropoiesis in the adults—a reversion to larval Hb synthesis mediated by reactivation of a larval erythropoietic site. Chegini *et al.* (1979) found that erythropoiesis occurs in both spleen and liver following phenylhydrazine-induced anemia in adult *X. laevis*, but the report did not indicate any attempts to correlate these sites with synthesis of particular Hb types.

Carver and Meints (1977), using an *in vitro* spleen culture system from adult frogs, obtained evidence for a serum erythropoietic factor that is enhanced by induced anemia. The factor(s) is nondialyzable, and may represent the amphibian equivalent of erythropoietin. The factor(s) needs to be purified and characterized, and studies on the role of such a factor in the metamorphic Hb transition need to be done. Using the same spleen culture system, Carver and Meints (1977) found no effect of mammalian erythropoietin, prolactin, or T_3 on iron uptake by the adult spleen cells.

Hormones such as pituitary polypeptides, steroids, prostaglandins, and others may be involved in regulating the metamorphic Hb transition. However, no systematic study of the effects of humoral agents on the metamorphic Hb switch, either *in vivo* or *in vitro*, has been reported.

Other factors could be involved in mediating the metamorphic Hb transition, such as changes in environmental temperature, local changes within the erythropoietic organs in O_2 and CO_2 tension and pH, and systemic changes that may trigger changes in hormonal secretions in the animals. These possibilities will not be discussed further since the Hb transition usually has not been examined in experiments in which these variables have been manipulated.

4.6. What Is the Origin of the Stem Cells That Give Rise to Larval and Adult RBCs?

Broyles and Frieden (1973) raised the question of whether the RBCs arising in *R. catesbeiana* larval kidneys differentiate from the same stem cell as RBCs arising from the larval liver. To date, no experiments have been done that directly answer the question. However, experiments with *R. pipiens* performed by Turpen and co-workers give strong support to the concept that all hemopoietic cells in larval and adult amphibians, including erythropoietic cells in larval liver and kidneys and in adult spleen and bone marrow, arise from a common pluripotent stem cell line that initially becomes determined in the embryo—in the anterior, dorsal portion of the area of the embryo that subsequently gives rise to the anlagen of the mesonephric kidneys (Turpen *et al.* 1979; Carpenter and Turpen, 1979; Turpen, personal communication).

Turpen and co-workers made intraspecific chimeras of diploid and triploid *R. pipiens* embryos and used the ploidy label (actually, the amount of Feulgen-stained DNA and the area of the nucleus) to trace the origin and migration of stem cells giving rise to the major blood cell classes shown in Fig. 4. The most recent work has shown the following: (1) The major erythropoietic organ in larval *R. pipiens* is the liver, the hemopoietic cells of the liver arising not from the ventral blood islands (where the first, embryonic RBCs form) but from an anterior, dorsal intraembryonic source (Turpen *et al.*, 1979). (2) The hemopoiesis that occurs in the pronephros of the larva (mainly granulopoiesis) is dependent on colonization of the organ by stem cells from the area of the embryo noted above (Carpenter and Turpen, 1979). (3) Reciprocal transplantation of ploidy-labeled mesonephric anlagen has shown that hemopoietic cells found in all of the blood-forming organs of the larva are derived from cells that initially differentiate in the same area as the anterior mesonephros (Turpen, personal communication).

The local source of pluripotent stem cells in later developmental periods, e.g., metamorphosis, has not been determined. What is the pluripotent stem cell compartment in larvae and in adults?

5. SUMMARY

Figure 5 summarizes the changes that occur in RBC and Hb populations during development of amphibians, with particular attention to *R. catesbeiana*. As emphasized in the preceding discussion, parts of this scheme need further investigation.

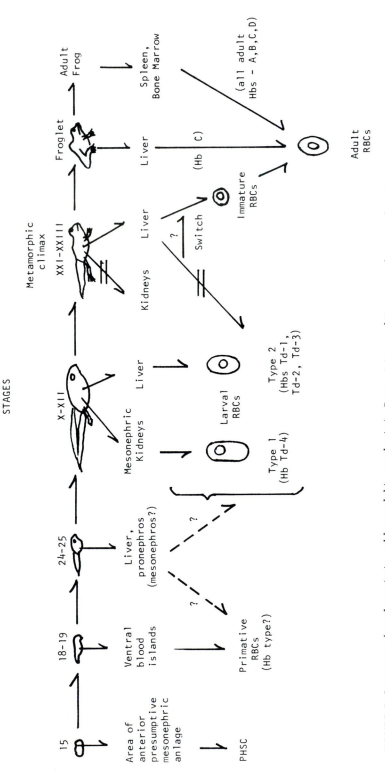

FIGURE 5. Ontogeny of erythropoiesis and hemoglobin synthesis in *Rana pipiens* and *Rana catesbeiana*. The information on embryonic events is taken primarily from results obtained with *R. pipiens* by Turpen and co-workers. See text for discussion, Fig. 1 for developmental stages, and Table 1 for nomenclature of larval Hb types. PHSC, pluripotent hemopoietic stem cells.

From a number of studies a general scheme of changes in erythro-poiesis and Hb synthesis during anuran development emerges, as follows.

1. The first RBCs arise in the ventral blood islands of the embryo (see Turpen et al., 1979). The type of Hb synthesized in these embryonic RBCs is not known.

2. As reviewed by Hollyfield (1966b), there is a considerable body of evidence that the kidneys (pronephric as well as mesonephric) are ery-thropoietically active in larvae and represent the second ontogenic site(s) of RBC formation. In R. pipiens, the pronephros becomes hemopoietically active at about Shumway stages 24–25, and is mainly a granulocyte pro-ducer during the larval period (Carpenter and Turpen, 1979). Erythro-poiesis constitutes less than 10% of the hemopoietic activity of the pro-nephros between TK stages I and VI. The relative number of different types of hemopoietic cells in the mesonephros at different stages of early development has not been published. However, at TK stages X–XII, the intertubular regions of the mesonephric kidneys are producing RBCs of defined morphology (type 1) containing larval Hb Td-4 (Broyles et al., 1981).

3. Liver erythropoiesis begins in R. pipiens at Shumway stages 24–25 (Turpen et al., 1979) and (in R. catesbeiana) continues through meta-morphosis (Maniatis and Ingram, 1971a). In R. pipiens, macrophages are the predominant hemopoietic cell type in liver from Shumway stage 22 through TK stage III; erythroblasts dominate from TK stages IV through X (Turpen et al., 1979). It is not known whether mesonephric kidney erythropoiesis and liver erythropoiesis begin simultaneously or at differ-ent developmental times. In R. catesbeiana tadpoles, liver erythropoiesis produces RBCs that differ in cell morphology (type 2) and Hb type from RBCs emanating from the kidneys (Broyles et al., 1981).

4. In R. catesbeiana, as development progresses toward metamor-phosis, the relative contribution of kidney and liver to the total eryth-ropoiesis changes—kidney erythropoiesis predominates at earlier stages, while liver predominates at later stages (Watt and Riggs, 1975).

5. During metamorphic climax (in R. catesbeiana, liver erythropoi-esis switches from larval to adult Hb production (Maniatis and Ingram, 1971c). Presumably, kidney erythropoiesis declines, but no data have been published that bear directly on this question. The molecular and cellular events that occur in the liver to mediate the metamorphic Hb switch are largely unknown. There is disagreement on whether both larval and adult Hbs are found in the same RBCs during the transition.

6. In froglets (postmetamorphic, juvenile frogs), the liver is the pri-mary erythropoietic site (Maniatis and Ingram, 1971c). The RBCs pro-duced are of uniform appearance (Benbassat, 1970) and appear to contain exclusively adult Hb (Maniatis and Ingram, 1971c). However, the results of Just and Atkinson (1972) indicate that there may be a low level of larval Hb in the circulating blood of froglets and adult frogs.

7. During maturation of the froglet into an adult frog, erythropoiesis shifts from the liver to another site(s). Jordan and Speidel (1923a) claimed

that the spleen is the main erythropoietic organ in adult frogs, with the bone marrow being active only in the spring of the year. Maniatis and Ingram (1971a) found that the bone marrow is the predominant site of erythropoiesis in adult *R. catesbeiana*, but they did not state at what time of year the adults were taken for study.

8. Induction of anemia in adult *R. catesbeiana*, *R. pipiens*, and *X. laevis* results in a partial reversion to larval Hb synthesis (Maclean and Jurd, 1971; Meints and Forehand, 1977). It is not known whether the larval and adult Hbs are in the same or separate cells in this situation. It is also not known if a larval erythropoietic site is reactivated during anemia.

The following aspects of erythropoiesis in bullfrog tadpoles are especially provocative: (1) Since it is known that erythropoietin is produced in mammalian fetal liver and that erythropoietin production in adult mammals involves the liver as well as the kidneys, it is interesting that the two primary sites involved in erythropoiesis in tadpoles are the liver and the kidneys. (2) Since Turpen and co-workers have found that the site of differentiation of pluripotent stem cells in *R. pipiens* embryos is in the area of the anterior mesonephric anlagen, the role of the mesonephric kidneys in erythropoiesis in *R. catesbeiana* tadpoles is all the more intriguing. The concentration of erythropoietic foci is greater in the anterior than in the posterior portion of the mesonephric kidneys of these animals (Broyles and Johnson, unpublished observations). Broyles and Frieden (1973) suggested that erythroid cells arising in the kidneys of these animals might migrate to the liver, where they are induced to synthesize larval Hbs different from the Hb arising in the kidney per se.

Each of the many changes occurring during amphibian metamorphosis is a complex series of events. This is especially true for the red blood cell–hemoglobin transition, as the foregoing discussion shows. A fundamental change occurs in the differentiation program of the erythroid cells that likely depends on cell-to-cell interactions within the erythropoietic organs, on changes in the milieu of those organs, on changes in the concentrations of circulating hormones, and is coordinated with other changes, such as the increase in serum ceruloplasmin, a change in transferrin, and the change in the location and structure of ferritin. Which of these changes are necessary for the change from larval to adult Hb synthesis and which directly influence the change in expression of globin genes are not clear, despite significant progress since previous reviews of this subject (Frieden, 1968; Frieden and Just, 1970).

ACKNOWLEDGMENTS.   I thank Dr. James B. Turpen and Dr. Austen Riggs for providing information about their current research and results, which in some cases is not yet in print. I also thank Dr. Russell H. Meints and Dr. Daniel M. Byrd for helpful suggestions. The skillful typing of Ms.

Cheryl Bontempi and Ms. Irene McMichael and the proofreading of Ms. Anna M. Parkinson are gratefully acknowledged. Some of the work discussed was supported in part by Grants AM 21386 and AM 21764 from the National Institutes of Health awarded to the author.

REFERENCES

Aggarwal, S. J., and Riggs, A., 1969, *J. Biol. Chem.* **244**:2372.
Araki, T., Kajita, A., and Shukuya, R., 1971, *Biochem. Biophys. Res. Commun.* **43**:1179.
Araki, T., Kajita, A., and Shukuya, R., 1973, *Nature New Biol.* **242**:254.
Baldwin, T. O., and Riggs, A., 1974, *J. Biol. Chem.* **249**:6110.
Benbassat, J., 1970, *Dev. Biol.* **21**:557.
Benbassat, J., 1974, *J. Cell Sci.* **16**:143.
Brown, J. E., and Theil, E. C., 1978, *J. Biol. Chem.* **253**:2673.
Broyles, R. H., and Deutsch, M. J., 1975, *Science* **190**:471.
Broyles, R. H., and Frieden, E., 1973, *Nature New Biol.* **241**:207.
Broyles, R. H., Johnson, G. M., Maples, P. B., and Kindell, G. R., 1981, *Dev. Biol.* **81**:299.
Cardellini, P., and Sala, M., 1979, *Comp. Biochem. Physiol.* **64B**:113.
Carpenter, K. L., and Turpen, J. B., 1979, *Differentiation* **14**:167.
Carver, F. J., and Meints, R. H., 1977, *J. Exp. Zool.* **201**:37.
Chegini, N., Aleporou, V., Bell, G., Hilder, V. A., and Maclean, N., 1979, *J. Cell Sci.* **35**:403.
Cline, M. J., and Golde, D. W., 1979, *Nature (London)* **277**:177.
DeWitt, W., 1968, *J. Mol. Biol.* **32**:502.
Dorn, A. R., and Broyles, R. H., 1978, *J. Supramol. Struct.* **1978**(Suppl. 2):164.
Ducibella, T., 1974a, *Dev. Biol.* **38**:175.
Ducibella, T., 1974b, *Dev. Biol.* **38**:187.
Feldhoff, R. C., 1971, *Comp. Biochem. Physiol.* **40B**:733.
Forman, L. J., and Just, J. J., 1976, *Dev. Biol.* **50**:537.
Foxon, G. E. H., 1964, in: *Physiology of the Amphibia* (J. A. Moore, ed.), pp. 151–209, Academic Press, New York.
Frieden, E., 1968, in: *Metamorphosis: A Problem in Developmental Biology* (W. Etkin and L. I. Gilbert, eds.), pp. 349–398, Appleton–Century–Crofts, New York.
Frieden, E., 1979, in: *Copper in the Environment* (J. O. Nriagu, ed.), pp. 241–283, Wiley, New York.
Frieden, E., and Just, J. J., 1970, in: *Mechanisms of Hormone Action*, Volume 1 (G. Litwack, ed.), pp. 1–52, Academic Press, New York.
Hamada, K., and Shukuya, R., 1966, *J. Biochem.* **59**:397.
Hamada, K., Sakai, Y., Tsushima, K., and Shukuya, R., 1966, *J. Biochem.* **60**:37.
Herner, A. E., and Frieden, E., 1961, *Arch. Biochem. Biophys.* **95**:25.
Hollyfield, J. G., 1966a, *J. Morphol.* **119**:1.
Hollyfield, J. G., 1966b, *Dev. Biol.* **14**:461.
Jordan, H. E., and Speidel, C. C., 1923a, *J. Exp. Med.* **38**:529.
Jordan, H. E., and Speidel, C. C., 1923b, *Am. J. Anat.* **32**:155.
Jurd, R. D., and Maclean, N., 1970, *J. Embryol. Exp. Morphol.* **23**:299.
Just, J. J., and Atkinson, B. G., 1972, *J. Exp. Zool.* **182**:271.
Just, J. J., Sperka, R., and Strange, S., 1977, *Experientia* **33**:1503.
Ledford, B. E., and Frieden, E., 1973, *Dev. Biol.* **30**:187.
McCutcheon, F. H., 1936, *J. Cell. Comp. Physiol.* **8**:63.
Maclean, N., and Jurd, R. D., 1971, *J. Cell Sci.* **9**:509.
Maclean, N., and Jurd, R. D., 1972, *Biol. Rev.* **47**:393.
Maniatis, G. M., and Ingram, V. M., 1971a, *J. Cell Biol.* **49**:372.

Maniatis, G. M., and Ingram, V. M., 1971b, *J. Cell Biol.* **49**:380.

Maniatis, G. M., and Ingram, V. M., 1971c, *J. Cell Biol.* **49**:390.

Meints, R. H., and Forehand, C., 1977, *Comp. Biochem. Physiol.* **58A**:265.

Miyauchi, H., LaRochelle, F. T., Jr., Suzuki, M., Freeman, M., and Frieden, E., 1977, *Gen. Comp. Endocrinol.* **33**:254.

Moss, B., and Ingram, V. M., 1968a, *J. Mol. Biol.* **32**:481.

Moss, B., and Ingram, V. M., 1968b, *J. Mol. Biol.* **32**:493.

Nieuwkoop, P. D., and Faber, J., 1967, *Normal Table of Xenopus laevis (Daudin)*, 2nd ed., North-Holland, Amsterdam.

Osaki, S., James, G. T., and Frieden, E., 1974, *Dev. Biol.* **39**:158.

Rosen, O. M., and Rosen, S. M., 1968, *Biochem. Biophys. Res. Commun.* **31**:82.

Rosenberg, M., 1970, *Proc. Natl. Acad. Sci. USA* **67**:32.

Saucier, W. J., 1976, Red blood cell differentiation in the bullfrog, *Rana catesbeiana*: The hemoglobin transition and erythropoietic sites during and following metamorphosis, M.S. thesis, University of Wisconsin, Milwaukee.

Shumway, W., 1940, *Anat. Rec.* **78**:139.

Sullivan, B., 1974, in: *Chemical Zoology* (M. Florkin and B. T. Scheer, eds.), pp. 77–122, Academic Press, New York.

Taylor, A. C., and Kollros, J. J., 1946, *Anat. Rec.* **94**:7.

Theil, E. C., 1973, *Dev. Biol.* **34**:282.

Turpen, J. B., Turpen, C. J., and Flajnik, M., 1979, *Dev. Biol.* **69**:466.

Watt, K. W. K., and Riggs, A., 1975, *J. Biol. Chem.* **250**:5934.

CHAPTER 15

Biochemical Characterization of Organ Differentiation and Maturation

SANDRA J. SMITH-GILL AND VIRGINIA CARVER

1. INTRODUCTION

During amphibian metamorphosis, many tissues and organs undergo dramatic biochemical changes associated with the physiological transition from an aquatic to a terrestrial environment. Because of the dramatic nature of many of these changes, and the implication of underlying hormonal control, they have been utilized as model systems in numerous investigations of the biochemical mechanism of action of thyroid hormone (TH) (reviewed by Cohen *et al.*, 1978, and Frieden, this volume). It is the purpose of this chapter to survey several organ systems and tissues whose changes during metamorphosis have been well characterized biochemically and to consider the experimental evidence for possible hormonal control of the changes. We will include the endodermally derived liver and intestine, the mesodermally derived kidney, and the ectodermally derived integument and eye. Biochemical changes in the blood are reviewed by Broyles (this volume). For all of these tissues, TH control has been implicated, but in many cases multiple actions of several other hormones are strongly suggested (see reviews by Dodd and Dodd, 1976,

SANDRA J. SMITH-GILL · Department of Zoology, University of Maryland, College Park, Maryland 20742. VIRGINIA CARVER · Department of Chemistry, The Florida State University, Tallahassee, Florida 32306.

and White and Nicoll, this volume). Morphological and cytological changes accompanying the metamorphosis of these tissues have been reviewed elsewhere in this volume (see especially Fox).

This discussion centers on changes occurring during anuran metamorphosis, with the principal organisms being ranid tadpoles and *Xenopus laevis* larvae. Unless otherwise indicated, the staging series of Taylor and Kollros and Nieuwkoop and Faber will be employed for the ranids and *Xenopus*, respectively; these series are reviewed and compared by Dodd and Dodd (1976). We will use the following terminology with respect to the major subdivisions of ranid metamorphosis: Premetamorphosis, from Taylor and Kollros stages I to X; prometamorphosis, from stages XI to XIX; and climax, beginning at stage XX.

2. THE LIVER

Thyroxin-induced metamorphosis of the liver is one of the more dramatic organ systems in which thyroxin-induced biochemical changes have been studied extensively by a number of laboratories. The changes accompanying metamorphosis encompass a panorama of biochemical differentiation events. The larval liver is a major site of proliferation of hematopoietic cells (reviewed by Broyles, this volume). Likewise, several groups of specialized proteins increase dramatically, including manyfold increases of the ornithine–urea-cycle enzymes associated with the animal's transition from ammonotelism to ureotelism (Fig. 1). Studies on the mode of action of TH on tadpole liver have concentrated on transcriptional effects, translational effects, and, to a lesser degree, changes in mitochondrial metabolism. Proliferative effects of thyroxin on tadpole hepatocytes have been described, and recently DNA metabolism has been investigated. However, the relationship of such proliferative effects to other metabolic changes, and to organelle biogenesis, remains to be elucidated.

This discussion attempts to synthesize studies made in many different laboratories on several species into an overview of the biochemical patterns emerging in the metamorphosis of the anuran liver. A detailed review of the molecular biology of TH action on the anuran liver may be found in Cohen *et al.* (1978) (see also Frieden, this volume).

2.1. DNA Synthesis, Cell Division, and Nuclear Morphology

An important question from the point of view of developmental and genetic control mechanisms is whether molecular changes accompanying liver metamorphosis occur in a fixed population of hepatocytes, or

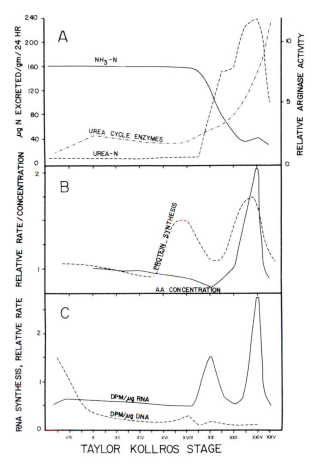

FIGURE 1. Theoretical composite of biochemical changes in ranid liver during spontaneous metamorphosis. (A) Nitrogen excretion, after Ashley *et al.* (1968); urea-cycle enzymes based on data from laboratories of Frieden, Cohen, and Smith-Gill. (B) Protein synthesis, based on unpublished data of A. Kistler, A. Miyauchi, and E. Frieden, and of Smith-Gill; amino acid concentration after unpublished data of Kistler *et al.* (C) RNA synthesis, based on unpublished data of Smith-Gill.

whether cell division or other population dynamic changes accompany or precede the developmental changes. Although metamorphosis of the anuran liver is commonly assumed to occur in a "fixed population of cells" (Cohen, 1970; Kistler *et al.*, 1975; VanDenbos and Frieden, 1976; Morris and Cole, 1978; Cohen *et al.*, 1978), several studies suggest considerable DNA synthesis and even cellular turnover accompany metamorphosis of the anuran liver (Atkinson *et al.*, 1972; Kistler and Weber, 1975; Smith-Gill, 1979; Smith-Gill *et al.*, 1979) (Fig. 2).

Early morphological studies on amphibian liver conflict with respect to the question of either mitosis or cellular turnover during spontaneous

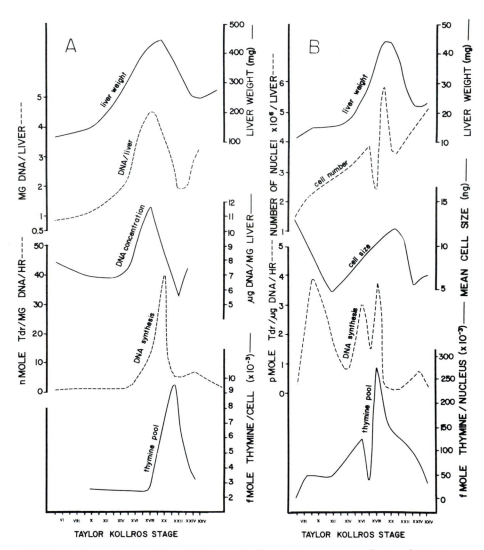

FIGURE 2. Comparison of liver DNA metabolism during metamorphosis of (A) *R. catesbeiana* (based on data of Atkinson *et al.*, 1972) and (B) *R. pipiens* (based on data of Smith-Gill *et al.*, 1979, and Smith-Gill, 1979).

metamorphosis. The number of liver cell plates doubles during spontaneous metamorphosis, transforming the liver from a muralium simplex to a muralium duplex. On the other hand, morphological studies reveal some liver degeneration but very little mitotic activity during spontaneous metamorphosis, and Bennett and Glenn (1970) concluded that very little mitosis accompanies liver differentiation.

Published morphological studies on induced metamorphosis of the anuran liver are inconclusive with respect to the question of either increased cellular mitosis or cellular turnover. Kistler and Weber (1975) concluded from a biochemical and morphometric analysis that a transient growth response, involving both hypertrophy and hyperplasia, accompanies thyroxin-induced metamorphosis of *X. laevis* liver. Thyroxin-treatment increases the percentage of liver nuclei labeled with [³H]-thymidine ([³H]-Tdr) to about 150% of controls, while incorporation of [³H]-Tdr into the cytoplasm increases about three-fold over controls during the 16 days of thyroxin exposure (Fig. 3D). Paik and Baserga did not measure mitotic indices, but since Champy (cited in Paik and Baserga) reported no change in mitotic index during metamorphosis, they concluded that thyroxin stimulates the synthesis of liver DNA without much change in the rate of cell division. However, since the amount of DNA per nucleus remains constant during metamorphosis (Atkinson *et al.*, 1972), some mitosis must accompany the increased DNA synthesis.

Two laboratories have examined in detail DNA metabolism during spontaneous and T_3-induced metamorphosis in the livers of *Rana catesbeiana* and *R. pipiens* (Atkinson *et al.*, 1972; Smith-Gill, 1979; Smith-Gill *et al.*, 1979). Both studies used [³H]-Tdr incorporation into the acid-insoluble fraction to measure rates of DNA synthesis. In order to estimate synthetic rates using radioactively labeled precursors as tracers, it is necessary to know both the rate at which the labeled precursor is incorporated into the synthetic product, and the specific activity of the actual pool of precursors in the animal, which will be diluted by the endogenous precursor when a labeled precursor is added. Both studies measured the amount of radioactivity in the acid-soluble fraction at the end of the labeling period as an estimate of the amount of label in the precursor pool. They then hydrolyzed all the acid-soluble thymine components of liver homogenates to measure the size of the endogenous pool and calculate the specific activity.

Although the details and timing of specific biochemical events differ in *R. catesbeiana* and *R. pipiens*, the overall pattern in both species is similar (Figs. 2–4). Both show maximal cell numbers, DNA synthesis, and thymine pools (on a per cell basis) at preclimax or early climax stages during spontaneous metamorphosis, with all three components falling significantly at or following climax (Fig. 2). Both species also show early (within hours) and later (several days) periods of enhanced nuclear DNA synthesis following T_3 treatment and preceding expression of biochemical differentiation (Figs. 3 and 4). Notably, all increases in DNA synthesis observed in *R. pipiens* are transient, while T_3-induced DNA synthesis in *R. catesbeiana* is greater than controls from Day 4 after T_3 treatment on. Whether these differences between *R. pipiens* and *R. catesbeiana* represent true species differences or are due to experimental conditions cannot be ascertained at this time.

On a per cell basis, the increase in acid-soluble thymine pool size during spontaneous metamorphosis of both species is transient and parallels increases in DNA synthesis (Fig. 2). Similarly, cellular thymine

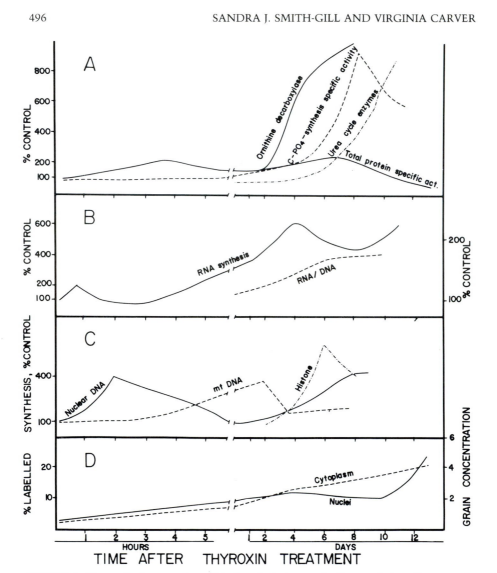

FIGURE 3. Theoretical composite of biochemical changes in *R. catesbeiana* liver accompanying Ty-induced metamorphosis. (A) Protein metabolism. General protein and ornithine decarboxylase synthesis are composite of data of Atkinson (1971), Fischer and Cohen (1974), Morris and Cole (1978). Urea-cycle enzymes after Wixom *et al.* (1972). (B) RNA synthesis and content, a composite of data of Eaton and Frieden (1969), Nakagawa *et al.* (1967), Kistler *et al.* (1975). (C) Rates of DNA synthesis after Atkinson *et al.* (1972); histone synthesis after Morris and Cole (1978). (D) Autoradiographic data after Paik and Baserga (1971).

pools also do not increase significantly during T_3-induced metamorphosis in either study. It may be concluded, therefore, that T_3 does not directly cause an increase in the cellular thymine component pool, and that the transient increase in the thymine pool observed during spontaneous metamorphosis is probably secondary to changes in nucleic acid turnover.

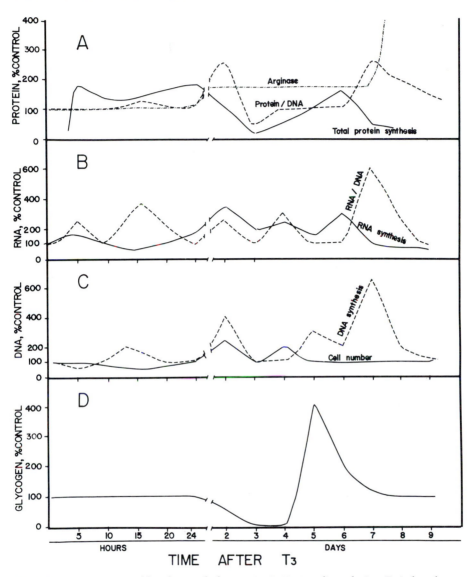

FIGURE 4. Composite of biochemical changes in *R. Pipiens* liver during T_3-induced met-amorphosis. (A) Arginase activity from Smith-Gill (1979); protein content and total protein synthesis based on unpublished data of Smith-Gill. (B) RNA synthesis and content, based on unpublished data of Smith-Gill. (C) DNA synthesis and cell number, from Smith-Gill (1979). (D) Total liver glycogen, based on unpublished data of Smith-Gill.

Cohen *et al.* (1978) have criticized the methodology used in the above studies because treating all the thymine components of the acid-soluble fraction as a single precursor pool ignores possible complications result-ing from Tdr metabolism, and they have pointed out that the kinetics in the study of Atkinson *et al.* (1972) do not conform to expected first-order kinetics. Under the conditions of the latter experiment, the counts in the acid-soluble fraction decrease significantly (severalfold) during the 90-min labeling period. The estimates of pool activity using soluble counts at the end of the labeling period may therefore be subject to considerable error, and it is not clear when the actual peak of DNA synthesis occurs in *R. catesbeiana* spontaneous metamorphosis. However, in the *R. pipiens* study the acid-soluble counts at the end of the labeling period do provide an accurate measure of the activity of the precursor pool, and the actual rates of change of DNA content fall within the 95% confidence limits of the synthetic rates calculated from [³H]-Tdr incorporation. Since the amount of DNA per nucleus remains constant during metamorphosis (Atkinson *et al.*, 1972; Smith-Gill *et al.*, 1979), and the total DNA per liver increases over fourfold between premetamorphosis and climax, there can be little doubt that both DNA synthesis and mitosis are occurring. However, precise estimates of rates are limited by the current state of technology of Tdr incorporation.

Conclusions based on the biochemical observations are corroborated by autoradiography (Smith-Gill, unpublished), which shows that [³H]-Tdr is in fact being incorporated into liver parenchymal cells, and labeling indices (percent [³H]-Tdr-labeled nuclei which are presumed to be synthesizing DNA) during spontaneous and induced metamorphosis parallel almost exactly biochemical incorporation data. Labeling indices in larval livers 6 days following T_3 injection are significantly increased over controls, in agreement with biochemical data. Collectively, these studies suggest that changes in DNA synthetic rates during both spontaneous and induced metamorphosis reflect changes in the number of cells engaged in DNA synthesis.

Morris and Cole (1978) interpreted their results to indicate no significant increases in the rate of histone synthesis during induced metamorphosis of *R. catesbeiana*, a result that would be inconsistent with elevated rates of DNA synthesis. In fact, their results show an increase of T_3-induced larvae over controls on Days 6 (190% of controls) and 8 (133% of controls) (Fig. 3C). Furthermore, Morris and Cole (1980) have recently reported enhanced phosphorylation of both H1 and H2a liver histones 2–8 days following T_4 treatment, with the greatest enhancement two- to fivefold) between 4 and 6 days. These two classes of histones are usually phosphorylated coincident with or preceding DNA synthesis. The results were not analyzed statistically in either study, but as presented they are consistent with the timing of DNA synthesis reported by Atkinson *et al.* (1972).

Van Denbos and Frieden (1976) estimated DNA turnover rates by a single injection of ¹²⁵I-labeled deoxyuridine to stage XVI *R. catesbeiana* larvae and a subsequent monitoring of changes in DNA specific activity. They assumed that any expansion of the cellular population would result in a decrease in DNA specific activity with time, but their predictions did not allow for more complicated population dynamics involving possible cell death. The abrupt 20% decrease in label recorded at metamorphic climax suggests some stage-specific loss of DNA, consistent with cellular turnover at climax (discussed below). Interpretation of biochemical long-term label turnover requires prior knowledge of the relationship of the labeled cell population to future cell death. When both turnover and synthesis are occurring simultaneously, a study such as that of VanDenbos and Frieden requires autoradiographic interpretation.

During spontaneous metamorphosis of both *R. pipiens* and *R. catesbeiana*, there are significant declines in liver DNA content at preclimax or climax stages while DNA synthetic rates are high. Similarly, in T_3-induced metamorphosis of *R. pipiens*, at no time after the first 24 hr following treatment do synthesis rates in the experimental group fall below control levels, yet over the same time intervals there are net decreases in total DNA content. Therefore, degradation rates in T_3 treated animals must be greater than in control levels. Notably, a very early event following T_3 treatment *in vivo* is a decrease in total liver DNA, and T_3 treatment of cultured liver cubes increases rates of central cell necrosis. These results suggest that rates of DNA turnover, including both synthesis and degradation, must play a significant role in determining net total DNA synthetic rates and fluctuation of DNA levels. The possibility that cell death, as well as DNA synthesis, may play a significant role in metamorphosis of the anuran liver has been suggested previously by Kistler and Weber (1975), whose combined biochemical and morphometric analysis of induced metamorphosis of *X. laevis* liver suggests that atrophy, including cell death, and macrophage activity are characteristic features of the late metamorphic response.

The significant decline in rates of DNA synthesis at climax correlates with morphological studies that show that liver nuclei, euchromatic at early metamorphic stages, become progressively heterochromatic during spontaneous and induced metamorphosis (Bennett and Glenn, 1970; Bennett *et al.*, 1970). The development of heterochromatin during spontaneous and induced metamorphosis could correlate either with withdrawal of cells from cycle or with an increased specialization of the liver cells. Heterochromatin is generally taken as evidence of genetic inactivity, and in this case its prime function may be to turn off genes for larval functions and for mitosis, genes that are no longer needed in the adult stage. On the other hand, heterochromatin is known to correlate with, and affect, patterns of cell division and cell growth, particularly cell and nuclear volume. In both spontaneous and induced metamorphosis, rates of DNA synthesis drop significantly following climax (Figs. 2–4). In the anuran liver, therefore, the development of heterochromatin may provide a useful cytological marker for withdrawal from the cell cycle and/or differentiation.

In conclusion, the general pattern suggested by available data is consistent with the interpretation presented by Smith-Gill *et al.* (1979) of the population dynamics of the anuran liver during metamorphosis. During pre- and prometamorphosis, the liver is basically a simple expanding population, with changes in DNA content corresponding to rates of DNA synthesis. DNA content reaches a maximum at metamorphic climax, when synthetic rates drop. At the same time, cell death and turnover rates increase, producing a net decrease in DNA content and cell number. The proliferation and turnover of liver cells at climax may involve a replacement of "larval" by "adult" cells. Following climax, cell turnover rates decrease, and the new cell population expresses adult differentiated characters, while becoming once more an expanding population.

The occurrence of turnover at metamorphic climax suggests a functional relationship between the population dynamics of the liver and differentiation events accompanying metamorphosis. However, critical experiments to test for such a relationship have not been performed, and the degree of DNA synthesis and possible cell turnover remains an area of controversy. Furthermore, TH-induced liver differentiation has been observed under experimental conditions where no DNA synthesis can be demonstrated (Cohen et al., 1978). Therefore, the relationship of DNA synthesis and possible cell turnover to other developmental changes during spontaneous metamorphosis of the liver still requires definition.

2.2. Protein Synthesis, RNA Metabolism, and Cytoplasmic Morphology during Metamorphosis

2.2.1. Protein Synthesis

During spontaneous and induced metamorphosis of ranid tadpoles, concentration and total amounts of liver protein increase (Fig. 5). Cellular protein increases in *Xenopus* also, but the response during induced metamorphosis appears to be only transient (Kistler and Weber, 1975).

Following TH administration, a specific group of proteins, including albumin, ceruloplasmin, and enzymes of the ornithine–urea cycle, increases dramatically (reviewed by Cohen et al., 1978; see also Broyles, this volume). The urea-cycle enzymes have been used by many workers as markers of biochemical differentiation in the anuran liver. The first systematic study of the enzymes of the urea cycle during amphibian metamorphosis was reported by Cohen's laboratory. They reported that the activities (measured both as units per gram liver and as specific activity or units per milligram protein) of carbamyl phosphate synthetase I (CPS-I), ornithine transcarbamylase (OT), and the arginine synthetase system remain low during pre- and prometamorphosis of *R. catesbeiana*, but activities increase dramatically (2- to 100-fold) at metamorphic climax and reach maximum levels in adults. The increases do not begin until just prior to metamorphic climax, after Taylor and Kollros stage XVIII. The rises in enzyme levels correlate with increased levels of urea excretion (Fig. 1A).

Early experiments from the laboratories of Frieden and Cohen demonstrated that TH induces increased activity of tadpole liver arginase and

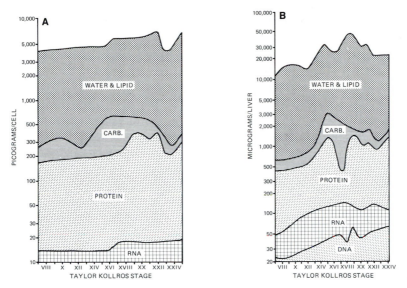

FIGURE 5. Cellular macromolecular content (A) and total liver macromolecular content (B) during spontaneous metamorphosis of *R. pipiens,* based on data from Smith-Gill *et al.* (1979) and Smith-Gill and Darling (unpublished manuscript).

of CPS-I. More recently it has been shown that all the enzymes of the urea cycle increase coordinately during TH-induced metamorphosis (Figs. 3 and 4).

The TH-induced increases in urea-cycle enzymes are generally agreed to involve *de novo* synthesis of the enzymes, although this remains to be conclusively demonstrated. In a detailed study of CPS-I induction, Cohen's laboratory concluded that the primary cause of increased activity is *de novo* synthesis. The recent demonstration (Mori *et al.,* 1979) that the level of *in vitro* translatable mRNA for immunoprecipitated CPS-I in bullfrog tadpole liver increases following T_4 treatment supports the hypothesis of *de novo* synthesis. *In vivo,* increases in enzyme activity are not detectable until several days after TH administration, but *de novo* synthesis can be detected immunologically within 24 hr, suggesting increased conversion of an inactive precursor to an active enzyme. Both CPS-I and OT are mitochondrial enzymes, located in the inner membrane matrix fraction of the mitochondria. Earlier studies (Cohen, 1970) suggested that CPS-I precursor subunits were synthesized extramitochondrially, transported into mitochondria, and assembled in the matrix space, but more recent evidence (Mori *et al.,* 1979) suggests that CPS-I is synthesized as a single large polypeptide. This polypeptide may be an inactive precursor that later is activated in the mitochondria. In addition, CPS-I has a longer half-life following TH treatment, and CPS-I turnover was found to be transcription dependent. Therefore, regulation of CPS-I activity by TH appears to involve coordinately increased synthesis, possible increased posttranslational activation, and increased enzyme stability.

Reports on total protein synthetic rates during spontaneous metamorphosis vary. Morris and Cole (1978) reported a doubling of lysine incorporation rates into both total liver and mitochondrial proteins, and a fivefold increase in total histone incorporation rates between "premetamorphic" and "late metamorphic" stages of *R. catesbeiana*. They measured precursor specific activities, but used a rather long labeling period (5 hr) for liver that may be undergoing substantial cellular turnover; since they did not show incorporation kinetics, their data are difficult to evaluate. Recent studies from the laboratories of Frieden and of Smith-Gill on the kinetics of several amino acids indicate that pool specific activity decreases within 30 min of incubation, while incorporation into the TCA-insoluble fraction is linear for only 90 min to 4 hr, depending upon the amino acid. The uptake of amino acid precursors into the acid-soluble pool varies over 10-fold during spontaneous metamorphosis. The concentration of leucine in the endogenous free amino acid pool of *R. catesbeiana* liver increases slightly during prometamorphosis, then increases transiently over fourfold at stage XXIV. If the endogenous leucine pool of *R. pipiens* liver varies similarly, then the overall pattern of liver protein synthesis in the two species during spontaneous metamorphosis is similar. Incorporation of [^{3}H]leucine into protein does not change substantially between pre- and mid-prometamorphosis. Transient but significant increases in protein synthesis are seen just prior to climax (stages XVII–XVIII), and again just following climax, between stages XXI and XXIV (summarized in Fig. 1B).

Rates of total protein synthesis *in vivo* increase up to 100-fold within 4 hr after TH injection (Figs. 3 and 4); Cohen's laboratory has reported that TH-induced increases in ornithine decarboxylase in *R. catesbeiana* parallel the increases in general protein synthesis and are similarly inhibited by cycloheximide. Later secondary increases in protein synthesis follow, but reports on the timing and duration vary. It should be noted that these variations may reflect differences in type of hormone treatment, as well as type of precursor used. In general, there appears to be some increased general protein synthesis, but it may be transient, while increases in specific activity of urea-cycle enzymes are much greater and are sustained (summarized in Figs. 3 and 4). The early increases in protein synthesis observed within hours after TH treatment may very well reflect a primary response to TH. However, later increases in protein synthesis that follow a lag of 4–6 days either may reflect a classica "lag" period characteristic of many TH responses (Tata, 1971; Frieden, this volume) or, alternatively, may be secondary. The relationship of changes in protein synthesis to TH action remains to be defined.

2.2.2. RNA Synthesis

Transcriptional events during metamorphosis have been investigated by several laboratories. The results are often difficult to compare, due to differences that include method and duration of TH administration, dosage and hormone used (T_3, T_4), temperature, species, developmental stages of animals used (often specific stages are not even reported), and possible seasonal differences. Furthermore, photoperiod is seldom reported, although diurnal rhythm can affect nucleic acid and protein metabolism, as well as cellular responsiveness to exogenous hormones. However, some generalizations are possible.

During spontaneous metamorphosis of R. pipiens at 22°C, total liver RNA content increases proportionally to liver weight (Fig. 5). Rates of [^{14}C]orotic acid incorporation, on a per cell basis, fall during spontaneous metamorphosis, although total RNA synthesis rises as the cell number increases (Fig. 1). The RNA/DNA ratio increases with metamorphic development. The nuclear to cytoplasmic distribution of total RNA also varies during metamorphic development. The increases in cellular RNA content and changes in the intracellular distribution may reflect increased accumulation as well as possible alterations in processing and/or transport of hnRNA associated with metamorphosis (discussed below).

Accompanying TH-induced metamorphosis, there have been reports of significant increases in total liver RNA (summarized in Figs. 3 and 4). In X. laevis, the increase may be only transitory (Kistler and Weber, 1975). Three different peaks of in vivo RNA synthesis have been reported for both R. catesbeiana (Fig. 3) and R. pipiens (Fig. 4): (1) the earliest increase is seen within 30 min in R. catesbeiana and after 4 hr in R. pipiens as an increased incorporation of precursors into RNA, which then falls to control levels within 1–2 hr after the increase; (2) a second increase is observed after 1–2 days in R. pipiens and 2–3 days in R. catesbeiana and also in X. laevis (Tata, 1971), when uridine and phosphate are each incorporated into both nuclear and mitochondrial RNA; (3) incorporation levels decrease to control levels, after which there is another sustained increase of precursor incorporation into RNA at 5–7 days in R. pipiens and 10–12 days in R. catesbeiana. Nuclear RNA polymerase activity increases following TH administration (Tata, 1971; Cohen et al., 1978).

The nature of TH-induced RNA synthesis has been studied in detail by the laboratories of Cohen (Cohen et al., 1978; Mori et al., 1979) and Tata (1971). Isolated chromatin from TH-treated larvae shows an increased template efficiency over controls. The increased precursor incorporation is reflected in both nuclear and cytoplasmic RNA. The GC content of the 30-min peak compares with values for tadpole DNA, while at later times the composition approaches that of tadpole RNA. Overall, the RNA base compositions differ from those of controls, but the rRNA is preferentially stimulated. The later increase is also accompanied by an accelerated turnover of ribosomes, heavier polysome aggregates, and an increased rate of microsomal phospholipid synthesis that coincides with both the appearance of microsomal RNA and an increased rate of protein formation in the microsomes. In addition, isolated liver mRNA fractions from T_4-treated larvae will translate immunoprecipitable CPS-I in *vitro* at a rate about twice that of untreated controls, suggesting a quantitative change in mRNA composition.

The early increases of total RNA synthesis correlate well with the early (4–5 hr) increases in protein synthesis, although a causal relationship has not been demonstrated. Notably, in *R. pipiens* the early increases in RNA and protein synthesis precede an increase in DNA synthesis, and the later two periods of RNA synthesis correlate well with periods of DNA synthesis. Patterns of RNA accumulation correlate with patterns of protein accumulation and increases in activity of the urea-cycle enzymes (Fig. 4). Furthermore, during spontaneous metamorphosis of *R. pipiens*, levels of RNA synthesis correlate highly with levels of DNA synthesis.

Actinomycin D inhibits RNA synthesis in both TH-treated and control *R. catesbeiana* tadpoles, but it does not inhibit the rise in CPS-I if given 5 or more days after TH treatment. If TH is administered for only 2 days, CPS-I levels continue to rise above controls, but enzyme levels remain below groups continuously treated with TH. It was concluded that intracellular mechanisms for the induction of CPS-I are activated during the first 2 days after exposure to TH and are probably associated with the 2- to 4-day increase in RNA synthesis (Cohen *et al.*, 1978).

The increase in rRNA synthesis correlates well with morphological studies showing proliferation of nucleoli accompanying spontaneous and thyroxin-induced metamorphosis. Whereas 80% of the nuclei observed prior to climax have at most one nucleolus, in late climax stages the predominant number is four (Bennett and Glenn, 1970; Bennett *et al.*, 1970). Thyroxin-induced supernumerary nucleoli have also been reported for cultured mammalian cells.

The stimulation of rRNA synthesis is accompanied by a turnover and redistribution of ribosomes; the new ribosomes are more tightly bound to cellular membranes and have a slow turnover (half-life 7–8 days). The breakdown of the old ribosomes is accelerated by T_3. The ribosomal turnover is accompanied by the formation of heavier polysome aggregates, and an increased rate of microsomal phospholipid synthesis that coincides with the appearance of microsomal RNA and an increased rate of protein formation in the ribosomes (Tata, 1971).

Cytologically, the biochemical changes of ribosomal turnover are reflected in a dramatic morphological reorganization of the cytoplasm. TH-treated livers are in general characterized by an increase in the number of ribosomes, most of which are attached to RER, and by a proliferation and increase in complexity of the RER (Bennett *et al.*, 1970; Tata, 1971). Prior to metamorphic climax, the RER characteristically is observed in stacks of 4–20 cisternae, which do not appear swollen and are located primarily in peripheral regions of the cell. At metamorphic climax there is a transition to a proliferation of RER, which is often in stacks but also frequently shows branching patterns throughout the cytoplasm. The cisternae appear swollen and the cisternal spaces stain more densely.

Furthermore, there is an increased occurrence of RER in a perinuclear location; both single and multiple cisternae are observed in close association with the nuclear envelope. Perinuclear RER is even more apparent in adult hepatocytes. An increased association of RER with mitochondria was also observed in postclimax livers (Bennett *et al.*, 1970; Bennett and Glenn, 1970; Tata, 1971).

Tata's laboratory (Beckingham-Smith and Tata, 1976) has suggested two possible developmental control mechanisms associated with this turnover in the ribosomal population and RER: (1) newly synthesized proteins associated with liver differentiation may be restricted to the TH-induced ribosomes, or (2) a compartmentalization of protein synthesis may exist within the cells, mediated by ribosomal attachment to specific endoplasmic reticulum structures. Either of these mechanisms represents exciting possibilities of structural mediation of posttranscriptional control, and merits further investigation.

Weber's laboratory (Hagenbuchle *et al.*, 1976) has also shown progressive accumulation of rRNA within the liver cytoplasm of larval *X. laevis*, and that TH alters pool turnover rates but not pool sizes for UTP and CTP. Their data also suggest TH may alter processing of nuclear hnRNA and pre-rRNA. Short-term labeling experiments show that in TH-treated livers, most of the 40 S hnRNA of the nucleus has moved into rRNA and mRNA fractions of the cytoplasm. In the controls, less label moves into cytoplasm, and a large proportion remains in the hnRNA fraction. These observations suggest that processing and/or transport of RNA may be altered by TH interaction with the cell.

2.2.3. Conclusions

TH stimulation of the anuran liver.results in the *de novo* synthesis of a group of specialized proteins that effect the differentiation of the larval liver into an adult liver. Both transcriptional and posttranscriptional controls are implicated in the TH-induced synthesis of these new proteins. At the transcriptional level, there is significant stimulation of hnRNA synthesis within 2 days after TH stimulation. TH also appears to stimulate increased processing and transport of hnRNA and pre-rRNA from the nucleus to the cytoplasm. Changes in the *in vitro* translating properties of isolated mRNA can be demonstrated within 1 day. Correlating with increased rRNA synthesis is the proliferation of hepatocyte nucleoli and changes in nuclear shape. Simultaneously, there is extensive morphological reorganization mediated by the synthesis and proliferation of new membranes to which the newly synthesized ribosomes bind more firmly. These membranes may originate in the perinuclear region. The new RER also associates intimately with the mitochondria, which may facilitate the transfer of newly synthesized mitochondrial proteins to the mitochondrial matrix for assembly. Both translation and posttranslational assembly of proteins are ultimately stimulated by TH. It is possible that the TH-induced structural and functional reorganization is mediated by the production of new generations of both molecules and cellular organelles, with increased turnover of old structures such that an "adult" population replaces a "larval" population as we have proposed for the hepatocytes themselves. The interrelationship of possible cellular turnover and turnover of subcellular organelles in TH-induced liver differentiation merits further investigation.

Although it is generally assumed that these changes in RNA and protein metabolism are specific responses to TH, it is possible that they represent, in part, nonspecific responses to stimulation of metabolic activity, such as cellular proliferation. The early sequence of RNA, protein synthesis, increased ornithine decarboxylase levels, and DNA synthesis are very similar to those that occur following partial hepatectomy of mammalian liver. Although the *in vitro* translating ability of isolated tadpole mRNA changes quantitatively following TH administration, to date there has been no demonstration of a qualitative change in mRNA, such as the appearance of a new class of mRNA in the TH-induced liver. The level of control of the change in RNA composition, and the relationship of the endpoints in gene expression (increases in a group of specialized proteins) to the early events remain to be defined. Even the urea-cycle enzymes may be under indirect TH control since prolactin, which is generally considered to be antagonistic to TH action, stimulates increases in arginase activity in premetamorphic liver at stages when no activity can be detected in untreated controls (Guardibassi *et al.*, 1970). Arginase activity in tadpole liver is also stimulated by rat growth hormone, hydrocortisone, insulin, and epinephrine. The interrelationships of the early TH-induced events to the later expressions of differentiation, as well as the possible roles of other hormones, must be defined if TH-induced metamorphosis is to be understood.

2.3. Liver Glycogen

Total liver carbohydrate, which would be primarily in the form of glycogen, accumulates during pre- and prometamorphosis and is then rapidly depleted during climax and postclimax stages (Hanke and Leist, 1971) (Fig. 5). The increases in total liver carbohydrate primarily reflect changes in cell number and, to a lesser degree, increased carbohydrate accumulation per cell. Cellular carbohydrate levels (pg/cell) peak at Taylor and Kollros stage XVI in *R. pipiens* declining to minimal levels by stages XXIII–XXIV. Total liver carbohydrate attains maximal levels between stages XVII and XIX; this latter rise is attributable to increases in cell number. Nearly all the liver glycogen is mobilized during metamorphic climax, but glycogen accumulation is resumed by stage XXV (Fig. 5). This pattern of glycogen mobilization is consistent with the energy requirements of nonfeeding climax stages.

Corticosteroids promote both gluconeogenesis and glyconeogenesis in larval amphibians, while TH promote glycolysis; glycogen is deposited in both liver and muscle during prometamorphic stages of *X. laevis* primarily in response to high glucocorticoid levels (Hanke and Leist, 1971). The decline in the levels of cellular glycogen in *R. pipiens* beginning at stage XVI is consistent with the observation that glucocorticoid levels fall dramatically at that stage, while circulating TH levels are beginning to rise (Dodd and Dodd, 1976; White and Nicoll, this volume). Undoubtedly the pattern of carbohydrate metabolism in the metamorphosing liver reflects the interaction of several hormones, including glucocorticoids and TH.

2.4. Mitochondrial Structure and Function during Metamorphosis

Among the organelles undergoing morphological reorganization during metamorphosis, the changes occurring in the mitochondria are quite evident and readily quantified. Mitochondria increase in number in TH-induced anuran livers (Beckingham-Smith and Tata, 1976), and morphological changes during spontaneous and TH-induced metamorphosis have been studied (Bennett *et al.*, 1970; Bennett and Glenn, 1970; Cohen, 1970).

The mitochondria of the pre- and prometamorphic liver of *R. catesbeiana* are characteristically small, rod shaped, and scattered throughout the cytoplasm. The matrix usually stains densely, and the cristae are prominent, tubulamellar in morphology. Preclimax (stage XVII–XIX) hepatocytes generally contain two morphological classes of mitochondria, one class small and similar to those found in younger larvae, and the other of larger tubular cristae. Mitochondria of climax larvae are significantly larger, medium staining, and their cristae less prominent. The climax mitochondria are more spherical in shape and double the volume of preclimax mitochondria (Bennett and Glenn, 1970). Mitochondria in TH-induced larvae undergo similar changes, such that they are more spherical, less oblong, and 7 times the volume of the controls. The matrix is slightly less densely staining than controls and has a "healthy" appearance, showing no evidence of "swelling" seen in TH-induced mammalian mitochondria. The cristae change only slightly (Bennett *et al.*, 1970). The appearance of the cristae is similar to that seen in normal adult rats. Mitochondria of climax livers are also more intimately associated with RER than in earlier stages (Bennett *et al.*, 1970; Bennett and Glenn, 1970).

Two enzymes of the ornithine–urea cycle, OT and CPS-I, are synthesized in the microsomes and transferred to the mitochondrial matrix (Cohen, 1970). It has been suggested that the intimate association of mitochondria with the RER facilitates this transfer (Beckingham-Smith and Tata, 1976). The increase in volume, without much change in structure of the cristae, may be associated with the accumulation of these enzymes in the matrix.

Cohen's laboratory (Cohen *et al.*, 1978) has used linear density gradient centrifugation analysis to define two subpopulations of mitochondria during metamorphosis. Their data indicate that the mitochondria of adult livers are characteristically of higher density than those of larval liver and that during metamorphosis a higher-density subpopulation, similar to that found in adults, replaces the lower-density larval subpopulation. Furthermore, in response to T_4 treatment, newly synthesized CPS-I, as detected by immunochemical methods, is preferentially incorporated into the higher-density subpopulation. The biochemical evidence for low-density larval and higher-density adult mitochondrial subpopulations correlates well with the morphological evidence for increases in mitochondrial volume during metamorphosis (Bennett *et al.*, 1970; Bennett and Glenn, 1970).

Whether mitochondrial enlargement is the result of accumulation of newly synthesized enzymes in the matrix or whether the generation of a new morphological subpopulation is a prerequisite to the enzyme accumulation remains to be determined. It has been suggested that mitochondrial proliferation is a prerequisite to mitochondrial metamorphosis, to produce a new generation of larger, adult-type mitochondria (Beckingham-Smith and Tata, 1976). The earliest response reported for mitochondria during TH-induced metamorphosis is mtDNA synthesis (Fig. 3C). A significant enhancement over control levels is seen within an hour after T_3 treatment, reaching a maximum at 3–4 days, and then declining. Cytoplasmic grains in autoradiographic experiments have been interpreted as indicating TH-induced mitochondrial proliferation (see Fig. 3D). Since the only available detailed fine-structural study of mitochondrial change reports only the endpoint (Bennett *et al.*, 1970), the relationship of the changes in mitochondrial structure to mtDNA synthesis remains to be determined.

The mitochondrion has been identified as a potential intracellular target site for TH action because of the well-known "calorigenic effect" of TH in mammals (reviewed by Tata, 1971). TH has been demonstrated to have reversible significant effects on both the structure and the function of mammalian mitochondria *in vivo* and *in vitro*. High-affinity, saturable binding sites for both T_4 and T_3, have been located on the rat mitochondrial membrane and have been implicated in TH effects on mammalian mitochondria. It has been suggested that TH has a dual site of action, acting on the nucleus to effect genetic changes associated with development and maintenance of the cell, and acting on the mitochondria to effect shorter-term effects on energy metabolism (discussed by Frieden, this volume).

However, the changes observed in tadpole liver mitochondria during metamorphosis are developmental changes of a permanent nature, rather than short-term energy effects. The growth observed during mitochondrial metamorphosis is true growth, and as such it is distinct from the swelling observed associated with TH-mediated energy changes in mammals. Furthermore, a calorigenic response to TH has not been demonstrated for tadpoles, although several enzymes involved in oxidative phosphorylation have been observed to increase in tadpole liver mitochondria during TH-induced metamorphosis (Cohen *et al.*, 1978). Similar changes in mitochondrial morphology have been observed during development of rat and chick liver.

The possible role of mitochondrial protein synthesis in the changes in mitochondrial morphology during metamorphosis has not been investigated. The relationship to protein synthesis on the RER also remains to be determined. It is possible that TH has multiple sites of action in the cell (Tata, 1975; see also Frieden, this volume), and dual sites of action on both the nucleus and the mitochondria may mediate mitochondrial biogenesis through coordinated cytoplasmic and mitochondrial protein synthesis (see Frieden, this volume).

In summary, liver mitochondrial biogenesis during amphibian metamorphosis is accompanied by coordinated morphological and biochemical changes. A larval subpopulation of mitochondria is gradually replaced, in response to TH, by an adult subpopulation characterized by larger volumes and higher densities. New enzymes of the urea cycle are synthesized on the microsomes and are transferred preferentially to the matrix of the adult-type, high-density mitochondria where they are assembled into functional enzymes. Accumulation of newly synthesized enzymes in the matrix correlates with mitochondrial enlargement. Biochemical evidence indicates that mtDNA synthesis and mitochondrial proliferation precede enlargement, but crucial experimental evidence of causal relationships between mitochondrial proliferation and biogenesis is lacking. Furthermore, the relationship of mitochondrial changes to other cellular changes requires definition, particularly with respect to the question of the site of TH action. The mitochondrial response may represent either a direct response to TH, a secondary response to other cellular, nuclear-mediated changes, or a coordinated response to multiple sites of action.

2.5. Temperature Effects on Liver Metamorphosis

Temperature alters cellular responsiveness to TH, such that high temperatures and TH synergistically accelerate metamorphosis, while at low temperatures cells become refractive to TH (Kollros, 1961). Larvae treated with TH at 5°C do not respond. However, if the larvae are then moved to a higher temperature, even long after hormone treatment, they respond rapidly, as measured by either biochemical or morphological response (Griswold *et al.*, 1972). In the liver of *R. catesbeiana* larvae, the length of the latent period, the rate of arginase increase, and the level at which arginase activity plateaus in response to TH are proportional to both TH dose and temperature. Pre- and prometamorphic arginase levels are temperature sensitive (Guardibassi *et al.*, 1970). Since the dosage of

TH needed for a given response increases at lower temperatures (Kollros, 1961), these results probably reflect changes in cellular responsiveness to TH.

When *R. catesbeiana* larvae are treated with labeled TH at 25°C, the hormone enters the cytoplasm and is rapidly transferred to the nucleus. However, when hormone is administered at 5°C, TH stays in the cytoplasm, associated with the microsomal, cytosol, and mitochondrial fractions. When larvae are moved from 5°C to 25°C, the hormone rapidly moves into the nuclear fraction. It has been suggested that the insensitivity to TH at low temperature is due to a temperature-dependent movement of the hormone to the nucleus (Griswold *et al.*, 1972).

Anura grown at low temperatures are larger at any given stage than those grown at higher temperatures (Berven *et al.*, 1979), but temperature regimes do not alter the allometric relationship between the liver and whole-body weight. Biochemical analysis of the liver indicates that low-temperature animals are larger because they have more cells and the cells are larger; hence there is an exponential increase of total body size with lowered temperatures. The cell enlargement is due to an actual accumulation of macromolecular components rather than to an accumulation of water, and hence represents true growth. On a per cell basis, protein and carbohydrate account for most of the increased liver cell growth (Fig. 6). In addition, temperature alters the content and intracellular distribution of RNA in liver cells, and the percent RNA in the nucleus is significantly elevated at lowered temperatures.

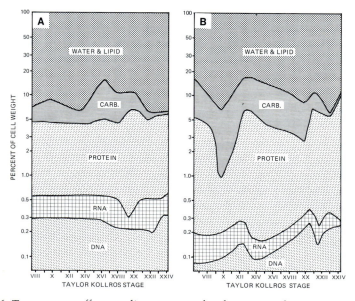

FIGURE 6. Temperature effects on liver macromolecular content during spontaneous metamorphosis of *R. pipiens*. Larvae raised continuously at (A) 22°C and (B) 18°C. From Smith-Gill and Darling (unpublished manuscript).

In summary, a number of important components of the metamorphic response in the liver, namely cellular responsiveness to TH, subcellular location of the hormone, rates of cell division, cellular growth, and the metabolism of RNA and other macromolecular components, are temperature sensitive. Temperature effects therefore may provide a useful tool to manipulate TH action on the liver, in order to possibly separate or change relative timing of the various morphological and biochemical events occurring within the cell. Furthermore, the effects of temperature at both the cellular and the organismal levels have a significant impact on metamorphic patterns in nature (Dodd and Dodd, 1976; Berven *et al.*, 1979), and therefore an understanding of temperature effects on metamorphosis is important both to the mechanisms of action of TH as well as to the natural ecology of metamorphosis.

2.6. Summary

Biochemical events accompanying thyroxin-stimulated metamorphosis of the anuran liver include DNA synthesis, synthesis and accumulation of a stable population of rRNA with an associated enumeration of nucleoli, lipid synthesis, elaboration of the RER, and a turnover of the mitochondrial population. Among the newly synthesized proteins are the enzymes of the ornithine–urea cycle, which develop in association with the animal's transition from ammonotelism to ureotelism. Regulation of the rates of protein synthesis appears to be at both transcriptional and posttranscriptional levels, but to date thyroxin-induced synthesis or accumulation of a new class of mRNA has not been demonstrated for anuran liver. In both spontaneous and TH-induced metamorphosis, periods of DNA synthesis precede the appearance of new proteins, and cell death may accompany liver metamorphosis. We have interpreted this to indicate that significant cellular turnover likely precedes biochemical differentiation, and that differentiation of adult function occurs in proliferating population of cells. However, a functional relationship between proliferation and liver metamorphosis has not been experimentally established.

The observation that many other agents, including prolactin, epinephrine, glucocorticoids, and temperature changes, can induce some or all of the macromolecular changes seen following TH stimulation, and that there are significant differences between TH-induced metamorphosis and spontaneous metamorphosis, suggest the possibility that multiple hormone interactions may be important in the overall metamorphic response. Possible candidates include pituitary growth-promoting hormones and the adrenal glucocorticoids. The thyroxin specificity of the metamorphic response remains to be demonstrated. Notably the relationships of late manifestations of liver differentiation (namely urea-cycle enzymes) to early macromolecular changes, and the thyroxin-specificity of both early and late events, must be established before the molecular mechanism of action of TH can be approached. Furthermore, if other hormones do interact with TH to bring about liver metamorphosis, their roles must be defined if the normal developmental processes occurring in the liver are to be understood.

3. THE KIDNEY

The kidney of the adult amphibian is a mesonephric kidney, while that of the embryonic and larval stages is a functional pronephros. The pro- and mesonephros coexist structurally and functionally over a considerable portion of the larval period, until the pronephros degenerates at metamorphic climax (reviewed by Fox, this volume). While the details of larval kidney function are poorly understood, several reports have suggested multiple roles of the kidney in development, and that its function is hormone responsive. For instance, the amphibian pronephros is a reservoir for hematopoietic stem cells during the larval period (reviewed by Broyles, this volume). A role for hormones is suggested by the observation that pronephric involution is more rapid in premetamorphic tadpoles when they are exposed to TH and by the reversible inhibition of pronephric involution by the TH antagonist thiourea. The kidney's role in the transition from ammonotelism to ureotelism is probably TH induced. More recently, a role for prolactin in the metamorphosing kidney has been postulated (White and Nicoll, 1979). The transition from ammonotelism to ureotelism and the possible roles of TH and other hormones in this transition will be examined below.

The transition from ammonotelism to ureotelism of most anurans involves not only changes in liver metabolism (Section 2), but also changes in kidney function (reviewed by Balinsky, 1970). While the frog actively excretes urea, probably in the proximal tubules located in the dorsal kidney, the larval kidney does not. The ability to do so first appears during metamorphosis and coincides with increased arginase activity in the dorsal kidney. Furthermore, adult frogs can utilize plasma urea for regulation of osmolarity. The ability to retain urea also appears at metamorphosis, for tadpoles do not accumulate urea in the blood.

The acquisition of urea excretion by the kidney is apparently dependent upon TH in both anurans and urodeles. Treatment of larvae with TH induces premature urea excretion, while thyroidectomy causes a decrease in percent urea excreted (Balinsky, 1970). Since both pro- and mesonephros function in the prometamorphic larva, the acquisition of urea excretion implies that kidney metamorphosis involves differentiation and subsequent change in function of the mesonephros, rather than mere persistence while the pronephros degenerates. The ability of the mesonephros to regulate urea excretion or retention develops coordinately with the increased ability of the liver to produce urea, the coordinating signal being TH. However, the mechanism of action of TH on the kidney and the possible interactions with other hormones remain to be defined.

The newly acquired urea-excreting ability correlates with the appearance of thyroxin-dependent prolactin (PRL) binding sites in the kidney (White and Nicoll, 1979). TH action on the mesonephros may therefore be indirect, for instance by increasing sensitivity to PRL or other factors. High-affinity, saturable PRL binding sites are low or undetectable in pre- and prometamorphic *R. catesbeiana* kidneys, but PRL binding becomes detectable and increases throughout metamorphic climax, reaching very high levels in adult kidneys. The increased binding corresponds to the stages when pronephric regression is greatest, and also when plasma TH levels are known to be rising. Increased PRL binding can be precociously induced by either immersion or injection of early prometamorphic larvae with thyroxin (White and Nicoll, 1979). White and Nicoll have suggested that the presence of thyroxin-dependent PRL receptors may be related to prolactin's role in controlling hydrostatic homeostasis. However, an interesting alternative hypothesis is possible, namely that PRL plays a role in the development of urea excretion and/ or retention by the kidney. Pro- and mesonephric tissues were not separated in the White and Nicoll (1979) study, but it would be of interest to know the specific localization of the receptors they reported.

In contrast to most anurans and urodeles, the aquatic *X. laevis* adult is ammonotelic. Significant quantities of ammonia can be detected in *X. laevis* blood, but not in adult ranid blood (Balinsky and Baldwin, 1962). Treatment of larval *X. laevis* with TH does not induce urea excretion (Balinsky, 1970). Deamination of L-amino acids by *X. laevis* kidneys is significantly higher than by ranid kidneys. *X. laevis* is apparently ammonotelic as an adult due in part to more active deamination in the kidney (Balinsky and Baldwin, 1962).

4. THE INTESTINE

In association with the transition from herbivorous to carnivorous feeding, the intestine of anuran amphibians undergoes a remarkable transformation during metamorphosis, essentially from a long, relatively simply organized longitudinal tube with brush border epithelial cell lining, few muscle cells, little connective tissue, and a serosa, to a typical multilayered villous vertebrate gut. Metamorphosis of the gut may be divided teleologically into two phases: (1) gut regression, which includes loss of the primary epithelium and shortening of the gut, and (2) postclimax gut development, including proliferation of the secondary epithelium with the formation of the villi and development of extraepithelial tissue and muscle. Any explanation of the phenomenon of gut metamorphosis must account for all of these events. A recent review of the development of the gut during spontaneous or induced metamorphosis

(Hourdry and Dauca, 1977) has addressed several of the above points. The present discussion summarizes the known changes in the intestine, as described for anurans, and considers primarily papers published since the above review.

4.1. Loss of Primary Epithelium and Shortening of the Gut

The most conspicuous feature of the larval anuran gut is the primary epithelium, which serves as the absorptive surface. The cells of the primary epithelium have been shown to incorporate [³H]-Tdr and divide during the feeding stages in *Xenopus* (Marshall and Dixon, 1978). Analysis of the incorporation of [³H]-Tdr injected 2 hr before sacrifice in premetamorphic *Alytes* indicated that 8% of the primary epithelium is labeled in the nucleus and that the specific activity (cpm/μg DNA) of the tissue is high. This observation is consistent with the hypothesis that a large portion of the gut is composed of primary epithelium undergoing rapid growth (Dauca and Hourdry, 1978a).

Prior to loss of the larval primary epithelium during spontaneous metamorphosis, a characteristic pattern of epithelial lysosomal activity and change is observed in several anurans. Histological and biochemical measurements have been correlated to confirm increases in the hydrolytic activity of several lysosomal hydrolases during prometamorphosis, including acid phosphatase, β-glucuronidase, aryl sulfatase, and *N*-acetyl-β-glucosamidase (Hourdry and Dauca, 1977; Hourdry, 1977). Both total and soluble activities of the ratio of two electrophoretically quantified *N*-acetyl-β-glucosamidase isoenzymes reach maxima at metamorphic climax (Hourdry, 1977).

Each of the above hydrolytic enzymes has been localized in the primary epithelium through cytochemistry (reviewed by Hourdry and Dauca, 1977). It is thought that the release of hydrolases into the cytoplasm initiates changes in the primary epithelium but the exact events occurring between the exposure of the intestine to TH and the increased hydrolase activity in primary epithelium are unknown. The primary lysosomes digest cytoplasmic material to form a residual body, which may be incorporated into a secondary (compound) lysosome. After a period of time a compound residual body is observed (Bonneville, 1963). The remains of the primary epithelium are subsequently liberated into the lumen of the intestine and egested (McAvoy and Dixon, 1977; Bonneville, 1963). This later event corresponds to metamorphic climax, a time when the animal has stopped feeding and the gut has begun to shorten and reduce its diameter (Carver and Frieden, 1977).

Alkaline phosphatase activity has been correlated with a functional epithelium in the feeding animal (Kaltenbach *et al.*, 1977). In feeding, growing larvae, alkaline phosphatase activity is high in the brush border of the primary epithelium posterior to the entrance of the bile duct and in the midintestine. With the onset of metamorphosis, alkaline phosphatase activity in these areas decreases in an antiparallel fashion to the increase in hydrolytic activity (Kaltenbach *et al.*, 1977; Hourdry *et al.*, 1978). The climactic increase in alkaline phosphatase activity, which occurs in the connective tissue of the duodenum and small intestine, may indicate that alkaline phosphatase also plays a role in the histogenesis of extraepithelial tissue during metamorphosis, as suggested by Kaltenbach *et al.*, (1977).

Since the metamorphosing intestine is a kaleidoscope of tissue destruction, proliferation, and differentiation, studies of enzymatic changes occurring in the gut during this period have necessarily relied heavily on histochemical data to yield significant results. Quantitative measurements of overall enzymatic total or specific activity would be of uncertain interpretation in such a tissue. For instance, one might have concluded in the above example that alkaline phosphatase activity did not change, or increased only slightly, during metamorphosis. Techniques recently developed for separation of anuran primary epithelial layer from the extraepithelial tissue may now make it possible to quantitate these hormone responses in the specific layers of the gut (Dauca and Hourdry, 1978b).

Morphological, histological, and enzymatic changes in the primary epithelium paralleling those seen during spontaneous metamorphosis have been demonstrated following immersion or injection of intact larvae with thyroxin. Decrease in the intestine length occurs within 1 day in *R. catesbeiana* (Carver and Frieden, 1977). Within 15–16 hr autolytic vacuoles are formed and acid phosphatase specific activity increases in *Discoglossus* (Hourdry and Dauca, 1977). Loss of primary epithelium in thyroxin-treated *Alytes* parallels the *in vivo* observations in *Discoglossus* (Dauca and Hourdry, 1977, 1978a). Therefore, TH appears to induce specific enzyme changes, shortening of the gut, and loss of the primary epithelium *in vivo*. Furthermore, thyroidectomy or hypophysectomy in prometamorphosis halts changes in the primary epithelium. TSH replacement therapy in a hypophysectomized animal results in autolysis of the primary epithelium (Hourdry and Dauca, 1977).

In vitro experiments support the hypothesis that TH acts directly on the epithelial tissue to induce regression of the gut. The autolytic changes in the primary epithelium of gut explants have been reported to be induced by 5×10^{-8} to 10^{-7} thyroxin g/ml after 2 days in medium supplemented by fetal calf serum (Pouyet and Hourdry, 1977). This is the first *in vitro* demonstration of thyroxin causing primary epithelial changes that mimic the *in vivo* changes occurring during spontaneous or TH-induced metamorphosis and therefore is an important report. The physiological significance is yet to be determined as the endogenous bound versus free serum levels of T_3 and T_4 were not reported (for a discussion of this problem see Samuels, 1978). Collectively, however, the above data suggest that primary epithelium is lost during spontaneous metamorphosis as a result of the direct action of TH on epithelial tissue.

4.2. Proliferation of Secondary Epithelium and Development of Extraepithelial Tissues

The secondary epithelium is formed by proliferation of stem cells at the base of the primary epithelium. The proliferation of the stem cells begins only after a lag of several days (typically 3–9) following initiation of changes in the primary epithelium during anuran spontaneous metamorphosis. The rapid proliferation of the secondary epithelium causes some folds in the tissue, which are modified into villi or longitudinal folds to increase surface area. For example, *Xenopus* larvae during metamorphic climax incorporate [³H]-Tdr DNA at the base of the longitudinal trough. Newly divided cells then migrate toward the crests of the longitudinal folds (McAvoy and Dixon, 1977). *R. catesbeiana*, *Alytes*, and *Discoglossus* form villi as a result of proliferation of the secondary epithelium (Kaltenbach *et al.*, 1977; Hourdry and Dauca, 1977; Dauca and Hourdry, 1978a).

Lysosomes are rare in the newly proliferated tissue, and enzymatic and lysosomal changes characteristic of the primary epithelium do not occur. However, alkaline phosphatase activity, associated with feeding, does increase in the connective tissue and muscle layers of the digestive tract during spontaneous metamorphosis. The brush border of the newly developed secondary epithelium increases in alkaline phosphatase activity late in metamorphosis (Kaltenbach *et al.*, 1977).

[³H]-Tdr incorporation into DNA in the secondary epithelium has been reported to occur in localized sections along the length of the intestine in *Bufo bufo*, while in *Alytes* it is uniformly distributed (Hourdry and Dauca, 1977; Dauca and Hourdry, 1978a). The secondary epithelium incorporates little [³H]-Tdr during premetamorphosis, but it incorporates 28% during prometamorphosis and 33% during the early climax stages. Although the number of secondary epithelial cells dividing in late metamorphic climax is high relative to total secondary epithelial cells, the specific activity (cpm/µg DNA) is low. This is consonant with the presence of a relatively large number of other cell types in gut tissue. Less than 10% of the secondary epithelium incorporates [³H]-Tdr at the end of metamorphic climax, the time when the secondary epithelium begins to leave the mitotic cycle and grow. The extraepithelial tissue has a low labeling index and fewer total cells than the secondary epithelium, but by the end of metamorphic climax the extraepithelial tissue occupies about half of the intestinal cross-sectional area (Dauca and Hourdry, 1978a).

Development of the secondary epithelium can be induced through treatment of intact animals. About 4 days after immersion of intact *Discoglossus* larvae in thyroxin, the secondary epithelium replaces the primary epithelium (Hourdry and Dauca, 1977). The events of DNA synthesis in response to thyroxin immersion in *Alytes* have also been carefully analyzed. After the primary epithelium degenerates, the specific activity of [³H]-Tdr incorporation can be accounted for by extraepithelial cell division at 7 days and by secondary epithelial cell replication at 9 days (Dauca and Hourdry, 1977, 1978c). Both the time frame and the sequence of events are similar to spontaneous metamorphosis.

Experiments with hormone replacement therapy in hypophysectomized or thyroidectomized animals have demonstrated only degeneration of the primary epithelium, as suggested by histological change and acid phosphatase activity increase (Hourdry and Dauca, 1977). Likewise, organ culture of *Discoglossus* intestine in serum containing medium supplemented with T_4 has resulted in cytological and acid phosphatase changes characteristic of primary epithelium (Pouyet and Hourdry, 1977). During *in vivo* treatment of intact animals the secondary epithelium proliferates only after a lag of several days following changes in the primary epithelium. The possibility that *in vitro* incubations must be continued over a longer period of time and under somewhat different conditions or that TH may exert its effects on the secondary epithelium and extraepithelial tissue indirectly or through other hormones should be considered (Kaltenbach *et al.*, 1977; Dauca and Hourdry, 1977, 1978c). Most of the papers discussed above have utilized T_4 to induce metamorphosis. Since T_3 also increases in concentration during metamorphosis and a role has been established for T_3 during induced metamorphosis (see White and Nicoll, this volume), the function of T_3 during metamorphic changes in the gut should be more carefully evaluated (Carver and Frieden, 1977).

Superimposed on the increases in plasma concentration of T_3 and T_4 during spontaneous metamorphosis are parallel increases in both growth hormone and PRL, and there are also changes in glucocorticoid levels (see White and Nicoll, this volume). Moog's laboratory has demonstrated that alkaline phosphatase in the mouse intestine is under pituitary–adrenal control, and only combined doses of TH growth hormone, and cortisone have been effective in producing normal growth in the intestine of the hypophysectomized, suckling rat (Yeh and Moog, 1978). Therefore, the possible role of other hormones, particularly the glucocorticoids, in development of the metamorphic anuran intestine deserves further investigation.

4.3. Summary

During metamorphosis, the anuran larval gut transforms from a long, simply organized, rapidly proliferating structure to a typical multilayered villous vertebrate gut. The first phase in the transition, gut regression, occurs during late prometamorphosis and metamorphic climax. It is characterized by a dramatic shortening of the gut and a loss of the primary epithelium. The latter appears to be initiated by mobilization of lysosomes and a release of hydrolases into the cytoplasm of the epithelial cells. During gut regression alkaline phosphatase activity, which is correlated with a functional larval epithelium, decreases. Administration of

TH to intact animals, and *in vivo* hormone inhibition and replacement, as well as *in vitro* experiments, support the hypothesis that gut regression is induced by direct action of TH on the gut primary epithelium.

Adult-type gut development is initiated by proliferation of stem cells at the base of the primary epithelium and development of the villi, as well as development of extraepithelial tissues, including muscle. *In vivo* hormone replacement in hypophysectomized animals and *in vitro* experiments have failed to induce normal development of secondary epithelium and extraepithelial tissues. Therefore, while the degenerative phase of gut metamorphosis appears to be under direct control of TH alone, the possibility that the regenerative phase results either by indirect action of TH or by interaction of TH with other hormones, such as the glucocorticoids, merits further investigation.

5. THE INTEGUMENT

5.1. The Pigmentary System

The larval amphibian develops a distinctive adult pigmentary pattern that becomes pronounced during and just after metamorphic climax. In anurans and urodeles this metamorphosis involves migrations and rearrangements of chromatophores as well as changes in chromatophore densities resulting both from differentiation of new chromatophores and from mitosis of already differentiated chromatophores (Smith-Gill, 1974; Bagnara *et al.*, 1978b). There are also changes in chromatophore morphologies (Smith-Gill, 1971; Bagnara *et al.*, 1978b, 1979), and the development of new chromatophore associations (Bagnara *et al.*, 1968). New classes of pigment are sometimes synthesized (Bagnara, 1976; Bagnara *et al.*, 1978b, 1979; Frost, 1978).

Many changes in pigment cells have been attributed to hormones. Relatively rapid, reversible, and transient changes are referred to as physiological color changes and are based primarily on intracellular movements of pigment granules. Longer-term changes, which involve quantitative and sometimes qualitative changes in the pigmentary content of the skin through pigment synthesis or breakdown, are referred to as morphological color changes (Bagnara and Hadley, 1973). This discussion will consider the longer-lasting morphological changes that occur during metamorphic development, ultimately effecting a developmental change in the type of pigment, pigment cells, or pattern in the integument.

Although variations occur, the skin of amphibians commonly contains several types of pigment cells, each with its own characteristic pigment and associated pigment organelle. The dermal and epidermal melanophores contain melanin polymerized over a protein matrix into a typical vertebrate melanosome pigment granule. The dermis also contains iridophores, in which purines are organized into reflective platelets, or refractosomes (Menter *et al.*, 1979). Additionally, the dermis contains yellow xanthophores or red erythrophores; these pigment cells usually contain pteridines organized in pterinosome organelles (Matsumoto, 1965) and carotenoids, generally distributed throughout the cytoplasm in membrane-bound vesicles (Bagnara *et al.*, 1968). The terminology, physiology, and morphology of amphibian pigmentation have been expeditiously reviewed (Bagnara and Hadley, 1973; Bagnara *et al.*, 1979).

In the adult, pigment cells may be arranged in disruptive patterns, such as dorsal spots, and in specific cell associations, such as the dermal chromatophore unit (Bagnara *et al.*, 1968) and the epidermal melanin unit. The adult chromatophore units have discrete morphological and functional relationships to the pigment cells. These cellular associations, as well as the disruptive chromatophore patterns typical of the adult, are not seen in the larva, but develop during metamorphosis when the dermis undergoes extensive morphological reorganization (see Fox, this volume).

The dermal chromatophore unit includes xanthophores, iridophores, and dermal melanophores (Bagnara and Hadley, 1973) (Fig. 7). Typically, the xanthophores are located below the basement membrane, with iridophores beneath the xanthopores. A dermal melanophore lies below each iridophore; processes from the melanophore extend upward around the iridophore, and from these processes fingerlike projections occupy

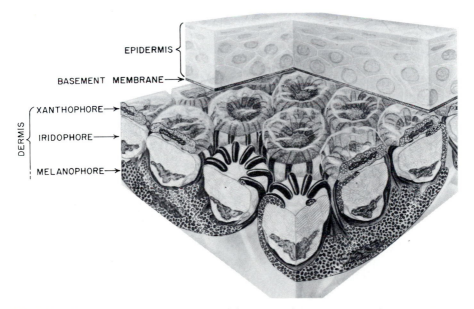

FIGURE 7. Diagrammatic representation of the anuran dermal chromatophore unit. A dermal melanophore lies below each iridophore. Dendritic melanophore processes extend upward around the iridophore to occupy the space between the iridophore and the xanthophore. Adaptation to a dark backbround is depicted, with melanosomes dispersed into the "fingers" of the melanophores. Reproduced with permission from Bagnara *et al.*, 1968.

the space between the iridophores and the xanthophores. The particulars of the arrangement of the pigment cells within the dermal chromatophore unit vary among species. The physiological and morphological states of the chromatophores determine the quantity and quality of light that will be refracted. Under appropriate hormonal stimulation, melanosomes may migrate into the "fingers," obscuring the light-reflecting capacity of the iridophores and thus modifying color. Depending on the species and location within the skin, iridophores may be dendritic and physiologically active, responding to appropriate hormonal stimulation by translocation of refractosomes, or may be permanently punctate and physiologically inactive (Butman *et al.*, 1979).

A second type of adult skin pigmentary unit is the epidermal melanin unit in which the dendritic epidermal melanocyte transfers cytocrine melanin to other, nonpigment-producing epidermal cells (Hadley, 1966). A functional epidermal melanin unit is present in both the adult and the larva (Bagnara and Hadley, personal communication).

Investigations of wild-type and several pattern mutants of *R. pipiens* have suggested that adult disruptive pigmentary patterning is correlated with three interrelated factors (as discussed in: Hadley, 1966; Bagnara and Hadley, 1973; Smith-Gill, 1973): (1) chromatophore densities; (2) local chromatophore patterns; and (3) cytophysiology of the chromatophores. The spatial patterning of the variations in chromatophore densities forms a permanent record of dorsal pigmentary patterning and is in itself sufficient to define disruptive patterning in the leopard frog. There is a concurrence of dermal and epidermal disruptive patterning, including a negative association of the dermal chromatophore unit with the epidermal melanin unit. Furthermore, studies utilizing mutants deficient in the production of one or more pigment types have established that disruptive pattern formation is independent of the differentiation of any particular chromatophore type, since blue frogs lacking xanthophore pigments, albino frogs lacking melanin pigments, and melanistic frogs lacking iridophore pigments all show normal spotting patterns.

5.1.1. Pattern Formation

Two aspects of amphibian pattern develop during metamorphosis. The first is the formation of local chromatophore associations, such as the dermal chromatophore unit. The second is the formation of disruptive patterning, which involves greater organization of chromatophore densities and morphologies. In both cases, the timing of pattern formation appears to coincide with metamorphic climax (Hadley, 1966; Bagnara *et al.*, 1968; Smith-Gill, 1974) and with the onset of dermal reorganization from an essentially one-dimensional structure to a multilayered, three-dimensional structure (see Fox, this volume). In *R. pipiens*, local differences in chromatophore densities characteristic of adult patterning do not differ between presumptive spot and interspot regions until metamorphic climax or later (Smith-Gill, 1974). Similarly, analysis of chromatophore associations during metamorphosis revealed no differences between spot or interspot regions until after climax stages (Smith-Gill, 1971).

While the timing of pattern development during metamorphosis can be specified, the mechanism of pattern formation has not been determined. In a detailed analysis of chromatophore patterning and cluster sizes during metamorphosis of mutant and wild-type *R. pipiens*, Smith-Gill (1971, 1974) concluded that one set of endocrine-independent cell–cell or cell–tissue interactions determines regional density differences, producing a prepattern and ultimately the adult spotting pattern, while a different set of cellular events determines the local chromatophore associations, such as mottling or pinpoint markings, superimposed upon the disruptive spot pattern. Whether yet another set of cell–cell associations determines the formation of the adult chromatophore unit remains to be examined. An exact analysis of factors influencing pattern formation is made difficult by the complex interactions of genes, cell–cell associations, and environmental and hormonal factors in pigment cell physiology and differentiation. Relatively few studies have attempted to experimentally analyze these interactions.

Early during development, prepattern is determined in *R. pipiens*, and those areas that will be spot or interspot regions can be recognized during premetamorphosis by the patterning of the larval fusiform epidermal melanocytes (Smith-Gill, 1974) (Fig. 8). Morphological analysis of wild-type and mutant *R. pipiens* suggested that the spotting prepattern is determined by the tissue environment of the integument, rather than being an inherent property of the chromatophores themselves (Smith-Gill, 1971, 1974). Prepattern can be heavily influenced by several factors, including collagen disposition in the larval basement lamella. Such deposition may even limit melanophore migration into the epidermis (Smith-Gill, 1974).

That specific pigmentary patterns are not hormonally determined has been demonstrated by interspecific neural fold transplants in urodeles, and by exchanges of neural folds among *R. pipiens* pattern mutants. In all cases the transplants express the phenotypic pattern of the neural fold donor. Volpe's (1964) experiments with pattern mutants have been interpreted as indicating that pattern is a cell-specific genetic property of the chromatophores themselves. However, in Volpe's experiments the transplanted pattern always appeared as discrete spots, rather than the diffuse pattern seen in other neural fold transplant experiments. Recent evidence supports an alternative interpretation: the tissue environment provided by the other neural fold derivatives included in the transplant may play an important pattern-determining role (Bagnara, personal communication).

Although pattern is not hormonally determined, some hormone action is required to mediate either chromatophore differentiation or pro-

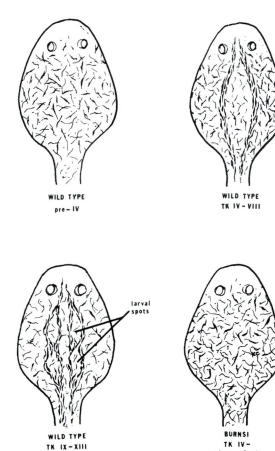

WILD TYPE
pre-IV

WILD TYPE
TK IV-VIII

larval
spots

WILD TYPE
TK IX-XIII

BURNSI
TK IV-
metamorphosis

FIGURE 8. Evidence for a pre-pattern in the patterning of fusiform epidermal melanophores in larval *R. pipiens*. Prior to stage IV, no discernible patterning is seen. After stage IV, parallel bands of anteroposteriorly oriented epidermal melanophores appear; these bands coincide with the location of larval spots, which become recognizable between stages IX and XIII and which later are transformed into adult spots at metamorphic climax through differentiation and mitosis of dermal chromatophores and epidermal melanocytes. Tadpoles carrying the *Burnsi* gene have no discernible patterning, correlating with the absence of spots in *Burnsi* adults. Reproduced with permission from Smith-Gill, 1974.

liferation for full expression of the adult pattern. One obvious candidate is TH. Subcutaneously implanted thyroxin–cholesterol pellets induce local precocious expression of the adult pigmentary pattern. In addition, MSH has been shown to influence both mitosis and differentiation of chromatophores. However, the adult pattern is not absolutely dependent on MSH as suggested by the faint adultlike pattern visible in hypophysectomized larvae induced to metamorphose by TH (for discussion see: Bagnara and Hadley, 1973; Bagnara, 1976).

Meticulous experiments by Pehlemann (1972) have convincingly demonstrated that the normal melanophore pattern of *Xenopus* develops as a result of MSH-coordinated melanophore mitosis and differentiation. Hypophysectomy is followed by a decrease in melanin synthesis, cessation of cell division, and partial depigmentation. Pituitary reimplantation stimulates mitosis and pigment synthesis. From this work the strong inference may be drawn that continued pituitary hormone stimulation, possibly by MSH, is required for melanophore division, maintenance, and possibly differentiation. The underlying biochemical mechanisms remain to be described.

The dermal melanophores of wild-type *R. pipiens* increase in number during prometamorphosis, probably as a result of mitosis. In addition, xanthophore densities increase while iridophore densities fall during prometamorphosis (Smith-Gill, 1974). Melanophores also proliferate just prior to metamorphic climax of *Pachymedusa dacnicolor*. The proliferation is accompanied by nuclear blebbing. MSH has also been implicated in both loss of heterochromatin and nuclear envelope blebbing (Bagnara *et al.*, 1978b). If chromatophore development in these two species is similar to that of *Xenopus*, these prometamorphic chromatophore density changes may be in response to circulating levels of MSH. This hypothesis could be tested if MSH levels could be measured directly by radioimmunoassay. Alternatively, repeated injections or infusion of MSH into prometamorphic animals might be used to evaluate the pigmentary response over extended time.

At metamorphic climax of *R. pipiens* there is an exponential increase in dermal melanophore densities; because of its brief time span, this is most likely due to the initiation of melanin synthesis in melanoblasts. The differences between spot and interspot densities arise from a differential increase in melanophore numbers between the two pattern regions, suggesting either a modulating effect of the tissue environment on the pigment cells or a differential cell-specific responsiveness of the cells localized within the two regions (Smith-Gill, 1974). Several observations suggest that functional subpopulations of chromatophores do exist in adult *R. pipiens* (discussed in: Hadley, 1966; Smith-Gill, 1973; Bagnara and Hadley, 1973; Butman *et al.*, 1979): (1) spot and interspot melanophores differ in melanin content; (2) spot and interspot chromatophores differ in their normal cytophysiological properties; (3) only interspot melanophores of northern frogs possess α-adrenergic receptors, while all dermal and epidermal melanophores possess β-adrenergic receptors; and (4) iridophores are physiologically heterogeneous, consisting of at least two populations differing in their MSH responsiveness. However, whether functional subpopulations are already differentiated in the larva, and hence differentially respond to endocrine influences during metamorphosis, or whether differential responses during metamorphosis are mediated through interactions with the local tissue environment, remains to be experimentally examined.

Possible multiple responses of larval chromatophores to MSH are suggested not only by the morphological responses of mitosis and pigment synthesis, but also by the observation that the larval integument is capable of hormonally mediated physiological responses to the background lighting conditions. These physiological responses of darkening or blanching are rapid and are superimposed on the long-term morphological responses. The ability of larvae to adapt to the background is acquired at a distinct time in development that is characteristic of the species

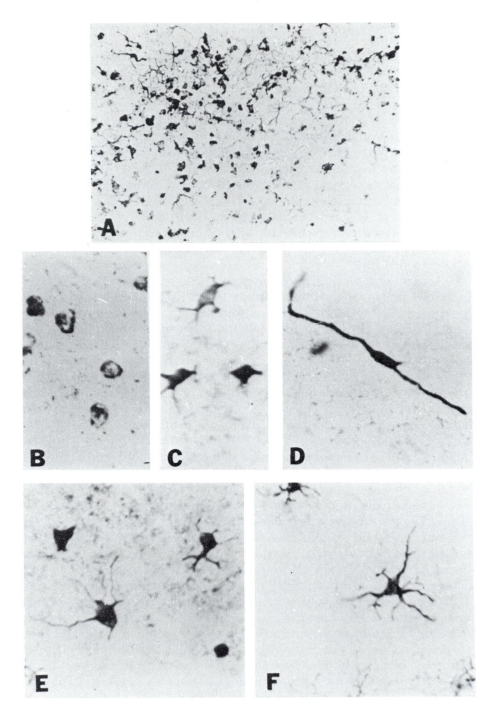

FIGURE 9. Morphogenesis of *R. pipiens* epidermal melanocytes. (A) Epidermis of stage XXI spot periphery, with many epidermal melanophores in various stages of differentiation. (B) Rounded melanoblasts with melanin granules in perinuclear position, abundant between

(Hadley, 1966; Bagnara and Hadley, 1973), and implies an ability of the larval chromatophores to distinguish between chronic and acute changes in MSH levels. Heterogeneous subpopulations, such as differential distribution of α- and β-adrenergic receptors, could underly such multiple responses. Alternatively, different molecular forms of MSH may be involved in control of the morphological and physiological responses. Furthermore, multiple hormonal control appears to underlie background adaptation in the larva. Although background adaptation in the adult appears to be under the control of MSH alone, in the larva a pineal hormone, probably melatonin, is strongly implicated in the blanching response. The responsiveness of larval melanophores to pineal extracts decreases with metamorphic development (discussed in: Hadley, 1966; Bagnara and Hadley, 1973; Bagnara, 1976).

Thus, not only may there be functional subpopulations within the integument at any stage of development, but the functional characteristics within and among these subpopulations may change with developmental time. Clearly, further investigation of these possibilities is necessary. In particular, it remains to be determined whether the functional subpopulations described in the adults are predetermined in the chromatoblasts and hence determine, for instance, the melanogenesis in the spot regions, or, alternatively, whether the properties of the surrounding extracellular environment permanently alter the differentiating chromatoblasts to mediate the development of functional subpopulations during the metamorphic transition to the adult pigmentary pattern.

TH has been experimentally associated with metamorphic effects on the pigmentary system (reviewed by Bagnara, 1976) and may specifically modulate the exponential increase in dermal melanophores that contributes to adult pattern (Smith-Gill, 1974; Bagnara, 1976). Convincing evidence confirming such a role for TH has yet to be presented. Although TH has been shown to affect pigment cells *in vitro* and *in vivo* by causing a physiological lightening of darkened skin (Bagnara and Hadley, 1973), it is not clear that TH has any true physiological lightening effects since the concentrations used in the above experiments were 10^{-6} to 10^{-5} M, with 10^{-6} M T_3 having much less effect. The demonstration of a direct effect on physiological skin lightening would be more convincing at concentrations that more nearly approach the physiological range (see White and Nicoll, this volume). Outside of the TH–cholesterol pellet implantations, little evidence exists concerning TH effects on morphological pigmentary changes.

The epidermal melanin unit is observed as cytocrine melanin is transferred from epidermal melanocytes to other epidermal cells. In *R. pi-*

---

stages XX and XXII. (C) Early branching. (D) Fusiform morphology, characteristic of larval epidermal melanophores up to stage XX, and also seen frequently between stages XXI and XXIV. (E) Early branching, seen after stage XXI. (F) Dendritic morphology, characteristic of epidermal melanocytes of juveniles and adults. Reproduced from Smith-Gill, 1971.

piens the larval epidermal melanocytes are fusiform in morphology, and transform to a dendritic morphology during metamorphic climax (Smith-Gill, 1971) (Fig. 9). Although the predominant morphological form of the larval epidermal melanocyte therefore differs from that of the adult, both morphological forms are apparently able to transfer cytocrine melanin (Bagnara and Hadley, personal communication). The morphogenesis of the functional epidermal melanin unit has not been described.

An interesting variation has been described in the metamorphic changes in the pigmentary system of tree frogs as exemplified by *P. dacnicolor*. The morphogenesis of the adult pigmentation during metamorphosis includes the loss of the overlying epidermal melanophores and subsequently the absence of the epidermal melanin unit in the adult. On the other hand, the dermal chromatophore unit has the same components as other amphibians. To form the dermal chromatophore unit, iridophores become punctate, xanthophores become positioned just below the basement membrane, and dermal melanophores move from the upper dermis to below the iridophores (Bagnara *et al.*, 1978b).

5.1.2. Pigment Synthesis

Expression of pigmentary pattern is dependent on pigment synthesis. This is exemplified by the melanoid phenotype in axolotls, which is controlled by the recessive melanoid gene and which can be induced in normal larvae by allopurinol (Bagnara *et al.*, 1978a; Frost and Bagnara, 1979a). In melanoid animals, pteridine synthesis is altered with concomitant reductions in xanthophores and iridophores and an increase in melanophores. The spotted or mottled phenotype of adults is thereby obscured, and the animal is characteristically a charcoal gray instead of a pattern on the usual dark-olive background.

Pigment synthesis appears to depend on a number of extrinsic factors. While some evidence exists that the pigment type of the propigment cells may be determined at the time of migration, specific tissue environments have often proved necessary for the differentiation of any particular chromatophore type. Furthermore, the chromatophore types may be naturally interconvertible (discussed below). Ectodermal associations are necessary for melanogenesis; the mesoderm also has melanin-promoting capacity (reviewed by: Bagnara and Hadley, 1973; Bagnara *et al.*, 1979).

Recently, a model describing the differentiation of all the vertebrate chromatophore types from a single pluripotent chromatoblast stem cell type has been presented (Bagnara *et al.*, 1979) (Fig. 10). An important component of this model is that propigment cells may not be determined to be any specific chromatophore type at the time of migration from the neural crest but rather that interactions with the local tissue and endocrine environment may be important in determining the differentiated expression at any given time. All pigment cells are proposed to contain a "primordial organelle" with a propensity to differentiate preferentially

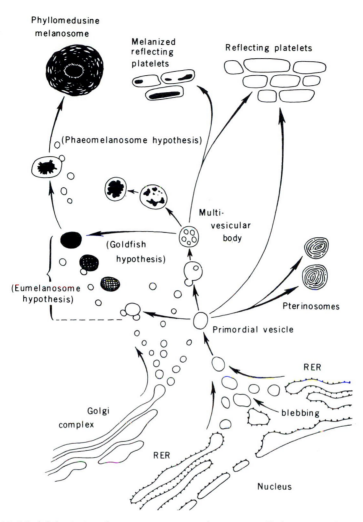

FIGURE 10. Model depicting the common origin of pigment cells from a single pluri-potent chromatoblast, and the common origin of pigment granules from a primordial organelle derived from the endoplasmic reticulum. Reproduced with permission from Bagnara *et al.*, 1979, copyright 1979, The American Association for the Advancement of Science.

into melanosomes, but also capable of differentiating into pterinosomes or refractosomes, depending on substrate availability. For example, it has been proposed that reduced xanthine dehydrogenase activity in the melanoid mutant, or following allopurinol treatment of wild-type axolotl larvae, blocks purine and pteridine metabolism, thereby passively enhancing melanophore differentiation (Frost and Bagnara, 1979a,b). That xanthophores and iridophores are potentially interconvertible has been demonstrated recently through a series of intriguing *in vitro* experiments from Ide's laboratory (reviewed by Bagnara *et al.*, 1979). The surprising discovery of mosaic pigment cells containing organelles characteristic of more than one type of pigment cell suggests that such multiple differentiations may occur naturally in intact animals (reviewed by Bagnara *et al.*, 1979).

The regulation of purine, pteridine, and melanin synthetic pathways may be interrelated, and low-molecular-weight compounds such as the purines and pteridines themselves, including pigment precursors, may act as regulatory molecules in pigment cell development (Frost and Bagnara, 1979a,b). These interrelationships remain largely unexplored, and present exciting possibilities for future investigations into the developmental regulation of cell differentiation.

An important role for hormones in the regulation of pigment cell differentiation is implied, although direct evidence is lacking. For instance, in hypophysectomized animals (i.e., lacking MSH) presumptive melanoblasts may differentiate as iridophores. Hypophysectomized larvae have fewer but more punctate melanophores and more but stellate iridophores; this is the expected physiological response when no MSH is present. If MSH is administered to these hypophysectomized larvae, the melanophores differentiate in areas where iridophores had been (for discussion see Bagnara and Hadley, 1973). Thus, MSH may play a role in modulating alternate pathways of pigment synthesis.

Few specific pigment changes associated with metamorphosis have been reported in any biochemical detail, although the differentiation of many new chromatophores accompanies metamorphosis. The melanin content of *R. heckscherii* skin almost doubles during metamorphosis, and skin tyrosinase increases severalfold at the completion of spontaneous metamorphosis. Thyroxin treatment elevates larval skin tyrosinase. During spontaneous metamorphosis, tyrosinase does not increase until tail resorption is nearly complete, long after climax when many of the pigmentary changes are occurring (Bennett and Frieden, 1962). Similarly, measurements of soluble tyrosinase and dopa oxidase activity in skin of metamorphosing *R. pipiens* fail to reveal any correlation between enzyme activity and stages of proposed melanin synthesis. Significant increases in activity are observed, however, in postclimax stages, after the completion of proposed melanophore differentiation, and activity continues to rise with juvenile development (Smith-Gill, 1971). Unfortunately, bound enzyme activity was not measured in either study. Since control of melanogenesis may involve translocation of the enzyme to the premelanosome (Bagnara *et al.*, 1979), any study of melanogenesis during metamorphosis must include measures of melanosome-bound as well as soluble tyrosinase or dopa oxidase activity if a functional relationship between enzyme levels and melanin synthesis is to be established.

The Mexican leaf frog, *P. dacnicolor*, has an unusual type of melanosome, the development of which has been reported only recently. The larva has a mature melanosome containing eumelanin pigment. At the end of metamorphosis a red pteridine dimer, pterorhodin, is deposited on the larval melanosome as a fibrous material to form a compound melanosome (Bagnara *et al.*, 1978b).

The close coordination of the morphological and biochemical changes in the *P. dacnicolor* melanosome with metamorphosis suggests that pterorhodin deposition may be hormonally mediated. Attempts to induce localized pterorhodin synthesis with T_4 implants have been reported on a limited but relatively unsuccessful scale (Frost, 1978). However, hypophysectomy followed by T_4 administration induced, in a single case, pterorhodin synthesis. Furthermore, skin of animals at or older than Taylor and Kollros stage XX will produce pterorhodin after a period of time, but isolated skin of younger animals, even in the presence of T_4 fails to produce pterorhodin (Ide, as cited in Frost, 1978). Apparently the pigment cells must be in some way preprimed by another hormone or factor before TH can exert its effect. Once the cells have been primed, TH may serve to coordinate several effects. The evidence for a role of TH in initiating pterorhodin synthesis is circumstantial but provocative. It is interesting that the epidermal melanophores, which apparently contain the same melanin pigment as the dermal melanophores, do not undergo a similar transition to a compound melanosome (Bagnara *et al.*, 1978), suggesting cell-specific responses to the stimuli initiating pterorhodin synthesis.

Qualitative and semiquantitative changes in pterorhodin and its precursors have been studied in *P. dacnicolor*. For instance, biopterin, a presumptive precursor to pterorhodin, is present as early as stage I but is not converted to pigment until completion of metamorphosis. Another intermediate, 7-methylxanthopterin, is present after stage XX, but pterorhodin is produced only at stage XXV and later (Frost, 1978; Frost and Bagnara, 1979b). The immediate factors that induce and mediate pterorhodin synthesis are not clearly established.

Pleurodeles blue, a pigment absent in young larvae of certain newts, is detected in the skin of older larvae but disappears at metamorphosis. Bagnara has demonstrated that thyroxin implants into larval skin induce a decrease in the concentration of Pleurodeles blue in the area of the implant (reviewed in Bagnara, 1976). Since this pigment is not found in the adult, the decrease in local concentration in the implant area is rather convincing evidence of TH action.

5.1.3. Conclusions

During metamorphosis, chromatophore densities, associations, and morphologies typical of adult pigmentary pattern develop. Complex interactions among pigment cells, their biochemical products, and their tissue and endocrine environments mediate pattern formation and pigment synthesis. Morphological transitions include migrations and rearrangements of existing chromatophores, mitosis of chromatoblasts and existing differentiated chromatophores, and differentiation of new chromatophores. Pigment synthesis and chromatophore differentiation are essential to pattern expression, and may involve the production of new pigments or the development of new organelles. Specific pigment classes may also be degraded.

Both TH and MSH have been suggested to be important in the development of the amphibian skin pigmentary pattern. Much of the evidence is circumstantial and is based on the temporal association of pigment changes with metamorphosis. Outside the thyroxin-implant experiments, there is little direct evidence for a role of TH in morphological chromatophore changes accompanying metamorphosis. Even in those experiments, the possibility of a relatively nonspecific mitogenic effect has not been excluded. The role of TH in the metamorphic development of the pigmentary system remains largely unexplored.

5.2. Nonpigmentary Changes

Many studies have reported the morphological changes in the integument during embryonic development and metamorphosis of amphibians (reviewed by Fox, this volume). The tadpole skin is protective, secretory, and respiratory in function, and with metamorphosis there are changes in cell proliferation, differentiation, and function. Some of the biochemical changes in skin that occur during metamorphosis are treated below.

5.2.1. Sodium Transport

The process of osmoregulation has been studied in adult frogs, while that in larval amphibians has received somewhat less attention. Tadpole gills are the primary regulators of Na^+ and Cl^- transport, but at metamorphosis this function must be assumed by other organs, such as the skin and the kidney (White and Nicoll, 1979). Recently PRL has been shown to be important to osmoregulation in the premetamorphic animal and to the climactic skin (Eddy, 1978, 1979; discussed in Eddy and Allen, 1979).

TH may have direct action on the integument via the development of the active-transport system. Although Na^+/K^+ ATPase activity is closely related to active sodium transport in frog skin, the development of Na^+/K^+ ATPase activity earlier than potential difference after immersion in T_4 suggests that other factors are required for complete development of the active sodium transport system. TH modulation of secretion of other hormones known to be involved in active sodium transport (i.e., PRL and aldosterone) cannot be excluded.

Transepidermal active sodium transport is a well-known phenomenon in frogs; however, tadpoles apparently do not actively transport Na^+ until late during metamorphosis. The active transport of Na^+ can be detected through the *in vitro* measurement of the short circuit current (SSC) or transepithelial potential difference (discussed by Eddy and Allen,

1979), measured between the outside mucosal and the inside serosal surfaces. This potential difference during metamorphosis develops such that stage XX animals have established a potential difference approaching that of *R. catesbeiana* adult ventral skin (Taylor and Barker, 1965). Furthermore, immersion of the larvae in T_4 (100 ng/ml) for 12 days also resulted in the development of the transepidermal potential difference. Though the effect was achieved with T_4, the long lag period may indicate that the hormone's effect on sodium transport is an indirect one.

The ventral skin of premetamorphic tadpoles has only modest quantities of Na^+/K^+ ATPase, and the lack of active transport of sodium in these tadpoles may be related to this deficiency. Direct measurement of Na^+/K^+ ATPase in T_4-treated *R. catesbeiana* tadpole skins has shown that adult levels of the enzyme are present as early as 4 days after T_4 treatment is initiated (Kawada *et al.*, 1972). Since development of the potential difference requires a longer time interval (12 days), other skin factors are apparently required for functional sodium transport (Taylor and Barker, 1965).

In vitro demonstration that PRL is responsible for the SSC current has also been made in *R. catesbeiana.* Isolated ventral skin of stage XXV animals responds to exogenous PRL with a net increase in SSC. Since the possibility of antidiuretic hormone (ADH) contamination in the PRL preparation was excluded and since the highest serum levels of PRL occur at stage XXIV, PRL appears to be specifically responsible for the *in vitro* increase in SSC (Eddy and Allen, 1979). A PRL-like factor is also important to premetamorphic osmoregulation, as has been demonstrated by decreases in Na^+ and H_2O content in other tissues of *R. catesbeiana* treated with PRL (Eddy, 1979). That PRL is present during early development and functions in osmoregulation is consistent with findings in other lower vertebrates (Nicoll, 1974). However, radioimmunoassay of premetamorphic levels of serum PRL has not been reported. Again, those cells responding to PRL at the end of metamorphosis might be newly divided or differentiated cells or cells with newly available receptors. Since exposure to T_4 *in vivo* produces increases in SSC only after a prolonged lag period (Taylor and Barker, 1965), it is possible that T_4 in some way modulates PRL production or availability. Endogenous T_4 would of course be available during the above experiments (Eddy and Allen, 1979) with spontaneously metamorphosing ventral skin exposed *in vitro* to various PRL treatments, but T_4 is not responsible for the observed effect since *in vitro* T_4 treatment does not stimulate active transport (Barker and Taylor, 1965). In any event, PRL is apparently involved in sodium transport in the ventral skin during metamorphic climax (Eddy and Allen, 1979).

A role for PRL in osmoregulation has also been proposed for urodeles and *Xenopus* (as discussed in Platt and Christopher, 1977). In these cases PRL increases water uptake or turnover, but the effect has not been specifically localized to the skin.

5.2.2. Keratinization

During the process of keratinization, a heterogeneous group of cellular proteins forms a keratin complex via sulfhydryl, hydrogen, and polar bonds (Spearman, 1977). While it has been known for some time that keratinization is initiated in amphibians at the time of metamorphosis (see Fox, this volume), demonstration that a specific protein is produced for the keratin complex has only recently been made. Reeves (1977) has described a system that utilizes an antibody to one keratin protein from *Xenopus* to identify the newly synthesized adult-type protein that first appears during metamorphosis. Larval skin also can be precociously induced by T_3, *in vivo* and *in vitro*, to produce that keratin protein. Furthermore, Reeves was able to isolate poly-A-containing messenger which produced immunoprecipitable keratin-type protein when translated *in vitro*. The *in vitro* demonstration of the same process substantiates the role for T_3 in the induction of the protein. However, the response varied with the age of the skin, and Reeves suggested that other factors may be important in bringing about the maturation of the skin. These factors may well be hormonal.

5.2.3. DNA Synthesis

$[^3H]$-Tdr incorporation into DNA in *R. pipiens* tadpole shank epidermis has been used to study cell population phenomena in spontaneous and T_4-induced metamorphosis. During spontaneous metamorphosis the outer layer of the shank epidermis ceases to divide, while the basal cells continue to proliferate (Wright, 1977). T_4 stimulates DNA synthesis in all layers of the epidermis (outer, intermediate, and basal) 48 hr after T_4 immersion of stage IX–XI tadpoles, but the later responses of the individual cells, while in the same direction, are not identical. Cells in all layers of the epidermis proliferate; subsequently, the outer epidermal cells virtually cease dividing as they differentiate into the stratum corneum, but some cells of the intermediate and basal layers remain in the proliferative pool. The data from spontaneous and T_4-induced metamorphosis are essentially in agreement. Wright has suggested that these data indicate that both the G_1 phase of the cell cycle the proliferative time have been shortened. A similar shortening of the G_1 phase has been reported for *Pleurodeles waltii* larvae treated with T_4 (as discussed in Wright, 1977).

5.2.3. Conclusion

While the proliferation of epidermal cells, Na^+/K^+ ATPase sunthesis, and at least one protein involved in keratinization appear to respond to TH, it is clear that other functions of the skin are influenced by PRL, such as osmoregulation. Development of the active sodium transport system and regulation of the system once it has developed may be mediated by different hormones. Only TH have been implicated in the development. Other hormones, e.g., PRL, ADH, and fluorocortisone (a synthetic mineralocorticoid), are known to stimulate transepithelial potential difference in the SCC once the system has developed (Eddy and Allen, 1979, and unpublished).

6. THE EYE

By early posthatching stages, the amphibian eye has already assumed histological structure characteristic of the adult (Wilt, 1959; detailed background on the anatomy and visual physiology of the amphibian eye can be found in several recent reviews: Gordon and Hood, 1976; Grüsser-Cornehls and Himstedt, 1976; Keating and Kennard, 1976) (Fig. 11). During larval development and metamorphosis, dramatic morphological changes take place in the accessory ocular structures. The external and internal larval corneas fuse to form the single cornea characteristic of the adult. The nictitating membrane, upper and lower eyelids, and conjunctival sacs form from epidermal folds. The extrinsic ocular muscles increase in size and shift in position. Intraorbitol implantation of TH in a cholesterol pellet causes precocious development of all these changes and a bulging of the eyes through apparent local, direct hormone action (Kaltenbach, 1953). The corneal reflex also develops at the onset of metamorphic climax, and in anurans its development is TH sensitive (see Kollros, this volume). The eye itself increases in size through growth and mitosis in all layers, but overall morphological organization of the lens

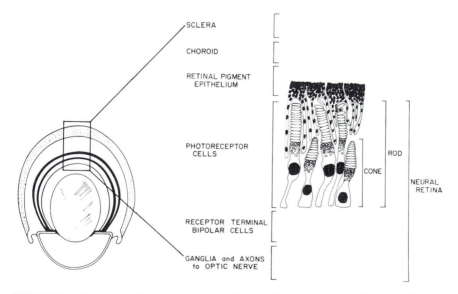

FIGURE 11. Diagrammatic representation of the cell associations in the amphibian eye.

and retina does not change. However, at the biochemical level several dramatic transitions occur, including changes in the lens crystallins and the visual pigments. This discussion considers the biochemical basis of lens and retinal growth, and the metamorphically correlated biochemical transitions in the lens, neural retina, and retinal pigmented epithelium.

6.1. The Lens

6.1.1. Lens Growth

Between stages X and XXV, the diameter and wet weight of *R. catesbeiana* lenses increase exponentially, and dry weight increases linearly. Overall growth, therefore, is due to both cellular growth and increases in protein, as well as hydration. Immersion of premetamorphic tadpoles in thyroxin mimics all these changes. The increase in hydration is most rapid between stages XXII and XXIV, corresponding to the stage when tadpoles first swim with their eyes above water. Since the refractive index increases as a result of increased hydration, these thyroxin-induced changes may be interpreted as an adaptation to the terrestrial environment (Polansky and Bennett, 1973).

Of the hormones that play a role in lens growth by cell division, those of the thyroid may have the most direct effect, although pituitary hormones also have been implicated as important mitogens in the adult lens. Implantation studies utilizing thyroxin pellets have shown that the hormone stimulates [³H]-Tdr incorporation and cell division in the lens epithelium of the treated eye in premetamorphic tadpoles of both *R. catesbeiana* and *R. clamitans* (Kaltenbach and Hobbs, 1972). Since the hormone was localized by a cholesterol pellet and the contralateral control eye with cholesterol only showed no increase in [³H]-Tdr incorporation, the response is thought to be a direct, peripheral response to thyroxin. *In vitro* demonstration of T_4 or T_3 induction of mitosis in tadpole lens epithelium has not been reported.

Even though the frequency of mitosis in the cells of the adult lens epithelium varies with season, a clear demonstration that the adult tissue retains any normal responsiveness to TH remains to be made (Worgul and Rothstein, 1974). Some pituitary factor appears to stimulate adult lens cell mitosis since the lens epithelium of hypophysectomized animals ceases to divide several weeks after pituitary removal. However, reconstitution of hypophysectomized animals with either pituitary extract, TSH, or growth hormone has been reported for each to correlate with increased [³H]-Tdr incorporation and cell division, which controls cell migration and possibly fiber formation (Hayden and Rothstein, 1978; Van Buskirk, 1977; Worgul and Rothstein, 1974). Further *in vitro* study is required to determine whether the above factors act directly or indirectly and whether they act as general growth stimulators or have cell-specific mitogenic effects.

6.1.2. Crystallin Transitions

Changes in the crystallin patterns have been demonstrated during metamorphic development of *X. laevis*. In amphibians, as for other vertebrates, γ-crystallins are characteristic of fiber development. The concentration of γ-crystallins decreases during metamorphosis, and one major band disappears at metamorphic climax. The concentration of α- and β-crystallins increases dramatically (Clayton, 1970; discussed in Doyle and Maclean, 1978).

Chemical thyroidectomy by immersion of *X. laevis* larvae in propylthiouracil does not retard the development of the adult crystallin pattern. Comparison of thyroidectomized larvae grown in crowded (growth inhibiting) and uncrowded (growth promoting, allowing development into giants) conditions indicated that the degree of developmental transition is correlated with larval growth and the associated increase in lens diameter (Doyle and Maclean, 1978). Since this species is aquatic as an adult, changes in the refractive index to accommodate air vision would not be necessary, and these results support the hypothesis (Clayton, 1970) that developmental transitions in crystallin proteins maintain the appropriate refractile properties as the lens increases in diameter. However, the mechanism of control has not been determined, and it may involve any of the endocrine factors influencing lens growth discussed above.

Quantitative and qualitative differences in the lens crystallin proteins have also been observed between tadpoles and adult *R. catesbeiana*. Like *X. laevis*, these changes do not appear to be thyroxin dependent, for immersion of tadpoles in thyroxine does not alter the electrophoretic patterns. In contrast to *X. laevis*, the shifts in the soluble lens protein profiles occur postmetamorphically (Polansky and Bennett, 1973). Since the emergence from water at metamorphic climax is not accompanied by changes in the crystallins, the pattern of development in *R. catesbeiana* further supports Clayton's (1970) hypothesis concerning the role of crystallins and suggests the following interpretation for functional changes in the lens at metamorphosis. The change in refractive index needed to accommodate the transition from aquatic to air vision is accomplished by a rapid hydration at metamorphic climax, probably coordinated by TH. As the juvenile frog grows and the lenses increase in diameter, later postmetamorphic changes in crystallin proteins may serve to maintain the new refractive index of the adult eye. The critical experiments to test this hypothesis and to determine the control mechanisms involved have not been reported.

6.2. The Retina

The morphology of the frog retina is like that of other vertebrate retinas and is composed of three nuclear layers and two plexiform layers (Fig. 11). The photoreceptor cells, the rods and cones, are located toward

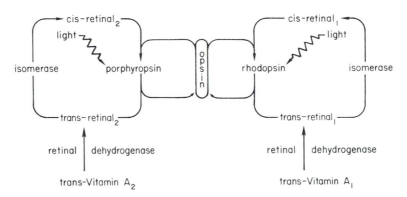

FIGURE 12. Photopigment interconversion. The *cis* isomer of retinal can react with opsin to produce rhodopsin or porphyropsin. Vision is initiated by the absorption of light by the photopigment, which isomerizes the molecule to the *trans* configuration. Following isomerization, the photopigment undergoes conformational changes to produce a series of photoproducts that yield *trans*-retinal. *trans*-Retinal may be recycled.

the back of the eye and abut against the absorptive retinal pigmented epithelium (RPE), which contains endogenous melanin pigment. The neural layers are toward the front of the eye; only light penetrating the neural layers is perceived by the photoreceptive cells (Gordon and Hood, 1976).

The rods and cones interdigitate with long strands of the RPE. The outer segments (OS) of the rods and cones contain the photopigments. The OS have the appearance of vertically stacked, membrane-bound discs, which form by invaginations of the plasma membrane. The discs of the OS continually turn over, with older discs moving up and eventually pinching off to detach from the OS and to be absorbed by the RPE. Invaginations of the plasma membrane at the base of the OS form new discs.

During larval development, the pigmented and neural retinas grow and cooperate in retinal disc phagocytosis and in the production of the larval and adult forms of the photopigments (Fig. 12). In addition, functional synaptic relationships of the neural retina develop, as reflected in electrical function, dendritic morphology, ultrastructural organization of the synapses, and remodeling of the axons and myelin sheaths of the optic nerve (reviewed by Keating and Kennard, 1976).

6.2.1. Cell Proliferation in the Neural Retina

During spontaneous metamorphosis, the retinal layers increase in size and stainability, primarily through increases in rod length, thickening of the inner plexiform layer, and mitosis of cells in the inner nuclear layer. In *X. laevis*, the retina develops a dorsal–ventral polarity that correlates with patterns of mitosis; in another anuran, *R. pipiens*, this correlation does not hold (Keating and Kennard, 1976). In early premeta-morphic *Xenopus* [^{3}H]-Tdr is incorporated into mitotic figures in approximately equal numbers of cells in the dorsal and ventral retinas, but between stages 54 and 63 mitosis in the ventral retina exceeds that in the dorsal retina. By metamorphic climax (stages 61–65), approxi-mately 10-fold more ventral cells have incorporated [^{3}H]-Tdr and entered mitosis than dorsal cells (Beach and Jacobson, 1979a). This difference is a function of both increased ventral retinal cell proliferation and de-creased dorsal retinal cell proliferation rather than a change in the length of the cell cycle. The transient increase in ventral labeled cells during prometamorphosis may indicate that ventral retina cells are sensitive to low levels of TH.

As early as stage 50, thyroxin injected into the vitreous humor of *Xenopus* eyes directly stimulates local precocious incorporation of [^{3}H]-Tdr into the retinal cells, similar to the incorporation seen during spontaneous metamorphosis. Dorsal retinal cells show a transient in-crease in [^{3}H]-Tdr incorporation 3 days after exogenous thyroxin treatment, but after 6 days, ventral retinal cells incorporate six-fold more [^{3}H]-Tdr than dorsal retinal cells. After metamorphosis, dorsal retinal cells do not respond to thyroxin. Ventral retinal cells in the same animals demon-strate a transient increase in [^{3}H]-Tdr incorporation (Beach and Jacobson, 1979b). The proliferation of the cells of the ventral retina may be associated with the population of cells that will produce new visual pigments (reviewed below).

The differential growth pattern in the retina might be expected to result in mismatching of retinotectal connections, which are necessary for normal image processing. However, incorporation of [^{3}H]-Tdr (mul-tiple, spaced injections) during the histogenesis of the retina and studies of the development of the tectum have suggested that retinal and tectal growth are complementary. Furthermore, matching is made possible by shifting of some nasal retinal fibers caudally to make appropriate con-nections (Longley, 1978). A role for hormones in coordinating these re-tinotectal connections is presumed but has not been experimentally de-fined.

6.2.2. The RPE and Retinal-Disc Shedding

Cytologically, the RPE is characterized by increasing basophilia as metamorphosis proceeds, correlating with increasing functional activity (Ohtsu *et al.*, 1964). The RPE cooperates with the neural retina by se-questering vitamin A_1 or converting A_1 to A_2 for the production of pho-topigments by the retina (Kinney and Fisher, 1978b; Reuter *et al.*, 1971). The RPE is necessary for photopigment regeneration by the rods; without RPE, rod pigment will not regenerate (reviewed by Gordon and Hood, 1976). In addition, the RPE functions to keep the area between the RPE and the ROS (retinal outer segments) free of debris by phagocytosing egg melanin released from the developing neural retina a well as older discs. The RPE is capable of phagocytosis during embryogenesis, and of endogenous melanin synthesis by the time the feeding stage is attained (Kinney and Fisher, 1978a), presumably the first time that vision is useful to the animal.

In *Xenopus*, the photoreceptor system has recently been studied by careful quantification of [³H]leucine incorporation into the ROS, COS (cone outer segments), RPE during various stages of development. [³H]-Leucine is spread diffusely in the COS, whereas that in the ROS is displaced outwards toward the sclera with time. The ROS grow at a constant rate per day until metamorphosis, when there is a dramatic increase in the rate of disc shedding in the ROS and/or phagocytosis by the RPE. Such studies have demonstrated that these events have the net result of shortening the length of the ROS without gross histological change in the cellular organization of the rods of RPE (Kinney and Fisher, 1978b). Although these events correlate with the onset of metamorphosis, it is yet to be established whether increased shedding is due to changes in T_3, T_4, pituitary or pineal hormones.

6.2.3. Photopigment Conversion

Vertebrate rods contain two systems of photopigments: a red photosensitive rhodopsin, in cycle with vitamin A_1; and a purple photopigment, porphyropsin, in a similar cycle with vitamin A_2. Each photopigment is a conjugated protein, opsin, containing as the prosthetic group retinal$_1$ or retinal$_2$, the aldehydes of vitamins A_1 and A_2, respectively (Fig. 12). The porphyropsin–A_2 system is considered to be evolutionarily more primitive and is characteristic of freshwater fish, while the rhodopsin–A_1 system is characteristic of marine fish and terrestrial forms (for a review of the evolution of photopigments, see Wald, 1952). For most anurans, the aquatic larva utilizes the porphyropsin–A_2 system, while the terrestrial adult utilizes the rhodopsin–A_1 system (Wald, 1952; Wilt, 1959). *X. laevis*, a fully aquatic anuran, is unusual in that both tadpoles and adults use porphyropsin as the visual pigment (Bridges *et al.*, 1977).

The transition from porphyropsin to rhodopsin occurs during anuran metamorphosis, and the retina of a partially metamorphosed larva contains a mixture of the two types of pigments (Wald, 1952; Wilt, 1959; Crim, 1975a). TH has been shown to induce rhodopsin synthesis in *R. catesbeiana* tadpoles (Wilt, 1959; Ohtsu *et al.*, 1964). Surprisingly, the TH-induced rhodopsin synthesis in *R. catesbeiana* is not antagonized by PRL, despite the demonstration of peripheral TH–PRL antagonisms in other tissues (Crim, 1975a).

The larval isomerase and 3,4-retinal dehydrogenase are not specific for retinal$_2$; furthermore retinal$_1$ may combine with tadpole opsin as well as with the usual adult opsin. These observations indicate that it is the prosthetic group rather than the enzyme specificity that changes during eye metamorphosis (Ohtsu *et al.*, 1964). Reuter *et al.* (1971) have suggested that vitamin A 3,4-dehydrogenase may be sequestered in the RPE, and thus only that form of vitamin A provided by the RPE is available for photopigment production However, there are some differences in vitamin A metabolism between tadpoles and adults. The eye of premetamorphic *R. catesbeiana* rapidly sequesters radioactive vitamin A_1 injected intraperitoneally, and this ability declines as metamorphosis proceeds. There is no difference between larvae and adults in the sequestering of vitamin A_2. Furthermore, isolated larval eyes have the ability to convert vitamin A_1 to $retinal_2$, but this ability declines with metamorphic development. Experiments utilizing radioactive vitamin A_1 precursors indicated that retinal rather than vitamin A is probably the direct intermediate in the conversion, suggesting the possibility of thyroxine-mediated changes in enzyme activity (Ohtsu *et al.*, 1964.

Since the adult bullfrog retina was shown to contain 30–40% porphyropsin sequestered in the dorsal zone of the retina (Reuter *et al.*, 1971), the presence of rhodopsin in the ventral zone may be associated with the visual sensitivity of the bullfrog, which positions itself with its eyes partially above water. If the vitamin A form donated by the RPE determines the form of visual pigment produced in the retina (Reuter *et al.*, 1971), one would also expect to find 3,4-retinal dehydrogenase exclusively in the dorsal zone. Reuter *et al.* (1971) were further able to demonstrate this point when they showed that *R. catesbeiana* dorsal neural retina (usually containing porphyropsin) generated rhodopsin when placed on the ventral RPE (usually adjacent to neural retina in which rhodopsin is found) and vice versa.

Notophthalmus viridescens is an interesting paradox of hormone-induced changes of photopigments. While the aquatic larval form has porphyropsin, the terrestrial eft has primarily thyroxin-induced rhodopsin, and the aquatic adult has primarily porphyropsin (Wald, 1952). PRL, which initiates the water drive in the second metamorphosis of the eft, has been shown to induce 3,4-dehydroretinal and subsequent formation of porphyropsin in the adult. Indeed, PRL induces porphyropsin synthesis irrespective of aquatic or terrestrial environment. The California newt (*Taricha torosa*), which does not undergo a second metamorphosis, has only A_1 in the retina. This condition is not altered by either PRL or thyroidectomy plus PRL treatments (Crim, 1975b). The European newts *Triturus cristatus* and *T. alperstris* also undergo a second metamorphosis and have pigment alternations similar to *Notophthalmus*. Additionally, they have a seasonal change in photopigments (reviewed in Grüsser-Cornehls and Himstedt, 1976).

Whether TH or other hormones subsequently induce a particular visual pigment during metamorphosis is therefore a function of prior evolutionary selection. Depending upon the life history of the specific organism, TH may serve as convenient devices to coordinate the change in larval photopigment to the adult photopigment.

7. SUMMARY AND CONCLUSIONS

We have reviewed several tissues that undergo dramatic biochemical transitions during metamorphosis. In many cases, such as the photopigments of the eye and the urea-cycle enzymes of the liver, transitions in specific biochemical functions can be directly related to the physiological transition from an aquatic to a terrestrial environment.

The tissues reviewed here present several contrasting patterns of cellular population dynamics accompanying the biochemical transitions. On the one hand, the eye increases in size with modest morphological reorganization. The integument undergoes extensive morphological reorganization at metamorphosis, with accompanying changes in cell associations, pigment cell patterns, and various physiological functions. In contrast, the internal kidney and the derivatives of the gut undergo cellular turnover. In the case of the kidney, an entire organ, the pronephros, degenerates. There is a dramatic turnover of the cells of the gut epithelium. There is also turnover of cells of the liver, although the extent and significance of this turnover require further definition. In all cases where it has been studied, however, some cell division accompanies the metamorphosis of each of these tissues, and the relationship of DNA synthesis and cell division to the biochemical transitions remains to be defined.

Many of the biochemical changes accompanying metamorphosis have been utilized as model systems for the study of TH action. Each of the organ systems so approached, however, is a heterogeneous population of cells with many biochemical changes occurring simultaneously, such that definition of the exact events leading from TH stimulation to a specific biochemical event is difficult. Despite extensive investigations, the mechanisms of TH action on these organ systems thus remains to be precisely defined (see Frieden, this volume). Many investigators favor the hypothesis of direct gene action as the mechanism of TH-induced changes in biochemical function. Yet only in two cases, keratin synthesis in the skin and CPS-I synthesis in the liver, has even a quantitative change in mRNA accumulation been demonstrated in association with metamorphosis and TH action, and the level of TH control of this accumulation requires further definition. To date, TH-induced synthesis of a new class of mRNA has not been demonstrated.

An additional complicating factor in the understanding of biochemical transitions during metamorphosis is incomplete definition of hormonal control. There are three necessary criteria that must be met *in vivo* to demonstrate specific hormonal control over a process: (1) addition of excess hormone should elicit a specific response; (2) removal of hormone should abolish the response; and (3) the normal state should be reproducible by removal of the hormone source(s) plus replacement therapy with physiological doses of hormone(s). Definitive demonstration of specific hormonal control requires *in vitro* studies, where the hormonal stimulus can be precisely controlled. These criteria have been met for few, if any, of the biochemical transitions reviewed in this chapter. Furthermore, TH treatment alone, whether by injection or immersion, seems insufficient to bring about all the changes seen during spontaneous metamorphosis. Examples of metamorphic changes reviewed above that cannot be completely induced by TH treatment of an intact larva include changes in the intestine, pigmentary changes, and lens crystallin transitions. A major feature of TH-induced metamorphosis not seen in spontaneous metamorphosis is the apparently cyclical increase in RNA, DNA, and protein synthesis in the liver.

Possible sources for the differences between spontaneous and thyroxin-induced metamorphosis are severalfold. First, a process that normally takes several months to several years, depending on the species and environmental conditions, is compacted into a week to 10 days. The juveniles resulting from such a process are often not normal, as the timing of developmental events both within and among tissues is upset. Second, the response of the tissue is likely to reflect the general hormonal milieu. Most experiments on induced metamorphosis have been performed on intact larvae at stages ranging from premetamorphosis to late prometamorphosis. Yet, undoubtedly both the cellular responsiveness and the tissue environment are very different than at later stages when climax changes normally occur. The normal metamorphic response may depend on interaction with other hormones. Even with TH-induced metamorphosis, the final biochemical indices of differentiation often do not occur until several days after hormone administration and could, in fact, represent secondary responses to the hormone. The relationship between the early events and later events remains to be established.

While the effects observed after administration of exogenous hormone are more likely to reveal information about primary actions of the hormone than long-term effects during spontaneous development, they are less likely to provide real insight into the normal developmental process and hormonal regulation thereof. It is desirable to understand both. The molecular mechanism of action of TH cannot be approached until the hormonal-specificity of the response has been established. The regulation of normal development during metamorphosis undoubtedly requires interaction of TH with many other hormones. These interactions must be defined before metamorphosis will be understood (for a further discussion of this topic, see Frieden, this volume).

ACKNOWLEDGMENTS. We would like to thank our colleagues who have critically read the manuscript, including Joseph Bagnara, Frank Carver, Michael Doyle, Lynne Eddy, Doug Gill, Harry Lipner, and Ted Williams. We are grateful to Joseph Bagnara for providing original illustrations used for Figs. 7 and 10. Unpublished data on R. pipiens are from a study supported in part by grants to S. J. Smith-Gill from the National Institutes of Health, AM 20518, and from the General Research Board of the University of Maryland. Preparation of this review was also supported in part by NIH Grant 01236 to E. Frieden.

REFERENCES

Ashley, H., Katti, P., and Frieden, E., 1968, *Dev. Biol.* **17**:293.

Atkinson, B. G., 1971, in: *Relationships of Endocrines to Growth and Development* (R. P. Breitenbach and A. D. Kenney, eds.), pp. 48–82, University of Missouri Press, Columbia.

Atkinson, B. G., Atkinson, K. H., Just, J. J., and Frieden, E., 1972, *Dev. Biol.* **29**:162.

Bagnara, J. T., 1976, in: *Physiology of the Amphibia*, Volume III (B. Lofts, ed.), pp. 1–52, Academic Press, New York.

Bagnara, J. T., and Hadley, M. E., 1973, *Chromatophores and Color Changes*, Prentice–Hall, Englewood Cliffs, N.J.

Bagnara, J. T., Taylor, J. D., and Hadley, M. E., 1968, *J. Cell Biol.* **38**:67.

Bagnara, J. T., Frost, S. K., and Matsumoto, J., 1978a, *Am. Zool.* **18**:301.

Bagnara, J. T., Ferris, W., Turner, W. A., Jr., and Taylor, J. D., 1978b, *Dev. Biol.* **64**:149.

Bagnara, J. T., Matsumoto, J., Ferris, W., Frost, S. K., Turner, W. A., Jr., Tchen, T. T., and Taylor, J. D., 1979, *Science* **203**:410.

Balinsky, J. B., 1970, in: *Comparative Biochemistry of Nitrogen* Volume 2 (J. W. Campbell, ed.), pp. 519–637, Academic Press, New York.

Balinsky, J. B., and Baldwin, E., 1962, *Biochem. J.* **82**:187.

Barker, S. B., and Taylor, R. E., Jr., 1965, in: *Current Topics in Thyroid Research* (C. Cassaon and M. Andreoli, eds.), Academic Press, New York.

Beach, D. H., and Jacobson, M., 1979a, *J. Comp. Neurol.* **183**:603.

Beach, D. H., and Jacobson, M., 1979b, *J. Comp. Neurol.* **183**:615.

Beckingham-Smith, K., and Tata, J. R., 1976, in: *The Developmental Biology of Plants and Animals* (C. F. Graham and P. F. Wareing, eds.), pp. 232–247, Saunders, Philadelphia.

Bennett, T. P., and Frieden, E., 1962, in: *Comparative Biochemistry*, Volume IV (M. Florkin and H. S. Mason, eds.), pp. 483–556, Academic Press, New York.

Bennett, T. P., and Glenn, J. S., 1970, *Dev. Biol.* **22**:535.

Bennett, T. P., Glenn, J. S., and Sheldon, H., 1970, *Dev. Biol.* **22**:232.

Berven, K. A., Gill, D. E., and Smith-Gill, S. J., 1979, *Evolution* **33**:609.

Bonneville, M. A., 1963, *J. Cell Biol.* **18**:579.

Bridges, C. D. B., Hollyfield, J. G., Witkovsky, P., and Gallin, E., 1977, *Exp. Eye Res.* **24**:7.

Butman, B. T., Obika, M., Tchen, T., and Taylor, J. E., 1979, *J. Exp. Zool.* **208**:17.

Carver, V. H., and Frieden, E., 1977, *Gen. Comp. Endocrinol.* **31**:202.

Clayton, R. M., 1970, *Curr. Top. Dev. Biol.* **5**:115.

Cohen, P. P., 1970, *Science* **168**:533.

Cohen, P. P., Brucker, R. F., and Morris, S. M., 1978, in: *Hormonal Proteins and Peptides* (C. H. Li, ed.), pp. 273–381, Academic Press, New York.

Crim, J. W., 1975a, *J. Exp. Zool.* **192**:355.

Crim, J. W., 1975b, *Gen. Comp. Endocrinol.* **26**:233.

Dauca, M., and Hourdry, J., 1977, *Wilhelm Roux Arch. Entwicklungsmech. Org.* **183**:119.

Dauca, M., and Hourdry, J., 1978a, *Biol. Cell.* **31**:277.

Dauca, M., and Hourdry, J., 1978b, *Biol. Cell.* **33**:85.

Dauca, M., and Hourdry, J., 1978c, *C. R. Acad. Sci. Ser. D* **286**:973.

Dodd, M. H. I., and Dodd, J. M., 1976, in: *Physiology of the Amphibia*, Volume III (B. Lofts, ed.), pp. 467–599, Academic Press, New York.

Doyle, M. J., and Maclean, N., 1978, *J. Embryol. Exp. Morphol.* **46**:215.

Eaton, J. E., and Frieden, E., 1969, *Gen. Comp. Endocrinol. Suppl.* **2**:398.

Eddy, L. J., 1978, *Fed. Proc. Fed. Am. Soc. Exp. Biol.* **37**:621.

Eddy, L. J., 1979, *Gen. Comp. Endocrinol.* **37**:369.

Eddy, L. J., and Allen, R. F., 1979, *Gen. Comp. Endocrinol.* **38**:360.

Fischer, M. S., and Cohen, P. P., 1974, *Dev. Biol.* **36**:357.

Frost, S., 1978, Developmental aspects of pigmentation of the Mexican leaf frog, *Pachymedusa dacnicolor*, Ph.D. thesis, University of Arizona.

Frost, S. K., and Bagnara, J. T., 1979a, *J. Exp. Zool.* **209**:455.

Frost, S. K., and Bagnara, J. T., 1979b, *Pigment Cell* **4**:in press.

Gordon, J., and Hood, D. C., 1976, in: *The Amphibian Visual System, A Multidisciplinary Approach* (K. V. Fite, ed.), pp. 29–86, Academic Press, New York.

Griswold, M. D., Fischer, M. S., and Cohen, P. P., 1972, *Proc. Natl. Acad. Sci. USA* **69**:1486.

Grüsser-Cornehls, U., and Himstedt, W., 1976, in: *The Amphibian Visual System, A Multidisciplinary Approach* (K. V. Fite, ed.), pp. 203–266, Academic Press, New York.

Guardibassi, A., Olivero, M., Campantico, R., Rinaudo, M. T., Giunta, C., and Bruno, R., 1970, *Gen. Comp. Endocrinol.* **14**:148.

Hadley, M. E., 1966, Cytophysiological studies on the chromatophores of *Rana pipiens*, Ph.D. thesis, Brown University.

Hagenbuchle, O., Schibler, U., and Weber, R., 1976, *Mol. Cell. Endocrinol.* **4**:61.

Hanke, W., and Leist, K. H., 1971, *Gen. Comp. Endocrinol.* **16**:137.

Hayden, J. H., and Rothstein, H., 1978, *J. Cell Biol.* **79**:19a.

Hourdry, J., 1977, *C. R. Acad. Sci. Ser. D* **284**:2023.

Hourdry, J., and Dauca, M., 1977, *Int. Rev. Cytol. Suppl.* **5**:337.

Hourdry, J., Chabot, J. G., Mendard, D., and Hugon, J. S., 1978, *Biol. Cell.* **32**:17a.

Kaltenbach, J. C., 1953, *J. Exp. Zool.* **122**:41.

Kaltenbach, J. C., and Hobbs, A. W., 1972, *J. Exp. Zool.* **179**:157.

Kaltenbach, J. C., Lipson, M. J., and Wang, C. H. K., 1977, *J. Exp. Zool.* **202**:103.

Kawada, J., Taylor, R. E., Jr., and Barker, S. B., 1972, *Endocrinol. Jpn.* **19**:53.

Keating, M. J., and Kennard, D., 1976, in: *The Amphibian Visual System, A Multidisciplinary Approach* (K. V. Fite, ed.), pp. 267–315, Academic Press, New York.

Kinney, M. S., and Fisher, S. K., 1978a, *Proc. R. Soc. London Ser. B* **210**:149.

Kinney, M. S., and Fisher, S. K., 1978b, *Proc. R. Soc. London Ser. B* **201**:169.

Kistler, A., and Weber, R., 1975, *Mol. Cell. Endocrinol.* **2**:216.

Kistler, A., Yoshizato, K., and Frieden, E., 1975, *Dev. Biol.* **46**:151.

Kollros, J. J., 1961, *Am. Zool.* **1**:107.

Longley, A., 1978, *J. Embryol. Exp. Morphol.* **45**:249.

McAvoy, J. W., and Dixon, K. E., 1977, *J. Exp. Zool.* **202**:129.

Marshall, J. A., and Dixon, K. E., 1978, *J. Exp. Zool.* **203**:31.

Matsumoto, J., 1965, *J. Cell Biol.* **27**:493.

Menter, D. G., Obika, M., Tchen, T. T., and Taylor, J. D., 1979, *J. Morphol.* **160**:103.

Mori, M., Morris, S. M., Jr., and Cohen, P. P., 1979, *Proc. Natl. Acad. Sci. USA* **76**:3179.

Morris, S. M., and Cole, D. R., 1978, *Dev. Biol.* **62**:52.

Morris, S. M., and Cole, D. R., 1980, *Dev. Biol.* **74**:379.

Nakagawa, H., Kim, K., and Cohen, P. P., 1967, *J. Biol. Chem.* **242**:635.

Nicoll, C. S., 1974, in: *Handbook of Physiology*, Section 7, Volume IV, Part 2 (E. Knobil and W. H. Sawyer, eds.), American Physiological Society, Washington, D.C., pp. 253–292.

Ohtsu, J., Naito, K., and Wilt, F. H., 1964, *Dev. Biol.* **10**:216.

Paik, W. K., and Baserga, R., 1971, *Exp. Cell Res.* **64**:190.

Pehlemann, F. W., 1972, in: *Pigmentation: Its Genesis and Biologic Control* (V. Riley, ed.), Appleton–Century–Crofts, New York, pp. 295–305.

Platt, J. E., and Christopher, M. A., 1977, *Gen. Comp. Endocrinol.* **31**:243.

Polansky, J. R., and Bennett, J. P., 1973, *Dev. Biol.* **33**:380.

Pouyet, J. C., and Hourdry, J., 1977, *Biol. Cell.* **29**:123.

Reeves, R., 1977, *Dev. Biol.* **60**:163.

Reuter, T. E., White, R. H., and Wald, G., 1971, *J. Gen. Physiol.* **58**:351.

Samuels, H. H., 1978, in: *Receptors and Hormone Action*, Volume III (L. Birnbaumer and B. W. O'Malley, eds.), pp. 35–74, Academic Press, New York.

Smith-Gill, S. J., 1971, A genetic and developmental analysis of the cytophysiology of the disruptive pigmentary patterning in *Rana pipiens*, Ph.D. thesis, University of Michigan.

Smith-Gill, S, J., 1973, *J. Morphol.* **140**:271.

Smith-Gill, S. J., 1974, *Dev. Biol.* **37**:153.

Smith-Gill, S. J., 1979, *Dev. Growth Differ.* **21**:289.

Smith-Gill, S. J., Reilly, J. G., and Weber, E. M., 1979, *Dev. Growth Differ.* **21**:279.

Spearman, R. I. C., 1977, *Symp. Zool. Soc. London* **39**:335.

Tata, J. R., 1971, in: *Hormones and Development* (M. Hamburgh and E. J. W. Barrington, eds.), pp. 19–40, Appleton–Century–Crofts, New York.

Tata, J. R., 1975, *Nature (London)* **257**:18.

Taylor, R. E., Jr., and Barker, S. B., 1965, *Science* **148**:1612.

Van Buskirk, R., 1977, *Experientia* **33**:438.

VanDenbos, G., and Frieden, E., 1976, *J. Biol. Chem.* **251**:4111.

Volpe, E. P., 1964, *J. Exp. Zool.* **157**:179.

Wald, G., 1952, in: *Modern Trends in Physiology and Biochemistry*, (E. S. G. Barron, ed.), pp. 337–376, Academic Press, New York.

White, B. A., and Nicoll, C. S., 1979, *Science* **204**:851.

Wilt, F. J., 1959, *Devel. Biol.* **1**:199.

Wixom, R. L., Reddy, M. K., and Cohen, P. P., 1972, *J. Biol. Chem.* **247**:3684.

Worgul, B. V., and Rothstein, H., 1974, *Cell Tissue Kinet.* **7**:415.

Wright, M. L., 1977, *J. Exp. Zool.* **202**:223.

Yeh, K. Y., and Moog, F., 1978, *Growth* **42**:495.

CHAPTER 16

The Dual Role of Thyroid Hormones in Vertebrate Development and Calorigenesis

EARL FRIEDEN

1. INTRODUCTION

Few areas of biological research have been more frustrating than the efforts to derive a rational explanation of the mechanism of action of the thyroid hormone(s) (TH). A huge and bewildering array of observations has been published and has generated more confusion than clarification. The difficulty of successfully integrating these data has been compounded by the use of a wide range of TH dosages, from physiological (10^{-10} to 10^{-8} M) to pharmacological (up to 10^{-6} M) to toxic (>10^{-6} M); by the variation in animal condition, from completely athyroid to euthyroid to hyperthyroid; and by the use of systems varying from pure enzymes to whole animals, frequently on variable diets. The object of this chapter will be to resolve the actions of TH into two major roles and trace their development in terms of the metabolic needs and developmental requirements of all classes of vertebrates with particular emphasis on their role in amphibian metamorphosis (Shellabarger, 1964). Our major thesis will be as follows:

EARL FRIEDEN · Department of Chemistry, Florida State University, Tallahassee, Florida 32306.

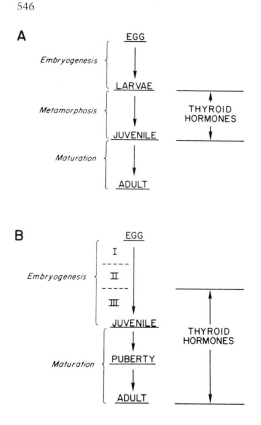

FIGURE 1. The effect of thyroid hormones on the developmental sequence in (A) a typical amphibian and (B) a typical mammal.

1. The role of TH was first manifested in most poikilothermic lower vertebrates as a hormone controlling essential aspects of differentiation, development, and maturation. The most crucial role of TH appeared in the transition from aquatic (larva) to terrestrial (juvenile) form in amphibians (Frieden and Just, 1970). All these responses are probably initiated in the nucleus, providing an irreversible imprint on genetic expression. The role of TH in the developmental sequence of a typical amphibian is shown in Fig. 1A.

2. Higher vertebrates have retained this sensitivity to TH in their developmental history. It is reflected in the key role of TH in brain, nerve, connective tissue, and endocrine maturation. Even in the postdevelopmental period of the mammal, the nucleus may require TH for the direction of certain highly sensitive "permissive" responses. This provides a baseline metabolic and endocrinological effect that serves as a prerequisite for late fatal development, growth, and maturation, and normal metabolism (Table 1). The effect of TH on the development of a typical higher mammal is illustrated in Fig. 1B.

In the homeotherm, in birds and mammals, TH became involved in temperature adaptation. In general, the homeotherm has at its disposal two methods of thermoregulation—physical methods concerned with the rate at which heat is exchanged by the animal and chemical methods

TABLE 1. THYROID HORMONE EFFECTS ON
THE DEVELOPMENT OF VERTEBRATE SYSTEMS

Animal	Developmental effect
Fish (teleosts)	Maturation of gonads, bone, CNS, skin pigments (metamorphosis)
Amphibians	Metamorphosis of larval forms (also intermediate stages)
Reptiles	Possible effects on embryo, hatching, growth
Birds	Development of embryo, bone, feathers, gonads
Mammals	Differentiation: Late fetal stages. Maturation of bone, brain, gonads

involving the rate at which heat is produced by the tissues. TH provides a major control of heat production in the homeothermic vertebrates. These calorigenic effects may be initiated in the nucleus, but are eventually reflected in mitochondrial function and seem to be relatively reversible.

2. THE ROLE OF IODINE IN INVERTEBRATES

Is it correct to ignore the possible role of TH in invertebrates? Organic iodine is found even in plants and, more commonly, in marine invertebrates. Virtually all of the larger phyla of invertebrates, except echinoderms and protozoans, synthesize iodoproteins to some extent. While further studies may yet identify TH as more than a curiosity in invertebrate systems, the presence of organic iodine does not establish the physiological presence and function of TH. Through the extensive work of Frantz, Morrison, Taurog, and their co-workers, we now know of several common enzymes that readily iodinate proteins. All that is needed is iodine, hydrogen peroxide, a peroxidase, and a tyrosine-containing protein:

$$\text{protein} - \langle \text{C}_6\text{H}_4 \rangle - OH + 2I^- + H_2O_2 \xrightarrow{\text{peroxidase}} \text{protein} - \langle \text{C}_6\text{H}_2\text{I}_2 \rangle - OH + 2H_2O$$

While diiodotyrosyl residues are the principal product, mild conditions can also yield some iodothyronyl peptides. Thus, organic iodination does not require the highly specialized machinery of a thyroid gland, and the presence of organic iodine in marine and other forms may be a fortuitous combination of favorable enzymatic events.

The most primitive system in which thyroxine (T_4) appears to be functional has been described by Spangenberg (1974). She has reported that TH (T_4 at 10^{-7} M) induces differentiation in an invertebrate coelenterate, the jellyfish *Aurelia aurite*. Jellyfish polyps, preconditioned at 19°C are induced to strobilate by iodide and certain organic iodine compounds. During strobilation (equivalent to metamorphosis), jellyfish polyps (strobilae) become segmented and evolve into several young medusae (ephyra) in a typical sequential order. Strobilating jellyfish synthesize organic iodine compounds, including T_4, which are concentrated at the distal differentiating regions. Goitrogenic agents (e.g., propylthiouracil, thiourea, KSCN) prevent the induction of strobilation by reducing iodide uptake and impairing synthesis of iodinated organic compounds. T_4 was found after 24–28 hr only in strobilating forms of *Aurelia*, suggesting that T_4 is involved primarily in the differentiation of new structures. Spangenberg's data constitute the most convincing evidence of an authentic role of organic iodine in the development of an invertebrate. All other credible reports, even those related to the protochordate tunicates, are confined to the vertebrates.

2.1. The Mystique of Iodine

How essential is iodine to TH? For almost a century, the presence of iodine has dominated considerations of TH structure and function. When it was found that replacement of the 3'-iodine of the thyronines by an alkyl group, e.g., methyl or isopropyl (Fig. 2), produced analogs with equal or greater thyromimetic activity, no serious challenge to the halogen mystique seemed to be involved. It was assumed that the 3 and 5 positions had to be filled by halogens, particularly iodine or bromine, in order to satisfy the minimum electronic and structural requirements for thyromimetic activity.

Then Jorgensen, Block, and their co-workers synthesized several halogen-free derivatives of 3,5-dimethyl-L-thyronine (Fig. 2). These alkyl thyronines have been shown to have significant thyromimetic activity

TABLE 2. THYROMIMETIC ACTIVITY OF METHYLTHYRONINES[a,b]

Thyronine	Tail decrease	Tadpole urea excretion	Disc regression	Rat antigoiter
L-T_4	100%	100%	100%	100%
TMe$_2$	<0.2	<0.2	0.08	<0.5
TMe$_3$	15	15	0.2	3
TMe$_4$	15	15	0.6	2
3'-iPr-TMe$_2$	25	20	3.0	18
T_0	<0.1	<0.1	<0.1	<0.1
T_2	5	6	1.3	5
L-T_3	1000	1000	900	600
DL-MB-T_3	1200	900	500	300
MB-Tetrac	150	250	90	—

[a] Data from Frieden and Yoshizato (1974).
[b] See Fig. 2 for representative structures of the thyronines.

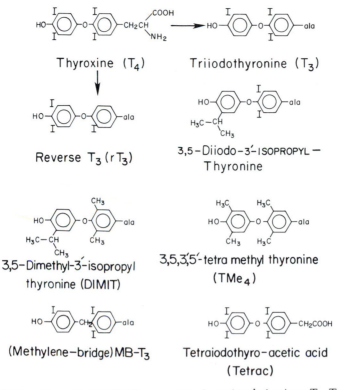

Thyroxine (T$_4$)   Triiodothyronine (T$_3$)

Reverse T$_3$ (rT$_3$)

3,5–Diiodo–3'–ISOPROPYL –
Thyronine

3,5–Dimethyl-3'-isopropyl
thyronine (DIMIT)

3,5,3',5'-tetra methyl thyronine
(TMe$_4$)

(Methylene–bridge) MB–T$_3$

Tetraiodothyro–acetic acid
(Tetrac)

FIGURE 2. Structure of key naturally occurring thyronine derivatives, T$_4$, T$_3$, rT$_3$, Tetrac, and important analogs including two halogen-free thyronines.

in the rat (Jorgensen, 1978) and bullfrog tadpole (Frieden and Yoshizato, 1974), especially 3,5-dimethyl-3'-isopropylthyronine, which had 20–25% of the activity of T$_4$. As shown in Table 2, other methylthyronines have appreciable TH activity. Thus, there is no absolute requirement for iodine or any halogen (or the ether oxygen) for TH activity in amphibians or mammals. The essential chemical features required for significant TH activity need to be redefined as a specialized central lipophilic core, provided by the thyronine moiety, sterically constrained by bulky 3,5–3' substituents with two specific anionic groups located at the distal ends of the molecule (Fig. 3).

These new data challenge two other important ideas related to TH activity. The comparable activities of the T$_3$ and T$_4$ analogs trimethyl-thyronine (TMe$_3$) and tetramethylthyronine (TMe$_4$; Table 2) argues against the proposal that T$_3$ is the active hormone produced by the metabolic 5'-deiodination of the prohormone, T$_4$. The very high activity of the methylene-bridge (MB) analogs (Table 2) also rules out the hypothesis proposed by Niemann and colleagues pertaining to the requirement for

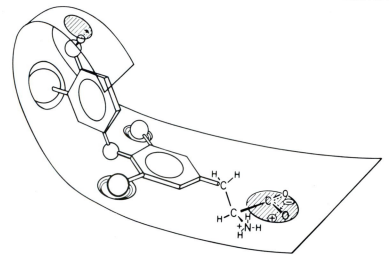

FIGURE 3. Three-dimensional representation of the steric and functional features of a hormonally active 3,5,5'-trisubstituted thyronine binding to the nuclear receptor. From Jorgensen and Andrea, 1976, p. 306.

the thyronine phenolic groups to undergo reversible oxidation to a stable, more useful quinoid form.

The presence of iodine in TH probably reflects favorable evolutionary circumstances in the origin of compounds with TH activity. The unique contribution of iodine to the thyroid hormones now appears to be due to its biosynthetic convenience. Since early animals originated in the sea, they had ready access to iodide, for which early trapping mechanisms were developed. The presence of iodide along with the H_2O_2–peroxidase systems afforded a convenient mechanism for introducing iodine *ortho* to the phenolic group of tyrosine in or out of proteins to form mono- and diiodotyrosines. Ample proteases were available for the hydrolytic release of iodothyronines from their peptide linkages. Specific binding proteins, e.g., thyronine-binding globulin (TBG), prealbumin, evolved to favor transport of these iodo amino acids. The ability to preferentially trap iodine permitted reutilization of the iodine after metabolic deiodination, further contributing to the economy of TH biosynthesis. All these circumstances combined to favor the biosynthesis of T_4 and T_3 with their unique geometry and structure.

3. THYROID HORMONE DOSE–RESPONSE RELATIONSHIPS

The recognition of the developmental and calorigenic roles of TH requires clarification of crucial dose–response relationships. Sensitivity to dose levels is a prime consideration in any endocrine system, particularly in the elucidation of hypo-, eu-, and hyperhormonal states. For TH there are the added complications of the occurrence of two distinct types of responses with different thresholds and extensive involvement of the hormone with pituitary function along with a correspondingly broad impact on the entire hormonal regulatory system. To clarify this situation we have delineated four TH dose–response categories.

1. *A developmental (subcalorigenic) range* (<0.5 μg T_4/100 g body wt, corresponding to a tissue concentration <10^{-8} M). This includes the maintenance of a minimal level of pituitary viability with a residual growth hormone (GH) secretion and peripheral tissue responsiveness to normal stimuli of growth and development, late prenatal and postdelivery effects in the mammal up to puberty and the premetamorphic or early prometamorphic period (possibly Taylor and Kollros stages X–XV) of the metamorphosing tadpole.

2. *A hypothyroid (pituitary support) range* (1–5 μg T_4/100 g, tissue concentration approximately 10^{-7} to 10^{-8} M). This refers to the TH dose necessary to establish normal pituitary trophic and other hormonal output in the thyroidectomized animal. This amount of TH suffices for the more sensitive calorigenic and feedback pituitary responses in addition to the specific developmental effects. It also reflects the level of the hormone achieved in the tadpole in late prometamorphosis or the early stages of metamorphic climax as the amphibian thyroid approaches maximal function.

3. *A normal regulatory (euthyroid) range* (5–10 μg T_4/100 g, approximately 10^{-7} M). This dose reflects the normal latitude in TH secretion, supportive of physiological and adaptive responses, centering around energy metabolism and thermogenesis. An intact endocrine system is also required for these functions.

4. *The hyperthyroid range* (>10 μg T_4/100 g, >10^{-7} M). This range varies from abnormally high basal metabolic rates and protein catabolism to frank thyrotoxicosis accompanied by numerous metabolic excesses and distortions of energy metabolism. Here we must emphasize that prolonged exposure to TH >10^{-6} M, either *in vivo* or *in vitro*, may not relate to normal hormonal function. Any claim that the effects induced by these pharmacological doses represent authentic hormone actions may be challenged.

The above four categories reflect the available data on dose–response relationships reported primarily for the rat and for the bullfrog tadpole. Evans *et al.* (1960) reported the minimum daily injection level of T_4 for the production of normal growth in thyroidectomized animals to be 0.5 μg T_4/100 g per day, known to be dependent on a minimal secretion of GH from the pituitary gland under TH stimulation. A dose up to 2 μg/100 g per day was required for a normal metabolic rate and erythropoietic and pituitary feedback functions (category 2 above). A clear experimental distinction was made between these two dose levels.

TH levels necessary for amphibian metamorphosis estimated from the data of Etkin (1968) and Yamamoto *et al.* (1966) suggest an inducing

dose of less than 10^{-8} M (in active tissues in a steady state) with metamorphic climax requiring a 5- to 10-fold increase in TH. Quantities of
T_3 or T_4 as low as 50 pg/g per day could induce metamorphosis in immature tiger salamander larvae.

The most sensitive range of TH concentrations, 10^{-8} to 10^{-9} M,
initiates an irreversible trigger action typical for developmental and maturational responses. It represents the hormone level necessary for the
early phases of metamorphosis in anurans and for proper development
in mammals of brain and pituitary cells and the maturation of nerve and
supportive tissues. Without this minimum TH at the appropriate time,
the tadpole remains a larva and the human infant becomes a cretin.

4. THYROID HORMONE RECEPTORS

The lowest range of TH concentration, which was first established
for a variety of in vivo biological responses, has been supported in recent
studies on T_3- and T_4-binding receptor proteins in the nuclei of rat liver
and tadpole cells. Dissociation constants, K_d, average about 10^{-10} M or
less. The receptor proteins, reported by DeGroot, Oppenheimer, and Samuels and their co-workers for rat liver nuclei and GH_1 (a rat pituitary
tumor cell line) nuclei, show the same number of sites for T_3 and T_4,
with the K_d for T_3 being one-tenth that for T_4. A possible distinctive type
of TH receptor has been observed in bullfrog tadpole liver and tail nuclei
by Kistler et al. (1975). The K_d values for T_3 and T_4 are comparable (10^{-10}
M). T_3 has twice the number of sites as T_4. In competition experiments,
T_3 competes with all T_4 sites, but T_4 competes with only about one-half
of the T_3 sites. Nuclei from tail cells at the onset of metamorphic climax
(stage XVIII–XIX) contained twice the number of binding sites as nuclei
from early prometamorphosis (stage X) (Yoshizato and Frieden, 1975).

5. THYROID HORMONE EFFECTS IN
POIKILOTHERMIC VERTEBRATES

Among the three major classes of poikilothermic vertebrates, fish,
amphibians, and reptiles, TH are frequently involved in numerous and
diverse developmental actions, but do not seem to be seriously concerned
with calorigenesis. The variability of reports of TH effects on the metabolism of poikilotherms may be understood better if dose levels are
taken into account as summarized in Table 2. For example, Lewis and
Frieden observed a calorigenic response to T_3 and T_4 in the bullfrog
tadpole, but this response required doses at least 10-fold greater than
those necessary to initiate metamorphosis. During spontaneous metamorphosis, Funkhouser has shown that oxygen consumption is significantly reduced. Packard (1976) reported small increases in mean oxygen
consumption of liver slices (29% in females, 11% in males) from adult
Rana pipiens injected with L-T_4 (2 μg/10 g body wt) daily for 5 days.

Thyroidectomy of the teleost parrot fish did not reduce oxygen consumption, nor did a modest dose of T_4 stimulate oxygen uptake.

In reptiles, ambient temperature is important in conditioning TH responses. At 21°C oxygen consumption of the lizard, *Anolis carolinensis*, is not affected by thyroidectomy, T_4, or TSH. But at 30°C, thyroidectomy reduces oxygen uptake and both T_4 and TSH stimulate oxygen utilization (Maher and Levedahl, 1959). Thus, in reptiles, elevated temperature may initiate susceptibility to a calorigenic response. In amphibians the induction of metamorphosis shows a remarkable temperature sensitivity. At 5–10°C virtually all anurans fail to respond to TH (Ashley *et al.*, 1968).

These data are merely representative examples of numerous and sometimes ambiguous reports that, in sum, suggest that the metabolic rate of poikilothermic vertebrates is independent of TH. However, under elevated ambient temperatures, or with pharmacological doses of TH, a calorigenic response can be elicited.

There is disagreement about the universality of metamorphiclike phenomena and the regulatory role of TH in the lower vertebrates, particularly fish. In their text Gorbman and Bern (1962) state that T_4-induced metamorphosis is not peculiar to the amphibians. Furthermore, Turner and Bagnara (1976) have noted that developmental changes (metamorphosis) in a teleost fish, the salmon, are accompanied by an increased metabolic rate and hyperactivity of the thyroid gland. These changes include transformation of the freshwater parr into the migratory marine smolt, which involves the silvering of the integument due to the deposition of guanine and electrolyte adaptations compatible to a marine existence. In contrast, the striking metamorphosis of the lamprey, a cyclostome, cannot be affected by administration of T_4 to the larvae.

The evolution of thyroidal function in fish has been traced by Sage (1973). Although the thyroid gland evolved from the gut, there is no evidence that TH functions as part of the gastrointestinal endocrine system or is involved in the control of glucose by the pancreatic islets. The control of the thyroid gland appears to have evolved from pituitary control of the gonads, suggesting an early role of TH in reproduction. This idea is supported by the presence of cycles of thyroid activity related to reproduction in both teleosts and elasmobranchs. Other effects of TH in teleosts, e.g., maturation of the nervous system and effects on behavior were acquired during the course of teleost evolution. The maturational effects of TH may have been extended to other morphogenetic actions on growth, development, metamorphosis, and on the integument. This may have marked the beginning of the multifunctional role of TH.

6. METAMORPHIC HORMONES IN AMPHIBIANS

The most spectacular role of TH occurs in amphibians, where it dominates the metamorphic transition from the aquatic larva to the juvenile frog. This process has become a model system for the study of comparative biochemistry, differentiation, and fetal-to-adult transitions. Within the phylum itself, particularly in the urodeles, there is a remarkable variety of responses; amphibian tissue provides a rich reservoir of biological systems that reveal a variety of hormonal responses to TH as well as to other endocrine components.

Best understood is the hormonal control of the thyroid gland in anurans, which has the same fundamental components as in mammals. A variety of stimuli can initiate neuroendocrine control in the hypothalamus, causing the release of TRH, which in turn stimulates the anterior pituitary to secrete TSH. TSH regulates the output of T_3 and T_4 by the thyroid gland. The ultimate response of the peripheral tissue depends on their receiving TH from the binding proteins in the serum and their sensitivity and capacity to respond. A modifying effect of prolactin in promoting growth phenomena and in delaying metamorphosis is discussed at length by White and Nicoll (this volume). Growth hormone has a similar but less acute effect.

Miyauchi et al. (1977) reported a spike of plasma T_3 and T_4 with a maximum at stage XX and declining at stage XXIV. Regard et al. (1978) also reported a sharp rise in T_3 and T_4, but the maximum levels were reported after stage XXII. The results of Mondou and Kaltenbach (1979) agree with those of Miyauchi et al. (1977) in that sharp elevations of plasma T_3 and T_4 are observed at stages XX and XXI. Cause and effect relationships suggest that the increase in plasma T_3 should precede its ability to produce climax. These data and their implications are considered at length in Chapter 11 on the hormonal control of amphibian metamorphosis.

Analytical difficulties have precluded the confirmation of the ebb and flow of the other relevant hormones such as TRH, TSH, PRL, and GH in even the most thoroughly studied amphibian order, Anura. The advent of specific radioimmunoassays and binding assays for TRH, TSH, etc. should fill this void in information.

While the hormonal complement appears to be complete in anurans, a variety of endocrine and metamorphic states have been reported for the other major amphibian order, Urodela (Fig. 4). Some urodeles bypass the aquatic larval or the terrestrial adult stages. These differences have culminated in neoteny, i.e., certain urodeles retain their larval body form throughout their entire life span; they become sexually mature and reproduce without a typical "adult" form. A defect can occur at any step in the metamorphic sequence of the urodeles. It is too soon to assess the role of hypothalamic hormones, although mammalian TRH has not been successful in inducing metamorphosis in any amphibian species. Pituitary inadequacy has been suggested for the neotenic state of the Mexican axolotl, *Ambystoma mexicanum*, which never undergoes metamorphosis in its normal habitat. When pituitary glands of *A. mexicanum* are transplanted into *A. tigrinum* (American axolotl), metamorphosis is not induced. In contrast, pituitary of the Mexican axolotl is ineffective or

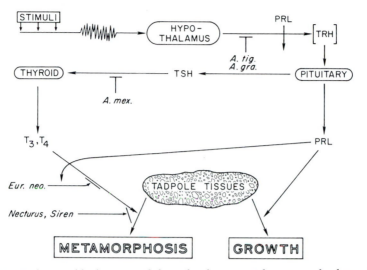

FIGURE 4. Endocrine blocks in amphibian development. The proposed scheme is based upon the current state of knowledge of the interactions of hormones affecting anuran metamorphosis. The long arrows designate the major axis and features of hypothalamo–pituitary–thyroid development. Small arrows and perpendicular lines represent experimentally demonstrated blocks in urodele development. Prolactin is shown both as promoting tissue growth and as interfering peripherally with the metamorphic action of thyroid hormones.

does not secrete a useful TSH. A block at the thyroid gland level appears to operate for salamanders such as *A. tigrinum* although considerable variability within this species is observed. Thyroid glands frequently are nonfunctional in many of the salamanders whose tissues are insensitive to TH.

Peripheral tissue response differences are found among many urodeles. In some salamanders (*Eurycea neotenes, E. tynerensis*), at least partial metamorphiclike responses can be forced by exposure to high dose of T_4 (10^{-5} to 10^{-6} M). Finally, a group of neotenic perennibranchiate salamanders (*Necturus, Siren, Amphiuma*) do not respond to any dose level of T_4. Obviously this variability in tissue response should be restudied with T_3 and some of its more active analogs.

7. THYROID HORMONE ACTION IN BIRDS

Birds comprise the other class of homeothermic vertebrates. It was hoped that birds might provide crucial information about the evolution of the role of TH in development and calorigenesis. However, the paucity of experimental information precludes this; only a few facts are available.

TH effects numerous developmental processes in birds. It is essential for gonadal maturation and for optimal growth and egg production. Molting and marked effects on feather growth are influenced by TH. Also it is one of a group of hormones controlling pigmentation in birds as well as in other vertebrates.

In some stages in the life of birds, temperature regulation is quite dependent on TH. In adults, thyroid secretion rate is inversely proportional to environmental temperature. Thermoregulatory processes involve TH, especially in the 0- to 7-day-old chick. T_3 and T_4 increase body temperature when the chick is held in a thermally neutral environment. TH reduces hypothermia arising from exposure to cold, but thyroidectomized chicks do not recover from hypothermia unless warmed; yet both the embryo and the adult chicken have been described as being refractory to injected TH.

8. METABOLISM AND CALORIGENESIS

TH regulate the metabolism of most of the adult bulk tissues—skeletal muscle, heart, liver, kidney—but do not normally affect the lungs, lymphatic system, gonads and accessory organs, nervous tissue, skin, smooth muscle, or thyroid gland in physiological doses. They have a depressant effect on the pituitary. In these affected tissues, T_3 and T_4 show no fundamental qualitative differences in their action, although T_3 acts sooner than T_4 and its effects are of shorter duration. This has been explained on the basis that T_3 is bound less firmly to serum proteins (especially the T_4-binding globulin) and is thus more rapidly released to tissues than is T_4. The metabolic effects of T_4 and T_3 are therefore referred to interchangeably. Though TH have a unique chemical structure, they induce numerous effects that may ultimately be initiated by a mechanism resembling that of other hormones. In common with GH, insulin, and adrenal glucocorticoids, TH have a general stimulatory metabolic effect on numerous tissues in the vertebrates (Frieden and Lipner, 1971).

Historically the action of TH in vertebrate postdevelopmental stages has been associated with its ability to increase oxygen utilization and stimulate calorigenesis in many key tissues and organs; in adult homeotherms its principal function has evolved toward thermogenesis. The rate of thyroid secretion is inversely related to the environmental temperature. Thyroidectomized animals are much more sensitive to the effects of cold; in human myxedema (thyroid deficiency in the adult), the heat-regulating mechanism is impaired. In fact, this calorigenic effect of TH is a major contributor to the maintenance of the body temperature of homeotherms in the 0–15°C range. An intact hypothalamo–pituitary system is also required for this temperature homeostasis. Cold induces increased TSH secretion and greater synthesis, secretion, and turnover of T_4 in the blood and tissues of the homeotherm.

The normal thyroid state of the animal provides the thermostatic setting for the metabolic rate. Other hormonal effects on metabolic rate are superimposed on this basal calorigenic level. However, in poikilo-

thermic animals such as amphibians, the responses to TH are not necessarily associated with calorigenesis, although an increase in oxygen
consumption can be induced with large doses of T_3 or T_4. In these animals,
the hormone induces and maintains the metamorphic changes leading
to maturation.

Calorigenesis does not account for all the effects of TH in restoring
the hypothyroid adult to the normal (euthyroid) state. Greater oxygen
utilization may be viewed as the result, rather than the cause, of these
metabolic events.

8.1. Hypo- and Hyperthyroidism

The essential role of TH in normal growth and development is demonstrated by the association of cretinism with thyroid deficiency. This
condition results from iodine shortage or congenital thyroid insufficiency
beginning at birth. The human cretin is a mental and physical dwarf with
severely inhibited growth and maturation of the skeletal and nervous
systems. The bones are infantile and abnormally calcified, and gonadal
development is juvenile. The cretin manifests mental retardation, apparently due to the lack of both brain development and normal nerve
growth. The basal metabolic rate is depressed by as much as 20–40% and
the growth rate is reduced to less than 10% of normal. The multiple
effects can be reversed by replacement therapy with T_3 or T_4, if administered sufficiently early in the growth period. Cretinism can be produced
experimentally by surgical or chemical thyroidectomy of postnatal animals.

In contrast to GH, TH will not stimulate growth beyond the normal
level. Although there is an upper limit in the growth response to TH,
excessive doses can produce an exaggerated effect on catabolic and oxidative metabolism with a large increase in oxygen consumption, a negative nitrogen balance, and a loss of weight. This is considered a thyrotoxic effect, distinct from the normal response. Certain diseases of thyroid
function reflect these thyrotoxic effects. Grave's disease is the most common form of thyrotoxicosis. It is characterized uniquely by a diffuse goiter
and exophthalmos, along with typical hyperthyroid symptoms of elevated
basal metabolic rate, bone demineralization, tachycardia, and negative
nitrogen balance. Other hyperthyroid disorders include toxic goiter, toxic
thyroid adenoma, and a variety of lesser characterized thyrotoxic states.

9. THYROID HORMONES AND DIFFERENTIATION

The induction of amphibian metamorphosis is a dramatic effect of
TH, illustrating its role in both differentiation and development. Typical
of amphibians, embryonic development results in a tadpole larval form
that differs greatly from the adult frog in both structure and habitat. This
transition of the tadpole to the frog during a discrete period of postem-

TABLE 3. BIOCHEMICAL SYSTEMS KNOWN TO BE EXTENSIVELY MODIFIED DURING ANURAN METAMORPHOSIS

Tissue, organ	Biochemical system	Effect	Role in metamorphosis
Respiration	Whole animal	No increase; decrease in some species	Calorigenic response not associated with metamorphosis
Liver	Urea production	Increase in urea-cycle enzymes	Transition from ammonotelism to ureotelism
	RNA and protein biosynthesis	Increased and different	T_3-mediated genetic expression via DNA
Tail	Hydrolytic enzymes of lysosomal type	Stimulation of the biosynthesis of cathepsin, phosphatase, RNase, DNase, β-glucuronidase, etc.	Accounts for tail resorption, adaptation to rapid movement
Blood, erythropoietic tissues	Erythrocytes: hemoglobin (Hb)	Repression of tadpole Hb, induction of frog Hb synthesis	Adaptive oxygen capture
	Serum protein biosynthesis (in liver)	Increase in serum albumin, ceruloplasmin, etc.	Improved homeostasis
Intestine	Hydrolytic enzymes	Increase in gut cells	Resorption of larval cells
	Proteases	Appearance of peptic activity in foregut	Digestion of animal tissue
Skin	Melanin	Increased synthesis	Protective coloring
	Na^+/K^+ ATPase	Increased activity	Maintenance of electrolyte balance
	Collagen	Change in deposition, breakdown	Adaptive skin properties
Pancreas	Enzyme, hormone secretion		Greater enzyme and hormone secretion
Acinar cells	DNA, RNA, proteins	Regression, then regrowth; increase in acid phosphatase, a new α-amylase	Preparation for carnivorous diet
Islet cells	Insulin	Stimulation of insulin secretion	
Eye	Light-sensitive pigments	Shift to rhodopsin	Repression of porphyropsin synthesis

bryonic change is known as metamorphic growth and maturation. Amphibian metamorphosis has become a model system for the study of comparative biochemistry and cellular differentiation, especially since it illustrates a series of remarkable biochemical and structural adaptations. In 1912, F. G. Gudernatsch showed that feeding thyroid glands to tadpoles initiated metamorphosis; this effect was eventually traced to T_4. Later, B. M. Allen demonstrated the pituitary control over the secretion of TH. More recently, W. Etkin and A. Voitkevitch showed that pituitary activity is in turn dependent on signals from the hypothalamus. If any of these tissues—thyroid, pituitary, or hypothalamus—is destroyed, normal metamorphosis is arrested. Metamorphosis can be induced by T_3, T_4, or TSH. These hormones can even induce similar transformations in amphibian species that do not metamorphose under normal conditions, e.g., the Mexican axolotl.

The extensive anatomical transformations during metamorphosis were described by biologists even before 1900. These changes reflect many millions of years of evolution. The modern bullfrog tadpole resembles a fish, with a streamlined body and tail, gills, lidless eyes, thin skin, and a mouth suited to feeding on aquatic plants; within a period of several months it loses its fishlike characteristics and changes to a young frog, a versatile land animal with lungs, eyelids, thick skin, large limbs with powerful muscles, and a mouth and tongue adapted for capturing insects.

The numerous metabolic and tissue systems that are extensively modified during anuran metamorphosis are summarized in Tables 3 and 4 (Frieden and Just, 1970). This compilation reflects the metabolic fate

TABLE 4. OTHER TISSUES MODIFIED DURING ANURAN
METAMORPHOSIS

Tissues, organ	Effect	Role in metamorphosis
Intestine	Extensive shortening; relocation, development of hydrolytic enzymes (e.g., pepsin in foregut)	Accommodation to diet
Limb buds	Development and growth of tissues (skin, nerve, etc.)	Locomotion on land
Skin	Collagenolysis; development of serous mucous glands; increase in serotonin, melanophores, Na transport	Protective adaption: color, etc.
Nervous	LM cells enlarged but fewer; initiation of corneal reflex; atrophy of Mauthner cells	Innervation of new structures
Reticuloendo-thelial system	Increase in lymphocytes and immune response	Improved immune mechanism
Bone	Stimulation of Ca transport and ossification	Formation of juvenile frog skeleton
Gills, kidney, spleen, etc.	Gills atrophy; other changes not defined	Preparation for land

of tissues that regress (e.g., gills, pancreas, tail) and those that undergo synthesis (e.g., limbs, muscle, lungs). Among the most vivid biochemical changes associated with metamorphosis are the transition from ammonotelism to ureotelism, a complete shift in hemoglobin population to lower oxygen-binding species, and a huge increase in the hydrolytic enzymes in tail tissues. These metabolic alterations occur spontaneously at metamorphic climax or they can be induced by 10^{-8} M T_3 or T_4 (Cohen et al., 1978).

As in the mammal, the basic initiating mechanism of TH action is under intensive investigation. The pattern may be similar to that of the steroid hormones, as Frieden and others have found strong evidence for T_3 and T_4 receptors in the nuclei of tadpole tail and liver cells. The T_3 receptors appear to increase during metamorphic climax and resemble many of the features of TH receptors from rat liver nuclei (Yoshizato and Frieden, 1975).

10. MECHANISM OF ACTION OF THYROID HORMONES

Hypotheses of TH action at the molecular level must encompass the following: (1) The developmental actions must be irreversible, while the calorigenic and many of the metabolic effects must be essentially reversible; (2) the action must be responsive to tissue hormone levels in the range 10^{-9} to 10^{-11} M T_3; (3) the action should show an order of responsiveness to analogs corresponding to their in vivo effects; (4) an essential role should be provided for the high affinity and low saturability of specific nuclear receptors; and (5) the response shoud be anabolic, permitting the cascade of metabolic responses that fit the 12- to 48-hr lag period and the multiplicity of TH effects. The breadth of TH effects is so great that there may not be a single unified mechanism of action. Sterling, Dratman, Edelman, and others have enumerated five possible mechanisms that may account for TH action at the cellular level (Table 5).

TABLE 5. PROPOSED MECHANISMS OF
THYROID HORMONE ACTION

1. Nuclear transcription; protein synthesis
2. Mitochondrial activation
3. Plasma membrane effects
   Amino acid transport
   Na^+/K^+ ATPase (sodium pump)
   Adenyl cyclase $\rightarrow$ cyclic AMP
4. Adrenergic receptor pathway
5. Incorporation into tyrosine pathways

10.1. Control of Nuclear Transcription and Protein Synthesis

The most widely accepted mechanism for the response to TH is the stimulation of specific mRNA and protein synthesis by the control of nuclear DNA-dependent RNA transcription (Fig. 5), as is well documented for several steroid hormone receptors. Extensive proof for the existence of powerful nuclear receptors in mammals and amphibians is at hand (Oppenheimer, 1979). However, T_3 and T_4 may be unique in that no significant cytoplasmic receptors or translocation reactions have been found. Nevertheless, there is a large body of circumstantial evidence that suggests that the basic unit of TH action is the T_3–nuclear receptor complex. The findings compatible with a receptor role for a nonhistone nuclear protein include the limited capacity and specific high affinity of binding, correlation between distribution and TH-responsive tissues, agreement between nuclear binding and thyromimetic effect, and correlation between nuclear occupancy and biological activity. An increase in nuclear activity after T_3 treatment has been shown in several predictable reactions including the formation of poly-A RNA, RNA polymerase activity, and in a few cases specific mRNA products, e.g., pituitary GH and hepatic α_2-μ-globulin. In the nucleus the T_3 receptor–protein(s) complex initiates a permanent activation of an unexpressed portion of the chromatin. This activation of transcription results in a number of events leading to specific mRNAs that migrate to the cytoplasm to code

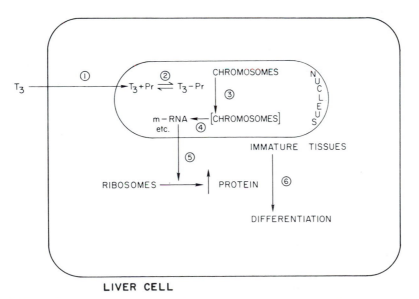

LIVER CELL

FIGURE 5. Schematic representation of the early events proposed for the action of thyroid hormones. Pr is the specific nuclear receptor protein that binds the thyroid hormones.

for new proteins, these events preceding the intracellular changes responsible for differentiation or changes in cellular activity. Except for the absence of significant cytosol receptors and the translocation step, this mechanism follows the same steps as observed for the steroid hormones.

10.2. Alternative Mechanisms of Thyroid Hormone Action

Several other possible actions of TH have been proposed in recent years (Sterling, 1979). These hypotheses provide reasonable extranuclear explanations for the reversible metabolic and calorigenic effects of T_3 and T_4, but do not offer much hope in accounting for developmental or differentiation responses. It is conceivable that a combination of these basic mechanisms might be involved in the action of TH. The four additional mechanisms will be considered in the order listed in Table 5.

The crucial role of TH in oxidative metabolism has focused on the mitochondria as a subcellular locus of hormone action. Mitochondrial oxidation and phosphorylation to ATP are normally tightly coupled. T_4 proved to be a physiological uncoupling agent, lowering the P:O ratio and increasing heat production. TH also shows a slow anabolic effect on mitochondrial protein synthesis and the number of respiratory assemblies. These effects have been difficult to demonstrate *in vitro* except when excessive levels of T_3 and T_4 are used. Moreover, the demonstration of an adequate number of mitochondrial T_3-receptor sites has not been possible.

The cell membrane has been proposed as a site of action of TH on certain transport processes. Increased intracellular uptake of amino acids has been observed in TH-sensitive cells, e.g., embryonic chick bone, rat cartilage. In regressing tissue such as tadpole tail cells, T_3 inhibited the uptake of certain amino acids.

The stimulation of the sodium pump in the cell membrane by T_3 and T_4 has been well documented. This is due to the increased activity of Na^+/K^+ ATPase in the membrane arising from an increase in the number of pump units, reflecting enhanced nuclear transcription. Since the sodium pump accounts for 40% of the oxygen consumption in many normal mammalian tissues, this effect offers a convenient rationale for calorigenesis.

cAMP mechanisms have also been proposed for several isolated actions of TH, but not for its crucial effects on calorigenesis or development.

There is considerable overlap and summation between the activities of TH and the catecholamines, which affect β-adrenergic receptors. T_3 does increase the number of β-adrenergic receptors. However, whenever the individual effects are dissected out, it seems likely that the two hormones have their own separate intrinsic metabolic actions and that the two may be additive in various thyroid states.

A direct involvement in the numerous branch points of tyrosine metabolism would provide for a significant role of T_3 and T_4 in protein and catecholamine biosynthesis. However, the trace amounts of hormone involved, especially T_3, seem to preclude a commanding role of T_3 in directly influencing a highly ubiquitous species such as tyrosine.

ACKNOWLEDGMENT.   This work was supported by NIH Grant HD-01236.

REFERENCES

Ashley, H., Katti, P., and Frieden, E., 1968, *Dev. Biol.* **17**:293.

Cohen, P. P., Brucken, R. F., and Morris, S. M., 1978, in: *Hormonal Proteins and Peptides*, Volume VI (C. H. Li, ed.), pp. 273–382, Academic Press, New York.

Etkin, W., 1968, in: *Metamorphosis: A Problem in Developmental Biology* (W. Etkin and L. I. Gilbert, eds.), p. 313, Appleton–Century–Crofts, New York.

Evans, E. S., Rosenberg, L. L., and Simpson, M. E., 1960, *Endocrinology* **66**:433.

Frieden, E., and Just, J. J., 1970, in: *Biochemical Actions of Hormones* (G. Litwack, ed.), p. 1–52, Academic Press, New York.

Frieden, E., and Lipner, H., 1971, *Biochemical Endocrinology of the Vertebrates*, Prentice–Hall, Englewood Cliffs, N.J.

Frieden, E., and Yoshizato, K., 1974, *Endocrinology* **95**:188.

Gorbman, A., and Bern, H. A., 1962, *A Textbook of Comparative Endocrinology*, Wiley, New York.

Jorgensen, E. C., 1978, in: *Hormonal Proteins and Peptides*, Volume VI (C. H. Li, ed.), pp. 108–204, Academic Press, New York.

Jorgensen, E. C., and Andrea, T. A., 1976, *Excerpta Med. Found. Int. Congr. Ser.* **361**:303.

Kistler, A., Yoshizato, K., and Frieden, E., 1975, *Endocrinology* **97**:1025.

Maher, M. J., and Levedahl, B., 1959, *J. Exp. Zool.* **140**:169.

Miyauchi, H., LaRochelle, Jr., F. T., Suzuki, M., Freeman, M., and Frieden, E., 1977, *Gen. Comp. Endocrinol.* **33**:254.

Mondou, P. M., and Kaltenbach, J. C., 1979, *Gen. Comp. Endocrinol.* **39**:343.

Oppenheimer, J. H., 1979, *Science* **203**:971.

Packard, G. C., 1976, *Gen. Comp. Endocrinol.* **28**:334.

Regard, E., Taurog, A., and Nakashima, T., 1978, *Endocrinology* **102**:674.

Sage, M., 1973, *Am. Zool.* **13**:899.

Shellabarger, C. J., 1964, in: *The Thyroid Gland*, Volume I (R. Pitt-Rivers and W. R. Trotter, eds.), Butterworths, London, Chap. 9, p. 187.

Spangenberg, D. B., 1974, *Am. Zool.* **14**:825.

Sterling, K., 1979, *N. Engl. J. Med.* **300**:117, 173.

Turner, C. D., and Bagnara, J. T., 1976, *General Endocrinology*, Saunders, Philadelphia.

Yamamoto, K., Kanski, D., and Frieden, E., 1966, *Gen. Comp. Endocrinol.* **6**:312.

Yoshizato, K., and Frieden, E., 1975, *Nature (London)* **254**:705.

Index